Advances in
Nuclear Science
and Technology

VOLUME 10

Advances in Nuclear Science and Technology

Series Editors

Ernest J. Henley
University of Houston, Houston, Texas

Jeffery Lewins
University of London, London, England

Martin Becker
Rensselaer Polytechnic Institute, Troy, New York

Advances in
Nuclear Science
and Technology

VOLUME 10

Edited by

Ernest J. Henley
University of Houston
Houston, Texas

Jeffery Lewins
University of London
London, England

and

Martin Becker
Rensselaer Polytechnic Institute
Troy, New York

PLENUM PRESS · NEW YORK AND LONDON

The Library of Congress cataloged the first volume of this title as follows:

Advances in nuclear science and technology. v. 1—
  1962—
New York, Academic Press.
  v. Illus., diagrs. 24 cm. annual.
  Editors: 1962—   E. J. Henley and H. Kouts.
  1. Nuclear engineering—Yearbooks. 2. Nuclear physics—Yearbooks.
  I. Henley, Ernest J., ed. II. Kouts, Herbert J., 1919-   ed.
TK9001.A3   621.48058   62-13039

Library of Congress Card Number 62-13039

ISBN 978-1-4613-9915-5   ISBN 978-1-4613-9913-1 (eBook)

DOI 10.1007/978-1-4613-9913-1

©1977 Plenum Press, New York
Softcover reprint of the hardcover 1st edition 1977
A Division of Plenum Publishing Corporation
227 West 17th Street, New York, N.Y., 10011

Preface

The editors are pleased to present to the nuclear community our new-look annual review. In its new look, with Plenum our new publisher, we may hope for a more rapid presentation to our audience of the contents for their consideration; the contents themselves, however, are motivated from the same spirit as the first nine volumes, reviews of important developments in both a historical and an anticipatory vein, interspersed with occasional new contributions that seem to the editors to have more than ephemeral interest.

In this volume the articles are representative of the editorial board policy of covering a range of pertinent topics from abstract theory to practice and include reviews of both sorts with a spicing of something new. Conn's review of a conceptual design of a fusion reactor is timely in bringing to the attention of the general nuclear community what is perhaps well known to those working in fusion - that practical fusion reactors are going to require much skillful and complex engineering to make the bright hopes of fusion as the inexhaustible energy source bear fruit. Werner's review of numerical solutions for fission reactor kinetics, while not exactly backward looking, is at least directed to what is now a well established, almost conventional field. Fabic's summary of the current loss-of-coolant accident codes is one realisation of the intensity of effort that enables us to call a light water reactor 'conventional.' The comprehensive coverage given to optimization theory by Terney and Wade is of course in an abstract form applicable to topics wider than fission or fusion reactors. The new material by Hesketh on thermodynamic theory, the 'spice' amongst the reviews, may seem even more divorced from nuclear technology until the problem of describing thermal material diffusion in fast reactor fuels at high temperature is contemplated. Sjöstrand's perceptive treatment of the apparently simple concept

of extrapolation distance is also a reminder of the valuable
work done in a number of smaller European countries.

The occasion of the new letter press is a suitable opportunity to acknowledge the work of the Editorial Board and the
changes that have occurred in recent years. We may quietly
thank those leaving for their support, given in characteris-
tically quiet and modest ways. We will shortly welcome to
the Board the new members from France, England, and the
United States in our endeavour to bring together the distillation of experience from all countries of the world; to
engineer the well known quotation from Donne - no nation is
a nuclear island. In similar spirit, the existing editors
have much.pleasure in welcoming Martin Becker to join them
in the preparation of further volumes on a basis of continuing
to secure this international coverage.

The application of nuclear power for land based electricity is going through a reappraisal in many countries,
part economic but part moral, though we forget sometimes the
immense experience in nuclear navies belonging to two super-
powers and to at least two other nations. We expect, of
course, that the series can contribute to the factual examination; in so far as it is a crisis of the mind we might
suggest the relevance of a quotation from Thomas Hardy in
The Mayor of Casterbridge: "But the ingenious machinery con-
trived by the Gods for reducing human possibilities of
amelioration to a minimum - which arranges that wisdom to do
shall come pari passu with the zest of doing - stood in the
way of all that."

Finally, but as always, our thanks are expressed on
behalf of this international constituency to our authors
for their endeavours in enriching their fellows.

                                        E. J. Henley
                                        J. Lewins
                                        M. Becker

Contents

Thermodynamic Developments
    R. V. Hesketh

Kinetics of Nuclear System: Solution Methods
            for the Space-Time Dependent Neutron
            Diffusion Equation
    W. Werner

Review of Existing Codes for Loss-of-Coolant
            Accident Analysis
    Stanislav Fabic

OPTIMAL CONTROL APPLICATIONS
IN
NUCLEAR REACTOR DESIGN AND OPERATION

W. B. TERNEY*
Institute für Neutronenphysik und Reactortechnik
Kernforschungszentrum Karlsruhe, Karlsruhe, West Germany
and
D. C. WADE
Applied Physics Division, Argonne National Laboratory
Argonne, Illinois

I. INTRODUCTION

A. Motivation

Modern control theory has developed over the past 15 years. The state variable representation of physical systems has come into wide use in dealing with nonlinear and time-varying systems. The classical calculus of variations (1,2,3) has been modified and extended (4-7) to yield procedures for determining the necessary conditions for optimality. The Pontryagin maximum principle (7) is a well-known example of such extensions. Dynamic programming, (8,9) an alternate approach to optimal control problems which has much in common with the variational and Hamilton-Jacobi formulations but is more amenable to stochastic systems and to straightforward solutions by digital computer, has also been developed. Mathematicians have applied functional analyses to optimization problems.

A proliferating literature describes recent research in optimal control theory; some excellent reviews and summaries of the methods (10-13) have been published.

Concurrent with the development of the theory, new, fast digital computers with large memory capacities have become

*Present address: Combustion Engineering, Inc., Windsor, Connecticut.

available. These make it possible to obtain numerical solutions to the equations obtained by an application of the
theory.

In the last decade, nuclear power has become competitive
with fossil energy sources and has developed into a mature
industry. As a result, increasing emphasis has been placed
on reducing design margins and improving operating procedures
so as to maximize return on investment. This economic incentive has motivated a growing interest in optimization
techniques and their application in the nuclear power reactor
field.

B.   Survey of Applications of Optimal Control Theory
     in Nuclear Engineering

Already there has been fairly widespread use of optimal
control methods in the nuclear field; both to nuclear design
problems and to nuclear plant operation problems. The areas
that have received attention in the past may be grouped into
the categories:
          1.   Reactor startup
          2.   Xenon Shutdown
          3.   Suppression of Xenon Spatial Oscillations
          4.   Shield Design
          5.   Beginning of Life (BOL) Design
          6.   Poison Management
          7.   Fuel Management
          8.   Plant Allocation

These applications are briefly described below as an introduction to the variety of problems, mathematical formulations,
and optimization theories that have been considered in the
nuclear engineering field.

   1. Reactor Startup *(14-17)*

The problem of bringing a reactor from a shutdown
condition to full power in minimum time is an example of a
reactor startup problem of interest in nuclear rocket technology. Other types of criteria (minimum coolant consumption, etc.) can be easily imagined. Constraints on the
allowable reactivity insertion and on the flux and power
trajectories may be included. Mohler *(14)* has treated a
number of these types of problems with the variational cal-

culus and Pontryagin's theory.* These applications are
summarized in the book (14) by Mohler and Shen.

Stacey (15,16) has solved similar problems for both
point and spatially dependent reactor models using a different technique. For point kinetics, the problem is formulated in terms of integral equations--a strategy which eliminates the problem of solving two-point boundary value
equations. The control is synthesized as a linear expansion
of a limited set of orthonormal functions. The unknown coefficients are obtained by applying the optimization condition. In the spatially dependent problem (16), the state
variables are also synthesized as linear combinations of the
eigenfunctions of the homogeneous system operator in order
to remove the spatial dependence. The procedure used in
the point kinetics case is then followed.

2. Xenon Shutdown (18-27)

In the xenon shutdown problem, an attempt is made
to control the power history of a reactor just prior to
shutdown in such a way as to limit the xenon buildup after
shutdown, thereby minimizing the control required for peak
xenon override. Various performance criteria are used ranging from minimizing the control time to minimizing the lost
flux time. In general, the problem is formulated so that
the reactor flux is the control variable and xenon and iodine
concentrations are the state variables. The state variables
are sometimes constrained. There is a considerable amount
of literature on this subject. The problem has been attacked
both by the Pontryagin approach (23) and by dynamic programming (24,25). The review article by Lewins and Babb (26)
gives a summary of the various approaches.

3. Xenon Spatial Oscillations (28-33)

In large thermal reactors, the delayed poisoning
effect of xenon can induce spatial power density oscillations
which may grow to unsafe amplitudes. These spatial oscil-

---

*A characteristic of the variational approach is that the
 Euler equations constitute a two-point time boundary value
 problem, having initial conditions on some variables and
 final conditions on the others; thus, a choice of an iterative solution technique is generally required.

lations can be suppressed by the manipulation of control
rods. There have been several attempts to optimize the rod
motion patterns to limit the oscillations.

Wiberg (28,29) first considered the problem using
the variational calculus. In order to gain numerical results,
he expanded the spatial flux in (difficult to compute) natural modes and linearized the equations. Stacey later treated
the problem using both dynamic programming (30) and variational (31) approaches. In the dynamic programming approach
the xenon transient in a 3D reactor model was treated as a
multi-stage decision process. Two alternate rod pattern
control schemes were proposed over three decision intervals
into which the control period was divided. The goal was to
minimize a weighted power peaking integral by the optimal
selection from this set of postulated rod patterns. By
working backward to decision points, where the state of the
reactor was the same for several trajectories, the effectiveness of the various rod programs could be compared, the
best option selected, all others discarded, and the procedure continued back to the beginning of the transient. In
this "tree pruning" manner the number of alternate transients
which had to be calculated were substantially reduced. In
the second approach, the reactor was nodulized (two nodes)
and the two-point boundary value equations arising from the
calculus of variations were solved by an iterative technique
which simultaneously improved the control and the assumed
missing initial conditions.

Recently, a simplified implementation of a variational approach (32) has been successfully tested on operating power reactors (33).

4. Shield Design (34-37)

The idea of determining the distribution and concentrations of materials in a shield to optimize its attenuation of radiation was considered very early in the development of reactor technology. Starting with Hurwitz (34) and
Troubetzkoy (35), attempts have been made to use the calculus
of variations and Pontryagin's theory to determine optimal
configurations. Lewins (36), using the variational theory,
gave a test for optimality. Terney and Fenech (37) developed
a numerical gradient procedure for iterating to the optimum
shield dimensions once a laminated material arrangement close
to the optimum configuration was selected.

5. BOL Design* (38-49)

In this type problem the goal is to arrange the
fissile and nonfissile materials in a critical reactor so as
to optimize the BOL performance according to a specified
criterion; e.g., minimum mass or minimum power peaking. In
the mid-fifties, Goertzel (38) first obtained a solution to
the minimum mass problem with no constraints on the fuel
density or power level using the variational calculus.
Russian workers (40) later used the Pontryagin principle for
this problem with upper limits on the control variable (in
this case, fuel density). Goldschmidt and Quenon (45) further extended the minimum mass problem by considering constraints on the state variables (fluxes) as well, again with
the Pontryagin formalism and a phase space interpretation to
determine the optimum trajectory or arrangement of fuel.
Russian authors (41,42) also used the Pontryagin principle
to consider minimum size and maximum power problems with
control variable constraints. Axford (43) extended the work
to additional problems with constraints on the flux (state
variable). Terney (46) applied the Pontryagin theory to
obtain the optimum k_∞ distribution which minimized a least
squares flux deviation criterion.

In these applications, the Pontryagin formalism
yields the necessary conditions which must be satisfied, and
thereby determines the allowable types of material regions;
e.g., maximum fuel, minimum fuel, etc. However, it does not
give "a priori" the way to synthesize or arrange the allowable regions. Wade and Terney (57) developed a gradient
procedure based on the Pontryagin formalism which iterates
to the optimum arrangement. A method based on linear programming was developed by Tzanos, et al. (49).

6. Poison Management (50-59)

The initial fuel loading pattern and placement and
movement of control poison determine the performance of a
core during its lifetime. Recently, there has been growing
interest and expanding efforts to optimize the reactor performance through the application of control theory.

---

* BOL = "Beginning of Life".

Haling and Crowther (50,51) proposed a method for minimizing the power peaking through the life of a core, by
selecting a desired end of life (EOL) power distribution
and manipulating the control poison to maintain this power
(and reactivity) distribution through the core life (no
formal control theory was used in this approach). Terney
and Fenech (54) formulated the problem of minimizing the
power peaking throughout life as a multi-stage decision
process in which the sequence of control actions at the
various decision points was to be optimized. Dynamic programming was used as a "tree pruning" device to eliminate
non-optimal paths and gain a solution for a simplified state
variable representation of the reactor.

Motoda and Kawai (55) have used the Pontryagin approach
to optimize the control poison program to maximize core lifetime. It was shown that the core lifetime alone is not a
sensitive performance criterion and that power peaking should
be considered. For simple reactor representations it was
possible to use phase space arguments to construct the
optimum trajectory from the resulting necessary Pontryagin
conditions. For certain cases it was shown that the Haling
mode is indeed optimal.

Wade and Terney (57,59) formulated the problem in terms
of the Pontryagin theory, using a generalized performance
index accounting for power peaking, lifetime, etc. They
used a multinode representation of the reactor with the
control affecting the nodal material buckling. The resulting Euler equations with mixed initial and final conditions
were solved by an iterative technique which does not rely
on guessing the missing initial conditions.

7. Fuel Management (60-69)

The choice of how to introduce new fuel and to
relocate partially depleted fuel at the end of each loading
cycle during the operation of a reactor plant can greatly
affect the nuclear fuel cost. The fuel management problem
has received considerable attention recently. In one of the
first attempts to optimize the fuel cycle using systematic
procedures, Wall and Fenech (60) considered the problem as
a multi-stage decision process and used dynamic programming
to optimize batch reloading and shuffling in a three-zone
core. Dynamic programming was used essentially as a "tree

pruning" operation to reduce the number of trajectories to be considered. Sesonske and Stover (61,62) extended this to consider partial batch and zone-wise scatter reloading strategies with the dynamic programming approach. More recently, Suzuki and Kiyose (65) and Sauer (66) have linearized the equations and used a linear programming algorithm to consider very general reloading and scatter schemes. Mélice (63) has used some of the preliminary variational results for BOL design to obtain an optimum core reactivity profile which is to be maintained by the proper introduction and shuffling of fuel.

8.   Plant Allocation (70-75)

There has been some interest in the use of control theory to determine the pattern of allocation and introduction of different reactor types in an expanding power industry to make better use of all available fuel types and to keep cost low. Young (70) considered the problem of using two different reactor types to meet the power demand so as to minimize the amount of fuel which must be bought. Nelson and Young (75) recently have carried this theory of Hinton arcs further to consider more complicated problems. Widner, Mason, and MacAvoy (73,74) consider the problems of selecting the proper enrichment at refueling time and also determining the short-term phasing in and out of a particular reactor in a system. Simple models are assumed and the results are obtained by the intersection of two curves obtained essentially by parameter studies. Märkl (71,72) has also investigated allocating different reactors to a system. He uses the linear programming formulation to determine the optimum parameters.

C.   Scope and Purpose of the Article

As illustrated above, there has been widespread application of optimal control theory to nuclear engineering problems in recent years. For most problems a variety of mathematical formulations have been tried. The system equations may be cast as differential or integral; the problem and process may be thought of as input-output, variational, or as a multi-stage decision process. Several theoretical approaches are available for use ranging from the calculus of variations (Pontryagin Maximum Principle) to linear and dynamic programming. Each method has its strengths and

weaknesses. With the variational approach, two-point boundary value equations arise which are difficult to solve. If
analytical solutions or phase space arguments cannot be used,
an iterative technique must be applied. Dynamic programming,
on the other hand, may require large amounts of computing
time and large memory requirements when realistic reactor
models are used.

The purpose of this article is to review the various
optimization approaches available and to illustrate their
application to typical problems reported in the literature
in an attempt to summarize and codify the choices available
to a nuclear engineer.

In the next section, the mathematical formulation of
optimization problems is presented. This is followed by a
very brief review of the variational and dynamic programming optimization theories and solution techniques. (A
more detailed treatment is presented in the appendix.) After
reviewing the theory, applications from the literature are
presented in detail for some of the areas of interest to
nuclear engineers. These include the Beginning of Life design, fuel management, and poison management problems. Each
problem is treated by several methods to highlight the advantages and disadvantages of each.

II. PROBLEM FORMULATION

A. A contrast of Traditional and Optimal Design Procedures

Given the task of designing a piece of equipment, engineers have traditionally proceeded by the following approach:

1. Select a tentative system configuration based on
   previous experience but leaving certain parameter
   values or parts of the configuration undetermined.

2. Write equations which describe the system behavior.

3. Select values for adjustable system parameters or
   select a tentative system configuration.

4. Solve the equations to see if the system performs
   as desired, then,

5.   if not, select new system parameters or alter the
     system configuration and proceed by iteration
     guided by experience until acceptable system per-
     formance is obtained.

The alternate, optimal control theory approach is as
follows:

1.   Select a tentative system configuration - leaving
     part of the configuration undetermined.

2.   Write down in mathematical form what is meant by
     "acceptable " system performance, i.e., quantify
     the performance by a single number - a "cost or
     payoff functional" or "performance index" which
     is always positive and is small for a good design
     and large for a poor design (cost index).

3.   Write equations which describe the system behavior
     and constraints.

4.   Then determine the values of the adjustable para-
     meters or the undetermined components of the sys-
     tem configuration by minimizing the performance
     index while simultaneously satisfying the system
     equations.

The distinctive feature of the optimal (or modern) control theory approach is that the system components or adjustable parameters obtained yield not just acceptable but
rather the best possible (optimal) system performance as
measured against the criterion of step 2. Furthermore, a
design is achieved with a single rather than multiple solution of the equations. However, the combined set of system
and optimality condition equations are more complicated than
the system equations alone.

The optimization of the performance index subject to
the design constraints is accomplished using mathematical
techniques which are recent extensions or derivatives of
the classical calculus of variations. The optimization
approach has historically depended on the form in which the
system is described. The most common forms of system representation are presented below. They are:

1. the differential form, and

2. the multistage decision process form.

B. Differential System Representation

1. System Equations

Many physical systems can be described by a set of coupled ordinary differential equations

$$f_i(\underline{x}, \underline{u}, t) - \frac{d}{dt} x_i(t) = 0 \qquad i = 1, 2, \ldots L \qquad (II.B.1)$$

and algebraic equations

$$f_i(\underline{x}, \underline{u}, t) = 0 \qquad\qquad i = L+1, L+2, \ldots n \quad (II.B.2)$$

where

t is the independent variable (such as time)

$\underline{x}(t) = [x_1(t) x_2(t) \ldots x_n(t)]^T$ is a vector of state variables

$\underline{u}(t) = [u_1(t) u_2(t) \ldots u_m(t)]^T$ is a vector of control variables which are to be determined to optimize system performance, and

$f_i(\underline{x}, \underline{u}, t)$ are in general nonlinear, time-varying functions of $\underline{x}$ and $\underline{u}$.

The state variables are subject to initial and final conditions:

$$\Psi_i [\underline{x}(t_0), t_0, \underline{x}(t_1), t_1] = 0 \quad i = 1, 2, \ldots r \leq 2n+2$$

$$(II.B.3)$$

The control variables u_i are under the designer's control and are to be picked to yield the desired behavior of the state variables. The set of equations, Equations II.B.1, II.B.2, and II.B.3 are assumed to be adequate to uniquely determine $\underline{x}$ given $\underline{u}$ and the initial and final conditions.

The control and state variables may be further constrained by inequality relations

$$f_i(\underline{x},\underline{u},t) \leq 0 \quad i=n+1, \ldots J, J-n \leq m \qquad \text{II.B.4)}$$

These express design constraints on the control and state variables.

Not all physical systems can be described by such a set of equations -- systems which are described by partial differential equations are an example. These can sometimes be described approximately in terms of such a set by discretizing all but one of the variables or by expanding in terms of basis functions on all but one of the variables. Systems described by integral equations also fall outside of this framework. Even though Equations II.B.1, 2, 3 and 4 are not completely general, they are adequate for describing many physical systems of interest, and will be used as a basis for most of the discussion in the following sections.

2.   Performance Index

One of the distinctive features of the optimal control approach to design is the quantification of the design goals. It is assumed that the design objectives can be cast in the form of a "cost functional" or "performance index" such as:

$$f_o(\underline{x},\underline{u}) = \int_{t_0}^{t_1} L(\underline{x},\underline{u},t) \, dt + \phi[\underline{x}(t_0),t_0,\underline{x}(t_1),t_1] \qquad \text{(II.B.5)}$$

L is a function of the state and control vectors on the interval t_0 to t_1 and ϕ is a function of the state at the initial and terminal times. For example, L, might represent the square of the deviation of certain state variables from some desired target value [$L = (x_i - x_{desired})^2$], and ϕ might represent the final time itself or the square of the deviation of the final state from the desired one. The performance index is so defined that it is non-negative and is small for desired system performance but large for poor system performance.

3.   Statement of the Optimal Control Problem

The optimal control problem is to choose the control vector

$$\underline{u}^T(t) = [u_1(t) \ldots u_m(t)]$$

so as to minimize the cost functional

$$f_o(\underline{x},\underline{u}) = \Phi\left(\underline{x}(t_0), t_0, \underline{x}(t_1), t_1\right) + \int_{t_0}^{t_1} L(\underline{x},\underline{u},t)\, dt$$

subject to the system equations and associated initial and final conditions,

$$f_i(\underline{x},\underline{u},t) - \frac{d}{dt}\, x_i(t) = 0 \qquad i = 1,2, \ldots L$$

$$f_i(\underline{x},\underline{u},t) \qquad\qquad = 0 \qquad i = L + 1, \ldots n$$

$$\Psi_i[\underline{x}(t_0), t_0, \underline{x}(t_1), t_1] = 0 \qquad i = 1,2, \ldots r \leq 2n + 2$$

in the state variables

$$\underline{x}^T(t) = [x_1(t) x_2(t) \ldots x_n(t)]$$

and subject also to inequality constraints on the state and control variables

$$f_i(\underline{x},\underline{u},t) \leq 0 \qquad\qquad i = n + 1 \ldots J; J-n \leq m$$

C.   Multistage Decision Process System Representation

The optimal control problem stated above is in differential form, and as will be shown in the following sections, leads naturally to a calculus of variations approach to the necessary conditions for optimality. However, some problems are inherently discrete, or may be so posed by finite differencing, and in such a case it may be natural to view the system and the determination of the optimal control as a multistage decision process. This discrete representation lends itself to the dynamic programming formulation for the necessary conditions for optimality.

Consider the process depicted in Figure II.C.1. At each of the finite set of stages, $k = 1, 2, \text{---} N$, the system is specified by the value of a state vector, $\underline{x}^{k-1}$. At the start of each stage a decision, $\underline{u}^k$, is to be chosen from a finite set of decisions $\underline{u}^k \varepsilon \Omega$, and the system is transformed to a new state $\underline{x}^k$ which constitutes the initial state for the next stage. It is assumed that the system is suitably defined such that a specification of $\underline{x}^{k-1}$ and $\underline{u}^k$ leads to a unique $\underline{x}^k$. The state transformation equation is represented by

$$x_i^k = f_i(\underline{x}^{k-1}, \underline{u}^k, k) \qquad i = 1, 2, \ldots L \qquad (II.C.1)$$

Just as in the differential form of the optimal control problem, the state of the system may be subject to specified initial and final conditions

$$\Psi_i [\underline{x}^0, \underline{x}^N] = 0 \qquad i = 1, 2, \text{---} r \leq n \qquad (II.C.2)$$

and to equality and inequality constraints

$$f_i(\underline{x}^k, \underline{u}^k, k) = 0 \qquad i = L+1, L+2, \text{---} n \qquad (II.C.3)$$

$$f_i(\underline{x}^k, \underline{u}^k, k) \leq 0 \qquad i = n+1, n+2, \text{---} J \qquad (II.C.4)$$

$$J - n \leq m$$

A performance index or cost functional may be defined to quantify the performance of the multistage decision process:

$$f_o = \sum_{k=1}^{K} \Delta f_o^k + \Phi(\underline{x}^0, \underline{x}^K) \qquad (II.C.5)$$

where Δf_o^k is the incremental addition made at the k-th stage and includes contributions from the initial and final states. The control problem is to determine the sequence $\{\underline{u}^k \mid k = 1, 2, \ldots N\}$, so as to minimize the performance index while at the same time satisfying the state transfor-

mation equations, the end conditions, and the constraints.
As an example, if three decision alternatives can be made at
each of five stages, each decision leading to a different
value of f_0^k but not necessarily a different state at the
subsequent stage, a total of 3^5 possible sequences is possible. Since the performance index associates a value with
each sequence, it is meaningful to ask which particular sequence of decisions leads to the minimum performance index.
The particular set of decisions leading to the minimum value
of the performance index is the optimal choice.

A straightforward evaluation of all possible sequences
of a multistage decision process is to be avoided if the
number of stages or the number of decisions or state variables
characterizing each state is large. The dynamic programming
approach provides a technique to determine the optimal control in multistage decision processes. It will be described
in detail in Section III.D, but rests on the intuitive principle of optimality stated by Bellman, "An optimal policy

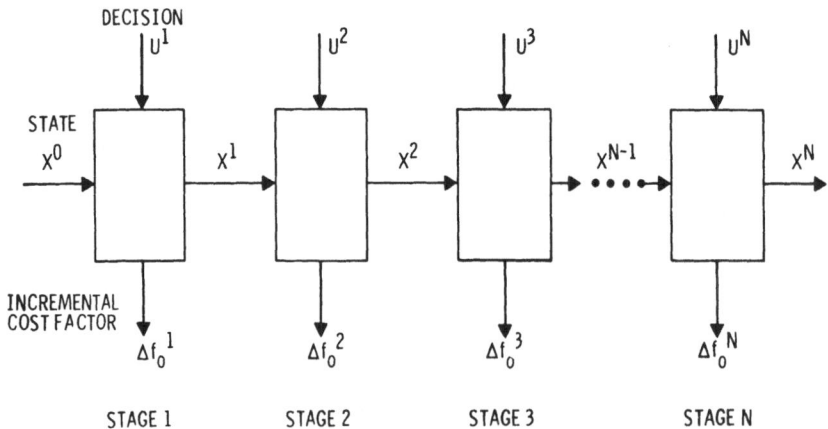

Figure II.C.1
A Multistage Decision Process

has the property that whatever the initial state and initial decision are, the remaining decisions must constitute an optimal policy with regard to the state resulting from the initial decision."[8]

Although it has been introduced here in a discrete form, the multistage decision process and the dynamic programming algorithms which are discussed later are not inherently restricted to discrete situations. The principle of optimality, in fact, provides an alternate path to the derivation of all results of the classical calculus of variations. The relationship between dynamic programming and the classical calculus of variations will be discussed in Section III.D.

III.   MATHEMATICAL APPROACHES TO THE OPTIMAL CONTROL PROBLEM

A.   Survey

The optimal control problem is to choose from the set of allowable control vectors, $\underline{u}(t)$, that particular one, $\underline{u}^*(t)$, which will minimize the performance index, $f_o(\underline{x},\underline{u})$ of Equation II.B.5 (II.C.5 for the multistage decision process). Three categories of approach to this problem have been used. They are based on:

a.   nonlinear and linear programming

b.   the calculus of variations

c.   dynamic programming

In the nonlinear programming approach, the definition of an optimum control is used directly; i.e., if $\underline{u}^*$ is the optimal control, then for all allowed controls

$$f_o(\underline{x},\underline{u}) - f_o(\underline{x}^*,\underline{u}^*) \geq 0$$

The procedure is to guess a $\underline{u}$, which may be the optimum; then guess a neighboring allowable choice for $\underline{u}$, solve the system equations for both cases, and evaluate the performance index, f_o, to determine whether it has increased or decreased. This is done repeatedly until a choice of $\underline{u}$ is found for which all feasible neighboring $\underline{u}$'s produce an

increase in f_o. Search techniques have been developed
which, after each iteration, are used to determine a
change in control that will decrease the value of f_o. For
very simple problems, geometric constructions may be used.
These involve representing f_o in the control phase space,
identifying the gradient direction and moving the control
point opposite the gradient direction. For higher dimensional problems, algebraic algorithms inspired by geometric
considerations are used. These go by the name "nonlinear
programming". If the system and constraint equations are
linear, special highly efficient "linear programming" algorithms are used.

In the calculus of variations approach, an augmented
performance index is created by adding the sum of Lagrange
multipliers times the constraint equations to the performance index functional. The resulting problem is closely
related to the classical "problem of Bolza". Examination
of the variation of the augmented functional leads to a
coupled set of Euler-Lagrange equations in the Lagrange
multiplier variables, the state variables, and the control
variables.* They are subject to initial conditions on some
of the variables and final conditions on others. As a result of the "two point" boundary values, an iterative solution scheme is required for all but the very simplest problems.

In the dynamic programming approach, given the constraints
on the state and control variables, the optimal value of the
performance index,

$$S(\underline{x},t) = \min_{\underline{u}}\{f_o\}$$

is recognized to depend only on the starting state, $\underline{x}$, and
starting time, t. The increment added to the value of f_o in
changing the initial state from $\underline{x}$ to $\underline{x} + \delta\underline{x}$ and the initial
time from t to $t + \underline{\delta}t$, $\delta t \geq 0$, is simply:

---

* For a certain class of problems the set of necessary
  conditions developed are the "Pontryagin maximum principle".

$$\int_{t}^{t+\delta t} f_o(\underline{x},\underline{u},\tau) \, d\tau$$

Using Bellman's "Principle of Optimality", a backward recursion algorithm may be written for the value of S as a
function of the starting state and time:

$$S(\underline{x},t) = \min_{\underline{u}} \left\{ \int_{t}^{t+\delta t} f_o(\underline{x},\underline{u},\tau) d\tau + S(\underline{x} + \delta\underline{x}, t + \delta t) \right\}$$

From this relationship it is possible to derive a partial
differential equation for $S(\underline{x},t)$. The solution of this
equation yields a "control law" which will specify the
optimal control to be used as a function of the state of
the system.

The equation shown above forms the basis for a recursive procedure for determining $S(\underline{x},t)$. Because of this
approach for solving the dynamic programming equations on a
digital computer, it is well suited for multi-stage optimization problems. Computer storage requirements often limit
its applicability to low dimensional systems, however. An
inherent strength of the dynamic programming approach is the
natural way in which it accommodates inequality constraints
on the state and control variables and (though not considered here) stochastic perturbations to the system equations.

As will be shown in the next section, inequality constraints introduce substantial complexity into the calculus
of variation approach to the optimal control problem. These
complexities can sometimes be circumvented by slightly relaxing the constraint and using the device of "penalty functions". Here the inequality constraint equations, Equation
II.B.4, are removed from the problem statement and the performance index, Equation II.B.5, is modified to include
terms (penalty functions) which go from very small to very
large as the constraint relations pass from being satisfied
to being violated.

In the next section, several extended problems of Bolza
are treated to develop the necessary conditions for optimality for the differential form of system representation
posed by Equations II.B.1 through II.B.5. Several numerical
solution techniques for the resulting two-point boundary
value equations are then discussed. Following that, a
simple discrete multistage decision process is optimized by
dynamic programming to illustrate the recursive solution
procedure for the multistage decision process system representation posed by Equations II.C.1 through II.C.5. This is
followed by a discussion of the connection between the dynamic programming and calculus of variation treatment of
optimization problems.

These brief descriptions provide a basis for proceeding
with the discussion of nuclear engineering applications of
optimal control theory presented in the later sections. In
an appendix, optimal control theories are reviewed in more
detail. The approach is heuristic rather than rigorous and
is intended to provide an overview which should be augmented
by specialized texts on each of the subjects.

B. Necessary Conditions for Optimality for the Differential
    Form of System Representation

The optimal control problem is to minimize the cost
functional:

$$f_o(\underline{x},\underline{u}) = \int_{t_0}^{t_1} L(\underline{x},\underline{u},t)\ dt + \Phi(\underline{x}\ (t_o),\ t_o,\underline{x}\ (t_1),t_1) \qquad \text{(III.B.1)}$$

which depends on the n element state vector, $\underline{x}$, and the
m-element control vector, $\underline{u}$, subject to the initial and
final conditions,

$$\psi_i(\underline{x}(t_o),\ t_o,\ \underline{x}\ (t_1),\ t_1) = 0 \qquad \text{(III.B.2)}$$

$$i = 1,2,\ldots r \le 2n+2$$

and the constraint relations,

$$f_i\ (\underline{x},\underline{u},t) - \frac{d}{dt}\ \underline{x}_i\ (t) = 0 \qquad i = 1,2,\ldots L{\le}n \quad \text{(III.B.3)}$$

$$f_i\ (\underline{x},\underline{u},t) = 0 \qquad\qquad\qquad i = L+1,\ldots n \qquad \text{(III.B.4)}$$

and to the inequality constraints,

$$f_i \ (\underline{x},\underline{u},t) \leq 0 \qquad\qquad i = n+1, \ \ldots J \qquad (III.B.5)$$
$$J-n \leq m$$

Consider the case where the control is a piece-wise continuous function of time with piecewise continuous first and second time derivatives and the functions, f_i, possess continuous first and second partial derivatives in their arguments. The optimal control will be denoted by $\underline{u} = \underline{u}^*(t)$ and the corresponding optimal system response will be denoted by $\underline{x} = \underline{x}^*(t)$. The system response will be thought of as a "trajectory" in the phase space of the state variables which is given parametrically by the components along each axis:

$$x_1(t), \ x_2(t), \ x_3(t), \ldots x_n(t).$$

Similarly, the constraint relations, Equations III.B.2 through III.B.5 will be thought of as surfaces in the phase space of the state and control variables. The trajectories must lie on the surfaces specified by Equations III.B.3 and .4, must begin and end on the surfaces specified by Equations III.B.2, and must lie inside (closer to the origin) or at most on the surfaces specified by Equation III.B.5.

The necessary conditions for optimality for three cases are tabulated and discussed below:

a.   no inequality constraints

b.   inequality constraints which depend explicitly on the control vector, $\underline{u}$, and

c.   inequality constraints in the state variables which depend only implicitly on the control vector $\underline{u}$.

The derivations of the necessary conditions for a and b are presented in the appendix where they are treated as extended problems of Bolza.

1.   No Inequality Constraints

The inequality constraints, Equations III.B.5 are dropped from the problem statement. Define the augmented functional:

$$G_1 = \int_{t_0}^{t_1} \left\{ L(\underline{x},\underline{u},t) + \sum_{i=1}^{L} p_i\left(f_i - \frac{d}{dt}x_i\right) \right.$$

$$\left. + \sum_{i=L+1}^{M} p_i f_i \right\} dt$$

(III.B.6)

$$+ \sum_{i=1}^{r} e_i \psi_i + \Phi\ (\underline{x}(t_0),\ t_0,\ x(t_1),\ t_1),\ t_1)$$

and the function:

$$H(t,\underline{x},\underline{p},\underline{u}) = L + \sum_{i=1}^{n} p_i f_i$$

(III.B.7)

where the Lagrange multiplier vectors:

$$\underline{p}(t) = [p_1\ (t)\ p_2\ (t)\ \cdots\ p_n\ (t)]^T$$

(III.B.8)

$$e = [e_1 e_2 \cdots e_r]^T$$

have been introduced. The necessary conditions for the optimal control are derived in the appendix by considering variations of the augmented functional, G_1.

These are the "Multiplier Rule" consisting of the Euler-Lagrange equations, the corner conditions, the transversality conditions, and the Weierstrass condition and are enumerated below.

Euler-Lagrange Equations (to be satisfied in the time intervals between discontinuities in u(t)).

$$\frac{\partial H}{\partial x_i} = -\frac{d}{dt}p_i \qquad i = 1,2,\ \ldots\ L$$

(III.B.9)

$$\frac{\partial H}{\partial x_i} = 0 \qquad\qquad i = L+1, 1+2, \ldots n \qquad (III.B.10)$$

$$\frac{\partial H}{\partial u_i} = 0 \qquad\qquad i = 1, 2, \ldots m \qquad (III.B.11)$$

<u>Corner Conditions</u> (which hold at any time $t_0 \leq t_2 \leq t_1$ at which the control is discontinuous).

$$\left. \begin{array}{l} H(t_2^+) = H(t_2^-) \\[2ex] \underline{p}(t_2^+) = \underline{p}(t_2^-) \end{array} \right\} \qquad t_0 \leq t_2 \leq t_1 \qquad (III.B.12)$$

<u>Transversality Conditions</u>

$$d\Phi + \underline{e}^T d\underline{\Psi} + \left[Hdt - \sum_{i=1}^{L} p_i dx_i \right] \Bigg|_{t_0}^{t_1} = 0 \qquad (III.B.14)$$

where
$d\Phi$ and $d\psi$ are total derivatives including both dt and dx_i components.

<u>Weierstrass Condition</u>

$$H(t,\underline{x}^*,\underline{p}^*,\underline{u}) \geq H(t,\underline{x}^*,\underline{p}^*,\underline{u}^*) \quad t_0 \leq t \leq t_1 \qquad (III.B.15)$$

i.e., at any point along the optimal trajectory, $\underline{x}=\underline{x}^*(t)$, any change in control from the optimal control, $\underline{u}=\underline{u}^*(t)$ will cause the Pontryagin function to increase in value.

In addition to Equations III.B.9 through III.B.15, the system equations III.B.2 through III.B.4 must be satisfied. The $\underline{x},\underline{u}$, and $\underline{p}$ vectors constitute 2n+m unknown functions of time. The system equations, Equations III.B.3 and .4, and the Euler-Lagrange equations, Equations III.B.9, .10 and .11, provide 2n+m differential and algebraic relationships between these unknowns at each instant of time. Note that while the system equations are independent of the Lagrange multipliers, p_i, the Euler-Lagrange equations are dependent on all of the vectors: $\underline{x}$, $\underline{u}$, and $\underline{p}$.

The corner conditions, Equations III.B.12 and .13 provide conditions which must apply at any discontinuity in the control, $\underline{u}(t)$ on the interval $t_o \leq t \leq t_1$.

The initial and final conditions, Equations III.B.2 and the transversality conditions Equation III.B.14 provide a complementary set of boundary conditions. Independent variations subject to the transversality condition are taken only in those variables not already fixed by Equation III.B.2.

The Weierstrass condition, which arises from a consideration of strong variations, eliminates controls which maximize f_o rather than minimize f_o. If the goal were to maximize f_o rather than minimize it, all the necessary conditions listed above would remain unchanged except that the sense of the inequality sign in the Weierstrass condition would reverse.

An additional fact may be noted. The total time derivative of H evaluated along a trajectory is

$$\frac{d}{dt} H = \frac{\partial H}{\partial t} + \sum_{i=1}^{n} \frac{\partial H}{\partial x_i} \frac{d}{dt} x_i + \sum_{i=1}^{M} \frac{\partial H}{\partial p_i} \frac{d}{dt} p_i$$

However, along the optimal trajectory the time derivatives of x_i and p_i are given by the system and Euler equations. The appropriate substitutions yield:

$$\frac{d}{dt} H \bigg|_{\substack{\text{optimal} \\ \text{trajectory}}} = \frac{\partial H}{\partial t} \qquad\qquad \text{(III.B.16)}$$

As a result, unless H depends explicitly on t, it takes on the same <u>constant</u> value when evaluated at any point along the optimal trajectory.

2. Inequality Constraints Explicitly Dependent on the Control

Consider again the optimal control problem with the inequality constraints, Equations III.B.5, restored. To transform the inequality constraints to equality constraints, define the "slack variables":

$$z_i^2 \geq 0 \qquad\qquad i = n+1,\ n+2,\ \ldots\ J \qquad (III.B.17)$$

such that the inequality constraints become

$$f_i\ (t,\underline{x},\underline{u},) + z_i^2 = 0 \qquad\qquad (III.B.18)$$

It is assumed that each inequality constraint contains an explicit dependence on the control, $\underline{u}$. Now define the augmented functional:

$$
\begin{aligned}
G_2 = \int_{t_0}^{t_1} & \left\{ L + \sum_{i=1}^{L} p_i \left(f_i - \frac{d}{dt}\, x_i \right) \right. \\
& \left. + \sum_{i=L+1}^{n} p_i f_i - \sum_{i=n+1}^{J} \upsilon_i (f_i + z_i^2) \right\} dt
\end{aligned}
$$

$$\qquad\qquad (III.B.19)$$

$$+ \sum_{i=1}^{r} e_i \Psi_i + \Phi(\underline{x}(t_0),\ t_0,\ \underline{x}(t_1),t_1)$$

and the function:

$$H(t,\underline{x},\underline{p},\underline{u}) = L + \sum_{i=1}^{n} p_i f_i$$

where, in addition to the $\underline{p}$ and $\underline{e}$ vectors defined above, an additional Lagrange multiplier vector,

$$\underline{\upsilon} = [\upsilon_{n+1}(t)\ \ \upsilon_{n+2}(t)\ \ \ldots\ \ \upsilon_J(t)]^T \qquad\qquad (III.B.20)$$

is defined.

By consideration of variations of the augmented functional, G_2, the following necessary conditions for optimality are obtained in the appendix.

Euler-Lagrange Equations

$$\left.\begin{array}{l}\text{either } \nu_i Z_i = 0 \\[1em] \text{or } \nu_i f_i = 0\end{array}\right\} \quad i = n+1, n+2, \ldots J \qquad (II.B.21)$$

$$\frac{\partial H}{\partial x_i} - \sum_{j=n+1}^{J} \nu_j \frac{\partial f_j}{\partial x_i} - \frac{d}{dt} p_i$$
$$\qquad\qquad\qquad\qquad i = 1,2, \ldots L \quad (III.B.22)$$

$$\frac{\partial H}{\partial x_i} - \sum_{j=n+1}^{J} \nu_j \frac{\partial f_j}{\partial u_i} = 0 \qquad i = L+1, \ldots n \quad (III.B.23)$$

$$\frac{\partial H}{\partial u_i} - \sum_{j=n+1}^{J} \nu_j \frac{\partial f_j}{\partial u_i} = 0 \qquad i = 1,2,\ldots m \quad (III.B.24)$$

Corner Conditions

$$\left.\begin{array}{l}H(t_2^+) = H(t_2^-) \\[1em] \underline{p}(t_2^+) = \underline{p}(t_2^-)\end{array}\right\} \qquad t_0 \leq t_2 \leq t_1 \quad (III.B.25)$$

Transversality Condition

$$d\Phi + \underline{e}^T d \underline{\Psi} + \left[Hdt - \sum_{i=1}^{L} p_i dx_i \right]\Bigg|_{t_o}^{t_1} = 0 \quad (III.B.26)$$

Weierstrass Condition

$$H(\underline{x}^*,p^*,u,t) \geq H(\underline{x}^*,\underline{p}^*,\underline{u}^*,t) \qquad\qquad (III.B.27)$$

These conditions are the same as in the previous case except for the ν_j terms in the Euler-Lagrange equations. Notice that from Equation III.B.21 the value of ν_i is required to be zero where $Z_i > 0$ (that is, where the inequality constraint is satisfied in the inequality sense). In this case, the remaining Euler equations reduce to those discussed above for no inequality constraints. On the other hand, if the inequality constraint is satisfied in an equality sense, Equation III.B.21 is satisfied for nonzero ν_i, and the ν_i terms enter the Euler equations to reflect the fact that variations are no longer independent, but must be taken in such a way as to keep the trajectory on the constraint surface.

Consider the special case of:

a. an integral performance index

b. differential system equations only

c. inequality constraints which depend on the control only

d. a fixed final state

$$f_o\,(\underline{x},\underline{u}) = \int_{t_0}^{t_1} L(\underline{x},\underline{u},t)\ dt \qquad\qquad (III.B.28)$$

$$\Psi_i(\underline{x}(t_0),\ t_0,\ \underline{x}(t_1),\ t_1) = 0 \qquad\qquad (III.B.29)$$
$$i = 1,2,\ldots r \le 2n+2$$

$$f_i\,(\underline{x},\underline{u},t) - \frac{d}{dt}\,x_i \quad\ = 0 \qquad\qquad (III.B.30)$$
$$i = 1,2,\ldots n$$

$$\chi_i(\underline{u}) \qquad\qquad\qquad \le 0 \qquad\qquad (III.B.31)$$
$$i = n+1,\ n+2\ldots J$$
$$J - n \le m$$

This special form of the case considered above is associated with the Pontryagin maximum principle. The necessary conditions for optimality are expressed compactly as:

$$\frac{\partial H}{\partial p_i} = \frac{d}{dt} x_i \qquad i=0, 1, 2,\ldots n \qquad (III.B.32)$$

$$\frac{\partial H}{\partial x_i} = -\frac{d}{dt} p_i \qquad i=0, 1, \ldots n \qquad (III.B.33)$$

$$H(\underline{x}^*,\underline{p}^*,\underline{u},t) \geq H(\underline{x}^*,\underline{p}^*,\underline{u}^*,t) \qquad (III.B.34)$$

where H and $\underline{p}(t)$ are continuous and $\underline{p}(t)$ is nonvanishing on the interval $t_0 \leq t \leq t_1$. Here, in conformance with usual notation, an additional element $x_o(t)$ has been added to the state vector such that the minimization of the performance index transforms into a minimization of the final value of x_o.

$$\frac{d}{dt} x_o(t) = L(\underline{x},\underline{u},t) \qquad (III.B.35)$$

$$x_o(t_1) = f_o = \int_{t_0}^{t_1} L(\underline{x},\underline{u},t)\, dt \qquad (III.B.36)$$

$$x_o(t_o) = 0$$

Also, a corresponding element $p_o(t)$ has been added to the Lagrange multiplier vector such that

$$p_o(t) = p_o = 1$$

and

$$H(\underline{x},\underline{p},\underline{u},t) = \sum_{i=0}^{n} p_i f_i \qquad (III.B.37)$$

Equation III.B.32 is a convenient way of writing the system equations; Equation III.B.33 are the Euler equations; the continuity of H and $\underline{p}$ correspond to the corner conditions; and the "minimum principle", Equation III.B.34 corresponds to the Weierstrass condition. Pontryagin's notation employed $p_o = -1$ in which case the sense of the inequality in Equation III.B.34 reverses to give a "maximum principle".

3. Inequality Constraints Implicitly Dependent on the
   Control

   Consider again the optimal control problem stated by
Equations III.B.1 through III.B.5. However, replace the
inequality constraints III.B.5 by the control-independent
state variable constraint:

$$\chi_i(\underline{x}, t) \leq 0 \qquad\qquad i = n+1, n+2...J \qquad \text{(III.B.38)}$$
$$J-n \leq m$$

   It was shown in the preceding discussion that when the
constraint is satisfied in the inequality sense, no additional
terms are required in the Euler equations and the necessary
conditions for optimality are given by Equations III.B.9
through III.B.15. Suppose, however, that Equation III.B.38
is satisfied in the equality sense for some particular i=k.
Here, unlike the previous case, the constraint equation cannot be used to relate the control and state variables on the
constraint surface because the control is missing from the
equation, $\chi_k(\underline{x}, t) = 0$.

   If the trajectory is to remain on the surface for a finite
time interval t_2 to t_3, the total derivative of χ_k evaluated
along a trajectory must be zero:

$$\frac{d}{dt} \chi_k = \chi_k^{(1)} = 0 \qquad\qquad\qquad \text{(III.B.39)}$$

$$\chi_k^{(1)} = \frac{\partial \chi_k}{\partial \tau} + \sum_{j=1}^{n} \frac{\partial \chi_k}{\partial x_j} f_j = 0 \qquad t_0 \leq t_2 \leq t_3 \leq t_1$$

   If this equation contains the control explicitly, it can
be used to relate the control and state variables on the
constraint surface. Suppose in general that it does not and
that the q_k-th total derivative of the k-th constraint surface evaluated along a trajectory is the first one to contain
an explicit dependence on the control. Then the equality
constraint:

$$\chi_k^{(q_k)}(\underline{x}, \underline{u}, t) = 0 \qquad\qquad t_2 \leq t \leq t_3 \qquad \text{(III.B.40)}$$

with the additional boundary conditions at the point of
entry (or exit) onto the surface

$$\underline{N}\ (\underline{x},t_2)_k \quad = \quad \left\{ \begin{array}{c} \chi_k\ (\underline{x},t_2) \\[6pt] \chi_k^{(1)}\ (\underline{x},t_2) \\ \vdots \\ \chi_k^{(q_k-1)}\ (\underline{x},t_2) \end{array} \right\} = 0 \qquad\qquad \text{(III.B.41)}$$

replace the original constraint, Equation III.B.38 and are equivalent to it in that they keep the trajectory on the surface once it enters at time t_2. By use of this tactic, each state variable constraint can be replaced by an equivalent one which displays an explicit dependence on the control, and the usual procedure for deriving the necessary condition employed.

Define the augmented functional:

$$
\begin{aligned}
G_3 = \int_{t_0}^{t_1} & \left\{ L + \sum_{i=1}^{L} p_i \left(f_i - \frac{d}{dt} x_i \right) \right. \\[6pt]
& + \sum_{i=L+1}^{n} p_i f_i + \sum_{i=n+1}^{J} \nu_i \chi_i^{(q_i)} \Bigg\} \ dt \\[6pt]
& + \Phi(x(t_0),t_0,x(t_1),t_1) + \sum_{i=1}^{r} e_i \Psi_i \\[6pt]
& + \left. \sum_{i=n+1}^{J} \Pi_i^T \underline{N}_i \right|_{t=t_2}
\end{aligned}
\qquad\qquad \text{(III.B.42)}
$$

where the Lagrange multipliers $\nu_i(t)$, $i=n+1,\ldots J$ have been introduced such that

$$\nu_i(t) = 0 \qquad\qquad \text{when } \chi_i < 0 \qquad\qquad \text{(III.B.43)}$$

$$\nu_i(t) \geq 0 \qquad\qquad \text{when } \chi_i = \chi_i^{(1)} = \ldots = \chi_i^{(q_i)} = 0$$

and the Lagrange multiplier vectors

$$\underline{\pi}_i = [\pi_1 \ \pi_2 \ \ldots \ \pi_{q_i}]^T \qquad i = n+1, \ldots J \qquad \text{(III.B.44)}$$

multiply the entry point initial conditions, $\underline{N}_i = \underline{0}$, specified by Equation III.B.41.

Define the function:

$$H = L + \sum_{i=1}^{n} p_i f_i$$

Then, by consideration of variations in the augmented functional G_3, identical to those illustrated in the appendix for the previous problem, it may be shown that the necessary conditions for the optimal control are:

Euler-Lagrange Equations

$$\sum_{i=n+1}^{J} \nu_i \ \chi_i^{(q_i)} = 0 \qquad \text{(III.B.45)}$$

$$\frac{\partial H}{\partial x_i} + \sum_{j=n+1}^{J} \nu_j \frac{\partial \chi_j (q_j)}{\partial x_i} = - \frac{d}{dt} p_i \quad i = 1,2,\ldots L \qquad \text{(III.B.46)}$$

$$\frac{\partial H}{\partial x_i} + \sum_{j=n+1}^{J} \nu_j \frac{\partial \chi (q_j)}{\partial x_i} = 0 \qquad i = L+1,\ldots n \quad \text{(III.B.47)}$$

$$\frac{\partial H}{\partial u_i} + \sum_{j=n+1}^{J} \nu_j \frac{\partial \chi_j (q_j)}{\partial u_i} = 0 \quad i = 1,2,\ldots m \quad \text{(III.B.48)}$$

Corner Conditions

When the trajectory intersects the constraint surfaces

$$H(t^-) = H(t^+) - \sum_{j=n+1}^{J} \underline{\Pi}_j^T \left(\frac{\partial N_i}{\partial t} \right) \tag{III.B.49}$$

$$p_i(t^-) = p_i(t^+) + \sum_{j=n+1}^{J} \underline{\Pi}_j^T \left(\frac{\partial N_j}{\partial x_i} \right) i = 1,2,\ldots n \tag{III.B.50}$$

and otherwise

$$H(t^+) = H(t^-) \tag{III.B.51}$$

$$p_i(t^+) = p_i(t^-) \qquad i = 1,2,\ldots n \tag{III.B.52}$$

Transversality Conditions

$$d\Phi + \underline{e}^T \, d\underline{\Psi} + [H \, dt - \underline{p}^T \, d\underline{x}] \Big|_{t_0}^{t_1} = 0 \tag{III.B.53}$$

Weierstrass Condition

$$H(\underline{x}^*, p^*, \underline{u}, t) \geq H(\underline{x}^*, p^*, \underline{u}^*, t) \tag{III.B.54}$$

As in the previous case, the inequality constraints introduce extra terms in the Euler Lagrange equations which are nonzero only when the trajectory lies on the constraint surface.

4. Iterative Solution Method for Two-Point Boundary Value Problems*

The necessary conditions for optimality shown in the previous section constitute a set of coupled nonlinear ordinary differential and algebraic equations subject to mixed boundary conditions -- the conditions on some variables

---

* The material of this section is based on a review made by Dr. Roger Rydin. An excellent discussion of the methods is presented in the book by Bryson and Ho (13).

apply at the initial time while others, as a result of the
transversality condition, apply at the final time. For very
simple problems, analytic solutions may be possible; alternate geometric constructions in phase space may be made.
These will be illustrated in Section IV.B where applications
to nuclear engineering problems are discussed. If all equations are linear, a formal matrix Riccatti equation procedure
exists for generating a solution to the equations.

In the general nonlinear case, however, a computer
numerical solution procedure is required. Standard numerical
integration routines for ordinary differential equations do
not apply here because of the absence of initial conditions
on some of the variables. There are three general methods
available for solving the two-point boundary value problem;
they are, (1) Boundary condition iteration, usually done
using the Newton-Raphson (or quasilinearization) method;
(2) Control signal iteration which is a gradient (steepest
descent or hill-climbing) technique using a first-order
variation of the functional to be minimized; and (3) Control
law iteration which is a gradient technique using first and
second order variations of the functional to be minimized.
A discussion of these methods follows.

a. Boundary Condition Iteration - The basic idea of
boundary condition iteration is to guess initial
values for the state variables for which only final
conditions are specified, solve all of the equations
simultaneously in the forward direction and to use
the resulting information to help choose an improved
set of initial conditions for the next iteration.
The selection of improved boundary conditions is
done by a quasilinearization technique. For example,
suppose 2n differential equations are to be solved:

$$\frac{d}{dt} \underline{x} = \underline{f}(\underline{x}, t); \quad \underline{x}(t_o) = \underline{x}_o,$$
(III.B.55)

and

$$\frac{d}{dt} \underline{p} = \underline{g}(x, \underline{p}, t); \quad \underline{p}(t_1) = \underline{p}_1.$$
(III.B.56)

Here it is assumed the Euler-Lagrange equations $\frac{\partial H}{\partial \underline{u}} = 0$

have been used to eliminate the control from the
equation set. (For many problems it will be impossible to express the control explicitly in terms of
the state and Lagrange multiplier variables. In that
case, one of the other two approaches must be used.)

Considering k to be an iteration index, and defining
$F_{\underline{x}}^k$ and $G_{\underline{x},\underline{p}}^k$ to be Jacobian matrices containing the
first-order partial derivatives of $\underline{f}$ and $\underline{g}$ with
respect to $\underline{x}$ and $\underline{p}$, a Taylor series expansion of
Equations III.B.55 and III.B.56 about the kth iterate,
keeping only first-order terms can be written as

$$\frac{d}{dt} \underline{x}^{k+1} = \underline{f}(\underline{x}^k, t) + F_{\underline{x}}^k (\underline{x}^{k+1} - \underline{x}^k) \qquad (III.B.57)$$

and

$$\frac{d}{dt} \underline{p}^{k+1} = \underline{g}(\underline{x}^k, \underline{p}^k, t) + G_{\underline{x}}^k (\underline{x}^{k+1} - \underline{x}^k)$$

$$+ G_{\underline{p}}^k (\underline{p}^{k+1} - \underline{p}^k) \qquad (III.B.58)$$

where all quantities with the superscript k can be
considered to be known from the previous iteration.
An improved guess for the initial condtions on $\underline{p}$ is
obtained by deriving a partitioned state transition
matrix for the homogeneous form of Equations
III.B.57 and III.B.58. This is accomplished by
solving the 2n homogeneous equations numerically,
i=1,n separate times, taking $x_{iH}(t_o) = 0$ and
$p_{jH}(t_o) = \delta_{ij}$; j=1...n. The term "partitioned" is
used because the unit initial conditions are applied
only to the equations in $\underline{p}$.

For each i, a row of the nxn (p only) state trans-
ition matrix $\underline{\emptyset}$ is obtained, where the jth term is
$p_{jH}(t_1)$. Once the full matrix $\underline{\emptyset}$ is available, the
new initial conditions are obtained by a matrix
inversion,

$$\underline{p}^{k+1}(t_o) - \underline{p}^{k}(t_o) = \underline{\emptyset}^{-1} \cdot (\underline{p}^{k}(t_1) - \underline{p}_1). \quad \text{(III.B.59)}$$

Then the inhomogeneous set of equations is numerically solved for the k+1th iterate; the entire process is repeated until convergence is obtained.

The primary advantage of the method is that is converges rapidly (convergence is quadratic if it converges at all). However, there is no guarantee that the method will converge, and it is sensitive to the initial set of guesses taken for $\underline{p}$. Further, none of the intermediate solutions are extremals, although an extremal is reached upon convergence. Finally, it must be possible to eliminate the control in terms of the other variables.

b. Control Signal Iteration - The control signal iteration method is based on the observation that the n system equations are only coupled to the n Euler-Lagrange equations through the control vector $\underline{u}$. Therefore, if one were to guess a value of $\underline{u}^{k}(t)$, where k is an iteration index, enough information would be available to completely solve the n system equations forward in time and subsequently to solve the n Euler-Lagrange equations backward in time using the just-obtained forward solutions. The problem then becomes one of computing a set of corrections $\delta\underline{u}^{k}$ which improves the control guess for the next iteration,

$$\underline{u}^{k+1} = \underline{u}^{k} + \delta\underline{u}^{k}. \quad \text{(III.B.60)}$$

The corrections are computed by a gradient technique which takes one of two general forms. In the first form, $\delta\underline{u}$ is chosen to be proportional to the negative of the gradient of the H function with respect to $\underline{u}$, such that the correction is always in a direction which reduces the value of H:

$$\delta u_i^{k} = -\varepsilon_i^{k} \frac{\partial H}{\partial u_i^{k}}; \quad i = 1, m, \quad \text{(III.B.61)}$$

where $\underline{\varepsilon}$ is a vector of undetermined constants (possibly time varying) which must be postulated or found by auxiliary means. If $\underline{\varepsilon}$ is too small, convergence is slow; if $\underline{\varepsilon}$ is too large, instability may result. Notice, in addition, that the magnitude of $\partial H/\partial \underline{u}$ is directly introduced into $\delta \underline{u}$, which may lead to problems of convergence or stability in regions having large gradients, or to insensitivity where the gradient is small. In the second form, a somewhat relaxed condition is used, namely

$$\delta u_i^k = - \varepsilon_i^k \; \text{sign}\left[\frac{\partial H}{\partial u_i^k}\right];$$

(III.B.62)

$$\text{sign}\,[z] = \begin{cases} 1, z>0 \\ 0, z=0 \\ -1, z<0 \end{cases}$$

At the expense of more slowly modifying the control vector $\underline{u}$, stability and convergence problems associated with using the magnitude of $\partial H/\partial \underline{u}$ in regions where the gradient is large are alleviated. The value of ε^k (which is usually chosen as a single constant ε instead of a vector), can be altered during the calculation (halved or doubled) by comparing successive values (k+1 vs. k) of the running performance index. An important feature of either form is the ability to directly include control constraints in the value of $\underline{u}^{k+1}$. Specifically, a test is made to see if the computed value of $\underline{u}^{k+1}$ violates a constraint, and if so, the new value is modified. Hence,

$$\underline{u} = \begin{cases} \underline{u}_{max} & \text{if } (\underline{u}^k + \delta\underline{u}^k) > \underline{u}_{max} \\ \underline{u}^k + \delta\underline{u}^k & \text{if } \underline{u}_{min} < (\underline{u}^k + \delta\underline{u}^k) < \underline{u}_{max} \\ \underline{u}_{min} & \text{if } (\underline{u}^k + \delta\underline{u}^k) < \underline{u}_{min} \end{cases}$$

(III.B.63)

An advantage of the method is its almost guaranteed convergence, even starting from a poor initial guess for the control. The primary disadvantage is that convergence is generally slower than in the other methods, especially in the vicinity of the optimum. To balance

this, however, it must be remarked that a near-optimal
solution would still be of value in a practical sense.

c. Control Law Iteration - The control law iteration method,
   sometimes called the second variation method, is similar
   to the control signal iteration method in philosophy.
   The primary difference is that second-order terms in
   the Taylor series expansion of H are retained; this
   leads to an additional 2n differential equations to
   be solved simultaneously with the original 2n system
   and Euler-Lagrange equations. The primary advantage is
   that the step size (equivalent to ε) is inherently determined. Convergence is rapid near the optimum, and
   hence computation time can be less than is required for
   the control signal method. The primary disadvantage is
   the increased programming complexity since many more
   equations and derivatives are involved.

For many practical problems, combinations of the above
approaches are required. The methods are not directly applicable in the cases of state variable or mixed constraints.

C.   Dynamic Programming for Multistage Decision Processes

    The dynamic programming approach to optimization derives
from the intuitive principle of optimality as stated by
Bellman, "An Optimal Sequence of decisions in a multistage
decision process has the property that whatever the initial
stage, state, and decision are, the remaining decisions must
constitute an optimal sequence of decisions for the remaining
problem which has as its initial condition the stage and state
resulting from the first decision."(8) This principle gives
rise to recursion formulas for solving a multistage decision
process. They have the effect of breaking an N-dimensional
optimization problem into a sequence of N one-dimensional
problems. The recursion formulas and the mechanics of
implementing them can proceed either in a stage (or time)-
increasing direction or vice versa, depending on which
boundary conditions are known for the state variables.

    The concepts are best presented by means of a simple
example. In the next section both forward and backward recursion solutions to a simple discrete multistage decision
process are presented. Following that, it is shown how, in
the limits of a continuous multistage decision process, the

recursion relations yield the necessary conditions for
optimality obtained in Section III.B by means of the calculus
of variations.

1.   A Discrete Dynamic Program Example

Consider the problem of minimizing the time to travel
from point 1 to point 2 when faced with a complex of one-
way streets, for each of which the travel time is known.
Figure III.C.1 shows the system, with the numbers along each
segment giving the travel time for that segment.

An i-j coordinate grid is set up as shown in the figure
and the position of the traveler within the network, or his
state, is represented by the coordinates i-j. Each horizontal
interval represents a stage; it is seen there are six stages.
At the start of stage 1 only one state is possible; it is

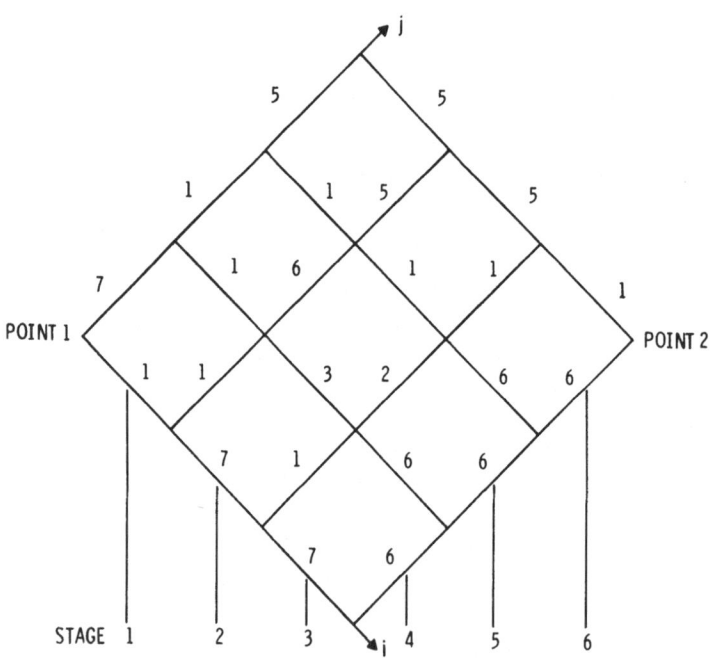

Figure III.C.1
A Discrete Network Problem

Point 1 or state (0,0). At the start of the second stage, two states are possible: (0,1) or (1,0). The state reached at stage 2 depends on the decision made during stage 1; whether to follow the path to the upper right or the path to the lower right.

The minimum time to Point 2 from i-j is a function of only the state i-j. Recognizing this, $S(i,j)$ may be defined:

$$S(i,j) = \text{Minimum time to point 2 from } i\text{-}j \qquad \text{(III.C.1)}$$

where
$$S(3,3) = 0 \qquad\qquad\qquad \text{(III.C.2)}$$

At each state, two paths lead to the next stage. Since the principle of optimality indicates that if there is more than one path leaving a state and leading eventually to the destination, only the best need be selected and retained for the rest of the calculations, the following backward recursion relationship can be written:

$$S(i,j) = \text{Min} \begin{bmatrix} \text{up:} & t(i,j;i,j+1) + S(i,j+1) \\ \text{down:} & t(i,j;i+1,j) + S(i+1,j) \end{bmatrix} \text{(III.C.3)}$$

Here, t is the travel time between states on the indicated path. For this backward algorithm, the starting point is Point 2. At each state, two paths lead to (different states at) the start of the next stage. By the simple addition and comparison indicated by Equation III.C.3, the "minimum time path to Point 2" is selected and the other path discarded from further consideration. Working backward from stage 6, each state of each successively earlier stage is labeled with the minimum travel time to Point 2 and an arrow indicating the minimum travel time path. When the start of stage 1 is reached, the minimum-time path from Point 1 to Point 2 may be identified by simply following the arrows. Figure III.C.2 shows the implementation of this algorithm.

This problem may also be solved with a forward algorithm. Recognizing that the principle of optimality indicates that if there is more than one path leading to a state from Point 1, only the best need be considered for the rest of the calculations, $S(i,j)$ may be defined:

$$S(i,j) = \text{Minimum time to } i\text{-}j \text{ from Point 1} \qquad \text{(III.C.4)}$$

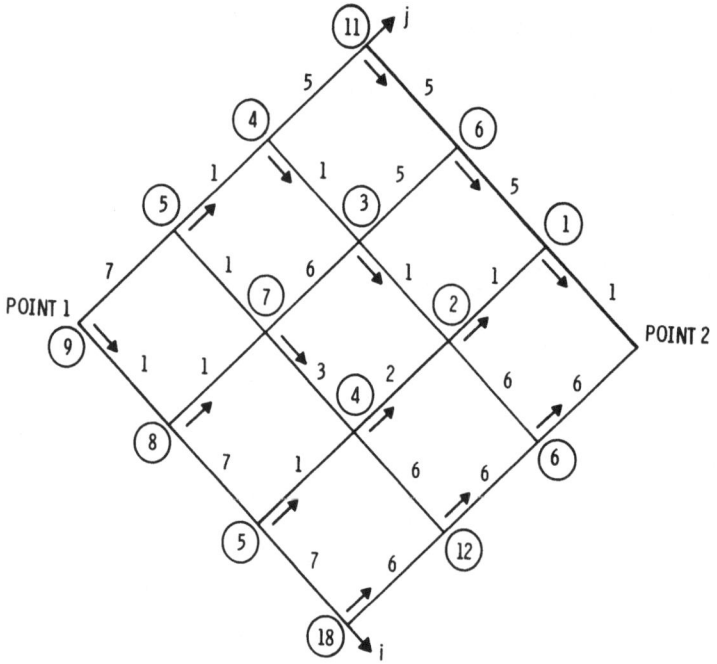

Figure III.C.2
Backward Dynamic Programming Algorithm

where

$$S(0,0) = 0 \qquad\qquad (III.C.5)$$

and the following recursion relationship written:

$$S(i,j) = Min \begin{bmatrix} up: & S(i,j-1) + t(i,j-1;i,j) \\ \\ down: & S(i-1,j) + t(i-1,j;i,j) \end{bmatrix} (III.C.6)$$

Here the algorithm is started at Point 1, and working forward, each state of each successive state may be labeled with the minimum time from Point 1 to it and a backward-pointing arrow along the minimum-time path leading to the

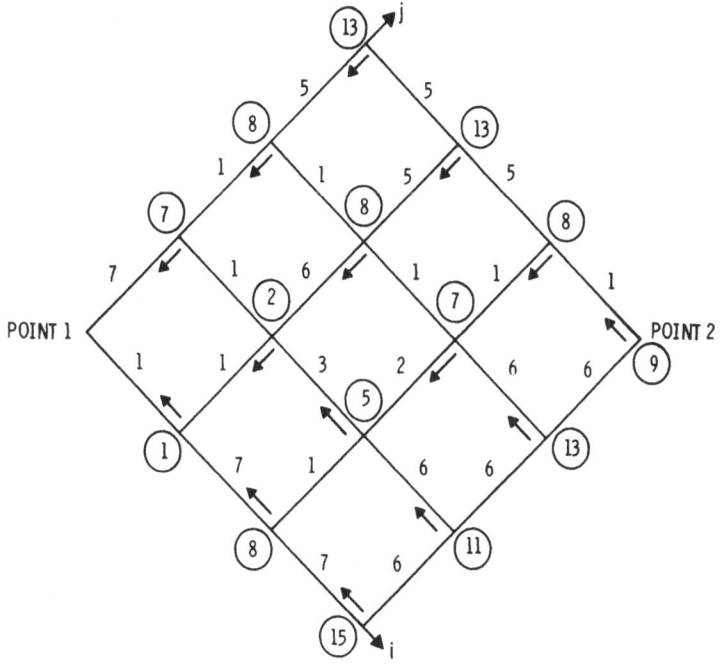

Figure III.C.3
Forward Dynamic Programming Algorithm

state. When Point 2 is reached, the arrows may be followed
backward to Point 1 along the optimal path. The implementation of this algorithm is shown in Figure III.C.3. Comparison with Figure III.C.2 shows the same optimal path is
determined by both algorithms. It passes through the states
0-0, 1-0, 1-1, 2-1, 2-2, 2-3, and 3-3. The travel time is
9 units.

This simple example displays many of the inherent features of the dynamic programming approach to optimization.
First, by eliminating paths which have no potential for being
optimal at the earliest possible stage, a significant reduction is achieved in the number of paths that must be evaluated.
This in effect is a "tree pruning" process, and it leads to a
reduction in the computer time and memory requirements needed
to find the optimal path over those required if all possible
paths are evaluated and compared only at the end.

Another feature of the dynamic programming approach is that it generates the optimal path not just for given initial or final states but for all states in between. This is in contrast to the variational approach which yields the optimal paths from the given initial state but no other. The forward and backward algorithms give the same optimal solution, but different information. The backward algorithm gives the best path from any state to the destination; whereas, the forward algorithm gives the optimal path from the initial state to every state. The choice of method in many cases depends on whether the initial or final state is known.

Note is taken of the ease with which the dynamic programming algorithm accommodates constraints on the control. In the sample problem, the control was the direction of leaving a state - either up or down. The control was constrained because of all possible directions only two were admissible. The dynamic programming algorithm accommodates constraints on the control by the simple procedure of selecting choices only from the allowed set of controls (which can be stage and state dependent).*

Finally, it is instructive to note the amount of information required to be saved in the dynamic programming algorithms. For each state of each stage the following data are required to be saved:

the stage identification

the state identification

the value of the optimal cost functional, S, and

the control decision of the state

Because of the way the backward and forward algorithms "fan out" from the final and initial states respectively, problems with many stages or high dimensionality in the state or control variables may require enormous amounts of data to be stored during the generation of the optimal solution.

---

*

Similarly, constraints on the state, while readily realized, have the advantage in dynamic programming of restricting the possible paths for evaluation still further.

2.   Dynamic Programming and the Calculus of Variations

The principle of optimality provides an alternative to the calculus of variations for deriving optimality conditions for systems represented in the differential form and provides a computing algorithm for determining the optimal control as well.

This section will present an outline of the derivation for a simple problem to illustrate the connection between dynamic programming and the calculus of variations. Consider a simple optimal control problem:

$$\underset{\underline{u}}{\text{Min}} \left\{ f_o(\underline{x},\underline{u}) = \int_{t_0}^{t_1} L(\underline{x},\underline{u},t)\ dt \right\} \qquad \text{(III.C.7)}$$

for a differential system:

$$f_i(\underline{x},\underline{u},t) - \frac{d}{dt}\ x_i = 0 \quad i = 1,2,\text{---}n \qquad \text{(III.C.8)}$$

subject to the initial conditions:

$$x_i(t_o) = x_i^o \qquad\qquad i = 1,2,\ \text{---}\ n \qquad \text{(III.C.9)}$$

Here, $\underline{x}$ is an n-component state vector and $\underline{u}$ is an m-component control vector.

It is recognized that the __minimum__ value of an integral of the form of f_o depends only on the initial time and position. Define

$$S(\underline{x}(t),t) = \underset{\underline{u}}{\text{Min}} \left\{ \int_{t}^{t_1} L(\underline{x},\underline{u},t)\ dt \right\} \qquad \text{(III.C.10)}$$

where $\underline{x}(t)$ denotes the state of the system at time = t.

It is clear that

$$S(\underline{x},t_1) = 0 \qquad \text{for all } \underline{x} \qquad (III.C.11)$$

Equation III.C.10 may also be written as:

$$S(\underline{x},t) = \underset{\underline{u}}{\text{Min}} \left\{ \int_t^{t+\Delta} L(\underline{x},\underline{u},t)\, dt + \int_{t+\Delta}^{t_1} L(\underline{x},\underline{u},t)\, dt \right\} (III.C.12)$$

where

$$\Delta \leq t_1 - t_0 \text{ is an arbitrary interval of time.}$$

The principle of optimality applied here states that along the optimal path from t to t_1, regardless of what path is taken from t to $t+\Delta$, the path from the resulting state at $t + \Delta$ to time t_1 is optimal for that starting state and time. Thus, Equation III.C.12 may be rewritten as

$$S(\underline{x},t) = \underset{\underline{u}}{\text{Min}} \left\{ \int_t^{t+\Delta} L(\underline{x},\underline{u},t)\,dt + S(\underline{x}(t+\Delta),\ t + \Delta) \right\} (III.C.13)$$

This computational algorithm may be compared with Equation III.C.3 for the discrete multistage decision process. It is clear that by dividing the time interval $[t_0, t_1]$ into N segments (or stages) of length Δ one could begin the computation at $t = t_1$ and successively work backward to time, $t_1 - \Delta$, $t_1 - 2\Delta$, etc. (using the system equations to determine the values of the state variables of each stage) in exactly the manner illustrated in Figure III.C.2.

Alternately, to derive the necessary conditions for optimality, Δ will be taken arbitrarily small, the mean value theorem will be used for the integral term and $S(t+\Delta)$ will be expressed as a Taylor series about $S(t)$. Neglecting second order terms, Equation III.C.13 becomes:

$$S(\underline{x},t) = \underset{\underline{u}}{\text{Min}} \left\{ L(\underline{x},\underline{u},t)\ \Delta + S(\underline{x},t)\ + \left(\frac{\partial S}{\partial t}\right)\Delta + \left(\frac{\partial S}{\partial \underline{x}}\right)^T \underline{f}(\underline{x},\underline{u},t)\Delta \right\}$$

which yields the condition for the optimal trajectory:

$$0 = \underset{\underline{u}}{Min} \left\{ L(\underline{x},\underline{u},t) + \frac{\partial S}{\partial t} + \left(\frac{\partial S}{\partial x}\right)^T \underline{f}(\underline{x},\underline{u},t) \right\} \qquad (III.C.14)$$

For piecewise continuous control functions, the minimization indicated by Equation III.C.14 may be achieved by requiring

$$\frac{\partial}{\partial u_i} \left\{ L(\underline{x},\underline{u},t) + \left(\frac{\partial S}{\partial x}\right)^T \underline{f} \right\} = 0 \quad i = 1,2,\ldots m \quad (III.C.15)$$

between discontinuities in $\underline{u}$. Beyond that, however, Equation III.C.14 requires that at each point along the optimal trajectory:

$$-\frac{\partial S}{\partial t} = L(\underline{x},\underline{u},t) + \left(\frac{\partial S}{\partial x}\right)^T \underline{f}(\underline{x},\underline{u},t) \qquad (III.C.16)$$

By imposing the definitions:

$$\underline{p} = \left(\frac{\partial S}{\partial x}\right) \qquad (III.C.17)$$

$$H(\underline{x},\underline{u},t,\underline{p}) = L(\underline{x},\underline{u},t) + \left(\frac{\partial S}{\partial x}\right)^T f(\underline{x},\underline{u},t) = \qquad (III.C.18)$$

$$L(\underline{x},\underline{u},t) + \underline{p}^T \underline{f}(\underline{x},\underline{u},t)$$

Equation III.C.15 assumes the form of the familiar Euler-Lagrange equation:

$$\frac{\partial}{\partial u_i} H(\underline{x},\underline{u},t,\underline{p}) = 0 \qquad i=1,2, \text{---} m \qquad (III.C.19)$$

At the "corners" where $\underline{u}$ is discontinuous, Equation III.C.14 is still required to hold.

Since $\frac{\partial S}{\partial t}$ has no dependence on $\underline{u}$, the two familiar corner conditions:

$$H(\underline{x},\underline{u},t^{+}) = H(\underline{x},\underline{u},t^{-})$$

$$\underline{p}(t^{+}) = \underline{p}(t^{-})$$

(III.C.20)

are necessary and sufficient to guarantee that

$$L(\underline{x},\underline{u},t) + \left(\frac{\partial S}{\partial x}\right)^{T} \underline{f}(\underline{x},\underline{u},t) = 0$$

as the corner is approached from either the positive or negative side.

To develop the remaining Euler-Lagrange equations, it is noted that by the definition of p (Equation III.C.17) and the chain rule:

$$\frac{d}{dt} p_{i} = \frac{d}{dt} \frac{\partial S}{\partial x_{i}} = \frac{\partial^{2} S}{\partial t \partial x_{i}} + \sum_{j=1}^{n} \frac{\partial^{2} S}{\partial x_{j} \partial x_{i}} f_{j} \quad i = 1,2, \ldots n$$

Differentiating Equation III.C.16 with respect to x_{i} yields

$$- \frac{\partial^{2} S}{\partial t \partial x_{i}} = \frac{\partial L}{\partial x_{i}} + \sum_{j=1}^{n} \frac{\partial^{2} S}{\partial x_{i} \partial x_{j}} f_{j} + \frac{\partial S}{\partial x_{i}} \frac{\partial f_{j}}{\partial x_{i}} \quad i = 1,2,\ldots n$$

When this expression for $\frac{\partial^{2} S}{\partial t \partial x_{i}}$ is substituted into the previous equation there results

$$\frac{d}{dt} p_{i} = - \frac{\partial}{\partial x_{i}} \left\{ L + \sum_{j=1}^{n} \frac{\partial S}{\partial x_{j}} f_{j} \right\} \quad i = 1,2, \ldots n$$

or, expressed in vector form

$$\frac{d}{dt} \underline{p} = \left(\frac{\partial H}{\partial x}\right)$$

(III.C.21)

it reduces to the familiar Euler-Lagrange equation.

Again, noting that $\frac{\partial S}{\partial t}$ has no dependence on $\underline{u}$, Equation III.C.14 is recognized as a statement of the Weierstrass condition:

$$0 = \underset{\underline{u}}{\text{Min}} \left\{ H(\underline{x},\underline{u},\underline{p},t) + \frac{\partial S}{\partial t} \right\}$$

Or

$$H(\underline{x}^*,\underline{u},\underline{p}^*,t) \geq H(\underline{x}^*,\underline{u}^*,\underline{p}^*,t) \quad t_0 \leq t \leq t_1 \qquad (III.C.22)$$

where

super * denotes the optimal trajectory.

Finally, since a final condition $\underline{x}(t_1)$ is not specified, the optimal value of the integral of f_o should be insensitive to its value:

$$\left(\frac{\partial S}{\partial \underline{x}} \right)\Bigg|_{t=t_1} = \underline{0}$$

With Equation III.C.17, this is recognized as the familiar natural boundary condition on the Lagrange multiplier variables:

$$\underline{p} \Bigg|_{t=t_1} = \underline{0} \qquad (III.C.23)$$

Alternately, if the problem statement were altered to require that the final state lie on a specified surface:

$$\underline{\psi}(\underline{x}(t_1),t_1) = \underline{x} - \underline{g}(t) = \underline{0} \qquad (III.C.24)$$

then the optimal value of the integral of f_o should be insensitive to the final time so long as the final state lies on the prescribed surface:

$$\frac{d}{dt} S \Bigg|_{\underline{x} \varepsilon \underline{g}(t)} = 0$$

This becomes

$$\frac{\partial S}{\partial t} + \left(\frac{\partial S}{\partial x}\right)^T \frac{d}{dt} g = 0$$

or, substituting for $\frac{\partial S}{\partial t}$ from Equation III.C.16 and for $\frac{\partial S}{\partial x}$

from Equation III.C.17, this becomes

$$-\left[H + \underline{p}^T\left(\frac{d\underline{g}}{dt}\right)\right]\Bigg|_{t=t_1} = 0 \qquad\qquad (III.C.25)$$

the transversality condition.

To summarize, it is seen that the principle of optimality leads to the requirement of Equation III.C.14 for the optimal trajectory. Examination of the implications of this equation lead to all the familiar results obtained in a completely different way in the calculus of variations including:

a.   the Euler-Lagrange equations III.C.19 and III.C.21

b.   the Weierstrass-Erdmann corner conditions, Equation III.C.20

c.   the Weierstrass condition, Equation III.C.22

d.   the transversality condition, Equation III.C.25

Further, a physical interpretation is given to the Lagrange multiplier variables, $p_i(t)$. Equation III.C.17 indicates that the value of $p_i(t)$ gives the amount by which the optimal value of the performance index changes per unit change in x_i at time t with the control held fixed.

IV.   NUCLEAR ENGINEERING APPLICATIONS
OF OPTIMAL CONTROL THEORY

A.   <u>Introduction</u>

In the following sections, selected applications of optimal control theory to nuclear engineering problems are discussed. These have been chosen to illustrate the character of physical and mathematical problems encountered in

problems of design interest and the features of the various
mathematical approaches which are employed in their solution.
Not all of the branches of nuclear engineering in which optimal control applications have been published:

    Reactor startup

    Xenon shutdown

    Suppression of Xenon Spatial Oscillations

    Shield design

    *Beginning of Life Design

    *Poison Management

    *Fuel Management

    Plant Allocation

are treated. The starred topics are treated here. Reactor
startup problems are discussed at length by Mohler and Shen(14).
The Xenon shutdown problem has been the subject of a review
by Lewins and Babb (26). The plant allocation problem is
considered outside the scope of this article which places its
emphasis on the physics problems of reactor design and operation.

    The next section contains a discussion of material
placement at BOL*. This problem, in one dimension, is well-
suited for a straightforward application of the variational
theory and is amenable to analytic solution. The treatment
of various combinations of control and state variable constraints is illustrated. (Shielding design optimization is
quite similar to the BOL material placement problems discussed and can be treated by the same methods).

    Fuel cycles are discussed following BOL material place-
ment. These problems introduce complexities due to the requirement of dealing with both space and time variations.
Fuel cycle problems can be conveniently represented as
multistage decision processes suitable for solution by
dynamic programming and linear programming techniques.

---

*   BOL denotes the "Beginning of Life" or initial loading of
    a nuclear reactor core.

Finally, poison management through one reload cycle is discussed. As with fuel cycles, space and time variations must be treated. Furthermore, the modeling of the neutron balance in the core must be relatively detailed. This subject is treated in detail as a result of the authors' interests and the timeliness of the topic. Suppression of spatial xenon oscillation problems are not discussed but are amenable to solution by the poison management procedures.

B.   Beginning of Life (BOL) Design

1.   Introduction

In this section the problem of determining fissile and nonfissile material distribution in a critical, BOL reactor to optimize some aspect of reactor performance is treated. There is an extensive literature (38-49) on this type problem. In all cases, the problem is formulated as a variational problem and the equations of Section III.B used to determine the necessary conditions for optimality. This approach leads to a determination of the allowed types of regions and conditions to be satisfied by the arrangement of these allowed region types. However, it does not indicate how many regions to use or the order in which to arrange them. The illustrative problems discussed below demonstrate different approaches for synthesizing the optimal design from the necessary conditions. Also, different classes of inequality constraints are treated.

In the first example, Terney (46) used a least squares performance index to minimize the deviation of the flux distribution from its average value by manipulating the k_∞ distribution. The difficulties in constructing the optimal reactor from the allowable regions are pointed out. In this problem, the inequality constraint involves the control only. In the second example, Goldschmidt and Quenon (45) used phase-space arguments to facilitate the "a priori" synthesis of the minimum critical mass fuel distribution. Here the inequality constraint displays an explicit dependence on both control and state variables. In the last problem, following Axford (43), the important class of problems involving state variable constraints which are only implicitly dependent in the control is illustrated. A table is included which summarizes the known analytic solutions for BOL material placement problems.

2.   The Flux Flattening Problem - Control Constraints

In this problem, Terney determined the k_∞ distribution to minimize the square deviation of the flux from its average value in a symmetric, 1-D slab reactor described by one group theory. The control was limited to lie in a range between a maximum and minimum k_∞.

The problem is

$$\text{Minimize } I_1 = \int_0^{t_f} (\phi - 1)^2 \, dt$$

subject to the normalization requirement

$$\int_0^{t_f} \phi \, dt = t_f$$

the flux equation

$$\frac{d^2}{dt^2} \phi(t) + \frac{k_\infty(t) - 1}{M^2} \phi(t) = 0$$

and the boundary conditions

$$\frac{d\phi(0)}{dt} = 0 \; ; \; \phi(t_f) = 0$$

at the centerline and known outer core boundary, $t = t_f$.

The control, $k_\infty(t)$, is bounded:

$$k_{\infty_{Min}} \leq k_\infty(t) \leq k_{\infty_{Max}}$$

Since the flux integral is fixed, the quantity to be optimized may be written as

$$\text{Maximize } I_2 = -\int_0^{t_f} \phi^2 dt$$

where

ϕ = The flux, is the state variable,

k_∞ = The infinite multiplication factor, is the control variable, and

M^2 = Migration area

t = Space dimension

Introducing the continuous variables, x_o and x_3, and letting x_1 be the flux and x_2 the current, and taking the control u as k_∞, the problem may be stated as

$$\underset{k_{\infty Min} \le u \le k_{\infty Max}}{\text{Maximize}} \quad x_o(t_f) \tag{IV.B.1}$$

subject to the state equations and boundary conditions

$$-x_1^2 - \frac{d}{dt} x_o = 0 \tag{IV.B.2}$$

$$x_2 - \frac{d}{dt} x_1 = 0 \tag{IV.B.3}$$

$$\frac{1-u}{M^2} x_1 - \frac{d}{dt} x_2 = 0 \tag{IV.B.4}$$

$$x_1 - \frac{d}{dt} x_3 = 0 \tag{IV.B.5}$$

$$x_o(0) = 0$$

$$x_1(t_f) = 0$$

$$x_2(0) = 0 \qquad \text{(IV.B.6)}$$

$$x_3(0) = 0$$

$$x_3(t_f) = t_f$$

This problem involves differential system equations only and an inequality constraint on the control alone. It is thus in a form suitable for solution by the Pontryagin maximum principle, Equations III.B.32 through III.B.34.

Define the Pontryagin function:

$$H = \sum_{i=0}^{3} \Psi_i(t)\, f_i(\underline{x},t) \qquad \text{(IV.B.7)}$$

$$= \psi_o(-x_1^2) + \psi_1(x_2) + \psi_3\left(\frac{1-u}{M^2}x_1\right) + \psi_3(x_1)$$

$$= -\psi_2\, x_1\, \frac{u}{M^2} + \left\{ \psi_1 x_2 - \psi_o x_1^2 + \left(\psi_3 + \frac{1}{M^2}\psi_2\right)x_1 \right\}$$

as the sum of products of Lagrange multipliers, ψ_i, times the nondifferential parts of the system equations. The Pontryagin function is known to be a constant since it is not explicitly dependent on t.

The Euler equations are given below with their corresponding external boundary conditions which arise from the transversality condition, Equation III.B.26.

$$\frac{d\Psi_o}{dt} = 0; \; \Psi_o = C_o = 1 \qquad \text{(IV.B.8)}$$

$$\frac{d\psi_1}{dt} = \frac{u-1}{M^2}\Psi_2 + 2\Psi_o\, x_1 - \Psi_3\; ; \; \Psi_1(o) = 0 \qquad \text{(IV.B.9)}$$

$$\frac{d\Psi_2}{dt} = -\Psi_1 \; ; \; \Psi_2(t_f) = 0 \qquad\qquad (IV.B.10)$$

$$\frac{d\Psi_3}{dt} = 0 \; ; \; \Psi_3 = C \qquad\qquad (IV.B.11)$$

The internal boundary conditions arising from the corner conditions on H and $\underline{\Psi}$ imply that

$$\Psi_2(t_i) = 0 \qquad\qquad (IV.B.12)$$

at interfaces (t_i) where u is discontinuous.

The Pontryagin optimality condition requires that

$$H(u^*,\Psi^*,x^*) \geq H(u,\Psi^*,x^*) \qquad\qquad (IV.B.13)$$

where u* is the optimal control.

Since the control is bounded above and below, three possibilities must be examined: u* lying at either boundary or u* lying between the boundaries. These possibilities give rise to three types of region for the makeup of the reactor. For example, if it is assumed that u* lies at the upper limit of its range, the only allowable variation in its value is for it to decrease. If $k_{\infty_{max}}$ is optimum, any change should not make H larger. From Equation IV.B.7, it is seen that it implies that ψ_2 must be less than zero. Similarly, for $u = k_{\infty_{min}}$ to be optimum, ψ_2 must be greater than zero.

Suppose u* lies between the boundaries. Then by considering small variations in a strong neighborhood of u*, it is clear that a necessary condition for optimality is

$$\frac{\partial H}{\partial u} = 0$$

From the definition of H, Equation IV.B.7, this implies that $x_1\psi_2 = 0$ over that section of core for which the control is not at either of its boundaries. Physical realizability rules out the option of zero flux, $x_1 = 0$; thus, ψ_2 is re-

quired to be zero over the region. Equation IV.B.10 then
requires that $\psi_1 = 0$ over this same region. These conditions
when used in Equation IV.B.9 with the conditions $\psi_3 =$ constant
from Equation IV.B.11 and $\psi_0 = 1$ yield

$$x_1 = \psi_3/2 = \text{constant flux}$$

This condition in Equation IV.B.3 shows that

$$x_2 = 0 \text{ zero current}$$

and when this condition is used in Equation IV.B.4, it yields

$$u = k_\infty = 1$$

Thus, the optimum reactor can be made up of only three
types of regions, namely:

$$u = k_{\infty \text{max}} \qquad \text{when } \psi_2 < 0 \qquad\qquad \text{(IV.B.14)}$$

$$u = k_{\infty \text{min}} \qquad \text{when } \psi_2 > 0 \qquad\qquad \text{(IV.B.15)}$$

$$u = 1 \qquad \text{when } \psi_2 = 0 \qquad\qquad \text{(IV.B.16)}$$

$$\psi_1 = 0$$

$$x_1 = \psi_3/2$$

$$x_2 = 0$$

It can be shown that a unity k_∞ region is not allowed
at all in the optimal reactor; the optimal reactor must be
made of only maximum and minimum k_∞ regions. This can be
seen by writing analytic solutions for the flux and Lagrange
multiplier equations, for each region type, exclusive of
boundary conditions, postulating various arrangements, and
comparing the number of unknown constants with the number of
boundary conditions. Only for combinations of maximum and
minimum k_∞ regions is a consistent solution to all the
optimization equations possible.

The nature of the outer region can be determined by examining the value of H at the outer boundary.

$$H\Big|_{t=t_f} \equiv \psi_1(t_f)x_2(t_f) \qquad \text{(IV.B.17)}$$

If the outer region is $k_{\infty min}$, ψ_1 must be positive from Equation IV.B.10 since ψ_2 is zero at the outer boundary and must be greater than zero within the region. Therefore, H has a negative value, since leakage is out of the core making $x_2(t_f)$ negative. Conversely, if the outer region is $k_{\infty max}$, H is positive. Since for the optimal reactor H must have its greatest value, it is concluded that the outer region must be one of $k_{\infty max}$.

From the foregoing arguments, the optimal reactor is known to consist of regions of minimum and maximum k_∞, with the outer region required to be $k_{\infty max}$. Remarkably,

these conclusions are obtainable by inspection. Progress becomes more tedious at this point, however. There is no formal way to determine how many region types there must be or how to arrange them. The number and arrangement depends on the magnitudes of $k_{\infty max}$ and $k_{\infty min}$. The optimal synthesis of allowed region types can be determined by postulating arrangements by trial and error until the Pontryagin conditions are satisfied and the Lagrange multiplier, ψ_2 has the proper sign in each region.

Analytic solutions were obtained for two and three region reactors (46) and were examined for a variety of values for $k_{\infty max}$ and $k_{\infty min}$. Numerical results were obtained for a reactor of half width 100 cm and M^2 of 50 cm^2. Figure IV.B.1 shows the flux and Lagrange multiplier distribution for a two-region reactor with maximum and minimum k_∞'s of 1.02 and 0.99 respectively. It can be seen that ψ_2 is negative in the region of $k_{\infty max}$ and positive in the region of $k_{\infty min}$ and switches at the interface as required by the Pontryagin principle. The flux distribution for a unity-maximum k_∞ reactor

is also given (this is the optimal solution to the maximum power problem $(41,42)$). The performance index for both is shown. For this case and the least squares performance index, the two-region reactor is optimal. When the maximum and minimum k_∞ values were changed to 1.04 and 0.98, respectively, the two-region reactor failed to satisfy the Pontryagin conditions since ψ_2 had the wrong signs. In this case, a three-region reactor satisfied the optimality conditions. Further changes in the k_∞ limits made the three-region reactor non-optimal and would require more regions.

This example demonstrates a difficulty associated with the variational approach that has been commented upon by many authors $(40-43)$. While the procedure specifies the necessary conditions for optimality and determines which

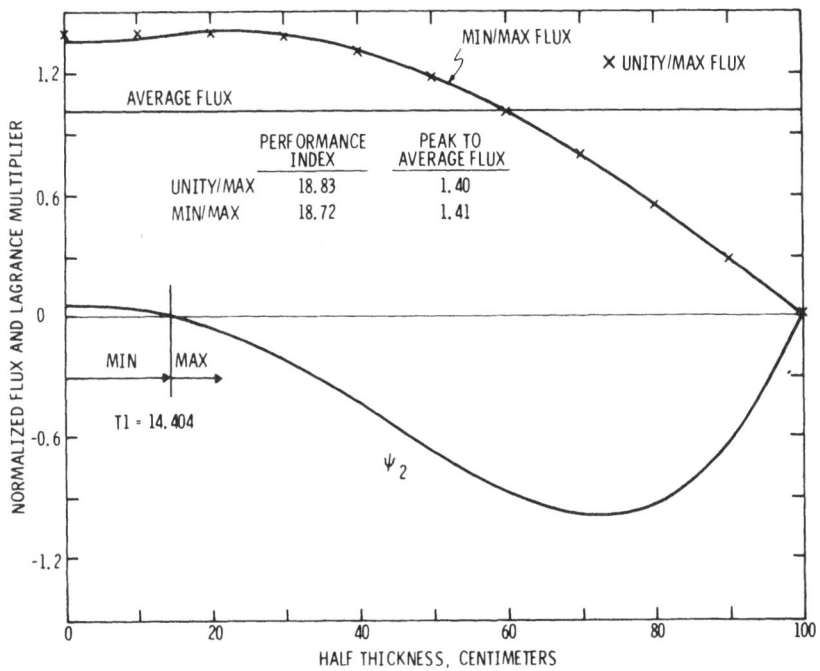

Fig. IV.B.1

Two-Region Min/Max Reactor

types of regions to use in constructing the optimal reactor
without requiring the solution of any equations, it does not
specify "a priori" a synthesis of the region types into an
optimal reactor. This can be a serious drawback if all possible combinations must be examined. The iterative technique
of Wade and Terney (57) described in Section IV.D was also
used on this problem, and converged to the optimum solution,
thus alleviating the difficulty of synthesis.

Another approach which can sometimes be used if the dimensionality of the problem is not large is illustrated by the
next example. In it, the use of phase-space concepts and
trajectories permits the optimal reactor to be synthesized
directly.

3.   The Minimum Critical Mass Problem-Mixed Constraints

For the problem of determining the fuel distribution
to minimize the critical mass in a one-dimensional slab, fast
reactor described by one group theory, Goldschmidt and Quenon
(45) carried previous work on the minimum mass problem (40)
further and used the concept of phase-space trajectories to
directly determine the optimum reactor. They first considered
the problem with no constraints on the local power density and
developed the phase-space interpretation. Then the local power
density was constrained; the optimal trajectory becomes obvious
in the phase-space representation, and the extension to this
important case is relatively easy. More recently, Goldschmidt
has extended this work to a two-group model (47,48).

The problem considered is to select the fuel density distribution (u) lying between upper and lower bounds which will
minimize the mass of the reactor for a given power output, P.
Letting x_1 be the flux and x_2 the current, the problem is

$$\text{Maximize } I = -\int_0^{t_f} u(t)\,dt$$

$$\text{for } u_{min} \leq u \leq max$$

or equivalently

$$\text{Maximize } x_o(t_f) \tag{IV.B.18}$$

where

$$\frac{d}{dt} x_o = u(t) ; \quad x_o(o) = 0 \qquad \text{(IV.B.19)}$$

where the flux satisfies

$$\frac{d}{dt} x_1 = - (\S_3 u + \S_4) x_2 ; \quad x_1(t_f) \qquad \text{(IV.B.20)}$$

$$= g \, x_2 \, (t_f)$$

$$\frac{d}{dt} x_2 = (\S_1 u - \S_2) x_1 ; \quad x_2(o) = 0 \qquad \text{(IV.B.21)}$$

and is normalized to

$$\int_o^{t_f} x_1 \, u \, dt - P = 0 \qquad \text{(IV.B.22)}$$

where

t_f = is the undetermined core boundary

$F(u) = \S_1 u - \S_2$ = Net Production Operation

$\dfrac{1}{D(u)} = \xi_3 u + \S_4$ = Inverse Diffusion Coefficient

g = a constant depending on the reflector properties

Since the normalization condition is merely a scaling condition on the flux, it can be satisfied after the optimal
distribution is found.

This again is a problem with differential state equations
and an inequality constraint dependent on the control only.
The Pontryagin principle thus applies. In this case, the
Pontryagin function may be written as

$$H = \sum_{i=0}^{2} \Psi_i f_i = (-\Psi_1 \S_4 x_2 + \Psi_2 \S_2 x_1 - 1)u - (\Psi_1 \xi_4 x_2 -$$

$$\psi_2 \xi_2 x_1) \qquad \text{(IV.B.23)}$$

where $\psi_o = 1$. The Pontryagin function is constant over the space since it has no explicit spatial dependence. The Euler equations, Equation III.B.33 for the Lagrange multipliers and the boundary equations obtained from applying the transversality conditions are

$$\frac{d\psi_1}{dt} = -(\S_1 u - \S_2)\psi_2 \; ; \psi_1(o) = 0 \qquad\qquad (IV.B.24)$$

$$\frac{d\psi_2}{dt} = (\S_3 u + \S_4)\psi_1 \; ; \; \psi_2(t_f) = -g\psi_1(t_f) \qquad (IV.B.25)$$

and

$$H(t_f) = 0 \qquad\qquad\qquad (IV.B.26)$$

Since H is a constant and is zero at $t = t_f$, it is zero everywhere. The internal boundary conditions arising from the corner conditions require H and ψ to be continuous, and imply that

$$\Phi = 0 \qquad\qquad\qquad (IV.B.27)$$

$$\chi = 0 \qquad\qquad\qquad (IV.B.28)$$

at the places where u is discontinuous since H must also always be zero. Here Φ and χ are defined in Equations IV.B.31 and IV.B.32 below.

Comparing the system and Euler equations, it can be seen that

$$\psi_1 = -\alpha x_2$$

$$\psi_2 = \alpha x_1 \qquad\qquad\qquad (IV.B.29)$$

where α is a normalization constant related to the flux normalization constant. Using these relations, H may be rewritten as

$$H = \Phi u + \chi \qquad\qquad\qquad (IV.B.30)$$

where

$$\Phi = \alpha x_2^2 \S_3 + \alpha \S_1 x_1^3 - 1 \qquad\qquad (IV.B.31)$$

$$\chi = \alpha\beta_4 x_2^2 + \alpha\beta_2 x_1^2 \qquad\qquad \text{(IV.B.32)}$$

The Pontryagin maximum principle requires that

$$H(\underline{u}^*,\underline{\psi}^*,\underline{x}^*) \geq H(\underline{u},\underline{\psi}^*,\underline{x}^*) \qquad\qquad \text{(IV.B.33)}$$

$$u_{min} \leq u^* \leq u_{max}$$

so that there are three ways to make up the reactor:

$$u = u_{max}; \quad \text{if } \Phi > 0$$

$$u = u_{min}; \quad \text{if } \Phi < 0$$

$$u_{min} < u < u_{max}; \quad \text{if } \frac{\partial H}{\partial u} = \Phi = 0$$

As before, there is no "a priori" way to tell how many regions should be used or where they should be located. However, again, the nature of the center-line region and outermost region can both be determined by examining the value of H and Φ at the boundaries. At the center line:

$$\Phi(0) = -1 + \alpha\beta_1 x_1^2(0)$$

but also

$$H(0) = 0 = \alpha[\beta_1 u(0) - \beta_2]\, x_1^2 - u(0)$$

so that

$$\alpha = \frac{u(0)}{[\beta_1 u(0) - \beta_2]\, x_1(0)^2}$$

and

$$\Phi(0) = \frac{\beta_2}{(\beta_1 u(0) - \beta_2)}$$

For a critical reactor, the production operator (the denominator of the above expression) must be positive, and thus $\Phi(0)$ is also positive as is α. Therefore, the central region must be a u_{max}. At the outer boundary:

$$\Phi(t_f) = -1 + \frac{(\beta_1 g^2 + \beta_2)u(t_f)}{(\beta_1 g^2 + \beta_3)u(t_f) + (\beta_4 - \beta_2 g^2)}$$

The sign of $\Phi(t_f)$ depends on the leakage parameter, g. When

$$g < (\S_4/\S_2)^{1/2} \; ; \; \Phi < 0 \text{ and } u = u_{min}$$

$$g > (\S_4/\S_2)^{1/2} \; ; \; \Phi > 0 \text{ and } u = u_{max} \qquad \text{(IV.B.34)}$$

To investigate the third possibility, $\Phi = 0$, we note that for Φ to remain zero over a finite region, not only Φ but its derivatives evaluated along a trajectory must remain zero.

$$\Phi = -1 + \alpha(\S_3 x_2^2 + \S_1 x_1^2)$$

$$\frac{d\Phi}{dt} = \frac{\partial\Phi}{\partial t} + \sum_{i=1}^{2} \frac{\partial\Phi}{\partial x_i} \frac{dx_i}{dt}$$

$$= -2\alpha(\S_1\S_4 + \S_2\S_3) x_1 x_2$$

$$\frac{d^2\Phi}{dt^2} = -2\alpha(\S_1\S_4 + \S_2\S_3) \; [(\S_1 u - \S_2)x_1^2 - (\S_3 u + \S_4) x_2^2]$$

It is instructive to examine the behavior of Φ, $\dfrac{d\Phi}{dt}$ and $\dfrac{d^2\Phi}{dt^2}$

in the phase space of the state variables x_1, x_1. Φ is represented by the equation of an ellipse and is negative inside the ellipse and positive outside. $\dfrac{d\Phi}{dt}$ is zero in the lines $x_1 = 0$ and $x_2 = 0$. In the interior of the ellipse, Φ is negative so that any optimal trajectory will have $u = u_{min}$. In that case, if $(\S_1 u_{min} - \S_2)$ is negative, $\dfrac{d^2\Phi}{dt^2}$ will be negative throughout the interior of the ellipse. Assume, however, that $(\S_1 u - \S_2) > 0$. In that case, the behavior of $\dfrac{d^2\Phi}{dt^2}$ is shown in Figure IV.B.2. Examination of the figure shows that there is no point in the phase space where Φ, $\dfrac{d}{dt} \Phi$, $\dfrac{d^2}{dt^2} \Phi$, are simultaneously zero. Thus, the third potential region type is eliminated.

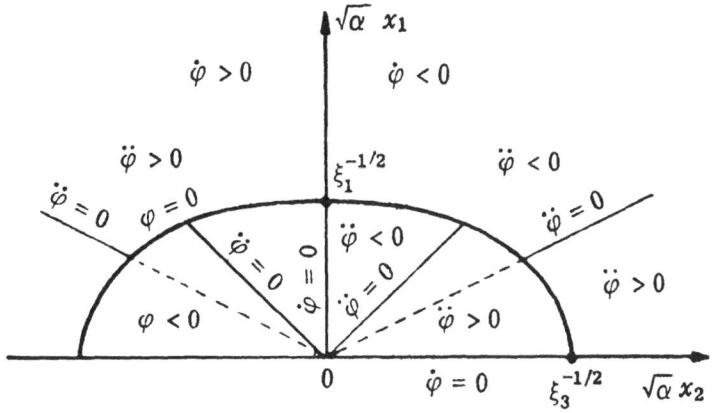

Figure IV.B.2

Function Φ, in Phase Space

From these arguments, it is seen that the optimal reactor consists of regions of maximum and minimum concentrations, with a maximum region in the center and either a maximum or minimum region at the outer core boundary depending on the effectiveness of the reflector; i.e., on the value of g. In what follows, g is assumed small enough to require a minimum region.

The number of regions and their arrangement is still not known except for the boundary regions. To determine the ordering, Goldschmidt and Quenon made further use of the $x_1 x_2$ phase space. Along the optimum trajectory, H = 0; i.e.,

$$\alpha(\S_{1}u - \S_{2}) \, x_1^2 + \alpha(\S_{3}u + \S_{4}) \, x_2^2 - u = 0 \quad \text{(IV.B.35)}$$

or since u is constant in each region,

$$\frac{\alpha x_1^2}{a} + \frac{\alpha x_2^2}{b} = 1 \qquad \text{(IV.B.36)}$$

where

$$a = (\S_1 - \S_2/u)^{-1/2}$$

$$b = (\S_3 + \S_4/u)^{-1/2}$$

$$(IV.B.37)$$

Equation IV.B.36 is the equation of the optimal trajectory in the flux-current phase space. The phase space is shown in Figure IV.B.3 as are the curves for $\Phi = 0$, the straight lines representing $\chi = 0$, and the external boundary conditions $x_1 = gx_2$. Since both χ and Φ must be zero when u is discontinuous, examination of the figure shows that there can be only one interface -- located at point B -- and only two zones. In the central region, $\Phi > 0$, $u = u_{max}$, and the trajectory is an ellipse (ANB). In the outer region, the shape depends on the magnitude of a. If

$\S_1 u_{min} - \S_2 > 0$ the trajectory is an ellipse(BC$_1$)

$\S_1 \dot{u}_{min} - \S_2 = 0$ it is a straight line parallel to the x_1, axis (BC$_2$)

$\S_1 u_{min} - \S_2 < 0$ it is a hyperbola (BC$_3$)

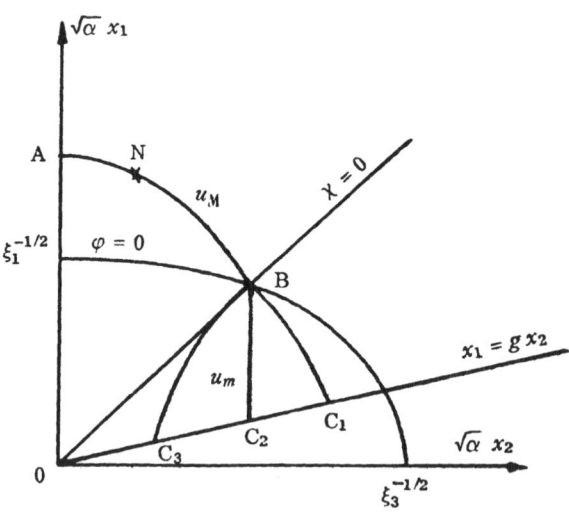

Figure IV.B.3

Optimal Trajectory without Constraint on the Power Density

It is also recalled that if g is large enough, no $u = u_{min}$
region exists at all as is evident in the figure. Therefore,
the optimum trajectory is ANB and then BC_i depending on the
nature of the material in the outer region. The core is
composed of two regions, maximum enrichment in the central
zone and minimum enrighment in the external region. The
dimensions can be obtained analytically, and the flux scaled
to meet the power requirement in accordance with Equation
IV.B.22. This thus completes the solution of the problem
with no constraint on power density.

Consider now the case where the local power density
is constrained:

$$ux_1 - p \leq 0 \qquad\qquad\qquad (IV.B.38)$$

and assume that the flux scaling, Equation IV.B.22, is such
that the constraint would first be violated at some given
point (denoted by N in Figure IV.B.3) in the reactor. The
inequality constraint, Equation IV.B.38, is seen to be explicitly dependent on both the control and state variables.
This problem thus falls into the class treated by Equations
III.B.21 through III.B.27 of the previous section. In this
case, while the system equations remain unchanged, the Euler-
Lagrange equations become:

$$\frac{d}{dt}\psi_1 = -(\S_1 u - \S_2)\psi_2 + \mu_2 u; \quad \psi_1(0) = 0$$

$$\frac{d}{dt}\psi_2 = (\S_2 u + \S_4)\psi_1 \qquad ; \quad \psi_2(t_f) = -g\psi_1(t_f)$$

where

$$\mu = 0 \text{ when } ux_1 - p \leq 0$$

$$\mu > 0 \text{ when } ux_1 - p = 0$$

When the constraint is satisfied in an inequality sense,
everything remains the same as for the case treated above
because $\mu = 0$. However, when Equation IV.B.38 is satisfied
in an equality sense, the Euler-Lagrange equation (Equation
III.B.24)

$$\frac{\partial H}{\partial u} - \mu x_1 = 0 \qquad\qquad\qquad (IV.B.39)$$

applies giving the necessary condition:

$$\Phi - \mu x_1 = 0 \qquad\qquad\qquad (IV.B.40)$$

Since the flux, x_1, is positive and μ is necessarily positive
for optimality, Equation IV.B.40 requires that the trajectory
in x_1, x_2 space join the constant power density constraint
surface at a place where $\Phi > 0$; i.e., where $u = u_{max}$. This
point is indicated as point N in Figure IV.B.3 and is the
point in the high fuel loading (u_{max}), part of the core
where the power density first reaches the limit. The fuel
loading in this new zone of the core is given by the requirement that the trajectory remain on the constraint:

$$u(t) = \frac{p}{x_1(t)}$$

The minimum mass reactor subject to a constraint on local
power density thus consists of three regions, an inner constant power density region, a maximum loading region and a
minimum loading region. The trajectory is shown in Figure
IV.B.4 as $x_1(0)NBC_3$.

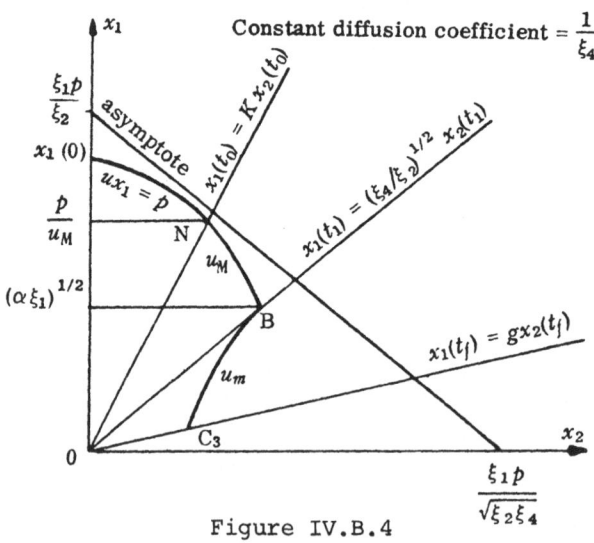

Figure IV.B.4

Optimal Trajectory with Constraint on the Power Density

This problem has illustrated how an examination of the phase space of the state variables can aid in the synthesis of an optimal design. The aproach is well suited, however, for problems of only a few state variables.

This and the previous problem have included constraints which are explicitly dependent on the control. The following example will illustrate the procedures for a state variable constraint; i.e., one having only an implicit dependence on the control.

4. The Maximum Power Problem - State Variable Constraints

In this section, the procedure and theory to be followed when a state variable constraint is present is illustrated. The work is taken from Axford (43), who solved a series of maximum power problems.

The problem is to maximize the power output of a fixed size, slab reactor described in a two-group diffusion model. The fuel concentration is bounded by a maximum limit and the fast flux is also limited by a maximum value.

The problem may be stated as

$$
\text{Maximize } f_o = \int_o^{t_f} dt \; u(t) \; x_3(t) = 0 \tag{IV.B.41}
$$

subject to the flux equations

$$
\frac{dx_1}{dt} = -\frac{1}{\tau^2} x_2 \quad ; \quad x_1(t_f) = 0 \tag{IV.B.42}
$$

$$
\frac{dx_2}{dt} = \eta u x_3 - \rho x_1 \quad ; \quad x_2(0) = 0 \tag{IV.B.43}
$$

$$
\frac{dx_3}{dt} = -\frac{1}{\ell^2} x_4 \quad ; \quad x_3(t_f) = 0 \tag{IV.B.44}
$$

$$\frac{dx_4}{dt} = \rho x_1 - (1+u) x_3 \; ; \; x_4(0) = 0 \qquad\qquad \text{(IV.B.45)}$$

and the control constraint

$$u - u_{max} \leq 0 \qquad\qquad\qquad \text{(IV.B.46)}$$

and the state variable constraint

$$S = x_1 - K \leq 0 \qquad\qquad\qquad \text{(IV.B.47)}$$

Here, x_1 and x_2 are the fast flux and current, respectively; x_3 and x_4, the thermal flux and current, respectively; u is a normalized fuel distribution and the other parameters have appropriate definitions.

The state variable constraint has no explicit dependence on the control and must be treated differently than in the previous section. The first step is to determine the "order" of the constraint by taking spatial derivatives until an explicit dependence on the control variable appears. As described in Section III.B, the resulting differential constraint expression is then used as a constraint explicitly dependent on the control with modified corner or jump conditions at interfaces where the constraint is satisfied. Taking spatial derivatives of the constraint:

$$S = x_1 - K$$

yields

$$S^{(1)} = \frac{dS}{dt} = \frac{dx_1}{dt} = -\frac{1}{\tau^2} x_2 \qquad\qquad \text{(IV.B.48)}$$

$$S^{(2)} = \frac{d^2 S}{dt^2} = -\frac{1}{\tau^2} \frac{dx_2}{dt} = -\frac{1}{\tau^2} [\eta u x_3 - \rho x_1] \qquad \text{(IV.B.49)}$$

The constraint is seen to be of second order, and Equation IV.B.49 is to be used in the same way as a control variable constraint.

For this case, the Pontryagin function is

$$H = [x_3 (1 + n\psi_2 - \psi_4)]u \qquad\qquad\qquad (IV.B.50)$$

$$-\left\{\frac{\psi_1 x_2}{\tau^2} + \rho x_1 (\psi_2 - \psi_4) + \psi_3 \frac{x_4}{\ell^2} + \psi_4 x_3\right\}$$

and the Euler equations for the Lagrange multipliers;

$$- \frac{d}{dt} \psi_i = \frac{\partial H}{\partial x_i} + \mu \frac{\partial}{\partial x_i} s^{(2)}$$

together with their boundary conditions from the transversality condition are:

$$\frac{d\psi_1}{dt} = \rho(\psi_2 - \psi_4) - \mu\gamma_1^2 \; ; \; \psi_1(0) = 0 \qquad\qquad (IV.B.51)$$

$$\frac{d\psi_2}{dt} = \frac{1}{\tau^2} \psi_1 \qquad\qquad ; \; \psi_2(t_f) = 0 \qquad\qquad (IV.B.52)$$

$$\frac{d\psi_3}{dt} = -u[1 + \psi_2 n - \psi_4] + \psi_4 - \mu\gamma_1 \frac{2nu}{\rho}; \psi_3(0)=0 \quad (IV.B.53)$$

$$\frac{d\psi_4}{dt} = \frac{1}{\ell^2} \psi_3; \; \psi_3(t_f) = 0 \qquad\qquad\qquad (IV.B.54)$$

where

$$\gamma_1^2 = \rho/\tau^2 \qquad\qquad\qquad\qquad (IV.B.55)$$

and

$$\mu = 0 \text{ if } x_1 - K < 0 \qquad\qquad\qquad (IV.B.56)$$

$$\mu > 0 \text{ if } x_1 - K = 0$$

The Pontryagin condition requires that the optimal H be greater than any other for allowable controls; i.e.,

$$H(u^*) \geq H(u) \; ; \; u \leq u_{max}; \; \frac{d^2 s}{dt^2} \leq 0 \qquad\qquad (IV.B.57)$$

This condition yields three possible regions:

1. u_o when $\dfrac{\partial H}{\partial u} = 0 = x_3[1 + \eta\psi_2 - \psi_4]$; $\mu = 0$ ((IV.B.58)

2. u_{max} when $0 < x_3[1 + \eta\psi_2 - \psi_4]$; $\mu = 0$ (IV.B.59)

3. $x_1 = K$ when $\dfrac{\partial H}{\partial u} + \mu \dfrac{\partial}{\partial u}\left(\dfrac{d^2 s}{dt^2}\right) = 0$;

$$0 < \mu = \dfrac{\rho}{\eta\gamma\hat{1}} [1 + \eta\psi_2 - \psi_4]$$

(IV.B.60)

The first occurs when u is completely free, the second where u has its maximum value, and the third when the fast flux has its maximum value.

The optimal synthesis of the allowed region types can be accomplished by the methods discussed in the previous sections. Obviously, it is impossible for the outermost region to be one of flat flux, since the flux must go to zero on the outer boundary and there must be leakage. At the outer boundary, $x_3[1 + \eta\psi_2 - \psi_4]$ is positive, so an outer maximum density fuel region is required for optimality. By postulating a region where u is free and Equation IV.B.58 must be satisfied, it can be shown that no solution exists for such a region. Therefore, the optimal reactor is constructed from maximum fuel density regions and flat fast flux regions, with the maximum fuel density region at the outside.

At the interface between the two region types, the ex-panded corner conditions , Equations III.B.49 and III.B.50 must be satisfied.   They are in vector notation:

$$H(t_i -) = H(t_i +) = \underline{\Pi}^T\left(\dfrac{\partial N}{\partial t}\right)$$

(IV.B.61)

$$\underline{\psi}(t_i -) = \underline{\psi}(t_i +) + \underline{\Pi}^T\left(\dfrac{\partial N}{\partial x}\right)$$

(IV.B.62)

where

$$\left(\dfrac{\partial N}{\partial t}\right) = \dfrac{d}{dt}\left\{\begin{array}{l} \underline{S}(x) \\ \dfrac{d}{dt}\underline{S}(x) \end{array}\right\}$$

(IV.B.63)

and

$$\frac{\partial N}{\partial x} = \begin{bmatrix} \dfrac{\partial S}{\partial x_1} & \dfrac{\partial S}{\partial x_2} & \dfrac{\partial S}{\partial x_3} & \dfrac{\partial S}{\partial x_4} \\[3ex] \left(\dfrac{\partial}{\partial x_1}\dfrac{dS}{dt}\right) & \left(\dfrac{\partial}{\partial x_2}\dfrac{dS}{dt}\right) & \left(\dfrac{\partial}{\partial x_3}\dfrac{dS}{dt}\right) & \left(\dfrac{\partial}{\partial x_4}\dfrac{dS}{dt}\right) \end{bmatrix} \qquad \text{(IV.B.64)}$$

and $\underline{\underline{\Pi}}$ is a two-component Lagrange multiplier vector introduced to ensure that S and $\frac{d}{dt}S$ are identically zero at the

interface. Since the S functions do not explicitly depend on t, Equation IV.B.61 reduces to the continuity condition:

$$H(t_-) = H(t_+) \qquad\qquad \text{(IV.B.65)}$$

In fact, since it has no explicit dependence on t, H is a constant throughout the optimal reactor. Further, expanding Equation IV.B.62 gives

$$\psi_{1_-} = \psi_{1_+} + \Pi_1$$

$$\psi_{2_-} = \psi_{2_+} - \Pi_2 \frac{\gamma_1^2}{\rho}$$

$$\qquad\qquad \text{(IV.B.66)}$$

$$\psi_{3_-} = \psi_{3_+}$$

$$\psi_{4_-} = \psi_{4_+}$$

so ψ_3 and ψ_4 are automatically continuous, and only ψ_1 and ψ_2 have the possibility of being discontinuous. Substituting Equation IV.B.50 into the requirement for continuity of H, it is found that H is continuous for $\psi_1 = \psi_2 = 0$, and continuous fuel density at the interface. Therefore, at interfaces between the maximum flux and maximum fuel density regions, the fuel concentration must be continuous. Physically, this requires the optimal reactor be composed of only two zones. The outer region will be of maximum fuel concentration, and the inner region will be of flat fast flux. The interface occurs where the fast flux reaches its maximum value. Axford

TABLE IV.B.1

Summary of Published BOL Optimization Results

Criterion	Reference	Control	Constraints Control	Constraints Other	Geometry Groups, Geometry, Boundary	Solution	Comments
Minimum Mass	Kochurov (40)	u-normalized fuel concentration	$0 \leq u$	None	2g, slab, $\phi = 0$	Two-region: Flat thermal flux $(u \lor \cos)$, $u = 0$	$u = 0$ must be in outer region
	Goertzel (38)	u-normalized fuel concentration	$0 \leq u$	None	$\geq$1g, slab, $\phi = 0$	Two-region: Flat thermal flux $(u \lor \cos)$, $u = 0$	Integral Formulation
	Kochurov	u-normalized fuel concentration	$0 \leq u \leq u_{max}$	None	2g, slab, $\phi = 0$	Three-region: $u = u_{max}$, flat thermal flux $(u \lor \cos)$, $u = 0$	MAX, MIN is not as good
	Goldschmidt (48)	u-fuel concentration	$u_{min} \leq u \leq u_{max}$	None	1g, slab, z condition	Two-region: $u = u_{max}$, u_{min}	No free u region; u_{max} must be at centerline, possibly u_{min} region, depending on condition; uses phase-space arguments
						Three-region: Flat power = p; $u = u_{max}$, $u = u_{min}$	u continuous at first interface
		u-fuel concentration	$u_{min} \leq u \leq u_{max}$	$\phi u - p < 0$, $\int \phi$ u=W	1g, slab, z condition	Two-region: Flat power = D, $u = u_{max}$	No free u region; u continuous. Equivalent to maximizing power with same constraints
Minimum Size	Zaritskaya (41)	u-normalized fuel concentration	$0 \leq u \leq u_{max}$	ϕ_{Th} u-D<0, $\int \phi$ u=W	2g, slab, $\phi = 0$		

TABLE IV.B.1

Criterion	Reference	Control	Constraints Control	Constraints Other	Groups, Geometry, Boundary	Solution	Comments
Maximum Power	Axford* (43)	u-normalized	$0 \leq u \leq u_{max}$	$\Phi_{Th}, u-Q_3 \leq 0$	2g, slab, $\Phi = 0$	Two-region: Flat power $= Q_3$, $u = u_{max}$	No free u region, u continuous
		u-normalized fuel concentration	$u_{min} \leq u \leq u_{max}$	$\Phi_{Th}, u-Q_3 < 0$	2g, slab, $\Phi = 0$	Three-region: $u = u_{min}$, Flat power $= Q_3$, $u = u_{max}$	
		u-normalized fuel concentration	$u \leq u_{max}$	$\Phi_{Fast} - K < 0$	2g, slab, $\Phi = 0$	Two-region: $\Phi_{Fast} = K, u = u_{max}$	
	Zaritskaya (42)	u-normalized fuel concentration	$u_{min} \leq u \leq u_{max}$	$\Phi' \leq$ No	1g, cyl, $\Phi' = 0$	Three-region: $u_{max}, u_{min}, u_{max}$	No free u region, outer region must be u_{max}, two-region doesn't work
Least Square Flux Deviation ($J=\int(\phi-1)^2$)	Terney (46)	k_∞	$k_{\infty min} \leq k_\infty \leq k_{\infty max}$	None	1g, slab, $\Phi = 0$	Maximum and Minimum k_∞ region. The number depends on magnitude of $k_{\infty max}$ and $k_{\infty min}$	No free u region; outer region must be $k_{\infty max}$

*Axford gives many other maximum problems with other state variables, constraints, etc.

obtained the analytic solutions for this reactor and showed
that it satisfied the optimality conditions.

In this problem, the class of problems with state variable constraints was illustrated. The approach is to turn
the state variable constraint into a "pseudo" control variable
constraint, and to impose higher order corner conditions.
Reasoning from the necessary conditions and physical arguments, it was possible to synthesize a plausible optimal
reactor, which, in fact, can be shown to be optimal.

5. Summary

Three examples of optimizing the beginning of life
material distribution in a reactor were illustrated in this
section. A sizable literature exists for this subject, with
problems formulated for a variety of performance indices and
reactor models. Table IV.B.1 summarizes the published results.

The examples showed that for a one-dimensional geometry
and few-group model, the Pontryagin condition:

$$H(u^*) \leq H(u)$$

combined with simple inspection of the system and Euler-
Lagrange equations permitted the identification of feasible
region types without tedious solution of equations. However,
they also demonstrated a difficulty associated with the
Pontryagin approach. While the formalism gives the necessary
conditions for optimality and specifies which types of regions
may be used in synthesizing the optimal reactor, it does not
indicate "a priori" the number of regions or how to arrange
them. Physical arguments can sometimes be used and other
times, trial and error must be relied on for the synthesis.
The phase-space approach is a useful way to determine the
optimal arrangement of feasible region types for low dimensional systems. However, its usefulness diminishes as the
dimensionality of the problem increases. The iterative technique of Wade and Terney to be discussed in Section IV.D is
another systematic way to obtain the optimal configuration.

C. Fuel Management

1. Introduction

The fuel management problem is concerned with the reduction of fuel costs by the adroit introduction of new fuel and rearrangement of the new and old fuel at the completion of each loading cycle comprising the life of a power plant. The selection of the fuel management scheme can significantly affect the cost of nuclear power because of the large cost of the fuel inventory.

Numerous procedures to determine optimal fuel management by the application of optimal control theory have been described. In all cases, the problem is visualized as a multi-stage decision process where at the end of each depletion cycle, decisions are made regarding which fuel to discharge, which to relocate, and the location and amount of new fuel. Generally, in the interest of tractability, the fuel properties are assumed to be known as a function of a single variable; for example, the burnup, and the core lifetimes, power peaking, etc., are also assumed to be known as functions of that variable. In addition, assumptions are generally made regarding the power shape and allowable power peaking, thereby decoupling the poison and fuel management problems from each other; i.e., it is assumed that whatever the fuel management scheme turns out to be, a poison management scheme will be found that produces acceptable core performance for each of the core loadings comprising the fuel cycle.

In this section, the dynamic programming approach of Wall and Fenech (60) and the linear programming work of Suzuki and Kiyose (65) are reviewed. Both approaches rely on the availability of large amounts of computer time and memory capacity for application to large, realistic systems.

2. Minimum Unit Power Cost; A Dynamic Programming Approach

Wall and Fenech (60) considered a three-zone, 1,000 MWe PWR reactor. The goal was to minimize the unit power costs by selecting the optimal refueling and zone-wise shuffling patterns to be used at each core refueling during the thirty-year life of the power plant. The process is visualized as a multistage decision process as shown in Figure IV.C.1. The n-th stage decision variable, k_n, determines where to

locate new fuel and what to do with irradiated fuel. Twenty-
eight unique decisions are considered at each refueling step.
These are enumerated in Table IV.C.1 and range from replacing
all the fuel (k=1) to introducing new fuel in the outer zone,
moving the irradiated fuel one zone inward, and discharging
the fuel from the central zone only (k=11). In all cases,
the alternatives comprise a partial batch reloading with
zone-wise batch reshuffling but not scatter reloading or
reshuffling.

With 28 separate decision variables and 10 reloading
times during the plant life, the number of possible core
histories is roughly 10^{28}, which is obviously too large a
number to calculate outright. Dynamic programming provides
a means to systematically reduce this number to a manageable
level.

It is assumed that the state of the reactor can be rep-
represented by the average burnup (MW days/tonne) of the fuel
in the three equivolume reactor zones.

$$\underline{\Theta}_n = [\Theta_n^1 \; \Theta_n^2 \; \Theta_n^3]^T$$

Here, $\underline{\Theta}_n$ refers to the end-of-cycle state for stage n. The
assumption was checked by showing that the zone-average uranium, uranium 235 and plutonium concentrations were (to within
a 3% tolerance) in a one-to-one correspondence with the burnup for a wide range of one-dimensional, four-group diffusion
theory depletion calculations. Figure IV.C.2 shows this to
be true. For a single fresh fuel enrichment, the state vector
at the beginning of the n-th core lifetime is a function of the
end of cycle state from the previous stage and the reloading
decision variable.

$$\underline{\phi}_n = T_1 \, (\underline{\Theta}_{n-1} \, , \, k_n) \tag{IV.C.1}$$

Here, $\underline{\phi}_n$ refers to the beginning of cycle state for stage n.

To decouple the poison management problem from the fuel
cycle problem, it was assumed that the stage n core lifetime,
δt_n, the power output, p_n, the power costs g_n, the maximum

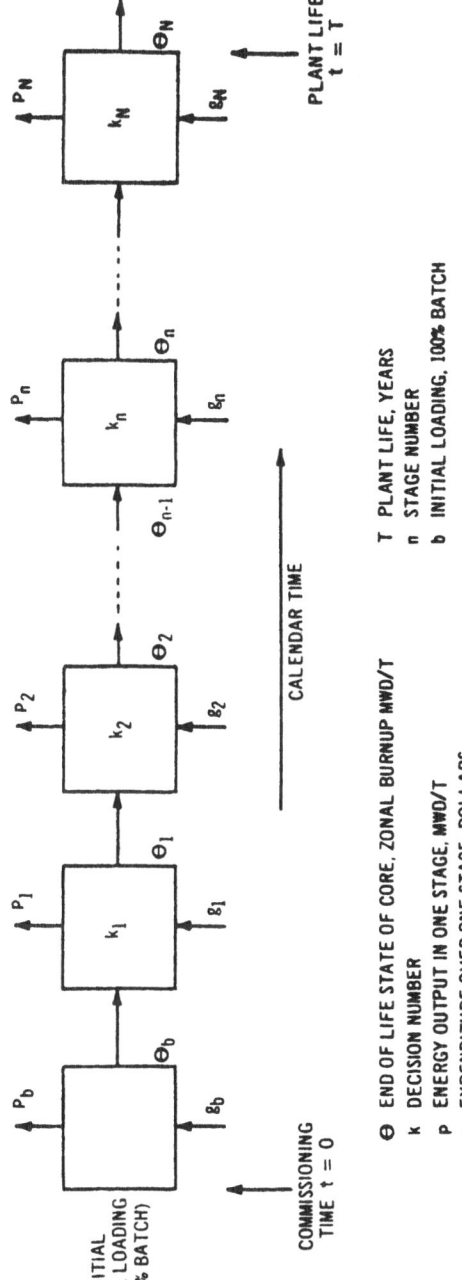

Figure IV.C.1

Block Diagram for Partial Batch Refueling

TABLE IV.C.1
Refueling Decisions and Reloading Patterns

| Decision | Zone | | | Reloading |
Number	1	2	3	Pattern
1	F	F	F	1, F F F
2	1	F	F	
3	2	F	F	2, O F F
4	3	F	F	
5	F	1	F	
6	F	2	F	3, F O F
7	F	3	F	
8	1	2	F	
9	1	3	F	
10	2	1	F	4, O O F
11	2	3	F	
12	3	1	F	
13	3	2	F	
14	F	F	1	
15	F	F	2	5, F F O
16	F	F	3	
17	1	F	2	
18	1	F	3	
19	2	F	1	6, O F O
20	2	F	3	
21	3	F	1	
22	3	F	2	
23	F	1	2	
24	F	1	3	
25	F	2	1	7, F O O
26	F	2	3	
27	F	3	1	
28	F	3	2	

F denotes fresh fuel
1, 2, 3 denotes previous zone of fuel.

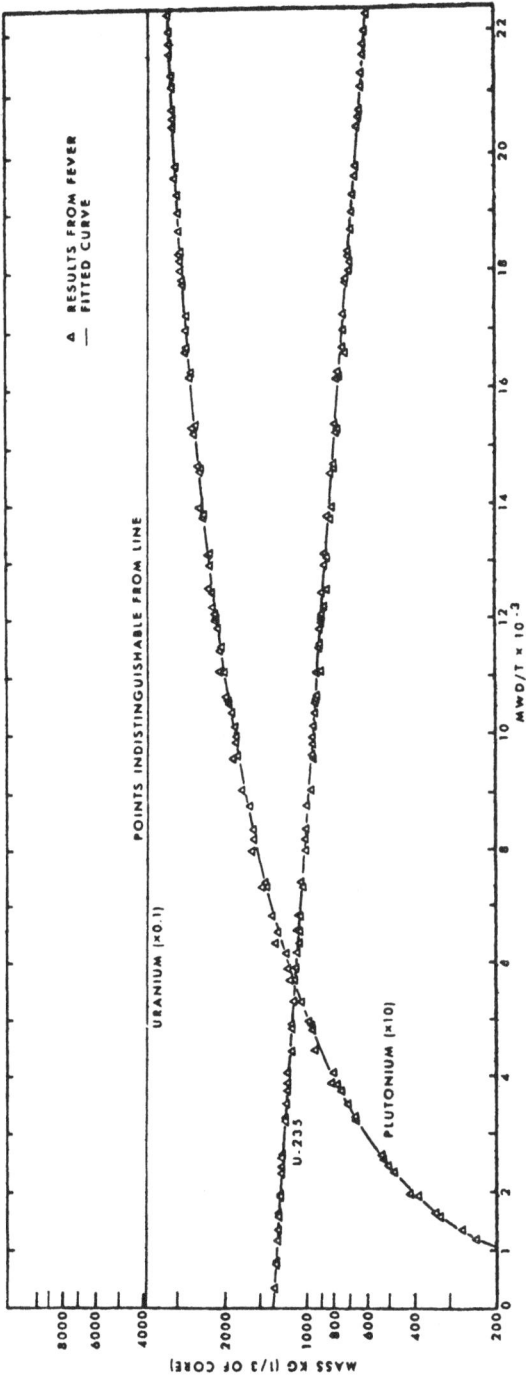

Figure IV.C.2

Change in Fuel Composition with Burnup

peaking, pp_n, and the end of cycle state were all unique functions of the beginning of cycle state. By virtue of Equation IV.C.1, this made them functions of the previous end of cycle state and the reloading decision:

$$\delta t_n = \delta t_n \ (\underline{\phi}_n) = \delta t_n \ (\Theta_{n-1}, k_n)$$

$$p_n = p_n \ (\underline{\phi}_n) = p_n \ (\Theta_{n-1}, k_n)$$

$$g_n = g_n \ (\underline{\phi}_n) = g_n \ (\Theta_{n-1}, k_n) \qquad (IV.C.2)$$

$$pp_n = pp_n \ (\underline{\phi}_n) = pp_n \ (\Theta_{n-1}, k_n)$$

$$\underline{\Theta}_n = \underline{\Theta}_n \ (\phi_n) = \underline{\Theta}_n \ (\Theta_{n-1}, k_n)$$

The assumption was checked by examining the results of numerous one-dimensional, four-group, diffusion theory depletion calculations* and found to be generally accurate to within 3%. For rapid calculation during the fuel cycle optimation, Equation IV.C.2 components were embodied in polynomial fits and tables dependent on the components of Θ. A different fit or table was used for each of the 28 control decisions.

As described above, the core lifetime was not fixed, but dependent on the criticality of the depleted core. The cycle cost increment, g_n, was based on δt_n and included fixed costs, reloading down time, etc., based on the data shown in Table IV.C.2.

The total power and total cost are given by:

$$P_N = p_b + p_1 + \ldots + p_n + \ldots + p_N \qquad (IV.C.3)$$

$$G_N = g_b + g_1 + \ldots + g_n + \ldots g_N$$

---

*   The calculations did not include a control rod representation; the assumption being that an appropriate poison management scheme could be found to guarantee the attainment of δt_n, p_n, etc.

TABLE IV.C.2
Basic Data on Reactor Model and Costs Basis

Reactor:

Heat output	3200 MWth
Net electrical output	1000 MWe
Load factor	0.8
Core loading	117 500 kg U
Fuel type, UO_2, stainless-steel cladding, 3.4 wt% enriched	
Core Geometry, cylinder	
radius	1.97 m
height	3.05 m

Costs:

Natural uranium	23.50 $/kgU
Separative work	30.00 $/kgU
UF_6 enrichment of zero value	0.2531 wt%
Conversion & fabrication of fuel elements	101 $/kgU
Separation plant charge, 972 kgU/day capacity	17 100 $/day
Uranium conversion, 1000 kgU/day capacity	5.6 $/kgU
Plutonium conversion	1500 $/kg Pu
Plutonium credit as metal	9500 $/kg Pu

Interest rates:

UF_6 use charges	4.75 %
Capital amortization	14 %
Fabrication interest	12 %

Capital Cost:	170 $/kWe
Operating Cost:	0.4 mills/kWhe

The goal of the optimization is to minimize the unit power costs.

$$\text{Min } J = \quad \text{Min } \frac{G_N}{P_N} \qquad\qquad \text{(IV.C.4)}$$

$$k_n; \; n=1,2,\dots N$$

while maintaining the required plant lifetime

$$\delta t_b + \sum_{i=1}^{N} \delta t_i \geq T \qquad\qquad \text{(IV.C.5)}$$

and satisfying the maximum allowed burnup requirement

$$\theta_i^n \leq \theta_{Max} \quad \begin{array}{l} i = 1,2,\ 3 \\ n = 1,2,\dots\ N \end{array} \qquad\qquad \text{(IV.C.6)}$$

while also satisfying the maximum allowed power peaking constraint

$$PP_n \leq PP_{Max} \qquad\qquad n = 1,\ 2,\ \dots\ N \qquad\qquad \text{(IV.C.7)}$$

For the dynamic programming concept, the ratio $G_n(\theta_n)/P_n(\theta_n)$ may be defined as

where

$G_n(\theta_n)/P_n(\theta_n)$ is the <u>minimum</u> power cost over n + 1 stages leading to an end state θ_n,

$G_n(\theta_n)$ is the total cost in dollars to the end of stage n

$P_n(\theta_n)$ is the total energy output in MW days to the end of stage n

and the optimization recursion relationship is written:

$$\frac{G_n(\theta_n)}{P_n(\theta_n)} = \underset{k_n}{\text{Min}} \; \frac{g_n(\theta_{n-1},\ k_n) + G_{n-1}(\theta_{n-1})}{P_n(\theta_{n-1},\ k_n) + P_{n-1}(\theta_{n-1})} \qquad \text{(IV.C.8)}$$

This relation states that knowing the best path to the various θ_{n-1} states, the best path to any particular n-th stage
state, θ_n, can be obtained by examining the way of getting
to that θ_n state from each of the alternate θ_{n-1} states and
selecting that one as optimum which produces the lowest total
unit cost. In other words, if there are several paths to the
state θ_n from the previous stage, the one producing the lowest
unit power cost not over the n-1st stage but over all stages
k = 1, 2, = n is selected and the others are discarded for the
remainder of the calculations. This eliminates paths which
have no potential for being the optimum and thus substantially
reduces the number of states and paths to be calculated.

A reduction in the number of calculations is achieved also
by considering the constraints. At a decision point, those
decisions which will lead to a constraint violation during the
next core lifetime are discarded. In general, the depletion
state generated by each path will be unique. However, in order to further reduce the number of calculations, absolute
identity of the various states, θ_n, at any stage may be relaxed. States that have identical burnup within a tolerance,
$\delta\theta = 250$ MWD/t, for example, are considered to be identical.

A program was written to solve recursion relationship,
Equation IV.C.8. At the end of the first core life, the
various decision possibilities of Table IV.C.1 are considered.
Those that will fail to satisfy the constraints of Equations
IV.C.6 and IV.C.7 during the next cycle are discarded, and
the others are used to calculate the reactor history over the
next stage. Then power and costs are calculated and compared
for any identical EOL states and the one of least unit cost is
chosen and recorded. The process is repeated for each stage
and the optimal path to each state recorded. During the first
part of the plant life, there will usually be unique end
states. As the process continues, an overlap of states within the $\delta\theta$ tolerance will occur and many paths will be discarded, thus reducing the number of potential paths, from
10^{28} to a manageable number.

For a maximum allowable burnup, $\theta_{max} = 22,000$ MWD/t, a
plant life of 30 years, and an allowable maximum power peaking

of pp_{max} = 1.75, the optimum policy as obtained by this procedure is shown in Table IV.C.3. The unit power cost is
6.115 mills/kwhr which is a significant reduction over the
value of 6.512 for 100% batch reloading, and is also lower
than the value of 6.14 obtained using the three-zone, out-
in policy (this latter turns out to be optimum if the allowable burnup is reduced to θ_{max} = 20750 with the same power
peaking limit). The number of states or paths that were
recorded at each stage is shown in Table IV.C.4. It is seen
that the number never exceeds 25.

TABLE IV.C.3

Optimal Policy and Performance

Item	Value
Maximum Permissible Burnup	22 MWD/kg
Maximum Permissible Power Peaking	1.75
Power Costs	6.115 mills/kWhr
Plant Life	30.43 y
No. Stages	26
Load Factor	0.774
Average Burnup	19.65 MWD/kg
Optimal Policy (decision number – Table IV.C.1 – by stage)	11,4,7(13,21,3)*,13,21

*7(13,21,3) denotes 7 cycles of the sequence: 13,21,3.

TABLE IV.C.4

Number of End States by Stage
Achieved by Dynamic Programming

Stage Number	Number of End States
2	3
3	8
4	16
5	22
6	23
7	22
8	23
9	21
10	21
11	22
12	21
13 through 23	20
24	21
25	20
26	14

Wall and Fenech ran a series of problems to investigate the effect of changes in the various constraint values and the value of $\delta\theta$ on the resulting optimal strategy. They found that $\delta\theta = 250$ MWD/t (about 1% of the discharge burnup) gave significant reductions in the number of calculations without altering the optimal policy. They also showed that variations of no more than 2.5% in the representations of the reactor neutronics, Equation IV.C.2, did not alter the optimal policy.

Wall and Fenech's work was restricted to a three-zone core with partial batch reloading and did not consider inter-zone, partial batch shuffling, such as roundelay schemes. Stover and Sesonske (61,62) used dynamic programming to study the case of zone-wise scatter reloading strategies with a state variable of 30 components and over 2,000 decision policies.

The dynamic programming approach provides a formal and systematic way to eliminate trajectories having no potential for being optimum at an early stage in the analysis, thereby eliminating unnecessary calculations. As illustrated above, it accommodates constraints on the control and state variables which often lead to difficulties in the variational optimization procedures. Indeed, by eliminating potential paths, the constraints actually decrease the amount of computation required in the dynamic programming approach.

The dynamic programming approach to fuel cycle analysis has several inherent shortcomings, however. For each state of each stage it is necessary to store the value of the optimal cost functional and the state and decision identification from the preceding stage which is on the optimal path to the state. Also, the state vector itself and perhaps other data as

$$\sum_{i=1}^{n} \delta t_i$$

is stored. Thus, as the number of states increases from stage to stage, or as the number of components of the state vector increases for more detailed reactor models, ever larger volumes of data must be stored. Also, the number of one-stage trajectories emanating from each state is equal to the number of allowed control policies (28 in the example discussed above). The number of one-stage trajectories to be considered thus also increases as the number of states increases from stage to stage. For problems with a large number of stages or states, or a large number of components in the state vector, the computation storage and time requirements become large.

Another difficulty is more fundamental. If there is only one path to a given state at a given stage, then no minimization of cost and elimination of trajectories is possible since there is only one path to the state. The average fuel burnup in a reactor zone is inherently a continuous variable so that in the fuel cycle problem as formulated above, each trajectory does in fact lead to a unique state and no application of dynamic programming is possible. This difficulty was overcome by defining as "identical" all states having burnups within a range of $\delta\theta = 250$ MWd/tonne, of the same value. The degree of coarseness which can be tolerated in the burnup grid without affecting the results will always be a point to consider in a dynamic programming formulation of the burnup problem.

3. Minimum Fuel Consumption; A Linear Programming Approach

Suzuki and Kiyose (65) considered a reactor of I equi-volume zones and minimized the fresh fuel consumption subject to constraints on maximum burnup and total plant energy output. The problem is cast as a multistage decision process, and the equations and performance index are formulated linearly (in terms of the numbers of fuel elements of various burnups) so that a linear programming solution is possible. They showed that to minimize fresh fuel consumption is equivalent to maximizing average burnup of the discharge fuel. Sauer (66) has also used the linear programming approach, although on a simpler neutronics model.

Figure IV.C.3 shows the multistage decision process representation of the problem. At each stage (k), the beginning-of-cycle (BOC) fuel loading is specified by the loading matrix, α^k

$$\alpha^k = \begin{bmatrix} \alpha^k_{11} & \alpha^k_{21} --- & \alpha^k_{I,1} \\ & & \\ \alpha^k_{1,J} & & \alpha^k_{I,J} \end{bmatrix} \quad ; \ k = 1, 2, \ldots K$$

$$\alpha^k_{ij} \geq 0$$

where α^k_{ij} denotes the number of fuel elements in zone i with burnup level e_j, and J is the maximum number of burnup indices in any zone. After constant power operation for a given time interval, (Δt), the fuel has depleted to an end-of-cycle (EOC) state, β^k:

$$\beta^k = \begin{bmatrix} \beta^k_{11} & \beta^k_{,I1} \\ & \\ \beta^k_{1,J'} & \beta^k_{IJ'} \end{bmatrix}$$

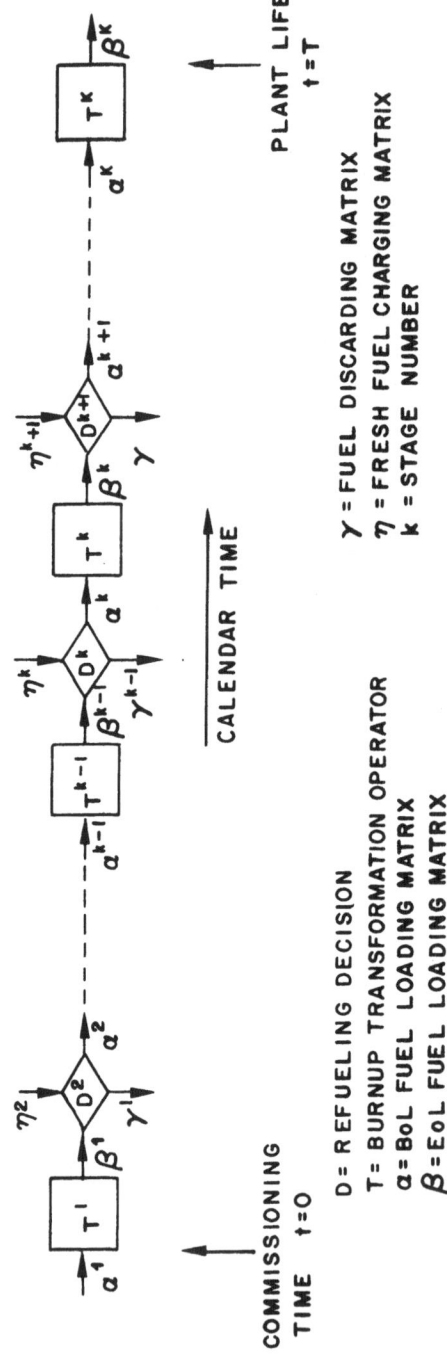

Figure IV.C.3

Multistage Decision Process for Refueling Strategies of an LWR

where

$$\beta_{ij}^{k} \geq 0 \; ; \; J' > J$$

At each decision point, fuel assemblies are discarded as specified by a matrix, γ^{k}, or reloaded into any of the zones, $i = 1, 2 \ldots I$. Also, new fuel is fed in as specified by a matrix, η^{k}:

$$\gamma^{k} = \begin{bmatrix} \gamma_{11}^{k} & --- & \gamma_{I1}^{k} \\ & & \\ \gamma_{1J'}^{k} & & \gamma_{IJ'}^{k} \end{bmatrix}$$

$$0 \leq \gamma_{ij}^{k} \leq \beta_{ij}^{k}$$

$$\eta^{k} = \begin{bmatrix} \eta_{11}^{k} & & \eta_{I1}^{k} \\ 0 & & 0 \\ 0 & & 0 \\ 0 & & 0 \end{bmatrix}$$

$$0 \leq \eta_{ij}^{k} = \alpha_{i1}^{k}$$

Denoting by N_i, the number of assemblies in zone i, the material conservation equations are:

$$\sum_{j=1}^{J} \alpha_{ij}^{k} = \sum_{j=1}^{J'} \beta_{ij}^{k} = N_i; \; k = 1, 2, \ldots K, i = 1, 2, --- I \quad (IV.C.9)$$

and

$$\sum_{i=1}^{I} \eta_{i1}^{k} = \sum_{i=1}^{I} \sum_{j=1}^{J'} \gamma_{ij}^{k-1} \qquad k = 2, 3, \ldots K \qquad (IV.C.10)$$

Thus far, only material balances have been considered and the operational requirements of the reactor core have been unaccounted for. In order to divorce the fuel and poison management problems, it is assumed that the reactor is operated according to the Haling principle (see Section IV.D.2) in which the fuel is distributed and the control manipulated so the power and reactivity distributions remain fixed throughout core life. The particular power and reactivity distribution chosen is one to supposedly maximize the burnup or minimize the critical mass for limited power peaking and is taken as an end-of-cycle rods-out distribution which is at the limit of the power-peaking constraint. Using the preselected end-of-cycle power and reactivity distributions, the average fluxes, the energy output, and the necessary reactivity for each of the I reactor zones are predetermined by means of separate depletion life study analyses. Furthermore, these analyses are used to determine δe_{ij}, the cycle energy output (or equivalently, the burnup increment) for each BOC loading and K_{ij}, the BOC reactivity contribution per unit BOC loading. Because by definition of the Haling principle, the power shape stays fixed throughout core life, the depletion transition operator, T, which transforms the BOC state α_{ij}^k to the EOC state β_{ij}^k in the kth stage after producing the allocated energy output δe_{ij}^k during Δt, can also be predetermined in the separate depletion lifestudies and is independent of both BOC state, α_{ij}^k and stage, k.

The neutronics representation of the core is thus reduced to a set of linear relations:

the burn cycle transition of fuel elements:

$$\underline{\beta}^k = T \, \underline{\alpha}^k \qquad\qquad\qquad\text{(IV.C.11)}$$

the generation of the required energy output by zone, E_i^k,

$$\sum_j \delta e_{ij}^k \, \alpha_{ij}^k = E_i^k \qquad \begin{array}{l} k = 1,2, \ \ldots \ K \\ i = 1,2, \ \ldots \ I \end{array} \qquad \text{(IV.C.12)}$$

and the reactivity requirement, k_i^k, by zone,

$$\sum_j K^k_{ij} \alpha^k_{ij} = k^k_i \qquad \begin{array}{l} k = 1, 2, \dots K \\ i = 1, 2, \dots I \end{array} \qquad \text{(IV.C.13)}$$

by the use of a set of separate, one-cycle depletion calculations.

The optimization problem is then to minimize the fresh
fuel loading:

$$\text{Minimize} \left\{ \sum_{k=1}^{K} \sum_{i=1}^{I} \alpha^k_{il} \right\} \qquad \text{(IV.C.14)}$$

subject to the conservation of fuel assemblies

$$\sum_j \alpha^k_{ij} = N_i \; ; \qquad \begin{array}{l} k = 1, 2, \dots K \\ i = 1, 2, \dots I \end{array} \qquad \text{(IV.C.15)}$$

and to the requirement that the number of fuel assemblies of
each burnup level, j', loaded at the (n+1)th stage is no
larger than the number of fuel assemblies of that burnup
available from the previous stage.

$$\sum_i \beta^k_{ij'} \geq \sum_i \alpha^{k+1}_{ij'} \; ; \; \begin{array}{l} k = 1, 2, \dots K-1 \\ j' = 2, \dots J' \end{array} \qquad \text{(IV.C.16)}$$

Further constraints are in the energy output requirement, the
reactivity requirement, and the burn-cycle transition relation,
Equations IV.C.11, IV.C.12, and IV.C.13. Also, the maximum
burnup constraint must be satisfied:

$$e^k_{iJ'} \leq e_{Max} \qquad \begin{array}{l} i = 1, 2., \dots I \\ k = 1, 2, \dots K \end{array} \qquad \text{(IV.C.17)}$$

This optimization problem is a standard linear programming
problem in which

$$\left\{ \alpha^k_{ij}; \quad \begin{array}{l} j = 1, 2, \dots J \\ i = 1, 2, \dots I \\ k = 1, 2, \dots K \end{array} \right\}$$

are the control variables and the performance index and all
constraint equations are linear in α^k_{ij}.

Suzuki and Kiyose solved the problem stated above for a five-zone (I=5), 460 MWe BWR over 10 reload cycles using a computer code written for the CDC-3600 computer. The number of decision variables was 8208 and the number of constraint equations was 1890. With the simplex method, a straightforward solution to the problem would have required tens of hours of computation time. The simplex method is broken into two parts: (1) seeking feasible solutions; and (2) choosing the best of these as optimal; most of the computation time is spent seeking feasible solutions. In order to decrease the computer time requirements for this problem, Suzuki and Kiyose first solved the stagewise optimization problem (requiring only 0.7 hour for all 10 stages); i.e., they optimized each stage, separate from the others. Then they used this result as a starting point for the overall optimization, which then required only an additional 4.3 hours of computation.

It was shown that the fresh fuel consumption for the stagewise and overall optimization differed only by one percent. However, the reloading patterns were different. The stagewise optimization approached an equilibrium cycle after about four cycles, whereas an equilibrium cycle was never established for the overall optimization problem. In general, the overall optimal policy required the removal of all fuel over a certain threshold burnup and its replacement with new fuel. The optimal policy reduced the fuel requirement by 10% from that required for a conventional one-fourth batch refueling policy.

The fuel cycle model solved by Suzuki and Kiyose was of realistic complexity. While the computer computation times were large, they were not unacceptable in view of the potential 10% reduction in fresh fuel requirements they identify. Several approximations were necessary to cast the problem into a form applicable for linear programming. In addition, the identity of the individual fuel assemblies is lost, since there is no geometric detail within a zone and the quantities treated are the numbers of elements of various types rather than the elements themselves.

4.   Summary

In this section, the problem of optimizing the fuel management for a reactor plant was viewed as a multistage

decision process. The two techniques well suited for this
type problem, namely, dynamic programming and linear programming, were demonstrated. In each case it was necessary
to decouple the fuel and poison management problems in order
to simplify the problem enough to make a solution feasible.

In addition, many simplifying assumptions in the reactor
model were required which limit the applicability of the results. Even so, in both methods, the computing times and
memory requirement become very large.

There have been other approaches also to the fuel management problem. Melice (66) used variational results to construct an optimal k_∞ profile which optimizes each stage of

the fuel management process. Hoshino (69) has used a heuristic learning technique on a fairly simple model in an
attempt to decide which, if any, of the stagewise conditions
might be used to effect an overall optimization.

A rather detailed neutronics model is necessary to adequately account for power-peaking considerations in a fuel
management study. To date, there have been no published results of fuel cycle studies for such a detailed model. In
addition, no attempt has yet been made to develop a formal
iterative procedure which parallels intuitive design techniques. Motoda's "Method of Approximation Programming" (68)
and Wade and Terney's (57) approach to poison management
might be applicable to this problem. This might be particularly true for PWR's where, with control in the form of
soluble boron, the power distribution changes continually
throughout the core life. Further, with quarter or eighth-
core symmetry, three zones, and no rods, the number of
variables might not be unreasonable for a FLARE type approach to the neutronics.

D. Poison Management

    1.   Introduction

        Considerable work has been published concerning the
problem of determining the poison management or rod programming scheme to achieve desired reactor performance through one
core lifetime; for example, to achieve maximum lifetime, subject to a constraint on power peaking. It has been found, as
will be shown below, that while the lifetime itself is not

highly sensitive to the poison management scheme, the power
peaking is quite sensitive and thus influences the lifetime.
Other criteria are minimum power peaking throughout core life,
or specified power profile versus life.

Several problem formulations and solution techniques have
been used for the poison management problem. Motoda and
Kawai (55) and Suzuki and Kiyose (53) considered the maximum
lifetime problem using the Pontryagin necessary conditions.
By considering a two-region reactor, they were able to use
geometric constructions in burnup phase space to determine
and interpret the results. Later, Suzuki and Kiyose (56)
generalized the results to a multiregion reactor using topo-
graphic space arguments. The results were used by Motoda (58)
to determine the interrelationships between rod programming
and the fuel management. The work of Motoda and Kawai (55)
for a simple, two-region reactor model is reviewed first to
gain insight into the problem. Then the work of Terney and
Fenech (54) is summarized. They formulated the problem of
optimizing rod motion to minimize the power peaking through
the core life as a multistage decision problem, and dynamic
programming was used to determine the optimal rod motion.

Finally, the generalized poison management problem for a
multiregion reactor treated by Wade and Terney (57) is discussed. They formulated a general design problem incorpor-
ating lifetime, burnup and power-peaking considerations, and
used an iteration technique to gain a numerical solution to
the resulting generalized Bolza equations.

2.   Maximum Lifetime Problem with Constrained Power
      Peaking; A Phase-Space Solution to the Pontryagin
      Equations

Motoda and Kawai (55) treated the problem of optimizing the poison management to maximize the core lifetime
while constraining the allowable power peaking to a maximum
value and requiring that the control or poison concentration
is never less than zero. They considered a two-region reactor
and were able to use phase-space arguments to aid in gaining
the solution.

Consider a two-region reactor which satisfies the neutron
balance:

$$\nabla^2 \phi + S\phi = 0 \qquad\qquad\qquad (IV.D.1)$$

and depletion equations:

$$\frac{d\Sigma_i}{dt} = D_i (\phi) \qquad\qquad i = 1, 2 \qquad\qquad (IV.D.2)$$

The state of the reactor is represented by the unpoisoned, normalized material buckling in each region, (Σ_i), and the control poison in each region, C_i, is chosen to make the reactor critical. The region material buckling is denoted by S_i

$$S_i = \Sigma_i - C_i \qquad\qquad\qquad (IV.D.3)$$

Physical realizability dictates that the control is non-negative:

$$C_i \geq 0 \qquad\qquad i = 1, 2 \qquad\qquad (IV.D.4)$$

For a two-region reactor, the neutron balance, Equation IV.D.1 can be used to obtain analytic expressions in terms of the state (Σ_i) and control (C_i) variables, for the criticality relationship:

$$C (\underline{S}) = 0 \qquad\qquad\qquad (IV.D.5)$$

the region average flux ratio,

$$g = \bar{\phi}_2 / \bar{\phi}_1 = G(\underline{S}) \qquad\qquad\qquad (IV.D.6)$$

and for the power peaking,

$$f = Max\ \phi / Avg\ \phi = F(\underline{S}) \qquad\qquad\qquad (IV.D.7)$$

Here, $\underline{S}$ denotes the vector of region material bucklings, S_i.

It is convenient to visualize the reactor depletion and control process in a burnup phase space in which the coordinates are the control-free region material bucklings, Σ_1.

For the two-region model under consideration, the space has only two axes, Σ_1 and Σ_2. The criticality relationship, Equation IV.D.6, when evaluated with the control set to zero $C_1 = C_2 = 0$, defines a criticality curve on which the reactor

must operate (see Figure IV.D.1). Each point on the criticality curve has a unique region average flux ratio, g, and a
unique power-peaking factor, f. Their values are shown parametrically along the critical curve in Figure IV.D.1.

The two components of the control, C_1 and C_2 are selected
to bring the region bucklings, S_i from the control-free point
(Σ_1, Σ_2) to the criticality curve on which the reactor must
operate. Then the region fluxes are determined and the reactor depleted according to Equation IV.D.2, thereby changing
the material bucklings and leading to a trajectory in the Σ
space. The line AE is such a trajectory.

If it is assumed that the control-free material buckling
varies monotonically with depletion or flux exposure (i.e.,
the term $D(\phi(s))$ in the depletion equation always has the
same sign), then the trajectories will always move monotonically to the lower left of the coordinate system. Note that
this assumption is not universally applicable and is violated
by the use of burnable poisons or fuels of certain enrich-
ments. In what follows in addition to the above, it is
assumed that the control-free material bucklings vary linearly
with burnup; as a result, the curves of depletion in the phase
space and the depletion trajectories are as depicted in Figure
IV.D.1.

Several typical modes of operation are shown in the figure.
In the first mode, the control is maintained the same in both
regions. This implies that the control vector is a 45° line
drawn to the criticality curve from the state point. The
gradient of the burnup trajectory is determined by the value
of flux ratio, g, corresponding to the material buckling, $\underline{S}$.
Two trajectories of this sort are shown, one starting from A
and the other from B. Notice that the power shape is continuously shifting, since the operating point moves along the criticality curve. As a result, the trajectories are curved.

A second mode is to operate the reactor at a constant
power shape as long as possible without allowing the controls
to go negative. This means that the operating point is fixed
on the criticality curve. In this case, the trajectories are
straight lines since the value of the flux ratio, g, is constant. Two groups of trajectories of this sort are shown for

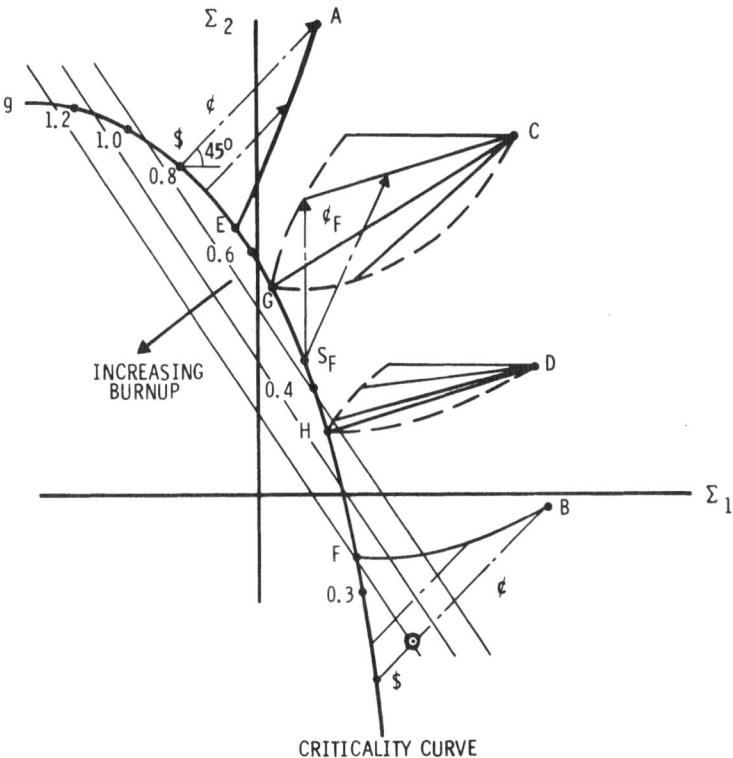

Figure IV.D.1

Depletion Trajectories in a Material Buckling Phase Space

initial points C and D, with the family of curves representing operation from the initial state at different fixed
points (i.e., values of g) on the criticality curve. The
dashed lines represent the end-of-life for the trajectory
at a given fixed power shape, since further operation at
this shape would require negative control.

Notice that for each initial state (C or D) there is a
unique trajectory and operating point which ends on the
criticality curve (i.e., with zero control) and gives the
maximum lifetime. The selection and maintenance of this
particular flux ratio throughout life corresponds to the
well-known Haling Principle (50) mode of operation. Whether
this mode is acceptable or not depends on the power peaking.
As is evident from the figure, this is determined by the
initial state.

From the figure, it can be seen that the maximum burnup
itself is not highly sensitive to the mode of operation, and
consideration should be given to the effect of control on
the power peaking. This may be done using the Pontryagin
formalism. The statement of the optimization problem is:

$$\text{Maximize } J = \int_{0}^{t_f} dt \qquad\qquad (IV.D.8)$$

with

$$c_i \geq 0$$

$$f \leq f_{Max}$$

and subject to the criticality and depletion equations,
IV.D.1 and IV.D.2 and to a required power output.

For a two-region cylindrical Boiling Water Reactor similar
to Vermont Yankee, Motoda and Kawai expressed the depletion
equations in the form:

$$\frac{d\Sigma_1}{dt} = \alpha\,\phi_1; \quad \frac{d\Sigma_2}{dt} = \alpha\,\phi_2 \qquad\qquad (IV.D.9)$$

where ϕ_1 and ϕ_2 are the average fluxes in regions 1 and 2 respectively normalized so that

$$\phi_1 + \phi_2 = 2 \tag{IV.D.10}$$

They further expressed the criticality relationship in the form

$$(\Sigma_2 - C_2) = \sum_{n=0}^{2} a_n (\Sigma_1 - C_1)^n \tag{IV.D.11}$$

and obtained an expression for the average flux ϕ_1 as

$$\phi_1 = \sum_{n=0}^{2} b_n (\Sigma_1 - C_1)^n \tag{IV.D.12}$$

Furthermore, using these, the constraints can be written as

$$u_1(\Sigma_1, \Sigma_2) \leq C_1(t) \leq u_2(\Sigma_1, \Sigma_2)$$

or more conveniently as

$$R(\Sigma_1, \Sigma_2, C_1) \geq 0 \tag{IV.D.13}$$

The optimization problem may be restated as

$$\text{Max } J = \int_{o}^{t_f} dt$$

subject to the depletion equations

$$\frac{d\Sigma_1}{dt} = \alpha \left\{ \sum_{n=0}^{2} b_n (\Sigma_1 - C_1)^n \right\} \tag{IV.D.14}$$

$$\frac{d\Sigma_2}{dt} = \alpha \left\{ 2 - \sum_{n=0}^{2} b_n (\Sigma_1 - C_1)^n \right\} \tag{IV.D.15}$$

and subject also to the constraint

$$R(\Sigma_1, \Sigma_2, C_1) \geq 0 \tag{IV.D.16}$$

and to the criticality relationship given by Equation IV.D.11.

The phase space is shown in Figure IV.D.2, where $\alpha = -34.4$. For $f_{max} = 1.40$, the maximum power peaking occurs in the outer region at point B on the criticality curve and in the inner region at C. Anywhere between the two points is an acceptable operating point on the criticality curve. The part of the phase space to the right and above the line A B C D represents the part of space for which the constraint $R(\Sigma_1, \Sigma_2, C_1) \geq 0$ can be satisfied when the operating point lies between B and C on the criticality curve. Point C represents the maximum attainable burnup within the power-peaking limits. Any initial states within G B C F can lead to an EOL state on the criticality curve while still satisfying the power-peaking requirements and the positivity requirements on the control.

The question of optimality can be solved by forming the Pontryagin function as

$$H = 1 + \psi_1 \, \alpha \left(\sum_{n=0}^{2} b_n (\Sigma_1 - C_1)^n \right)$$

$$+ \psi_2 \, \alpha \left(2 - \sum_{n=0}^{2} b_n (\Sigma_1 - C_1)^n \right)$$

or

$$H = 1 + \alpha (\psi_1 - \psi_2) \left(\sum_{n=0}^{2} b (\Sigma_1 - C_1)^n \right) 2\alpha\psi_2 \quad \text{(IV.D.17)}$$

The necessary conditions for optimality are

$$\mu R = 0; \quad \mu \geq 0 \tag{IV.D.18}$$

$$\frac{d\psi_i}{dt} = -\left(\frac{\partial H}{\partial \Sigma_i} + \mu(t) \frac{\partial R}{\partial \Sigma_i} \right); \quad i = 1, 2 \tag{IV.D.19}$$

$$\frac{\partial H}{\partial C_1} + \mu \frac{\partial R}{\partial C_1} = 0 \tag{IV.D.20}$$

$$H(C_{opt}) \geq H(C) \tag{IV.D.21}$$

together with the transversality conditions

$$H(t_f) = 0 \qquad\qquad\qquad\qquad\qquad\qquad (IV.D.22)$$

$$\underline{\psi(t_f)} \quad . \quad d\Sigma(t_f) = 0 \qquad\qquad\qquad\qquad (IV.D.23)$$

and the system equations, IV.D.14 and IV.D.15.

From Equation IV.D.18 it is seen that there are two possible modes of operation: either operate at the constraints
on the power peaking or control, in which case R = 0, or
satisfy these constraints in an inequality sense, in which
case μ is zero. This latter case corresponds to the unconstrained problem in which the operating point lies between
but not at points B and C on the criticality curve and the
control vector is neither horizontal nor vertical. Let us
consider the latter mode of operation first. From Equation
IV.D.20 we obtain

$$\frac{\partial H}{\partial C_1} = 0 = -\,\alpha(\psi_1 - \psi_2)\left\{ b_1 + 2b_2(\Sigma_1 - C_1)\right\} (IV.D.24)$$

which implies either

$$\text{(case a)} \quad \psi_1 \;=\; \psi_2$$

or $\qquad$ $$\text{(case b)} \quad b_1 \;+\; 2b_2(\Sigma_1 - C_1) = -\,\frac{\partial\phi_1}{\partial_1} = 0$$

i.e., $\qquad\qquad\qquad\qquad (\Sigma_1 - C_1) = -\,\frac{b_1}{2b_2} = \text{constant}$

Case a is clearly not a physically realizable situation since
with the second of the transversality conditions it would
imply

$$d\Sigma_1\Big|_{t_f} \;=\; -\,d\Sigma_2\Big|_{t_f}$$

Case b requires constant power shape operation on the asymp-
totes of the criticality curve (it is noted that $\frac{\partial\phi_1}{\partial C_1}$ can
be zero only where either all or none of the flux is in

node 1). Such a case is physically unrealistic. The unconstrained mode is thus eliminated from consideration.

The other possibility (case c) for satisfying the necessary conditions is to operate at the limits of the power peaking and/or control constraints $R \equiv 0$, $\mu \geq 0$. In this case, the Euler equations are given by

$$\frac{d\psi_1}{dt} = -\alpha(\psi_1 - \psi_2)\left(b_1 + b_2(\Sigma_1 - C_1)\right) - \mu \frac{\partial R}{\partial C_1} \quad \text{(IV.D.25)}$$

$$\frac{d\psi_2}{dt} = -\mu \frac{\partial R}{\partial \Sigma_1} \quad \text{(IV.D.26)}$$

where μ is given by

$$\mu \frac{\partial R}{\partial C_1} = \alpha(\psi_1 - \psi_2)\left(b_1 + b_2(\Sigma_1 - C_1)\right) \quad \text{(IV.D.27)}$$

and C_1 is given by

$$R(\Sigma_1, \Sigma_2, C_1) = 0$$

The optimal trajectories for case c are shown in Figure IV.D.2 and are discussed below. The optimal control procedure is to operate at constant power shape as long as possible; then, if it is still possible to go critical without requiring negative control or exceeding the peaking limits, to move along the criticality curve with C_2 zero and C_1 being adjusted to achieve criticality. Note that during this latter phase of operation, $f < f_{max} = 1.4$; however, the control constraint $C_2 = 0$ is active.

From initial states in the region A B C, the reactor is operated at point B at a peaking limit of 1.40 and an average flux ratio, $g = 0.81$, until the trajectory hits the vertical line AB. This defines EOL, since it is no longer possible to go critical without requiring C_1 to be negative. Similarly, from initial states in region FCD, the reactor is operated at the constant power shape point C, corresponding to $f = 1.40$ and $g = 0.53$, until the horizontal line CD is reached.

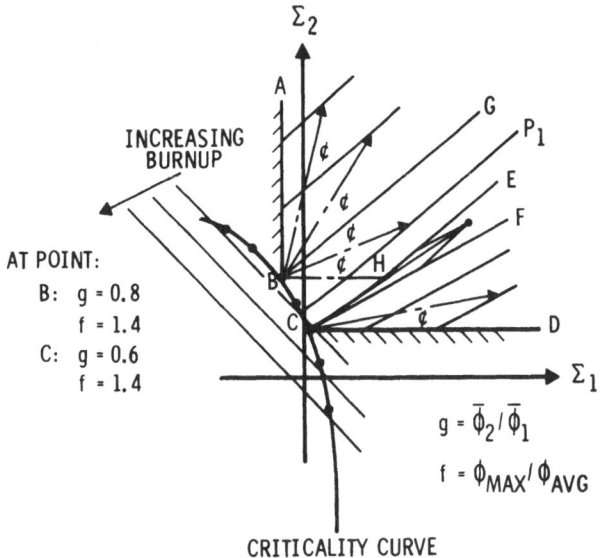

Figure IV.D.2

Optimal Trajectories in a Material Buckling Phase Space

After that, it is no longer possible to go critical without requiring C_2 to be negative.

From initial states in the region GBCE, the reactor is operated at the constant power shape corresponding to B until C_2 becomes zero (shown by horizontal line BH). Then the reactor is brought critical on C_1 with C_2 remaining zero, and the operating point moves down along the criticality curve until EOL. Within this region only an initial state along EHC will reach the maximum attainable burnup. From initial states within the region ECF, the trajectories are not unique except along the bounding curves. The reactor may operate at either B or C, the mode can switch around within the bounds, and all final states will be at C.

An examination of the above possibilities shows that the trajectories for case c will in all cases lead to greater burnups at EOL than can be obtained by trajectories for the

(case a) constant power shape trajectories without hitting
a power peaking or control constraint.

Further, note that only trajectory FC corresponds to the
Haling Principle (50) for giving the absolute maximum lifetime with the reactor operated at the maximum power peaking
and EOL occurring with all poison out. This can be achieved
with the Haling mode only for BOL states lying on FC.

This simple model shows the application of the Pontryagin
formalism and phase-space arguments to the maximum burnup
problem. The optimal EOL state should occur when all the
control is removed if this is consistent with the constraints
on the control or power peaking. Whether this is attainable
or not depends on the initial state of the reactor. Also,
the uniqueness of the trajectory depends on the initial state.
The results obtained by Motoda and Kawai have been generalized
by Suzuki and Kiyose (56) using topographical arguments for a
multiregion reactor, and by Fadilah and Lewins (76).

3. Minimum Power Peaking Through Core Life; A Dynamic Program Solution

In the preceding section, the problem of optimizing
poison management to maximize the core lifetime was investigated by means of the Pontryagin formalism. It was found that
power-peaking considerations were highly important and had the
effect of limiting the usable core life. Another approach to
the depletion problem is to optimize the poison management so
as to minimize the greatest power peaking which occurs through
the core life. The work of Terney and Fenech (54) in applying dynamic programming to this problem is reviewed here.

The reactor depletion process may be viewed as a multistage decision process (as illustrated in Figure IV.D.3)
where at the end of each depletion step a decision is made
concerning the pattern of rod withdrawal to keep the reactor
critical. The process begins at BOL, at which time the state
of the reactor is S_o. A rod withdrawal decision, Δy_1 is
made, and the core depleted to state S_1 with a maximum power
peaking of pp_1. Then another decision, Δy_2, is made and the
process repeated until end of life is reached.

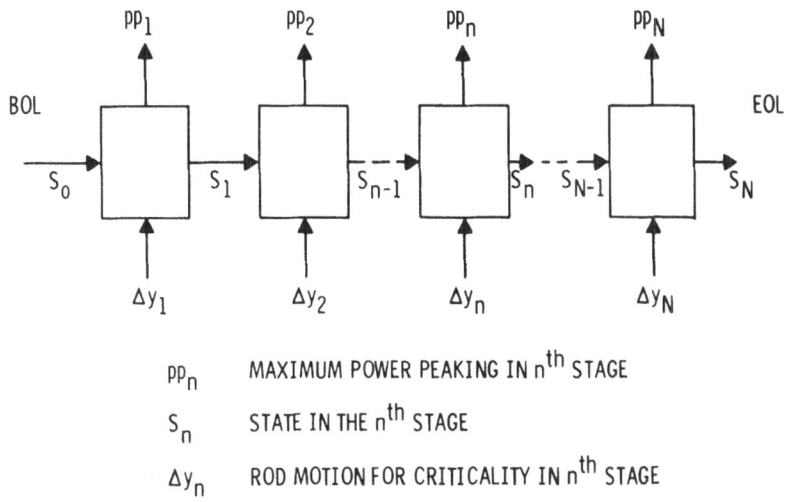

pp_n MAXIMUM POWER PEAKING IN n^{th} STAGE

S_n STATE IN THE nth STAGE

Δy_n ROD MOTION FOR CRITICALITY IN n^{th} STAGE

Figure IV.D.3

Control Rod Programming

The problem is to determine the optimal sequence of allowable decisions $\{\Delta y_1, \Delta y_2, \text{---} \Delta y_n\}$ which leads to the lowest maximum power peaking over the life of core.

To determine the optimal control sequence, a variety of tentative rod withdrawal decisions, Δy_i, must be made at each stage and the resulting peaking evaluated. The approach of factorially evaluating all possible depletion histories can quickly become overwhelming. If there were 10 stages and only three rod motion choices at each decision point, a total of 3^{10} trajectories would have to be evaluated. However, with the problem formulated as a multistage decision process, the dynamic programming "principle of optimality" may be used to stem this proliferation of trajectories.

As each stage is reached in the computation, all the trajectories leading to each state are examined. By discarding all trajectories except the single one which leads to the

state with minimum overall power peaking from BOL, one win-
nows out those few trajectories which retain the potential
for being optimal. As a result, an enormous reduction is
achieved in the number of trajectories which must be calculated.

Defining

pp_n = Maximum power peaking in the n'th stage

$r_n(S_n)$ = Minimum largest power peaking during
the n stages from BOL to stage n

$$r_1(S_1) = \text{Minimum } [pp_1 (S_0, \Delta y_1)] \\ \Delta y_1 \epsilon \qquad \qquad \text{(IV.D.28)}$$

the principle of optimality takes the form of a recursion
relationship for each state at a given stage:

$$r_n (S_n) = \text{Min} \quad \{\text{Max}[r_{n-1}(S_{n-1}), pp_n(S_n, \Delta y_n)]\}$$
$$\Delta y \epsilon R \qquad \qquad \text{(IV.D.29)}$$

The meaning of the recursion relationship is that all the
paths with allowable controls, $\Delta y \epsilon R$, leading to a state, S_n,
at stage n from any and all states, S_{n-1}, at stage n-1 are
examined and only one is retained: the one having lowest
overall power peaking from BOL to the stage n state under
consideration.

The constraints on the control

$\Delta y \epsilon R$

incorporate the allowed rod motion patterns, the criticality
requirement and could include limits on maximum allowed power
peaking and burnup. Constraints are easily handled in this
approach, and actually aid in reducing the number of trajectories to be considered. If a constraint is violated, that
path may be eliminated from future consideration.

In practice for the reactor depletion problem, the states
seldom are truly identical. However, by defining as identical
those states which have nearly the same properties, i.e.,

$$\left| S_n - S'_n \right| \leq \Delta S \qquad\qquad (IV.D.30)$$

the principle of optimality can be used to further eliminate trajectories. The larger the tolerance, ΔS, in identity of states, the more trajectories can be eliminated. However, care must be taken that ΔS is not taken so large that the optimal policy is affected.

Terney and Fenech implemented the above procedure for the reactor model of Figure IV.D.4. The control rod motion pattern was limited to three possible choices:

$$\left\{ \Delta y \epsilon R \; \middle| \; \begin{pmatrix} \Delta z_1 \\ 0 \end{pmatrix}, \begin{pmatrix} 0 \\ \Delta z_2 \end{pmatrix}, \begin{pmatrix} \Delta z \\ \Delta z \end{pmatrix} \right\} \qquad (IV.D.31)$$

consisting of moving the inner bank a distance of Δz_1; the outer bank, Δz_2; or moving both a distance of Δz. Initial criticality was achieved on the outer bank alone to avoid excessive initial power peaking. The state of the reactor was represented by the height of the two control rod banks H_1 and H_2, and the average burnups θ_1 and θ_2 in the unrodded inner and outer radial zones*

$$S_n = \left\{ \begin{matrix} \theta_1 \\ \theta_2 \\ H_1 \\ H_2 \end{matrix} \right\}_n \qquad\qquad (IV.D.32)$$

---

* The validity of the assumption that the state of the fuel is adequately characterized by the burnup was discussed in Section IV.C.2, where the application of dynamic programming to the fuel cycle problem was illustrated.

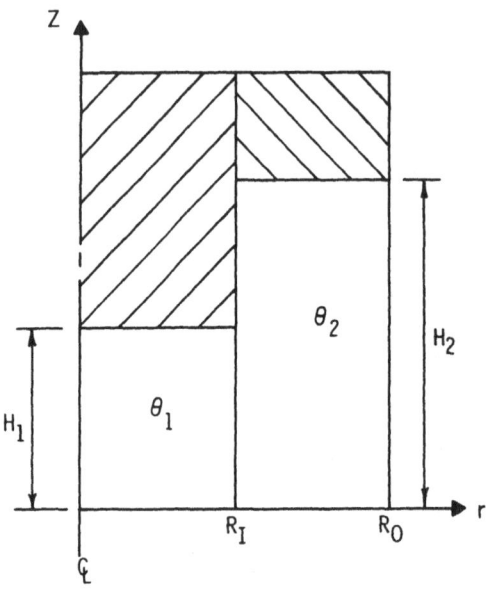

CHARACTERISTICS

POWER: 485 MWt   BARE HEIGHT: 243.2 cm
LOADING: 20.88 MTU OUTER RADIUS: 95.8 cm
ENRICHMENT: 3.4%   INNER RADIUS: 47.9 cm
                REFLECTOR SAVINGS 7.5 cm

Figure IV.D.4

Reactor System for Rod Programming Example

The tolerances on the burnup and bank heights were varied to determine how large they could become without affecting the optimal policy. A flux synthesis procedure was employed to solve the neutron balance equations for a reactor model using material compositions and macroscopic cross-sections in the form of fitted polynomial functions of region burnup, θ_i obtained from previously-computed depletions.

The optimal policy is shown in Figure IV.D.5. The maximum power peaking of 3.93 for the optimal policy occurs at EOL, with all rods out. This is consistent with the Haling principle discussed in the previous section. The maximum

power peaking is smaller than that which would be obtained by
other modes of operation. Straight bank operation gives a
peaking maximum of 6.4; a policy of removing the outer bank
completely before withdrawing any of the inner bank gives a
value of 4.30. The rationale behind the optimal policy is to
deplete the inner region somewhat before allowing the power
to shift into the central zone.

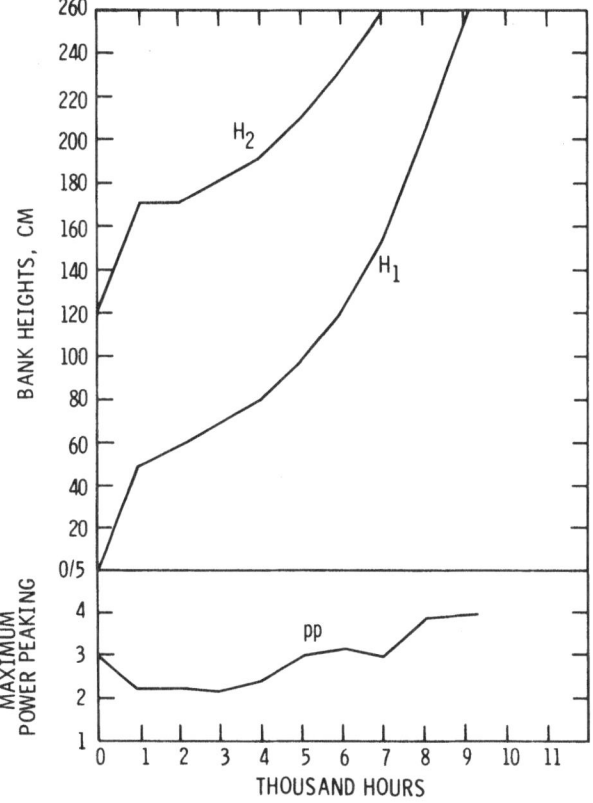

Figure IV.D.5

Optimal Rod Program

TABLE IV.D.1

EFFICACY OF DYNAMIC PROGRAMMING FOR CONTROL ROD PROGRAMMING

Depletion Time Step = 1000 h

Tolerances for Combining States: Burnup = 0.5 Mwd/k_g

Bank Height: 0.1 m

Stage No.	Theoretical Upper Limit on No.States (3^{n-1})	Actual No.States	No.States Same as One or More Other States	No.States That May Be Discarded	No. States To Be Carried Along
1	1	1	0	0	1
2	3	3	0	0	3
3	9	9	3	2	7
4	27	21	9	6	15
5	81	38	19	11	27
6	243	68	47	31	37
7	729	88	67	47	41
8	2187	88	79	63	25
9	6561	45	35	26	19[a]
10	19683	23	17	11	12[b]

[a] 7 at EOL.

[b] All essentially at EOL.

That dynamic programming reduced the required number of calculations is shown in Table IV.D.1. With three rod motion options, at the end of 10 stages the factorial approach would give as many as 20,000 states, which would mean 20,000 trajectories to calculate. By eliminating all but the optimal path leading to each state at each stage in the calculation, the principle of optimality reduces this such that the number of states at any stage never exceeds 90, and of these, at most 40 must be carried along for one more stage. It was found that reducing the state tolerances beyond 500 MWD/t in burnup and 10 cm in rod bank location did not affect the optimal policy while increasing the tolerance did.

The problem illustrates the use of dynamic programming on the lifetime power-peaking problem. As already discussed in Section IV.D.3, the disadvantage of dynamic programming is that it requires large amounts of computer memory storage and time to calculate even a reduced number of feasible paths. For this simple problem, there was no difficulty. All studies ran within several minutes on an early IBM OS-360 machine. However, a more realistic reactor model would require a larger number of state variables (e.g., burnup representation in many nodes) and more control options. In such a case, dynamic programming would not be feasible. For realistic problems and models, an iterative procedure appears desirable. The next section describes such an approach.

4. The Generalized Optimal Poison Management Problem;
    An Iterative Solution to the Pontryagin Equations

In the preceding sections, two aspects of optimizing the depletion behavior of a reactor were investigated; i.e., maximizing lifetime and minimizing power peaking. In this section, a generalized optimal depletion control problem is considered for a realistically-complex, spatially-nodalized reactor representation. The design goals are formulated as a least-squares deviation integral performance index. It is minimized by the optimal choice of a generalized control that affects the spatially dependent material bucklings. The generalized Bolza problem necessary conditions discussed in Section III are used to formulate the optimal control problem and give the necessary conditions for optimality. An iterative

procedure is presented to solve the resulting equations. The
description follows the work of Wade and Terney (57).

In general, the goals of the design and operation of a
fixed fuel, water-cooled power reactor are:

a.   The reactor should produce a large, integrated
     power release over its lifetime.

b.   A substantial burnup of fuel should occur in all
     parts of the reactor during core life.

c.   The critical heat flux and peak central temperature
     limitations should not be exceeded in any reactor
     configuration (e.g., peak xenon, no xenon, steady
     state xenon) during core life.

d.   Reactivity to override peak xenon should exist
     throughout life.

e.   Sufficient control should exist to shut the reactor
     down at all times in life.

These goals may be met by the judicious choice of the
control variables; e.g., initial material distribution within
the core, the control rod sequencing strategy followed during
core life, the selective replacement of fuel cells during
depletion, etc. The design objectives are quantified in a
"performance index" which is non-negative and small when the
objectives are met, and large when they are not. The problem
is then to minimize this performance index while satisfying
the system equation by the optimal choice of the control variables mentioned above.

In modified, one-group theory for a multinode reactor, the
system equations are given by the flux equation

$$[M + B^2]^\nu \underline{P} = 0 \qquad \nu = 0, 1, 2, \text{etc.} \qquad \text{(IV.D.33)}$$

where $\underline{P}$ is the vector of nodal power densities,

$$\underline{P} = [P_1 \ P_2 \ \text{---} \ P_N]^T \qquad \text{(IV.D.34)}$$

normalized with the nodal volumes, $\underline{V}$, such that:

$$\underline{V}^T \underline{P}^\nu = V_{core} \qquad\qquad (IV.D.35)$$

and the depletion equation

$$\frac{d}{dt} \underline{u} = [D] \underline{P}^{\nu=0}; \qquad \underline{u}(t = 0) = \underline{1} \qquad (IV.D.36)$$

where $\underline{u}$ is the vector of nodal fuel fraction remaining.

$$\underline{u} = [u_1 \text{ --- } u_N]^T \qquad\qquad (IV.D.37)$$

These equations must be satisfied for each of several reactor configurations, ν, considered at each instant of core life (e.g., steady state xenon, no xenon, etc.). By definition, $\nu = 0$, is the steady state xenon configuration used in the depletion equation.

The matrix $[M]$ contains the internodal coupling, the matrix $[D]$ is a diagonal matrix of fuel burnup rate per unit power, and the matrix $[B^2]$ is a diagonal matrix of nodal material buckling of the form

$$B^2_{ii} = \frac{\dfrac{k_{\infty i}}{k} - 1}{M^2_i \Sigma_{f_i}} \qquad\qquad (IV.D.38)$$

Table IV.D.2 defines the notation,

The nodal control variables, C_i, modify the nodal material bucklings

$$B^2_{ii}(t) = B^2_{ii_0}(t) + \left.\frac{\partial B_{ii}}{\partial C_i}\right|_t C_i(t) \qquad (IV.D.39)$$

and may be subject to constraints on their values and relationships which are indicated in symbolic form:

$$\underline{C} \in \Omega \qquad\qquad (IV.D.40)$$

All the material properties, except the control, are assumed known as functions of the fraction of fuel remaining; i.e.,

$$k_{\infty i} = k_{\infty i}(u_i); \quad M^2_i = M^2_i(u_i); \quad \Sigma_{f_i} = \Sigma_{f_i}(u_i) \quad (IV.D.41)$$

and are usually taken as third or fourth order polynomial fits in $(1 - u_i)$. These dependencies can be generated by independent cell depletion calculations.

The design objectives are quantified by defining a positive integral performance index of the form

$$PI = L_0(\underline{u}, T) + \int_0^T dt \; L_1(\underline{P}, \underline{u}) \qquad (IV.D.42)$$

which is to be minimized by the proper choice of control. The first term is:

$$L_0(\underline{u}, T) = W4 \langle u(t) \rangle \Big|_{t=T} \qquad (IV.D.43)$$

where $\langle u \rangle$ is the core-average fuel fraction remaining. The integrand is given by:

$$L_1(\underline{P}, \underline{u}) = \sum_\nu L_1^\nu(\underline{P}^\nu, \underline{u}) \qquad (IV.D.44)$$

where

$$L_1^\nu(\underline{P}^\nu, \underline{u}) = \sum_{channels, \kappa} (\beta_\kappa^\nu - \beta_\kappa^{\nu}) \; {}^2FR_\kappa^{\nu^2} W1_\kappa^\nu$$

$$+ \sum_{nodes, i} (P_i^\nu - P_i^{\nu d})^2 W2_i^\nu \qquad (IV.D.45$$

$$+ \sum_{nodes, i} (u_i - \langle u \rangle)^2 W3_i \delta_{\nu, 0}$$

$$+ \left[\left(\frac{1}{k}\right)^2 - \left(\frac{1}{k}\right)^{\nu d^2} \right] W5$$

The performance index is made up of the sum of the non-negative term L_0, which is small when the EOL average fuel fraction remaining is small and a time integral over the reactor life of a sum over reactor configurations at each time in core life of non-negative terms involving square deviations of (a) power profile shape factors, β, from their

target values β^d; (b) the power density profile from the desired profile, $\underline{P}^d$; (c) the depletion profile from uniformity;
and (d) the eigenvalue from the desired eigenvalue, k^d. The
weights, W1 to W5, are used to indicate the relative importance
to the reactor design of the various components of the performance index and the Kronecker, delta, $\delta_{\nu,\nu'}$ is used to indicate which terms apply only to certain configurations.

The first and second terms in L reflect the goals of
satisfying the critical heat flux* and peak central temperature requirements. These are not treated as hard constraints.
Rather, a penalty function approach is taken in which deviations
above the limits contribute significantly to the performance
index.

---

*   A rule of thumb for critical heat flux thermal perfor-
     mance is that the energy input into a coolant channel
     integrated from inlet to outlet should be small in
     channels for which the heat flux profile along the
     channel is excessively skewed toward either the inlet
     or outlet of the channel. The integrated energy input
     may be expressed in terms of the parameter:

$$FR_{channel\ \kappa} = \sum_{nodes\ i\epsilon\kappa} P_i V_i / \sum_{nodes\ i\epsilon\kappa} V_i$$

     and the skewedness of the heat flux profile is parameter-
     ized by the normalized centroid of the power density pro-
     file along the coolant channel:

$$\beta_{channel\ \kappa} = \sum_{nodes\ i\epsilon\kappa} Z_i P_i V_i / \sum_{nodes\ \epsilon k} P_i V_i$$

     where κ denotes the set of nodes through which the cool-
     ant channel passes and Z_i is the distance from the cen-
     ter of the node to the channel outlet normalized to the
     core height.

TABLE IV.D.2

DEFINITION OF TERMS

$M = (M_{ij})$

$$M_{ij} = \mu_{i \leftarrow j}/\Sigma_{f_i}$$

$$M_{ii} = - \sum_{j \neq i} \mu_{i \leftarrow j}/\Sigma_{f_i} - \mu_{e \leftarrow i}/\Sigma_{f_i} \quad \text{(e denotes the exterior of the core)}$$

$B = (B_{ii})$ diagonal matrix

$$B_{ii} = \frac{(k_{\infty i}/k) - 1}{M_i^2 \Sigma_{f_i}}$$

$D = (D_{ii})$ diagonal matrix

$$D_{ii} = -(1 + \alpha) <PD(t)>/N_{0_i} \quad \text{(α is the capture to fission ratio; <PD> denotes the core average power density; N_{0_i} is the initial nodal fuel concentration)}$$

$B_u^2 = (B_{u_{ii}}^2)$ diagonal matrix

$$B_{u_{ii}}^2 = \frac{\partial B_{ii}^2}{\partial u_i} P_i$$

TABLE IV.D.2

$$B_k^2 = (B_{k_{ii}}^2) \text{ diagonal matrix}$$

$$B_{k_{ii}}^2 = \frac{\partial B_{ii}^2}{\partial \, 1/k} = \frac{k_{\infty i}}{M_i^2 \Sigma_{f_i}}$$

$$B_c^2 = (B_{c_{ii}}^2) \text{ diagonal matrix}$$

$$B_{c_{ii}}^2 = \frac{\partial B_{ii}}{\partial C_i} P_i$$

$$F_u = (F_{u_{ii}}) \text{ diagonal matrix}$$

$$F_{u_{ii}} = \frac{1}{\Sigma_{fi}} \frac{\partial \Sigma_{f_i}}{\partial u_i} P_i$$

$$\left\{ \frac{\partial L_1}{\partial P} \right\} = \left\{ \frac{\partial L_1}{\partial P_i} \right\} \text{ vector}$$

$$\frac{\partial L_1}{\partial P_i} = 2W2_i (P_i - P_i^d)$$

$$+ (\Sigma_{\text{channels } l\epsilon\Gamma_i} \frac{(\beta_1 - \beta_1^d)}{(Z_i - \beta_1^d) V_i} FR_1 2Wl_1/Vol_1)$$

Γ_i = set of channels which include mode i

TABLE IV.D.2

$$\left\{ \frac{\partial L_1}{\partial U} \right\} = \left\{ \frac{\partial L_1}{\partial u_i} \right\} \quad \text{vector}$$

$$\frac{\partial L_1}{\partial u_i} = 2W3_i(u_i - \langle u \rangle)$$

$$- \frac{V_i}{V_{core}} \Sigma_i 2W3_i(u_i - \langle u \rangle) \quad (\langle u \rangle \text{ denotes core average fuel fraction remaining})$$

$$\frac{\partial L_1}{\partial \ 1/k} \quad \text{scalar}$$

$$\frac{\partial L}{\partial \ 1/k} = 2W5 \left(\frac{1}{k} - \frac{1}{k^d} \right)$$

The goal is to minimize the performance index (Equation IV.D.42) by the proper choice (within physically realizable bounds) of control which satisfies the state equations, IV.D.33 - IV.D.36. The control may be taken as the initial fuel or poison load in a node (in which case it is time independent), the presence or absence of control poison in a node, the values of the constants that define the functional behavior of the nodal properties, k_∞, M^2, and Σ_f, versus depletion (see Equation IV.D.41), etc. The following derivations assume only that the node i control influences the node i buckling term, B^2_{ii}, only.

The augmented functional and Pontryagin function may be formed as

$$J = L_0 + \int_0^T dt \left\{ L_1 + \sum_\nu \left\{ \underline{\Lambda}^{\nu T}[M+B^2(k^\nu)]\underline{P}^\nu \right. \right. \quad \text{(IV.D.46)}$$

$$\left. \left. + \xi^\nu(\underline{V}^T\underline{P}^\nu - V_{core}) + \underline{\lambda}^T\left[\frac{d}{dt}\,\underline{u} - [D]\underline{P}^\nu\right]\delta_{\nu,0} \right\} \right\}$$

and

$$H = L_1(\underline{P},\underline{u}) + \sum_\nu \left\{ \underline{\Lambda}^{\nu T}[M+B^2(k^\nu)]\underline{P}^\nu \right. \quad \text{(IV.D.47)}$$

$$\left. + \xi^\nu(\underline{V}^T\underline{P}^\nu - V_{core}) + \lambda^T[D]\ \underline{P}^\nu\delta_{\nu,0} \right\}$$

where

$$\underline{\Lambda}^\nu = [\Lambda_1^\nu(t)\ \Lambda_2^\nu(t)\ ---\ \Lambda_N^\nu(t)]^T \quad \nu=0,1,2,- \quad \text{(IV.D.48)}$$

$$\underline{\lambda} = [\lambda_1(t)\ \lambda_2(t)\ ---\ \lambda_N(t)]^T$$

are vectors of nodal Lagrange multipliers.

As stated, this is an extended Bolza problem with no inequality constraints. The necessary conditions for optimality are given by Equations III.B.9 to III.B.15, and consist of the Euler-Lagrange equations:

$$\frac{\partial H}{\partial \underline{P}} = 0 \quad ; \quad \frac{\partial H}{\partial 1/k} = 0 \quad ; \quad -\frac{d}{dt}\ \underline{\lambda} = \frac{\partial H}{\partial \underline{u}} \quad \text{(IV.D.49)}$$

subject to the corner conditions:

$$H(t-) = H(t+) \qquad\qquad\qquad\qquad (IV.D.50)$$

$$\lambda(t-) = \lambda(t+)$$

and the transversality conditions:

$$\left[\left(\frac{\partial L_0}{\partial u}\right)^T \delta \underline{u}\right]^T_0 + \left(H\delta t - \underline{\lambda}^T \delta \underline{u}\right)\Big|^T_0 = 0 \qquad (IV.D.51)$$

For constant power depletion, H does not depend explicitly on time and is therefore constant as well as continuous (as shown in Equation II.B.16). Furthermore, the Weierstrass condition requires that at each point along the optimal trajectory the Pontryagin function is minimized when the control is the optimal control.

$$H(\underline{P},\underline{u},1/k,\underline{\Lambda},\underline{\lambda},\xi,\underline{C}) \geq H(P,\underline{u},1/k,\underline{\Lambda},\underline{\lambda},\xi,\underline{C}_{opt}) \qquad (IV.D.52)$$

Evaluation of the terms in Equations IV.D.49 and IV.D.51 yields the Euler-Lagrange equations:

$$[M + B^2]^T \underline{\Lambda} = -\xi\underline{V} - [D]^T \underline{\lambda}\delta_{v,0} - \left(\frac{\partial L_1}{\partial \underline{P}}\right) \qquad (IV.D.53)$$

$$\underline{P}^T [B^2] \underline{\Lambda} = -\left(\frac{\partial L_1}{\partial 1/k}\right) \qquad\qquad (IV.D.54)$$

$$-\frac{d}{dt}\underline{\lambda} = \sum_v [B_u^2 - F_u M^T] \underline{\Lambda} + \left(\frac{\partial L_1}{\partial \underline{u}}\right) \qquad (IV.D.55)$$

subject to the final condition:

$$\lambda(T) = \left(\frac{\partial L_0}{\partial u}\right) \qquad\qquad\qquad (IV.D.56)$$

The matrices are defined in Table IV.D.2.

For realistic reactor models, the state equations IV.D.33, IV.D.35 and IV.D.36, and the Euler-Lagrange equations,

IV.D.53 - IV.D.56 present a system too complex for analytic
solution, and a numerical, iterative solution is appropriate.
Since the state equations are subject to known initial conditions:

$$\underline{u}(t = 0) = 1$$

while the Euler-Lagrange equations are subject to known final
conditions:

$$\underline{\lambda}(T) = \left(\frac{\partial L_0}{\partial \underline{u}}\right)$$

the control signal iteration strategy discussed in Section
III.B.4 is used.

a. A tentative choice of control is made and the state
    equations are solved forward in time as in a normal
    design depletion calculation. The state variables
    are saved and a partial performance index is cal-
    culated and saved.

$$PI_t = L_0 + \int_0^t L_1 \, dt \qquad\qquad (IV.D.57)$$

b. Then, using the tentative control and the resulting
    state variables, the Euler-Lagrange equations are
    solved backward in time, starting from their known
    final conditions.

c. An improvement in the tentative control is selected
    and the steps repeated iteratively until no further
    reduction in the performance index can be achieved.
    If, during any step, the partial performance index
    at time t exceeds the final performance index from
    the previous iteration, the control improvement will
    be known to have failed and an alternate one must be
    chosen.

The key step in the above procedure is the selection of
the improved tentative control. This is achieved by use of
the Weierstrass condition, Equation IV.D.52. It requires
that at each point along the optimal trajectory, the Pontryagin

function, H, must achieve its minimum value. To first order
the change in H for a small change, δc^{ν}, in control in each
spatial node at a given instant of time is given by:

$$\delta H(t) = \sum_{\nu} S^{\nu}(t)^T \, \delta c^{\nu}(t) + \sigma \, (\delta c^{\nu}{}^T \cdot \delta c^{\nu}) \qquad \text{(IV.D.58)}$$

Here

$$S_i^{\nu}(t) = \frac{\partial H(t)}{\partial c_i^{\nu}(t)} = \Lambda_i^{\nu}(t) \, \frac{\partial B_{ii}^{2\nu}(t)}{c_i^{\nu}(t)} \, P_i^{\nu}(t) \qquad \text{(IV.D.59)}$$

is called the control switching signal and is dependent on
the known values of nodal power density and Lagrange multiplier, Λ_i. It is clear that taking the change in control

proportional to the negative of the switching signal:

$$\delta c^{\nu}(t) = -\text{SCALE} * S^{\nu}(t) \qquad \text{(IV.D.60)}$$

will lead to a first order decrease in the Pontryagin function,
H. If the value of SCALE is too large, the linearity assumption inherent in Equation IV.D.58 will be violated and an
increase in H and PI may result. In such a case, the value
of SCALE is successively reduced until eventually the linearity assumption holds.

The control improvement procedure indicated by Equation
IV.D.59 is a "gradient" technique. It can easily accommodate
constraints on the control alone (see Equation IV.D.40) by
requiring of δc that:

$$c^{(n)} = (c^{(n-1)} + \delta c) \, \varepsilon \Omega$$

where super (n) is an iteration index. Convergence of the
iterative search to the optimum can sometimes be acceler-
ated by determining the control step direction as a relaxed
linear combination of the directions indicated by the current
switching signal and the previous step:

$$\delta c^{\nu(n)} = -b \, \text{SCALE} * S^{\nu}(t) + (1 - b) \, \delta c^{\nu(n-1)} \qquad \text{(IV.D.61)}$$

where b is the relaxation constant.

It is noted that if a control variable applies over a range of time, configuration, or space, the switching signal multiplying that control increment in Equation IV.D.60 should be integrated over the appropriate time, configuration or space interval.

The gradient technique for choosing changes in the control is only one way of utilizing the Weierstrass condition and Equation IV.D.58 in the iterative solution of the optimal control equations. Alternately, for control constraints that are complex but are linear functions of the control, the control improvement step in the iterative procedure is recognized as a linear programming problem:

$$\underset{\underline{\delta c}}{\text{Min}}\left\{\delta H(t)\right\} = \underset{\underline{\delta c}}{\text{Min}}\left\{\sum_{\nu} \underline{S}^{\nu T}(t) \ \underline{\delta c}^{\nu}(t)\right\} \qquad \text{(IV.D.62)}$$

subject to

$$\underline{A} \ \underline{\delta c}^{\nu}(t) \leq \underline{Q} \qquad \text{(IV.D.63)}$$

Here, possible constraints on the control, including constraints in control or step size and core criticality, are expressed by Equation IV.D.63, where $\underline{A}$ and $\underline{Q}$ are a control-independent matrix and vector, respectively. Standard linear programming techniques can be applied to selecting $\underline{\delta c}$ subject to Equations IV.D.62 and IV.D.63.

To test the procedures described above, Wade and Terney developed a computer code for treating one-dimensional, modified one-group reactor models of up to 50 nodes, three configurations at each time step, and 20 depletion steps. The minimum mass problem and the flux-flattening problems (Sections IV.B.1 and IV.B.2, respectively) were solved to verify the code and study convergence properties of the iteration scheme. Both gradient and linear programming control improvement algorithms were used. In both tests, the known optimal solutions

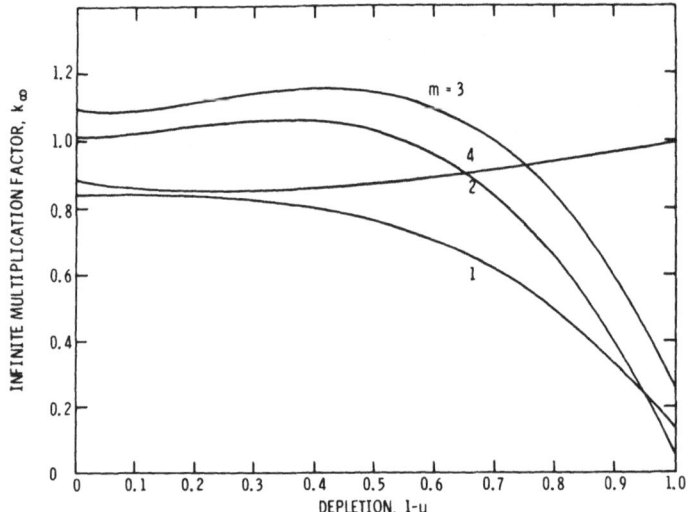

Figure IV.D.6, Part 1

Material Properties Versus Depletion
(Reactor zones m = 1 through 4 are identified
in Figure IV.D.7)

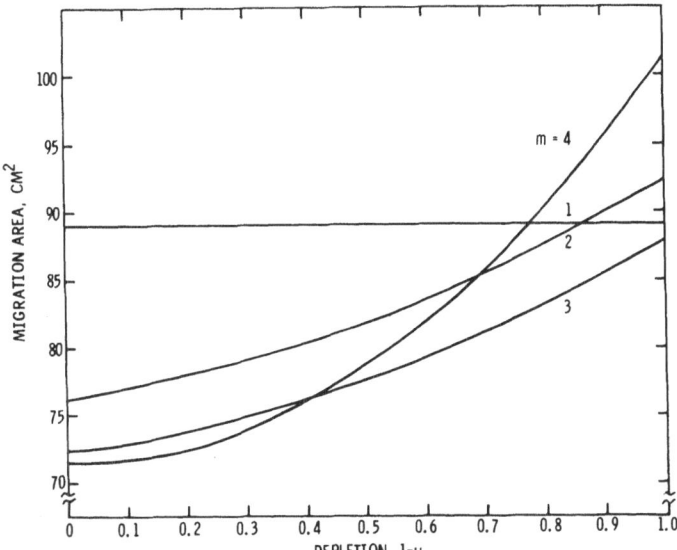

Figure IV.D.6, Part 2

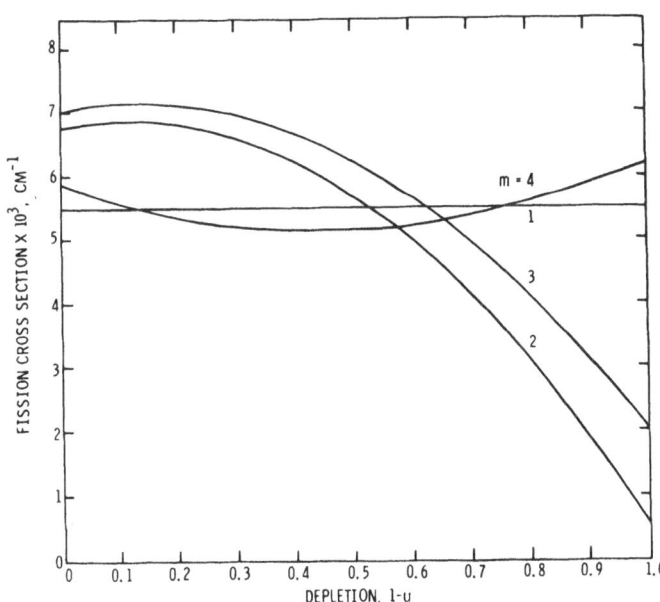

Figure IV.D.6, Part 3

were obtained or closely approached after only a few iterations. The iteration procedure markedly slowed down as the
optimum was approached. However, the vicinity of the optimal
solution was quickly attained and the added improvements in
achieving the absolute optimum were small.

The procedure was applied to a sample design problem of
representative complexity where the control was the initial
loading, or equivalently, the BOL level of the k_∞ versus

depletion curves given by Equation IV.D.41 and displayed in
Figure IV.D.6. The reactor model is shown in Figure IV.D.7,
together with the performance index data. Three reactor
configurations, each with 25 nodes, were considered. They
represented the no condition, steady state, and peak xenon
conditions; the effect of xenon was represented by uniform
decrements in k_∞ of 0.03 and 0.05 for steady state and peak

xenon, respectively. Three different material zones were
considered, with the top zone being rodded, partially rodded,
or unrodded, depending on the configuration. The control was
the initial level of the k_∞ versus $(1 - u)$ curves for each

of the four material types, an additional constraint being
that the initial rod worth or difference in the BOL k_∞ values

for k_∞ curves 1 and 2 was to be maintained at 0.177 Δk_∞.

The optimization procedure terminated after 11 iterations
when no further improvements were obtained. The optimized
results are compared to the unoptimized reactor performance
in Table IV.D.3. The performance index decreased 29% on the
first iteration and then decreased slowly on each successive
iteration until it was finally about 45% of its initial value.
The power peaking was reduced at all times in life and in all
configurations achieving a maximum value of 2.15. The eigenvalues were maintained within a few percentage of the target
values, and at EOL where criticality decreases because of
burnup, the eigenvalues were greatly improved.

A second calculation was made for the same reactor model
and performance index. In this case, however, not only the
BOL intercept of the k_∞ versus depletion curve was adjusted

but the shape was controlled as well. For this problem, the
initially-assumed rod worth versus depletion was maintained;
i.e., k_∞ curves 1 and 2 were constrained to maintain the

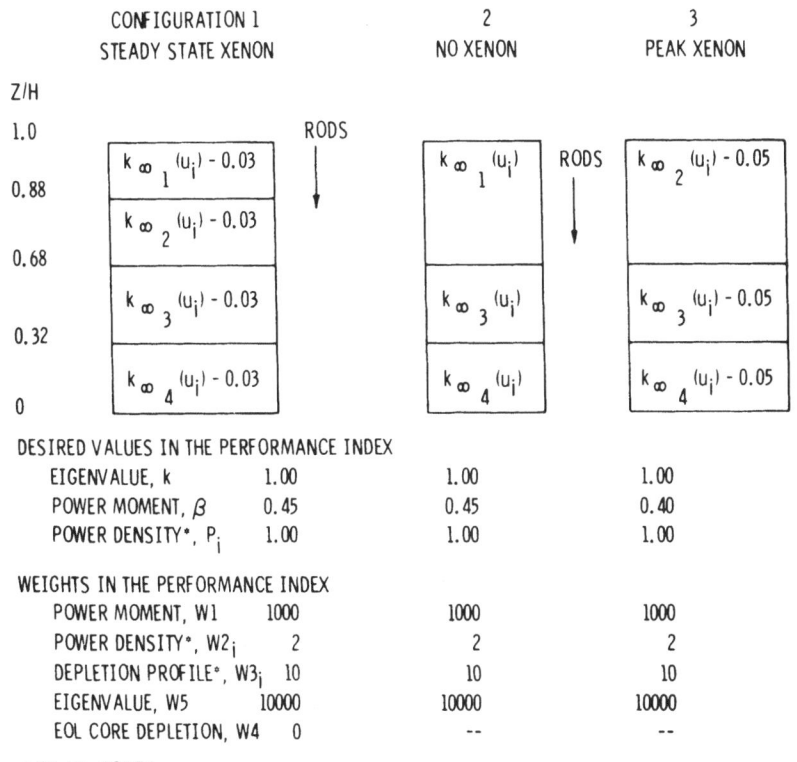

Figure IV.D.7
Sample Design Problem

initially-assumed separation versus depletion shown on Figure
IV.D.6. The control improvement algorithm was modified (Ref.
56) to treat this type of problem.

The optimized k_∞ versus depletion curves are shown in
Figure IV.D.8, and the optimized reactor performance is indicated by the last three columns of Table IV.D.3. The performance index was 15% smaller than that obtained in the previous
optimization problem, the worst power-peaking factor was 1.89,
and the largest eigenvalue deviation was about 0.015 Δk.
These results show improvement over the case when only the
BOL levels were optimized.

TABLE IV.D.3

Sample Design Problem Initial Case and Optimal Control Results

Time (hr)	Item	Initial			BOL Level Control			k_∞ versus Depletion with Fixed Rod Worth		
		SS XENON	NO XENON	PK XENON	SS XENON	NO XENON	PK XENON	SS XENON	NO XENON	PK XENON
0	K	1.01429	1.03143	0.99637	0.98888	1.00925	0.97469	1.00097	1.01460	0.98875
	PK/AVE (P)	2.194	2.507	2.105	1.389	1.878	1.381	1.587	1.886	1.571
	RMS (P)	76.340	89.089	71.009	36.482	61.083	31.600	46.948	61.124	46.040
	BETA (P)	0.456	0.504	0.439	0.517	0.627	0.446	0.464	0.582	0.402
	PK/AVE (U)	1.	-	-	1.000	-	-	1.000	-	-
1K	K	1.01322	1.02973	0.99540	0.98371	1.00090	0.97141	1.00093	1.01409	0.98878
	PK/AVE (P)	2.152	2.456	2.058	1.432	1.821	1.492	1.581	1.867	1.563
	RMS (P)	74.706	87.011	69.237	37.858	58.744	39.004	46.152	59.997	45.432
	BETA (P)	0.453	0.502	0.435	0.490	0.615	0.414	0.464	0.580	0.401
	PK/AVE (U)	1.052	-	-	1.043	-	-	1.041	-	-
2K	K	1.02809	1.04438	1.01003	0.98819	1.00152	0.97647	1.00094	1.01335	0.98887
	PK/AVE (P)	2.166	2.448	2.085	1.540	1.794	1.551	1.578	1.856	1.555
	RMS (P)	75.321	86.743	70.413	42.431	57.577	44.162	45.490	59.311	44.859
	BETA (P)	0.452	0.498	0.437	0.467	0.595	0.399	0.463	0.580	0.401
	PK/AVE (U)	1.110	-	-	1.088	-	-	1.086	-	-
3K	K	1.04633	1.06224	1.02808	0.99915	1.00881	0.98708	1.00103	1.01240	0.98903
	PK/AVE (P)	2.161	2.423	2.091	1.627	1.832	1.607	1.577	1.850	1.548
	RMS (P)	75.480	86.134	71.128	46.995	58.228	47.100	45.006	58.892	44.409
	BETA (P)	0.453	0.596	0.439	0.452	0.576	0.395	0.462	0.580	0.400
	PK/AVE (U)	1.175	-	-	1.136	-	-	1.137	-	-

4K	K	1.05475	1.06968	1.03649	1.01196	1.01916	0.99926	1.00122	1.01131	0.98931
	PK/AVE (P)	2.076	2.337	2.009	1.678	1.883	1.634	1.580	1.847	1.542
	RMS	73.113	83.621	68.843	49.656	59.789	48.230	44.715	58.568	44.139
	BETA (P)	0.453	0.498	0.439	0.446	0.562	0.396	0.461	0.578	0.400
	PK/AVE (U)	1.249	–	–	1.190	–	–	1.195	–	–
5K	K	1.04186	1.05477	1.02400	1.02057	1.02720	1.00752	1.00153	1.01017	0.98974
	PK/AVE (P)	1.874	2.210	1.797	1.665	1.904	1.609	1.585	1.844	1.538
	RMS	66.720	78.377	61.782	49.296	60.596	47.078	44.603	58.202	44.063
	BETA (P)	0.451	0.505	0.433	0.448	0.559	0.401	0.458	0.576	0.398
	PK/AVE (U)	1.333	–	–	1.250	–	–	1.262	–	–
6K	K	1.00434	1.01372	0.98829	1.01898	1.02738	1.00623	1.00201	1.00909	0.99037
	PK/AVE (P)	1.667	2.178	1.608	1.571	1.891	1.511	1.592	1.839	1.534
	RMS	55.892	71.082	50.766	44.897	59.971	42.737	44.623	57.690	44.156
	BETA (P)	0.443	0.525	0.410	0.461	0.571	0.408	0.456	0.571	0.396
	PK/AVE (U)	1.430	–	–	1.319	–	–	1.339	–	–
7K	K	0.95266	0.95691	0.94483	1.00425	1.01736	0.99278	1.00273	1.00824	0.99124
	PK/AVE (P)	1.608	2.319	1.935	1.435	1.879	1.348	1.598	1.831	1.530
	RMS	46.199	66.650	58.890	36.238	60.178	35.200	44.701	56.947	44.364
	BETA (P)	0.417	0.565	0.340	0.489	0.602	0.416	0.453	0.565	0.394
	PK/AVE (U)	1.540	–	–	1.408	–	–	1.429	–	–
8K	K	0.90449	0.90597	0.91570	0.98141	1.00220	0.97141	1.00373	1.00782	0.99239
	PK/AVE (P)	2.060	2.516	2.973	1.649	2.147	1.531	1.599	1.814	1.523
	RMS	53.555	73.674	97.642	34.759	67.894	28.905	44.743	55.896	44.624
	BETA (P)	0.372	0.629	0.245	0.552	0.654	0.427	0.449	0.556	0.392
	PK/AVE (U)	1.667	–	–	1.506	–	–	1.537	–	–

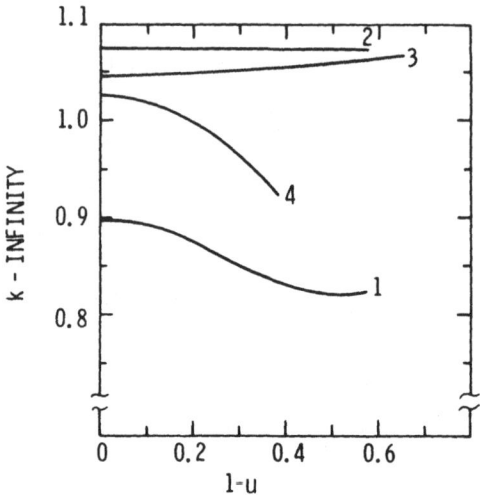

Figure IV.D.8

Optimal Mismatch Curves for Fixed Rod Worth

This section has described the formulation of a generalized optimal poison management problem for a reactor depletion
model of realistic complexity. An iterative solution technique
for the resulting two-point boundary value equations was developed. Test calculations have verified that the iterative solution quickly converges to the vicinity of the optimal solution. One attraction of the method is its applicability to
complex reactor models. Another is the nature of the iterative solution technique. Not only does each iteration yield
a better design than the preceding one, but each iteration is
a usable reactor design study for the control that was assumed.
Thus, the designer can intervene at any time to either termin-
ate the iterations with a suboptimal (unconverged) yet acceptable design, or alternately to speed the convergence by exer-
cising judgment and intuition not expressible as an algorithm.
That is, the iterative process follows the standard method of
designing reactors. A design is tried, evaluated and then
modified. The optimal control approach takes the guesswork
out of how to modify the design to gain improvement.

In the formulation presented above, a penalty function
approach was taken to include the design goals of conformance
to critical heat flux and peak central temperature constraints.

In this way, the complexities (see Section III.B.3) introduced by state variable constraints having no explicit dependence on the control were avoided. Further work is needed to
incorporate the state variable thermal performance constraints
into the iterative approach.

5. Summary

In this section, three approaches to the problem of
determining the poison or rod management scheme to optimize
some aspect of the reactor performance during a core lifetime were examined. A review of the Pontryagin approach with
phase-space arguments used by Motoda and Kawai to maximize
core lifetime aided in visualizing the process and in extracting physical interpretations. It also demonstrated that the
lifetime in itself is not highly sensitive to the poison
management scheme, but that the power peaking is and a constraint on it can fix the maximum attainable lifetime. Their
conclusion is the intuitively satisfying result that the maximum lifetime occurs when all the control poison can be removed at EOL, consistent with the power-peaking requirements.
This result has been generalized by Suzuki and Kiyose for a
multiregion reactor.

The dynamic programming approach of Terney and Fenech
demonstrated the capacity of dynamic programming to accommodate constraints. However, this method is limited by computer memory and computation time requirements. For realis-
tically-detailed reactor models and control options, these
requirements soon exceed limits of computational feasibility.

For realistically complex reactor models, iterative techniques appear quite promising. The approach of Wade and Terney
illustrated this point. It has the added attraction of following the normal design approach of successively iterating to
better and better designs. At present, its usefulness is
limited somewhat by the fact that an ability to handle state
variable or mixed control/state variable constraints has not
been demonstrated. To date, some success has been achieved
by a penalty function approach.

Motoda (68) has recently proposed a method called "modified
approximation programming" which also looks promising. It consists of iterating to the solution by solving a succession of
linear programming problems, where a new linearization is made
after each optimization step.

The future in this area would seem to be in the area of iterative techniques rather than in direct or analytic solutions, because of the complexity of realistic problems. Those methods, coupled with the Pontryagin approach, are very powerful, and probably more effective than direct search methods. In addition, methods utilizing the simple linear programming formalism may be very useful in conjunction with iterative methods.

V.   CONCLUSIONS

In this survey, the formulation and basis of the most common optimization techniques have been summarized. The application of the techniques to a variety of nuclear engineering problems has been critically reviewed.

For any particular application, a variety of problem formulations and solution techniques present themselves. The field is not yet mature enough for clearly-defined "best" approaches to have been identified. Certain trends are noticeable, however.

Historically, among the first applications of optimization techniques to nuclear engineering problems were the generation of analytic solutions to the calculus of variations or Pontryagin formulations of a problem. In some cases, phase-space constructions were used to develop a solution. However, as problems become less idealized and the reactor models more complex, progress by this technique becomes more difficult because analytic solutions are no longer trivial and constructions in multidimensional phase space are difficult.

Dynamic programming was also applied at an early historical stage to reactor optimization problems that were cast in a multistage decision process framework. Again, as reactor models become more detailed, progress by this technique becomes difficult because of requirements for excessive computer memory and calculational times.

One approach to overcoming these difficulties has been to utilize the linear programming technique. However, in some cases, the requirement for a linear representation of the system equations introduces an approximation that is too gross. Recent efforts have been made to employ linear programming iteratively; successively linearizing the nonlinear system equations around the point gained by the linear optimization.

Another approach is based on an iterative numerical solution to the nonlinear system equations and corresponding Euler-Lagrange equations which result from the generalized Bolza or Pontryagin formalism. The control modifications at each iterative step are chosen to further decrease the Pontryagin function in accordance with the Weierstrass condition. This technique has not yet been applied to a state variable or mixed state variable/control constraint problem, although a penalty function approach to constraints has been moderately successful.

With increasingly larger and faster computers, the dynamic programming method might again gain in popularity. However, the future trend probably will be toward iterative techniques. Those coupled with the Pontryagin formalism are very powerful. The iterative methods using linear programming also look very promising.

In view of the achievements to date, it can be expected that optimization procedures will be applied increasingly in the nuclear engineering field. This is particularly true for the poison and fuel management and plant allocation problems. In these cases, an improvement in performance of only a few percent can produce an economic payoff that is quite large, thereby economically justifying the cost of optimization studies.

REFERENCES

1. Bliss, G.A., Calculus of Variations, Open Court Pub-
    lishing Co., LaSalle, Illinois, 1962.

2. Gelfand, I.M. and Fomin, S.V., Calculus of Variations,
    Prentice-Hall, Englewood Cliffs, New Jersey, 1963.

3. Elsgolc, L.E., Calculus of Variations, Addison Wesley,
    Reading, Massachusetts, 1962.

4. Berkovitz, L.D., Journal of Mathematical Analysis and
    Applications, 3, PP 145-169, 1961.

5. Berkovitz, L.D., Journal of Mathematical Analysis and
    Applications, 5, PP 488-498, 1962.

6. Balakrishnan, A.V. and Newstadt, L.W., Editors,
    Computing Methods in Optimization Problems, Academic
    Press, New York, 1964.

7. Pontryagin, L.S., et al., The Mathematical Theory of
    Optimal Processes, Interscience, New York, 1962.

8. Bellman, R., Dynamic Programming, Princeton University
    Press, Princeton, New Jersey, 1957.

9. Bellman, R., and Dreyfus, S., Applied Dynamic Program-
    ming, Princeton University Press, Princeton, N.J., 1962.

10. Leitman, G., Editor, Optimization Techniques, Academic
    Press, New York, 1962.

11. Merriam, C.W., Optimization Theory and The Design of
    Feedback Control Systems, McGraw-Hill, N. Y., 1964.

12. Sage, A.P., Optimum Systems Control, Prentice-Hall,
    Englewood Cliffs, New Jersey, 1968.

13. Bryson, A.E., Jr. and Ho, Y.C., Applied Optimal Control,
    Blaisdell, Waltham, Massachusetts, 1969.

14. Mohler, R.R. and Shen, C.N., Optimal Control of Nuclear
    Reactors, Academic Press, New York, 1970.

15. Stacey, W.M., Jr., Nuclear Science and Engineering, 33, PP 249-267, 1968.

16. Stacey, W.M., Jr., Nuclear Science and Engineering, 39, PP 226-230, 1970.

17. Malan, G.F. and Koen, B.V., Nuclear Science and Engineering, 46, PP 385-393, 1971.

18. Ash, M., Bellman, R. and Kalaba, R., Nuclear Science and Engineering, 6, PP 152-156, 1959.

19. Rosztoczy, Z.R. and Weaver, L.E., Nuclear Science and Engineering, 20, PP 318-323, 1964.

20. Roberts, J.J. and Smith, H.P., Jr., Nuclear Science and Engineering, 22, PP 470-478, 1965.

21. Roberts, J.J. and Smith, H.P., Jr., Nuclear Science and Engineering, 23, PP 397-399, 1965.

22. Lewins, J., Nuclear Science and Engineering, 23, PP 404-406, 1965.

23. Roberts, J.J. and Smith, H.P., Jr., Nuclear Science and Engineering, 24, Page 95, 1966.

24. Ash, M., Nuclear Science and Engineering, 24, PP 77-86, 1966.

25. Ash, M., Optimal Shutdown and Control of Nuclear Reactors, Academic Press, New York, 1966.

26. Lewins, J. and Babb, A.L., "Optimum Nuclear Reactor Control Theory," Advances in Nuclear Science and Technology, 4, Academic Press, New York, 1968.

27. Salo, S., Nuclear Science and Engineering, 50, PP 46-52, 1973.

28. Wiberg, D.M., Optimal Feedback Control of Spatial Xenon Oscillations in a Nuclear Reactor, Ph.D. Thesis, California Inst. of Technology, Pasadena, Calif., 1965.

29.   Wiberg, D.M., Nuclear Science and Engineering, 27,
       PP 600-604, 1967.

30.   Stacey, W.M., Jr., Nuclear Science and Engineering, 33
       PP 162-168, 1968.

31.   Stacey, W.M., Jr., Nuclear Science and Engineering, 38,
       PP 229-243, 1969.

32.   Christie, A.M. and Poncelet, C.G., Nuclear Science
       and Engineering, 51, PP 10-24, 1973.

33.   Bauer, D.C. and Poncelet, C.G., Nuclear Technology, 21,
       1974.

34.   Hurwitz, H., Jr., "Note on a Theory of Minimum Weight
       Shields," KAPL-1441, 1957.

35.   Troubetzkoy, E.S., "Minimum Weight Shield Synthesis,"
       UNC-5017A, 1962.

36.   Lewins, J., Trans. Am. Nuclear. Society, 8, Page 492,
       1965.

37.   Terney, W.B. and Fenech, H., Nuclear Applications, 3,
       PP 46-52, 1967.

38.   Goertzel, G., Journal of Nuclear Energy, 2, PP 193-
       201, 1956.

39.   Wilkins, J.E., Jr., Nuclear Science and Engineering, 6,
       PP 229-232, 1959.

40.   Kochurov, B.P., Atomnaya Energiya, 20, PP 243-247, 1966.

41.   Zaritskaya, T.S. and Rudik, A.P., Atomnaya Energiya, 22,
       PP 6-10, 1967.

42.   Zaritskaya, T.S. and Rudik, A.P., Atomnaya Energiya, 23,
       PP 218-222, 1967.

43.   Axford, R.A., "Constrained Optimal Programming Problems
       in Reactor Statics," LA-4267, Los Alamos, 1969.

44.   Gandini, A., Salvatores, M. and Sena, G., Journal of Nuclear Energy, 23, PP 469-476, 1969.

45.   Goldschmidt, P. and Quenon, J., Nuclear Science and Engineering, 39, PP 311-319, 1970.

46.   Terney, W.B., Nuclear Science and Engineering, 45, PP 226-230, 1971.

47.   Goldschmidt, P., Nuclear Science and Engineering, 49, PP 263-273, 1972.

48.   Goldschmidt, P., Nuclear Science and Engineering, 50, PP 153-163, 1973.

49.   Tzanos, C.P., Gyftopoulos, E.P. and Driscoll, M.J., Nuclear Science and Engineering, 52, Page 84, 1973.

50.   Haling, R.K., "Operating Strategy for Maintaining an Optimum Power Distribution Throughout Life," Proc. ANS Topical Meeting on Nuclear Performance of Power Cores in San Francisco, California, TID-7672, PP 205-210, 1963.

51.   Crowther, R.L., Trans. American Nuclear Society, 6, Page 276, 1963.

52.   Crowther, R.L., "Extensions of the Power Control Method for Solution of Large Inhomogeneous Reactor Problems," Proceedings of the Conference on the Application of Computing Methods to Reactor Problems, Argonne National Laboratory, ANL-7050, PP 487-502, 1965.

53.   Suzuki, A. and Kiyose, R., Trans. American Nuclear Society, 11, Page 441, 1968.

54.   Terney, W.B. and Fenech, H., Nuclear Science and Engineering, 39, PP 109-114, 1970.

55.   Motoda, H. and Kawai, T., Nuclear Science and Engineering, 39, PP 114-118, 1970.

56.   Suzuki, A. and Kiyose, R., Nuclear Science and Engineering, 44, PP 121-134, 1971.

57.   Wade, D.C. and Terney, W.B., Nuclear Science and
       Engineering, 45, PP 199-217, 1971.

58.   Motoda, H., Nuclear Science and Engineering, 46,
       PP 88-111, 1971.

59.   Terney, W.B. and Wade, D.C., "Determination of Optimal
       K-Infinity Mismatch Curves Via Modern Control Theory,"
       Proceedings American Nuclear Society National Topical
       Meeting on New Developments in Reactor Physics and
       Shielding, CONF-720901, Kiamesha Lake, New York,1972.

60.   Wall, I. and Fenech, H., Nuclear Science and Engineer-
       ing, 22, PP 285-297, 1965.

61.   Stover, R.L. and Sesonske, A., Trans. American Nuclear
       Society, 11, Page 442, 1968.

62.   Stover, R.L. and Sesonske, A., Journal of Nuclear
       Energy, 23, PP 673-682, 1969.

63.   Mélice, M., Nuclear Science and Engineering, 37,
       PP 451-477, 1969.

64.   Fagan, J.R. and Sesonske, A., Journal of Nuclear
       Energy, 23, PP 683-696, 1969.

65.   Suzuki A. and Kiyose, R., Nuclear Science and Engin-
       eering, 46, PP 112-130, 1971.

66.   Sauar, T.O., Nuclear Science and Engineering, 46,
       PP 274-283, 1971.

67.   Kawai, T. and Kiguchi, T., Nuclear Science and Engin-
       eering, 43, PP 342-344, 1971.

68.   Motoda, H., Nuclear Science and Engineering, 49,
       PP 515-524, 1972.

69.   Hoshino, T., Nuclear Science and Engineering, 49,
       PP 59-71, 1972.

70.   Young, G., Nuclear News, 7, No. 11, Page 23, 1964.

71.   Märkl, H.,"Untersuchungen Über langfristig optimale
      Reaktorstrategien unter Berücksichtung des begrenzten
      Umfangs der Uranreserven, Teil I, Beschreibung des
      Optimierungsmodells," Nukleonik, 10, Page 207, 1967.

72.   Finnemann, H., Gulgeshell, W. and Märkl, H., Nukleonik
      12, Page 263, 1969.

73.   Widner, H., Mason, E.A. and MacAvoy, P.W., Trans.
      American Nuclear Society, 13, Page 785, 1970.

74.   Widner, H., Mason, E.A. and MacAvoy, P.W., Trans.
      American Nuclear Society, 13, Page 784, 1970.

75.   Nelson, P., Jr. and Young, G., Nuclear Science and
      Engineering, 46, PP 140-146, 1971.

76.   Fadilah, S.M. and Lewins, J., "Optimal Control Rod
      Programs in Power Reactors," Annals of Nuclear
      Energy 2, Page 443, Pergamon, Oxford, 1975.

Acknowledgements

We express our appreciation to the publishers of
Nuclear Science and Engineering for permission to reproduce
the following tables and figures:

    Terney, Fig. 1, NSE 45, p. 226
    Goldschmidt & Quenon, Figs. 1, 2, 3, NSE 39, p. 311
    Wall & Fenech, Figs. 1, 2, 3, and Tables 1, 2, 9,
      NSE 22, p. 288
    Suzuki & Kiyose, Fig. 1, NSE 46, p. 112
    Motoda & Kawai, Figs. 3, 5, NSE 39, p. 114
    Terney, Figs. 1, 2, 4, and Table 1, NSE 39, p. 109
    Wade & Terney, Figs. 4, 5, 6, 7 and Tables 1, t,
      NSE 45, p. 199

APPENDIX

OPTIMAL CONTROL THEORIES

A. INTRODUCTION

Geometric concepts and notation are useful in understanding the significance of steps in the optimal design
procedure. For example, in the more familiar problem of
finding the point, $t=t*$, on the real line that minimizes
the value of a real valued-function of t, $g_0(t)$, a well-
known necessary condition for a minimum in the case of no
constraints on the value of t is that:

$$\left. \frac{dg_0}{dt} \right|_{t*} = 0$$

Figure A.1 depicts the geometric interpretation. From
the figure, it is also seen that when a constraint, $t \geqslant 1$,
is added to the problem, the character of the necessary condition changes from the one given above to the condition:

$$dg_0 \geq 0 \qquad \text{for all allowed changes, dt}$$

Furthermore, if there are auxiliary constraint equations:

$$g_i(t) = 0 \quad i = 1,2, \ldots K$$

a necessary condition is that:

$$\left. \frac{d}{dt} \left[g_0 + \sum_{i=1}^{K} \lambda_i g_i \right] \right|_{t*} \quad 0$$

where λ_i are constants called Lagrange multipliers.

The concepts of a function space and a functional defined on that space make it possible to cast the optimal
control problem into a geometric form analogous to this
familiar problem in the calculus of functions. A function
space X corresponds to the real line; the elements of the

function space, $\underline{x}$ and $\underline{y}$, correspond to the values of the
independent variable, t, which comprise the real line; and
the functional, $f_0(\underline{x},\underline{u},t)$, defined in the function space
corresponds to the function $g_0(t)$ defined on the real line.

In the next several sections, which consist mostly of
definitions, these concepts are introduced. Then necessary
conditions for the minimization of a functional and their
geometric interpretations are discussed. This forms a background for a discussion of the results of the classical calculus of variations and its application (via the Bolza problem) to optimal control theory. The approach is heuristic
rather than rigorous and is intended to provide an overview
which should be augmented by specialized texts on each of
the subjects.

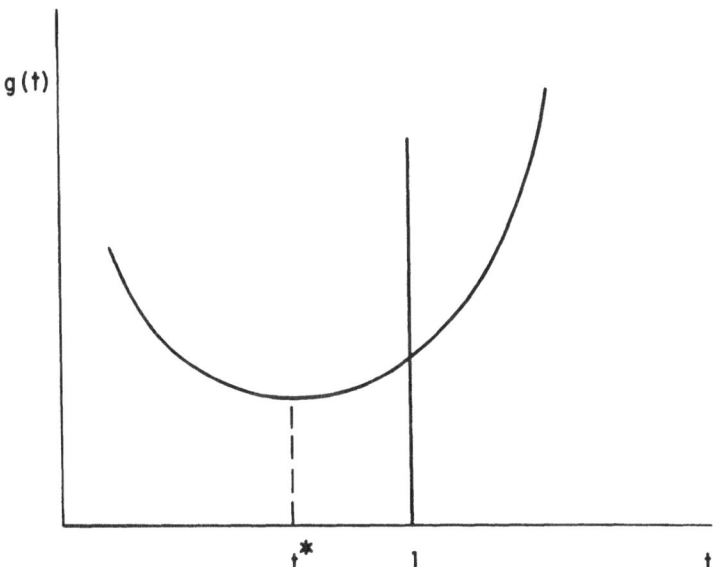

Figure A.1
A Minimization Problem in the Calculus of Functions

B.   UNDERLYING CONCEPTS AND NONLINEAR PROGRAMMING

   1. Concept of a Function Space

      It is useful to temporarily regard the combined state
function, $\underline{x}(t)$, and control function $\underline{u}(t)$, belonging to some
allowed class and denoted by $y(t)$:

$$x(t) \text{ and } u(t) \rightarrow y(t)$$

as representable by an element or "point" of a function
space, X. The function space, X, is a collection of functions, $y(t)$; each point in the space corresponds to some particular choice for $y(t)$. The choice of the space depends on
the problem at hand. For example, if the performance index
is of the form $f_0 = \int L(t,y,y')\, dt$, it is natural to choose
the space, X, as the set of continuous functions with continuous first derivatives*. If the performance index is of the
form $f_0 = \int L(t,y,y',y'')\, dt$, the space might be further restricted to include only functions having continuous second
derivatives as well. Each element in the space, $y \epsilon X$, represents some particular choice of the functions, $x_i(t)$ and
$u_i(t)$.

      Inequality and equality constraints on the state and con-
trol variables are viewed as defining an allowed subset in
the space X over which choices of the functions may be considered. Figure B.1 illustrates the concept for the case:

$$f_0 = \int_0^1 (x(t) + u(t))\, dt$$

$$f_1 = x + a\,u - \frac{dx}{dt} = 0$$

$$f_2 = \int_0^1 u^2(t)\, dt - 1 \leq 0$$

---

* Here, y' denotes a derivative with respect to t.

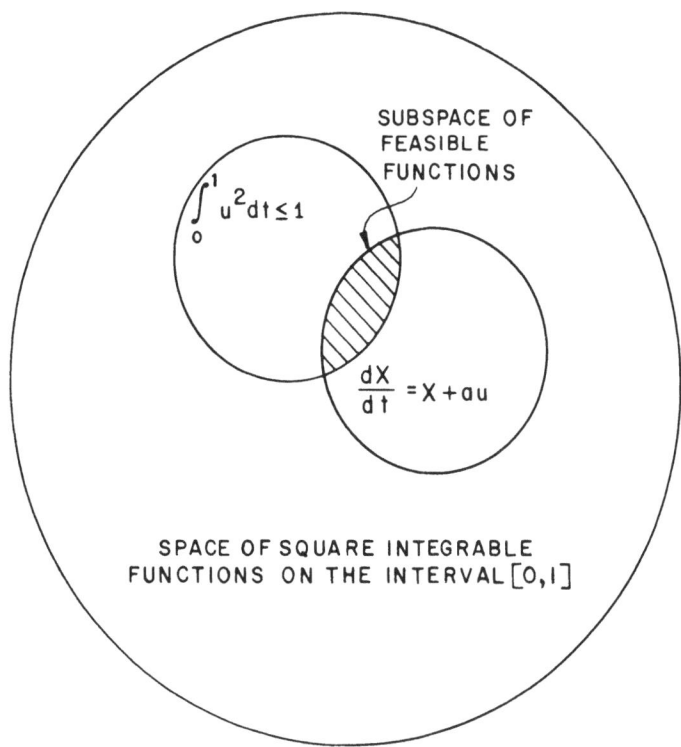

Figure B.1
A Function Space

For many practical problems, the space X can be taken
as an "inner product linear vector space". That assumption
is made in all that follows. For a linear vector space,
(being a collection of elements, $y_1, y_2, \ldots$), the operations of addition and multiplication by (real) numbers
$\alpha, \beta, \ldots$ are defined and obey the following axioms:

1. $y_1 + y_2 = y_2 + y_1 \varepsilon \ X$

2. $(y_1 + y_2) + y_3 = y_1 + (y_2 + y_3)$

3. there exists an element $0 \in X$ such that $y + 0 = y$ for any $y \in X$

4. For each $y \in X$ there exists an element, $-y$ such that $y + (-y) = 0$

5. $1 \cdot y = y$ (B.1)

6. $\alpha(\beta y) = (\alpha\beta) y$

7. $(\alpha+\beta) y = \alpha y + \beta y$

8. $\alpha(y_1 + y_2) = \alpha y_1 + \alpha y_2$

These axioms permit the familiar algebraic manipulations among real numbers to be applied to the elements of the function space X.

The inner product, denoted by (y_i, y_j), is defined as a real-valued function of pairs of elements of the linear vector space X which satisfies the axioms:

1. $(y_1, y_2) = (y_2, y_1)$

2. $(\alpha y_1, y_2) = \alpha(y_1, y_2)$ (B.2)

3. $(y_1+y_2, y_3) = (y_1, y_3) + (y_2, y_3)$

4. $(y,y) \geq 0$; $(y,y) = 0$ if, and only if, $y = 0$

The inner product embodies the concepts of both distance and angle between elements of the space:

"distance" between y_1 and $y_2 = \left\| y_1 - y_2 \right\|$ (B.3)

$= (y_1-y_2, y_1-y_2)^{\frac{1}{2}}$

$$\cos(\text{angle between } y_1 \text{ and } y_2) = \frac{(y_1, y_2)}{\left\| y_1 \right\| \cdot \left\| y_2 \right\|}$$ (B.4)

thus, y_2 is "perpendicular" (orthogonal) to y_1 if $(y_1, y_2) = 0$.

Example:

Let X be the space of continuous functions of time with continuous derivatives. Define the inner product:

$$(y_i, y_j) = \int_{-\frac{\pi}{2}}^{+\frac{\pi}{2}} y_i(t) \, y_j(t) \, dt$$

This definition is seen to satisfy the axioms. Let $y_1 = \cos t$ and $y_2 = \sin t$ be members of X. They are orthogonal, since:

$$\cos (\text{angle between } y_1 \text{ and } y_2) = \frac{(y_1, y_2)}{\sqrt{(y_1, y_1)(y_2, y_2)}}$$

$$= \frac{\displaystyle\int_{-\frac{\pi}{2}}^{+\frac{\pi}{2}} \sin(t) \cos(t) \, dt}{\sqrt{\displaystyle\int_{-\frac{\pi}{2}}^{+\frac{\pi}{2}} \sin^2(t) \, dt \int_{-\frac{\pi}{2}}^{+\frac{\pi}{2}} \cos^2 t \, dt}}$$

$$= \frac{0}{(\pi/2)(\pi/2)}$$

The distance between them is

$$\text{distance} = \sqrt{(y_1, y_1) + (y_2, y_2) - 2(y_1, y_2)}$$

$$= \sqrt{(\pi/2) + (\pi/2)}$$

$$= \sqrt{\pi}$$

2. Concept of a Functional

In the function-space context, the performance index
can be viewed as a transformation between the space X and the
real line, R. Specifically, examination of Equation II.C.1
shows that given the functions, $\underline{x}$(t) and $\underline{u}$(t), which combined
are an element of space X, the performance index produces a
single real number $f_0 = \alpha$ -- an element of the real line, R.
A rule (or mapping, or transformation).

$$f_0 \; : \; X \to R$$

which assigns a real number, α to each element y belonging
to a subset, D, of X, is called a functional. The subset D
of X is called the domain of the functional and the subset
of the real line, R, into which all α fall is called the
range.

Example:

X is the set of all continuous square integrable functions
over the interval [0,1]; i.e., for all

$$\left\{ y \; \epsilon \; X \; \middle| \; \int_0^1 y^2(t)dt \; < \; \infty \right\}$$

then the following are all functionals.

a. $f_0 = \displaystyle\int_{.5}^{.7} y(t) \; dt$

b. $f_0 = y(.2)$

c. $f_0 = \max \; y(t)$

$$0 \leq t \leq 1$$

On the linear vector space X, a functional which obeys
the superimposition principle:

$$f_0(\alpha y_1 + \beta y_2) = \alpha f_0(y_1) + \beta f_0(y_2)$$

$$y_1, \ y_2 \varepsilon D \varepsilon X \hspace{3cm} \text{(B.5)}$$

$$\alpha, \beta \text{ are real numbers}$$

is called a <u>linear functional</u>. In addition, the functional is said to be <u>continuous</u> at the point $y_1 \varepsilon D$ if for any reason $\varepsilon > 0$, there is a $\delta > 0$ such that:

$$\left| f_0(y_1) - f_0(y_2) \right| < \varepsilon$$

if

$$\left\| y_1 - y_2 \right\| < \delta$$

For a continuous linear functional "a small distance between the functions y_1 and y_2 in the function space produces a small difference in the value of the functional."

3. Concepts of the Variation of a Functional and the Functional Derivative

Suppose a small change is made in the point y_1 in space X, $(y_1 \rightarrow y_1 + h)$; one may ask about the resulting change in the value of a linear, continuous functional, f_0.

$$\Delta f_0 = f_0(y_1 + h) - f_0(y_1)$$

The functional, f_0, is said to be differentiable at point $y_1 \ \varepsilon \ X$ if there exists a <u>linear, continuous</u> mapping, (df_0/dy_1): $X \rightarrow X$ such that

$$\Delta f_0 = f_0(y_1 + h) - f_0(y_1) = \left(\frac{df_0}{dy_1}, h \right) + \S_1(h) \quad \text{(B.6)}$$

where

$$\frac{\S_1(h)}{\|h\|} \rightarrow 0 \text{ as } \|h\| \rightarrow 0 \hspace{3cm} \text{(B.7)}$$

for all $y_1 + h \ \varepsilon \ D$. The transformation, (df_0/dy_1), is called the functional derivative. It is the part of the incremental change, Δf_0, in the functional which is of first order in the change, h, in the function.

It is seen that for a small enough change of the point in function space, the change in the value of the functional is given by the first order term:

$$\Delta f_0 \underset{\sim}{\approx} \left(\frac{df_0}{dy_1} , h \right)$$

as

$$\| h \| \to 0 \tag{B.8}$$

This first order change, denoted by δf_0,

$$\delta f_0 = \left(\frac{df_0}{dy_1} , h \right) \tag{B.9}$$

is called the first variation of the functional. That is, the first differential or variation of the functional, δf_0, is the part of the change, $\Delta f_0 = f_0(y_1 + h) - f_0(y_1)$ in the functional which is for the first order in the change, h, in the function. It is given by the inner product of the change and the functional derivative.

Equation B.6 resembles a Taylor series expansion. In line with that notation, it is convenient to view the functional derivative as a gradient in function space. The change in the value of the functional between two points, y_1 and $y_1 + h$ in function space is given to first order by the inner product of the functional derivative or gradient and the change in position.

The extension to higher order functional derivatives and their geometric interpretation is straightforward. In particular, when the second order terms are represented explicitly:

$$\Delta f_0 = f_0(y_1 + h) - f_0(y_1)$$

$$= \delta f_0 + \delta^2 f_0 + \S_2(h) \tag{B.10}$$

where

$$\frac{\S_2(h)}{||n^2||} \to 0 \text{ as } ||h|| \to 0 \tag{B.11}$$

for all $y_1 + h \in D$. The term, $\delta^2 f_0$ is a quadratic form in the elements of the change, h, and is called the second variation.

Example:

X is the space of continuous square integral functions over the interval $[0,1]$. Assume given a functional,

$$f_0 = \int_0^1 y^2(t) \ dt = (y,y).$$

To find the variation of f_0 at $y = y^*$, note that:

$$\Delta f_0 = f_0(y^* + h) - f_0(y^*)$$

$$= \int_0^1 (y^{*2} + 2y^*h + h^2 - y^{*2}) \ dt$$

$$= \int_0^1 2y^*(t)h(t) \ dt + \int_0^1 h(t)^2 \ dt$$

The part of a Δf_0 which is of first order in h is

$$\int_0^1 2y^*(t)h(t) \ dt = (2y^*,h)$$

This clearly displays the linearity property (see Equation B.2). It also displays the continuity property since:

$$\left| \int_0^1 2y^*(t)h_1(t) \ dt - \int_0^1 2y^*(t)h_2(t) \ dt \right| < \varepsilon$$

if δ is chosen as

$$\delta = \varepsilon/\max\ (2y^*(t))$$

$$0 \le t \le 1$$

for any ε. Thus, the functional derivative of f_0 is

$$\frac{df_0}{dy^*} = 2y^*$$

and the first variation of the functional around the point y^* is

$$\delta f_0 = \int_0^1 2y^*(t)h(t)\ dt$$

The second functional derivative is

$$\frac{d^2f_0}{dy^{*2}} = 1$$

and the second variation is

$$\delta^2 f_0 = \int_0^1 h(t)lh(t)\ dt$$

4. <u>Traditional Forms of Variation From the Classical Calculus of Variations</u>

The classical calculus of variations out of which control theory has evolved is concerned with a particular function space X -- the space of continuous functions of one variable t, with piecewise continuous first derivatives, and with a particular type of functional -- functionals which may be expressed as definite integrals:

$$f_0 = \int_{t_0}^{t} L(t,y,y')\ dt \tag{B.12}$$

defined on that space.

In considering the change in the value of f_0 accompanying a change in the minimizing point of space X, a distinction has traditionally been made between "strong" and "weak" variations in the element of X:

a.   A "strong neighborhood" of a point $y^*(t)$ in space X is the collection of elements y for which:

$$\left\| y(t) - y^*(t) \right\| \le \epsilon \qquad (B.13)$$

for $\epsilon > 0$. A different strong neighborhood applies for each value of ϵ.

b.   A "weak neighborhood" of a point $y^*(t)$ in space X is the collection of elements of X for which:*

$$\left. \begin{array}{l} \left\| y(t) - y^*(t) \right\| \le \epsilon \\[2ex] \left\| y'(t) - y^{*\prime}(t) \right\| \le \epsilon \end{array} \right\} \qquad (B.14)$$

The weak neighborhood of y^* contains only those functions, y, which not only lie close to y^* but in addition, whose derivatives lie close to the derivative of y^*. Figure B.2 illustrates the difference. Since no condition is imposed on the derivatives in defining the strong neighborhood, it contains the entire weak neighborhood as a subset.

For functionals of the form of Equation B.12, weak variations about the function $y^*(t)$ whose derivative $y^{*\prime}(t)$ are often constructed as

$$y(t) = y^*(t) + \epsilon\eta(t)$$

$$y'(t) = y^{*\prime}(t) + \epsilon\eta'(t) \qquad (B.15)$$

$$\eta(t_0) = \eta(t_1) = 0$$

when $\eta(t)$ is a continuous function of time with a continuous first derivative and ϵ is a scalar. When such a variation is

---

*   Here, y' denotes a derivative with respect to t.

made for the case where L possesses first partial derivatives
in its arguments and where there are no discontinuities in y'
between t_0 and t_1, the first variation and functional derivation are obtained as follows:

$$\Delta f_0 = f_0(t,y^* + \varepsilon\eta, \; y^{*\prime} + \varepsilon\eta') - f_0(t,y^*,y^{*\prime})$$

$$= \int_{t_0}^{t_1} \left\{ L(t,y^*,y^{*\prime}) + \frac{\partial L}{\partial y}\varepsilon\eta + \frac{\partial L}{\partial y'}\varepsilon\eta' \right.$$
$$\left. + \theta((\varepsilon\eta)^2,(\varepsilon\eta')^2) - L(t,y^*,y^{*\prime}) \right\} dt$$

$$= \varepsilon \int_{t_0}^{t_1} \left(\frac{\partial L}{\partial y}\eta + \frac{\partial L}{\partial y'}\eta' \right) dt + \theta\left((\varepsilon\eta)^2, \; (\varepsilon\eta')^2 \right)$$

Integrating the second term by parts yields

$$\Delta f_0 = \left. \varepsilon\eta\frac{\partial L}{\partial y'} \right|_{t_0}^{t_1} + \int_{t_0}^{t_1} \left(\frac{\partial L}{\partial y} - \frac{d}{dt}\frac{\partial L}{\partial y'} \varepsilon\eta \right) dt$$
$$+ \theta\left((\varepsilon\eta)^2, (\varepsilon\eta')^2 \right)$$

which, because $\eta(t_0) = \eta(t_1) = 0$ reduces to

$$\Delta f_0 = \left. \right|_{t_0}^{t_1} \frac{\partial L}{\partial y} - \frac{d}{dt}\frac{\partial L}{\partial y'} \varepsilon\eta \; dt + \sigma \; (\varepsilon\eta)^2,(\varepsilon\eta')^2$$

In view of the definitions above, the functional derivative is

$$\frac{df_0}{dy} = \frac{\partial L}{\partial y} - \frac{d}{dt}\frac{\partial L}{\partial y'} \tag{B.16}$$

and the first variation is the familiar form:

$$\delta f = \int_{t_0}^{t_1} \left(\frac{\partial L}{\partial y} - \frac{d}{dt}\frac{\partial L}{\partial y'} \right) \varepsilon\eta \; dt$$

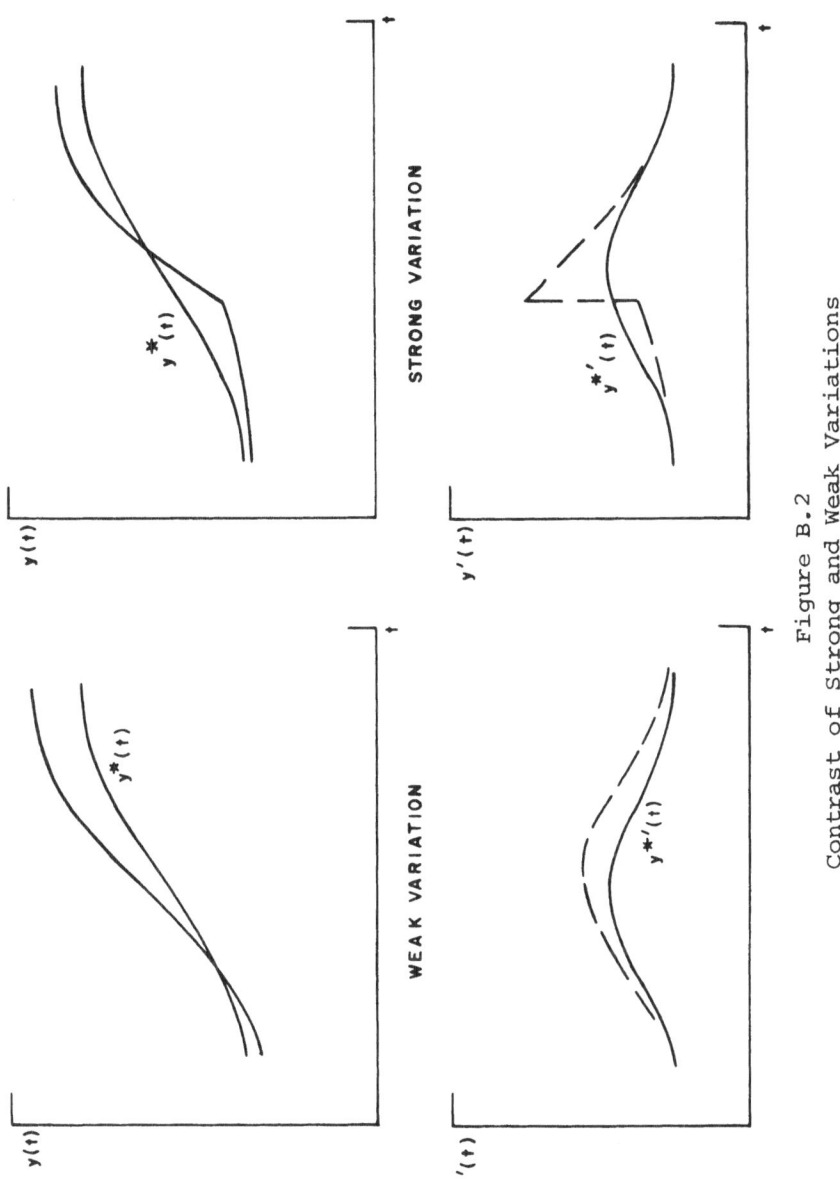

Figure B.2

Contrast of Strong and Weak Variations

5. Constraint Relations Viewed as Transformations

The performance index, f_0, was seen to be interpreted as a transformation between elements of space X and the real line. Similarly, the lefthand sides of the constraint relations, Equations II.B.1, II.B.2 and II.B.4 of the optimal control problem:

$$f_i(\underline{x},\underline{u},t) - \frac{d}{dt}x_i = 0 \qquad i = 1, 2, \text{---} L$$

$$f_i(\underline{x},\underline{u},t) = 0 \qquad i = L + 1, L + 2, \text{---} n$$

$$f_i(\underline{x},\underline{u},t) \leq 0 \qquad i = n + 1 \text{---} J$$

may also be viewed as transformations which map elements of space X onto the space of functions of t. For example, given a choice for $\underline{x}(t)$ and $\underline{u}(t)$ which, combined, are an element of X, the transformation $\bar{f}_n(\underline{x},\underline{u},t)$ produces a function of time only; which in general, does not equal 0 at all values of time. It is therefore clear that the constraint relations, which require time functions that are zero or less than zero, limit the elements of space X which may be considered in finding the one which minimizes f . The notation:

$$\left\{ y \in \Omega \left| \begin{array}{ll} f_i(\underline{x},\underline{u},t) - \frac{d}{dt}x_i = 0 & i = 1,2, \text{---} L \\ f_i(\underline{x},\underline{u},t) = 0 & i = L + 1 \text{---} n \\ f_i(\underline{x},\underline{u},t) \leq 0 & i = n + 1 \text{---} J \end{array} \right. \right\} \qquad (B.17)$$

denotes the allowed subset of space X from which the minimizing function, y* may be chosen.

Just as a small change in the position of the element in space X produces a change in the value of f_0, so does it produce a change in each of the time functions mapped by the lefthand sides of the constraint equations. Generalized derivatives may be defined here also such that

$$f_i(y^* + h) = f_i(y^*) + \frac{df_i}{dy^*}(h) + \S(h) \qquad (B.18)$$

where $\dfrac{\S(h)}{||h||} \to 0$ as $||h|| \to 0$

Here, the derivative $(df_i/dy^*)(h)$ maps the element, h, of X onto the space of functions of time.

6. Function-Space Interpretation of the Optimal Control Problem

With the concepts built up in the preceding sections, the optimal control problem has a simple interpretation.

Given a functional, $f_0 : X \to R$ which maps the elements y of the function-space X onto the real line, find the element y^* belonging to Ω, a subset of X which minimizes the value of the functional.

If there are no auxiliary (constraining) relationships that y must satisfy (e.g., no system and auxiliary equations II.B.1, II.B.2, II.B.3 and II.B.4) then Ω is unrestricted (an open set). In the usual case, additional relations must be satisfied and Ω is restricted (a closed set).

7. A Necessary Condition for a Minimum in an Unconstrained Problem-Geometric Interpretation

If the performance index is minimized by the choice $y = y^* \varepsilon \Omega \varepsilon X$, then by definition:

$$f_0(y^*) \le f_0(y) \text{ for all } y \varepsilon \Omega \varepsilon X. \qquad (B.19)$$

Choose an arbitrary h and an arbitrary positive scalar, $\mu > 0$, such that $y^* + \mu h \varepsilon \Omega$. By assumption, there are no constraints on the problem, so that Ω is an open subspace in X and there are no restrictions on the choice of h. Then

$$\left.\begin{aligned}
f_0(y^*) &\le f_0(y^* + \mu h) \\
&\le f_0(y^*) + \left(\frac{df_0}{dy^*}, h\right) + \S(\mu h) \\
0 &\le \mu\left(\frac{df_0}{dy^*}, h\right) + \S(\mu h)
\end{aligned}\right\} \qquad (B.20)$$

where use has been made of the derivative of f_0 at y^*, which is assumed to exist. Let $\mu \to 0$ and note that

$$\frac{\S(\mu h)}{\mu} \to 0$$

Then the component of the change, h, on the gradient of the performance index must be non-negative:

$$\left(\frac{df_0}{dy^*} , h \right) \geq 0 \tag{B.21}$$

but since h is arbitrary, this condition can be met only if the gradient is identically zero:

$$\frac{df_0}{dy^*} = 0 \tag{B.22}$$

This is a necessary condition for the function of y* to provide a minimum for the functional, f_0. The analogy to the minimization of a function of a real variable is obvious. When combined with Equation B.16 for weak variations of the form of Equation B.15, this condition is recognized as the familiar Euler equations.

8. A Necessary Condition for a Minimum for the Constrained Problem; Geometric Interpretation

Suppose y* is an optimum point; then $f_0(y) \geq f_0(y^*)$ for all $y \in \Omega$. By hypothesis, the problem is one in which auxiliary constraint equations apply. Figure B.3 shows an example situation involving one equality and four inequality constraints. At y*, some of the inequality constraints are inactive (f_2 and f_3 in the figure), while others (f_4 and f_5) are constraining the changes, μh, which can be made about y* to produce $y = y^* + \mu h$. All of the equality constraints (f in the figure) are active. This difference between the unconstrained (Ω an open subspace) and constrained (Ω a closed subspace) cases gives rise to very fundamental differences between the approaches required to obtain the optimum point, y*.

Suppose the first n constraints on the problem are equality constraints, the next K-n constraints are active inequality constraints; and the final J-K constraints are inactive inequality constraints; then consider as possible changes in the point y about y* the changes such that the direction of the change causes the new point to lie within the allowed domain.

$$h \in V(y^*) \quad \epsilon y + \mu h \quad \epsilon \ \Omega$$

and the amplitude is controlled by $\mu > 0$.

Then if y^* is an optimal point, $f_0(y^*) \leq f_0(y^* + \mu h)$

$$f_0(y^*) \leq f_0(y^*) + \left(\frac{df_0}{dy^*}, \mu h\right) + \xi(\mu h)$$

$$0 \leq \left(\frac{df_0}{dy^*}, n\right)\mu + \frac{\xi(\mu h)}{\mu}$$

Taking μ positive and small, the last term can be made to vanish, leaving the necessary condition:

$$0 \leq \left(\frac{df_0}{dy^*}, h\right) \qquad \text{for all } h\epsilon V(y^*) \tag{B.23}$$

that is, all allowable - consistent with the constraints - changes in the element of space X make an angle with the gradient of f_0 which lies between $-90°$ and $+90°$. Thus, a necessary condition that y^* be a minimizing point in the constrained case is that allowable change about y^* causes the value of the performance index to increase over its value at y^*. This is one form of the Kuhn-Tucker theorem (6) for nonlinear programming. It requires an additional subtle condition on the set $V(y^*)$ of allowable variations which was not discussed here.

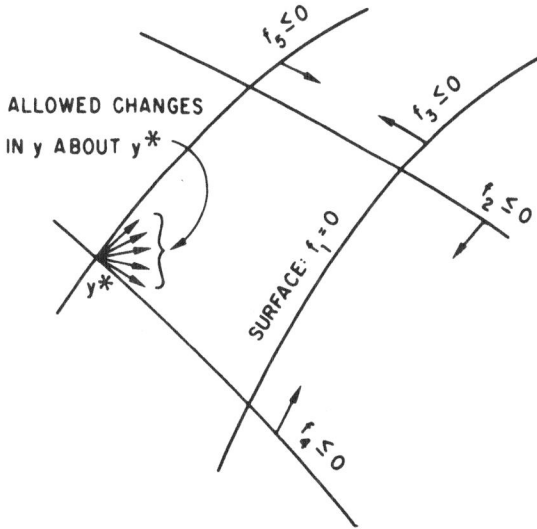

Figure B.3

Feasible Variations for a Constrained Problem

9. Nonlinear Programming Algorithms

The function-space geometric interpretation of the necessary conditions for an optimum suggests algorithms for finding the optimum point y*. These algorithms are called "nonlinear programming".

For the case of no constraints, the "method of steepest descent" proceeds roughly as follows:

a. A choice $y_i = y_o$ is selected as a possible minimizing function; here, sub i denotes an iteration index.

b. The direction $d_i = \dfrac{df_o}{dy_i}$ is determined. If $d_i = 0$, a local minimum has been found; if not,

c. an amplitude, $\alpha*$, is found for which

$$f_o(y_i + \alpha*d_i) \leq f_o(y_i + \alpha d_i)$$

for all $\alpha \geq 0$

d. if $f_o(y_i + \alpha*d_i) = f_o(y_i)$ stop

if $f_o(y_i + \alpha*d_i) < f_o(y_i)$ then set

$y_{i+1} = y_i + \alpha*d_i$,

return to step b, and iterate.

In the case where there are constraints, step b is modified to choose from among the allowed directions

$$d\epsilon V$$

that direction which has the maximum component on $-\dfrac{df_o}{dy_i}$.

An alternate nonlinear programming algorithm, the Newton-Raphson method, is based on finding the functions for which $\dfrac{df_o}{dy} = 0$ by a method analogous to Newton's method for functions of a real variable.

In the case where f_o and all f_i are linear in y, a minimizing function, if it exists, will be found at one of the constraint boundaries. Such problems, called "linear programming" problems, are solved by successively examining the vertices of the allowed domain Ω and evaluating the performance index. Since there are only a finite number of vertices, the method must eventually arrive at the optimum if it exists. This is the basis of the common simplex method.

10. Concept of a Lagrange Multiplier

Consider the special case in which the constraint relations are all expressed as functionals; i.e., they map elements of function space X into elements of the real line. The necessary conditions for an optimum discussed in Section B.8 still apply of course.

For a constrained problem that has no inequality constraints and n equality constraints, the arguments of Section A.B.8 show that the necessary conditions for a minimum are:

$$\left(\frac{df_o}{dy^*} \, , \, h \right) \geq 0$$

for all h for which

$$\left(\frac{df_i}{dy^*} \, , \, h \right) = 0 \qquad i = 1,2, \text{ --- } n$$

The latter set of n equations is the definition of the space of allowed variations, $h \epsilon V(y^*)$; the allowed changes are all h which are orthogonal to the set of vectors, $\frac{df_i}{dy^*}$, = 1,2, --- n; i.e., all vectors about y^* which lie simultaneously on all the constraint surfaces. It is noted that if a vector lies in a surface, its negative also lies on the surface. The first equation requires that $\frac{df_o}{dy^*}$ have a positive or zero component on every such h in the allowed space of variations; but since for any h, -h is also allowed, it is clear that the component must never differ from zero.

Thus, the vector $\dfrac{df_o}{dy*}$ is required to be orthogonal to all vectors which are simultaneously orthogonal to all $\dfrac{df_i}{dy*}$, $i =$ 1,2, --- n. One may conclude therefore that $\dfrac{df_o}{dy*}$ can be constructed from a linear combination of the vectors $\dfrac{df_i}{dy*}$ $i = 1,2, \ldots$ n; i.e., there exist numbers, λ_i, $i = 1,2,\ldots$ n not all zero such that

$$\frac{df_o}{dy*} + \sum_{i=1}^{n} \lambda_i \frac{df_i}{dy*} = 0$$

This is an equivalent form of the necessary conditions for a minimum and is called the "Multiplier Rule". The constants, λ_i are called Lagrange multipliers. Since the functional de- rivatives, $\dfrac{df_i}{dy*}$, depend on the choice of the point y* in function space, so do the values of the Lagrange multipliers.

The equation has the following interpretation: "in order to move in a direction that will decrease f_o, one would have to violate at least one of the equality constraints."

Next, consider the case where there are K-n active in- equality constraints only (i.e., no equality constraints). The necessary conditions are

$$\left(\frac{df_o}{dy*} , h \right) \geq 0$$

for all h which satisfy

$$\left(\frac{df_i}{dy*} , h \right) \leq 0 \qquad i = 1,2, \ldots \text{ K-n}$$

The second set of equations define the space of allowable variations; all vectors h which make an angle between 90° and 270° with every one of the vectors, $\dfrac{df_i}{dy*}$, $i = 1,2, \ldots$ K-n. The first equation requires that $\dfrac{df_o}{dy*}$ make an angle between -90° and $+90^o$ with all such vectors, h. One can conclude that $\dfrac{df_o}{dy*}$ can be constructed as a linear combination of the negatives of the vectors $\dfrac{df_i}{dy*}$, $i = 1,2, \ldots$ K-n; i.e., there exist $\mu_i \geq 0$ such that

$$\frac{df_o}{dy*} + \sum_{i=1}^{K-n} \mu_i \frac{df_i}{dy*} = 0$$

$$\mu_i \geq 0$$

This "Multiplier Rule" is an equivalent form of the necessary conditions for a minimum. It has the following interpretation: "in order to move in a direction that will decrease f_o, one would have to violate at least one of the inequality constraints."

Similar arguments are used in the case where both n equality and K-n inequality constraints are active, yielding the general Multiplier Rule

$$\frac{df_o}{dy*} + \sum_{i=1}^{n} \lambda_i \frac{df_i}{dy*} + \sum_{i=n+1}^{K} \mu_i \frac{df_i}{dy*} = 0 \qquad \text{(B.24)}$$

$$\mu_i \geq 0 \quad i = n+1, n+2, \ldots K$$

Let us consider the number of equations and unknowns represented in the Multiplier Rule. The vector y is being used to represent the combined state vector $\underline{x}$ which has n components and the control vector $\underline{u}$ which has m components; a total of n+m unknowns. The Multiplier Rule, Equation B.24, consists of one equation for each component of y*; i.e., n+m equations; however, it also contains the n unknowns, λ_i and the K-n unknowns, μ_i for a total of K more unknowns. To determine these, the K active constraint equations:

$$f_i \ (\underline{x},\underline{u},t) \qquad = 0 \qquad i = 1,2, \ldots M$$

$$f_i \ (\underline{x},\underline{y},t) \qquad \leq 0 \qquad i = n+1 \ldots K$$

are adjoined to the Multiplier Rule, providing the required K additional equations.

The geometric interpretation discussed here was for the case where the constraints are functionals and the Lagrange multipliers are scalar constants. The case of differential constraints in which the Lagrange multipliers are functions of time is discussed in Section C.6 of this Appendix.

11. Concept of a Hamiltonian

The Multiplier Rule, Equation B.24

$$\frac{df_o}{dy^*} + \sum_{i=1}^{n} \lambda_i \frac{df_i}{dy^*} + \sum_{i=n+1}^{K} \mu_i \frac{df_i}{dy^*} = 0$$

$$\mu_i \geq 0 \quad i = n+1, \ldots K$$

gives a necessary condition for the optimal choice of vector y^* to minimize the functional f_o, subject to the equality constraints:

$$f_i(y^*) = 0 \quad i = 1,2, \ldots n$$

and the active inequality constraints:

$$f_i(y^*) \leq 0 \quad i = n+1 \ldots K$$

By defining

$$\mu_i > 0 \text{ if } f_i(y^*) = 0 \quad \text{an active inequality constraint}$$

$$\mu_i = 0 \text{ if } f_i(y^*) < 0 \quad \text{an inactive inequality constraint}$$

the second summation can be extended to include both the active and inactive inequality constraints; i.e., the summation is from $i = n+1$ to J.

The multiplier Rule can be written in a more compact way by the definition of a combined functional and noting the linearity property of the functional derivative:

$$H = f_o + \sum_{i=1}^{n} \lambda_i f_i + \sum_{i=n+1}^{J} \mu_i f_i$$

The multiplier rule is then

$$\frac{\partial H}{\partial y*} = 0$$

The combined functional H is called the Hamiltonian.

The concepts of function space, neighborhood, functional derivative, and variation introduced in the previous sections form the foundation and a geometrical interpretation for the predominately analytical approach of the classical calculus of variations to be discussed in the next section.

C. THE CALCULUS OF VARIATIONS

1. Overview of the Classical Calculus of Variations

The classical calculus of variations is concerned with the space X of continuous functions of time with piecewise, continuous, first-time derivatives and with functionals of the form:

$$f_0 = \int_{t_0}^{t_1} L(y,y',t) \, dt$$

The "simplest problem" in the classical calculus of variations is to find that continuous function with piecewise continuous derivative that yields a relative minimum value for the functional, f_0. The end points, t_0 and t_1 may be fixed or may be defined parametrically as specified curves. Considerations of weak first variations of the function y about the minimizing curve y* give rise to well-known necessary conditions consisting of the Euler-Lagrange equations, the transversality condition, and the Weierstrass-Erdmann corner conditions. Consideration of strong variations* about the minimizing curve gives rise to the Weierstrass condition. Considerations of weak second variations give

---

* The distinction between strong and weak variations was discussed in Sections B.3 and B.4.

rise to additional necessary conditions: the Legendre and
the Jacobi conditions. These necessary conditions are derived and discussed below for the case of a real-valued
function, y.

The introduction of constraint conditions which the minimizing curve must satisfy gives rise to other well-known
problems in the classical calculus of variations:

The "isoperimetric problem": find the curve y which
minimizes f_0, subject to the constraint:

$$c = \int_{t_0}^{t_1} G(y,y',t)\ dt$$

and

The "problem of Lagrange": find the curve y which
minimizes f_0, subject to the constraint:

$$f_1\ (y,y',t) = 0$$

These side conditions lead to the introduction of Lagrange
multipliers. In the case of the isoperimetric problem, the
Lagrange multiplier is a constant; its geometric significance was discussed in Section B.10, above. In the case of
the problem of Lagrange, the multiplier is a function of
time. This case will be discussed in Section C.6.

When an additional term which contributes at the initial
and terminal time is added to the functional,

$$f_0 = \int_{t_0}^{t_1} L(y,y',t)dt + \Phi(t_0,t_1,y(t_0),y(t_1),y'(t_0),y'(t_1))$$

and the side conditions

$$f_i\ (y,y',t) = 0$$

are appended, the problem is called the "problem of Bolza".
In recent years, inequality constraints have also been considered in the problem of Bolza. Pontryagin's principle for

optimal control problems results from consideration of a
special form of the problem of Bolza. If the Φ term is present above in the functional with $L \equiv 0$, the problem is called
a Mayer problem. The various problems of Mayer, Lagrange and
Bolza are transformed one into the other by proper changes of
variable (1).

In the following sections the derivation of the first
variation of a functional expressible as a definite integral
is reviewed first. Then the necessary conditions for a minimum are deduced for the simplest problem of the calculus of
variations with a single state variable. Then follows a consideration of the problem of Bolza and the interpretation of
the optimal control problem as a Bolza problem.

2. Weak Variation of a Functional Expressible as a
   Definite Integral

Take as given the functional:

$$f_0 = \int_{t_0}^{t_1} L(t,y,y') \, dt \qquad (C.1)$$

defined on the space, X, of continuous real-valued functions,
$y(t)$, of one variable with piecewise continuous first derivatives. The function $L(t,y,y')$ is assumed to possess first
partial derivatives with respect to its arguments.

Figure C.1 shows two neighboring elements of space X.
The end points for curve y are at (t_0,y_0) and (t_1,y_1), while
the end points for the neighboring curve, $y + h$, are at
$(t_0 + \delta t_0, y_0 + \delta y_0)$ and $(t_1 + \delta t_1, y_1 + \delta y_1)$. For this
reason the curves are extended onto a common interval
$(t_0, t_1 + \delta t_1)$, by drawing tangents to the curves at their
end points. The function h is a continuous function of time
with a piecewise, continuous time derivative, h'.

A possible corner (value of t where $y'(t)$ jumps in value)
is at t_2 for curve y and $t_2 + \delta t_2$ for curve $y + h$. The
functional f_0 is expressed as:

$$f_0 = \int_{t_0}^{t_2^-} L(t,y,y') \, dt + \int_{t_2^+}^{t_1} L(t,y,y') \, dt$$

$$= f_{0_-} + f_{0_+}$$

(C.2)

By definition, the variation, δf_0, is to be linear in h, δt_0, δt_1, δy_0, δy_1 and is to differ from $[f_0(y + h, y' + h') - f_0(y,y)]$ by terms, §, of order 2 or greater in h and the end points.

Now,

$$\Delta f_0 = f_0(y+h,y'+h') - f_0(y,y') = \Delta f_{0_-} + \Delta f_{0_+}$$

and

$$\Delta f_{0_-} = \int_{t_0 + \delta t_0}^{t_2^- + \delta t_2^-} L(y,y+h,y'+h') \, dt - \int_{t_0}^{t_2^-} L(t,y,y') \, dt$$

$$= \int_{t_0}^{t_2^-} [L(t,y+h,y'+h') - L(t,y,y')] \, dt$$

$$+ \int_{t_2^-}^{t_2^- + \delta t_2^-} L(t,y+h,y'+h') \, dt$$

$$- \int_{t_0}^{t_0 + \delta t_0} L(t,y+h,y'+h') \, dt$$

Using a Taylor expansion and the mean value theorem

$$\Delta f_{0_-} = \int_{t_0}^{t_2} [L_y(t,y,y')h + L_{y'}(t,y,y') \, h'] \, dt$$

$$+ L(t_{2_-},y,y') \, \delta t_{2_-} - L(t_0,y,y') \, \delta t_0 + \S$$

where the notation $L_y = \dfrac{\partial L}{\partial y}, \ L_{y'} = \dfrac{\partial L}{\partial y'}$ is employed.

Integrating by parts to eliminate h' yields

$$\Delta f_{0_-} = \int_{t_0}^{t_2} (L_y - \frac{d}{dt} L_{y'}) h \, dt + L \bigg|_{t_{2_-}} \delta t_{2_-} + L_{y'} h \bigg|_{t_{2_-}}$$

$$- L \bigg|_{t_0} \delta t_0 - L_{y'} h \bigg|_{t_0} + \S$$

From the figure, it is seen that to terms of first order in h:

$$h(t_0) = \delta y_0 - y'(t_0) \delta t_0$$

$$h(\bar{t_2}) = \delta y_2 - y'(\bar{t_2}) \, \delta t_2$$

so that the part of Δf_{0_-} which is linear in h, δt_0, δy_0, δt_2 and δy_2 is:

$$\delta f_{0_-} = \int_{t_0}^{\bar{t_2}} (L_y - \frac{d}{dt} L_{y'}) h(t) \, dt + L_{y'} \, \delta y \bigg|_{t_0}^{\bar{t_2}}$$

$$+ (L-L_{y'} y') \delta t \bigg|_{t_0}^{\bar{t_2}}$$

A similar result is obtained for f_{0_+} so that

$$\delta f_0 = \int_{t_0}^{t_2} (L_y - \frac{d}{dt} L_{y'}) h(t) dt + \int_{t_2}^{t_1} (L_y - \frac{d}{dt} L_{y'}) h(t) dt$$

$$+ L_{y'} \delta y \Big|_{t_0}^{t_1} + (L - L_{y'} y') \delta t \Big|_{t_0}^{t_1} \\ + L_{y'} \delta y \Big|_{t_2^+}^{t_2^-} + (L - L_{y'} y') \delta t \Big|_{t_2^+}^{t_2^-} \Bigg\} \qquad (C.3)$$

which is therefore the first weak variation of functional f_0.

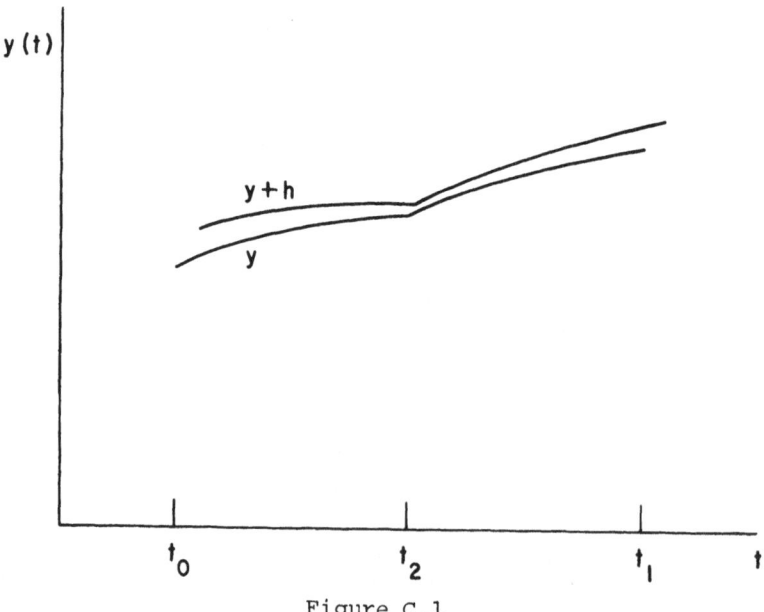

Figure C.1

A Curve and Its Weak Variation

3. The Euler-Lagrange, Transversality, and Weierstrass-Erdmann Necessary Conditions for a Minimum

If $y^*(t)$ is a minimizing curve for the functional, f_0, then all changes:

$y \rightarrow y^* + h$

$t_0 \rightarrow t_0 + \delta t_0$

$y_0 \rightarrow y_0 + \delta y_0$, etc.

of the type considered in the previous section should yield an increase in the value of f_0. As the amplitude of the changes in the curve becomes small, the increment, Δf_0, approaches the first variation, δf_0, so it is clear that δf_0 should exceed zero for all small changes in the curve from the minimizing curve, y^*. However, because by assumption, there were no constraints on the problem, the changes h, δt_0, etc., are of arbitrary sign; therefore, to guarantee that δf_0 is non-negative for all changes requires that the coefficients of these changes be identically zero.

The integrals in Equation C.3 will contribute nothing to δf_0 only if

$$L_y - \frac{d}{dt} L_{y'} = 0 \qquad (C.4)$$

These conditions are the wellknown Euler-Lagrange equations of the calculus of variations. Since they are second order partial differential equations, their integral contains two undetermined constants. These are uniquely determined by the requirement that the boundary terms in Equation C.3 are identically zero:

$$\left[L_{y'} \delta y + (L - L_{y'} y') \delta t \right]_{t_2}^{t_1} = 0 \qquad (C.5)$$

The form of the boundary terms depends on the type of problem. If the initial and final times are specified, $\delta t_1 = \delta t_0 = 0$, then only the term $L_y, \delta y \Big|_{t_0}^{t}$ remains.

To ensure that $\delta f_0 = 0$, this term must be zero; which can be guaranteed only if:

$$\left. L_{y'} \right|_{t_0} = \left. L_{y'} \right|_{t_1} = 0$$

since $\delta y(t_0)$ and $\delta y(t_1)$ are arbitrary. These two conditions are sufficient to uniquely determine the two constants in the solution of the Euler equation. Other combinations of boundary conditions arise; for example, if the starting time, t_0, and condition $y(t_0)$, are specified, but the terminal point is free, then it is necessary that

$$L_{y'}(t_1) = 0$$

$$L(t_1) - y'(t_1) L_{y'}(t_1) = 0$$

in order to ensure that $\delta f_0 = 0$. If it is required that the terminal point be on a curve $y = \phi(t)$, then $\delta y(t_1)$ and δt_1 are not independent ($\delta y = \frac{d\phi}{dt} \delta t$), giving rise to the "transversality condition".

$$L + (\phi' - y') L_{y'} = 0.$$

At corners where y' makes a jump in value, Equation C.3 shows that the conditions

$$\left(L_{y'} \delta y + (L - L_{y'} y') \delta t \right) \Bigg|_{t_2^+}^{t_2^-} = 0$$

will guarantee that $\delta f_0 = 0$. Since $\delta y(t_2^+) = \delta y(t_2^-)$ and $\delta t_2^+ = \delta t_2^-$ are arbitrary, this requirement gives rise to the Weierstrass-Erdmann corner conditions:

$$\left. L_{y'} \right|_{t_2^+} = \left. L_{y'} \right|_{t_2^-}$$

$$\left. (L - L_{y'} y') \right|_{t_2^+} = \left. (L - L_{y'} y') \right|_{t_2^-} \tag{C.6}$$

4. Strong Variations and the Weierstrass Necessary Condition

Consider next a different form of variation in the function y as shown in Figure C.2. Curve y which has the end values y_0 and y_1 at the end points t_0 and t_1, respectively, is assumed to satisfy the Euler-Lagrange equations on the interval (t_0,t_1) and the transversality conditions at the endpoints, t_0 and t_1. At the point 4, between 0 and 1 on y, a transverse curve, $\tilde{\varepsilon}$ is drawn, which, within a distance, ℓ, intersects some particular member of a one-parameter family of curves, $\bar{\varepsilon}$. connecting point 3 on $\bar{\varepsilon}$ to point 0 on curve y (see the figure). Each member of the family of curves, $\bar{\varepsilon}$, is also assumed to satisfy the Euler-Lagrange equations on (t_0,t_1), and the initial condition $y(t_0) = y_0$. As the distance, ℓ is shrunk to zero, curve $\tilde{\varepsilon}$ intersects successive members of the family $\bar{\varepsilon}$ and point 3 approaches point 4 until they meet, at which point, curves $\bar{\varepsilon}$ and y are the same curve. It is assumed that no corners occur on y or $\bar{\varepsilon}$ between points 0 and 1. The derivatives are denoted by y' for curve y and y' for curve $\bar{\varepsilon}$.

The curve 0 to 3 on $\bar{\varepsilon}$, 3 to 4 on $\tilde{\varepsilon}$ and 4 to 1 on y denoted by y+h is considered as a variation about the curve 0 to 1 on y. As the distance ℓ approaches zero, the two curves do not approach each other; thus, the variation is a "strong variation". Now the change in f_0 evaluated for the two curves is:

$$\Delta f_0 = \int_{t_0}^{t_3} L(t,(y+h),(y+h)') - L(t,y,y')\, dt$$

$$+ \int_{t_3}^{t_4} L(t,(y+h),(y+h)') - L(t,y,y')\, dt$$

By using a Taylor expansion and integrating by parts just as in the previous section, the first integral becomes

$$\int_0^{t_3} \left(\frac{\partial L}{\partial y} - \frac{d}{dx}\frac{\partial L}{\partial y'} \right) \Bigg|_{\substack{\text{curve} \\ y}} h\, dt \;+\; \frac{\partial L}{\partial y'}\,(t,y,y')\,\delta y \Bigg|_0^{t_3}$$

However, because curve y satisfies the Euler equations, the integral vanishes. Denoting $t_3 = t_4 - \ell$ and neglecting second and higher order terms, the first integral becomes:

$$\left. \frac{\partial L}{\partial y'} \right| \begin{matrix} \delta y \\ \text{curve} \\ y \end{matrix} \Bigg|_{(t_4 - \ell)}$$

The second integral is

$$\ell \left[(L(t,y+h,Y') - L(t,y,y') \right]_{t_4} + \theta\,(h^2)$$

which, when neglecting second and higher order terms and noting that y+h = y at t = t_4, this becomes:

$$\ell \left[L(t,y,Y') - L(t,y,y') \right]$$

The variation in y at time t_3 is given by

$$\delta y \Big|_{t_4 - \ell} = \left[y \Big|_{\substack{\text{curve} \\ \tilde{\varepsilon}}} - y \Big|_{\substack{\text{curve} \\ y}} \right]_{t_4 - \ell}$$

$$= y(t_4) - \ell Y' - y(t_4) - \ell y'$$

so that

$$\Delta f_0 = \left\{ L(t,y,Y') - L(t,y,y') - \frac{\partial L}{\partial y'}\,(Y' - y') \right\} \ell$$
$$+ \theta\,(h^2)$$

The quantity in braces is called the Weierstrass function

$$E = L(t,y,Y') - L(t,y,y') - \frac{\partial L}{\partial y'}\,(Y' - y') \quad \text{(C.7)}$$

and it is seen that it must be non-negative

$$E \geq 0 \qquad\qquad\qquad\qquad \text{(C.8)}$$

at all points t_4 between t_0 and t_1 in order to guarantee that the functional, f_0, achieves its minimum value for curve y.

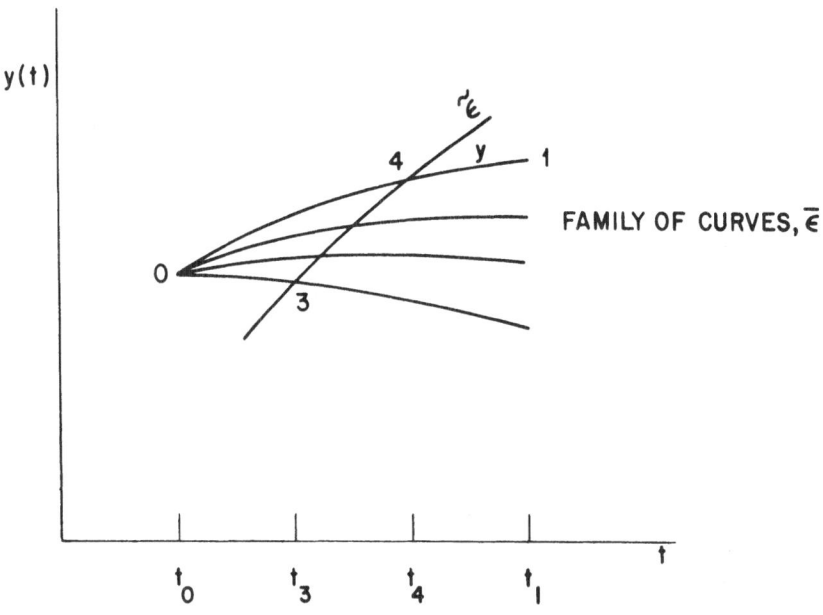

Figure C.2

A Curve and Its Strong Variation

5. Second Variations and the Legendre and Jacobi
    Necessary Conditions

Examination of the definition of the Weierstrass
function, Equation C.7, shows that it is, in fact, the difference between the increment Δf_o and the first order terms
δf_o when the derivative terms in f_o change from y' to Y'

$$E(t,y,y',Y') = f_o(t,y,Y') = f_o(t,y,y') - \frac{\partial f_o}{\partial y'} (Y'-y')$$

Thus, using a Taylor series expansion:

$$E(t,y,y',Y') = \frac{1}{2} \frac{\partial^2 f_o}{\partial y'^2} (y'-y')^2 + \frac{1}{3} \frac{\partial^3 f_o}{\partial y'^3} (Y'-y')^3 + \ldots$$

For the special case where the variation in y' is taken as

$$Y' = y' + \varepsilon C$$

with C a constant and ε allowed to vary arbitrarily in sign and become arbitrarily small, the Weierstrass condition requires that

$$\frac{\partial^2 f_o}{\partial y'^2} \varepsilon^2 + 0(\varepsilon^3) \geq 0$$

so that as $\varepsilon \to 0$ one may deduce the Legendre necessary condition:

$$\frac{\partial^2 f_o}{\partial y'^2} \geq 0 \qquad\qquad (C.9)$$

for a minimum.

While the necessary conditions based on the first variation (the Euler-Lagrange conditions, the transversality condition, and the corner conditions) eliminate from consideration all curves except those yielding a stationary value of f_o; i.e., either a minimum or a maximum, the Legendre and Weierstrass conditions further eliminate those curves that yield a maximum.

Examination of the second variation integral for weak variations lead to still another necessary condition. Seeking conditions guaranteeing the non-negativity of the second variation led Jacobi to a consideration of the equation:

$$\frac{\partial^2 f_o}{\partial y'^2} v''(t) + \left(\frac{d}{dt} \frac{\partial^2 f_o}{\partial y'^2} \right) v'(t) +$$
$$\left(\frac{d}{dt} \frac{\partial^2 f_o}{\partial y \partial y'} - \frac{\partial^2 f_o}{\partial y^2} \right) v(t) = 0 \qquad (C.10)$$

For a problem with fixed righthand end point, it turns out that if there exists a nontrivial solution $v(t)$ to Jacobi's equation which is zero, not only at the righthand end point but also at a point in the interval (t_0, t_1), then the curve y under consideration cannot minimize f_o, even though the Euler-Lagrange and Legendre necessary conditions are satisfied.

Sections C.2, C.3, C.4 and C.5 have presented a brief sketch of the derivation of necessary conditions for minimizing an integral functional of a scalar function of time possessing piecewise, continuous, first-time derivations. The necessary conditions consist of:

Euler-Lagrange Equations	Equation C.4
Weierstrass-Erdmann Corner Conditions	Equation C.6
Transversality Conditions	Equation C.5
Weierstrass Condition	Equation C.8
Legendre Condition	Equation C.9
Jacobi Condition	Equation C.10

The basic procedures described above for the simplest problem of the classical calculus of variations can be applied to obtain the necessary conditions for the Lagrange, Mayer and Bolza problems (which include subsidiary constraint relations). In addition, for these problems, it is useful to introduce the concept of a Lagrange multiplier. The next section discusses Lagrange multipliers. Then the problem of Bolza and its application to the optimal problem is outlined.

6. Differential Constraints and the Concept of Lagrange Multiplier Functions

In the section on nonlinear programming, it was shown that when the functional f_o is minimized by a curve, $y*$, subject to constraint relations which are expressed in the form of functionals, Lagrange multiplier constants are introduced such that the Hamiltonian composed of f_o and the Lagrange multipliers times the constraint functionals is also minimized by the curve, $y*$.

A similar situation applies when the constraint relations are not expressed as functionals. To illustrate this idea, suppose that in the functional of Equation C.1:

$$f_o = \int_{t_0}^{t_1} L(\underline{y}, \underline{y'}, t) \, dt$$

is to be minimized by the optimal choice of a vector-valued
function:

$$\underline{y}(t) = [y_1(t) \ y_2(t)]^T$$

Further assume that $\underline{y}(t)$ is constrained to lie on the surface defined by

$$g_1(\underline{y},\underline{y}',t) = 0 \qquad\qquad (C.11)$$

Suppose $\underline{y}(t) = \underline{y}^*(t)$ is an extremal for f_o subject to the
constraints, and let t_2 be an arbitrary value of time between
t_0 and t_1. Suppose a weak variation is made in each component
of $\underline{y}$ and that it is localized to a neighborhood of t_2 and has
the form:

$$y_1(t) \rightarrow y_1^*(t) + \eta_1(t)$$

$$y_1'(t) \rightarrow y_1^{*\prime}(t) + \eta_1'(t)$$

$$y_2(t) \rightarrow y_2^*(t) + \eta_2(t)$$

$$y_2'(t) \rightarrow y_2^{*\prime}(t) + \eta_2'(t)$$

Here, $\eta_1(t)$ and $\eta_2(t)$ are nonzero only in a neighborhood of
t_2 such that

$$\int_{t_0}^{t_1} \eta_1(t)dt = h_1$$

$$\int_{t_0}^{t_1} \eta_2(t)dt = h_2$$

The change in f_o induced by the variation in $\underline{y}$ is

$$\Delta f_o = \int_{t_0}^{t_1} \left\{ \left(\frac{\partial L}{\partial y_1} - \frac{d}{dt}\frac{\partial L}{\partial y_1'} \right)\eta_1 + \left(\frac{\partial L}{\partial y_2} - \frac{d}{dt}\frac{\partial L}{\partial y_2'} \right)\eta_2 \right\}dt + \S_1$$

$$\Delta f_o = \left(\frac{\partial L}{\partial y_1} - \frac{d}{dt} \frac{\partial L}{\partial y_1'} \right)_{t_2} h_1$$

$$+ \left(\frac{\partial L}{\partial y_2} - \frac{d}{dt} \frac{\partial L}{\partial y_2'} \right)_{t_2} h_2 + \S_2$$

(C.12)

where the functions $\S_i$ denote functions of order two or more in the variations and the sub t_2 indicates the derivatives are evaluated at t_2.

It is required that the new curve still satisfy the constraint relation:

$$g(\underline{y}+\underline{n}, \ \underline{y}'+\underline{n}', t) = 0$$

Subtracting Equation C.11 from the above and integrating from t_0 to t_1 yields

$$\left(\frac{\partial g}{\partial y_1} - \frac{d}{dt} \frac{\partial g}{\partial y_1'} \right)_{t_2} h_1 + \left(\frac{\partial g}{\partial y} - \frac{d}{dt} \frac{\partial g}{\partial y_2'} \right)_{t_2} h_2 + \S_3 = 0$$

or

$$h_2 = - \frac{\left(\frac{\partial g}{\partial y_1} - \frac{d}{dt} \frac{\partial g}{\partial y_1'} \right)_{t_2}}{\left(\frac{\partial g}{\partial y_2} - \frac{d}{dt} \frac{\partial g}{\partial y_2'} \right)_{t_2}} h_1 + \S_3$$

The interpretation of this is clear: the variations in y_1 and y_2 cannot be taken independently but must be related as shown in order that the varied curve remain on the constraint surface, $g = 0$. When this relationship is used in Equation C.12, it yields

$$\Delta f_o = \left\{ \left(\frac{\partial L}{\partial y_1} - \frac{d}{dt} \frac{\partial L}{\partial y_1'} \right)_{t_2} - \left(\frac{\partial L}{\partial y_2} - \frac{d}{dt} \frac{\partial L}{\partial y_2'} \right)_{t_2} \times \right.$$

$$\left. \frac{\left(\frac{\partial g}{\partial y_1} - \frac{d}{dt} \frac{\partial g}{\partial y_1'} \right)_{t_2}}{\left(\frac{\partial g}{\partial y_2} - \frac{d}{dt} \frac{\partial g}{\partial y_2'} \right)_{t_2}} h_1 \right\} + \S_4$$

Since h_1 is of arbitrary sign and t_2 can be taken anywhere

in the interior of the range t_0 to t_1, a necessary condition
for $\underline{y}^*$ to minimize f_o, subject to the constraint of Equation
C.11 is that the coefficient of h_1 be identically zero. This
yields the condition:

$$\frac{\left(\dfrac{\partial L}{\partial y_1} - \dfrac{d}{dt}\dfrac{\partial L}{\partial y_1'}\right)}{\left(\dfrac{\partial g}{\partial y_1} - \dfrac{d}{dt}\dfrac{\partial g}{\partial y_1'}\right)} = \frac{\left(\dfrac{\partial L}{\partial y_2} - \dfrac{d}{dt}\dfrac{\partial L}{\partial y_2'}\right)}{\left(\dfrac{\partial g}{\partial y_2} - \dfrac{d}{dt}\dfrac{\partial g}{\partial y_2'}\right)} \quad t_0 \le t \le t$$

The common value of these ratios will be a function of t; let
it be denoted by $-\lambda(t)$. The necessary condition then is expressed as the pair of equations:

$$\left(\frac{\partial L}{\partial y_1} - \frac{d}{dt}\frac{\partial L}{\partial y_1'}\right) + \lambda(t)\left(\frac{\partial g}{\partial y_1} - \frac{d}{dt}\frac{\partial g}{\partial y_1'}\right) = 0$$

$$\text{(C.13)}$$

$$\left(\frac{\partial L}{\partial y_2} - \frac{d}{dt}\frac{\partial L}{\partial y_2'}\right) + \lambda(t)\left(\frac{\partial g}{\partial y_2} - \frac{d}{dt}\frac{\partial g}{\partial y_2'}\right) = 0$$

These are recognized as Euler-Lagrange equations for an augmented functional of the form:

$$G = f_o + \int_{t_0}^{t_1} \lambda(t)\, g\, (\underline{y},\underline{y}',t)$$

$$G = \int_{t_0}^{t_1} \left\{ L(\underline{y},\underline{y},t) + \lambda(t)\, g\, (\underline{y},\underline{y}',t) \right\}\, dt \qquad \text{(C.14)}$$

Investigation of the other necessary conditions verifies
that to minimize a functional, f_o, subject to constraint relations is equivalent to minimizing an augmented functional
consisting of the original functional plus the sum of products of Lagrange multipliers times the constraint relations.
This approach is used below in the discussion of the Bolza
problem and its transformation to the optimal control problem.

7. The Classical Problem of Bolza

We now consider the classical "problem of Bolza" and
obtain the necessary conditions for a minimum. In the next

section, it will be shown how these necessary conditions
can be translated into necessary conditions for the optimal
control problem.

The classical problem of Bolza is to find the functions
$y_i(x)$, $(i = 1,---n, t_0 \leq t \leq t_1)$, which are piecewise, continuous, and minimize

$$f_o + \phi(t_o, \underline{y}(t_o), t_1, \underline{y}(t_1)) + \int_{t_0}^{t_1} L(t, \underline{y}, \underline{y}') dt \qquad (C.15)$$

while satisfying the system equations

$$g(t, \underline{y}, \underline{y}') = 0 \qquad\qquad i = 1, 2, ---m < n \qquad (C.16)$$

and the end conditions

$$\psi_i(t_0, \underline{y}(t_0), t_1, \underline{y}(t_1)) = 0 \quad i = 1, 2, ---r < 2n+2 \quad (C.17)$$

This problem introduces several complexities only briefly
mentioned above:

  a.   the state variable, $\underline{y}$, is vector valued so the phase
        space is multidimensional

  b.   the performance index depends on the beginning and
        terminal states

  c.   equality differential side conditions must be
        satisfied along the trajectory

  d.   the starting and ending states must lie on prescribed
        surfaces in the phase space.

To solve this problem, Lagrange multipliers:

$$\underline{p}(t) = [p_1(t) \; p_2(t) ---p_m(t)]^T$$

$$e \quad = [e_1 e_2 ---e_r]^T$$

are used to multiply the auxiliary constraint equations and
added to f_o to form an augmented functional, G, which, of

course, is equal to f_o when the constraint equations are

satisfied; i.e.,

$$G_1(t,\underline{y},\underline{y}',\underline{p},\underline{e}) = \phi + \underline{e}^T\underline{\psi} + \int_{t_0}^{t_1} (L + \underline{p}^T\underline{g})\,dt \tag{C.18}$$

where vector notation has been introduced with super T denoting a transpose. Notice that the end points are not necessarily fixed, and that corners are permitted in the solution. To make this clearer, G_1 may be rewritten as

$$G_1 = \phi + \underline{e}^T\underline{\psi} + \int_{t_0}^{t_2} F\,dt + \int_{t_2}^{t_1} F\,dt \tag{C.19}$$

where the integrand is denoted by F

$$F = L + \underline{p}^T\underline{g} \tag{C.20}$$

and t_2 is located where a jump can occur in $\underline{y}'$.

Now consider the changes induced in G_1 when the minimizing curve y is replaced by a different curve in its weak neighborhood. The total increment in G_1 is

$$\Delta G_1 = d\phi + \underline{e}^T d\underline{\psi} + \Delta\left(\int_{t_0}^{t_2} F\,dt + \int_{t_2}^{t_1} F\,dt\right) \tag{C.21}$$

First, examine the variation of the integrals where t_0, t_2 and t_1 are not fixed. This is done as in section C.2 with appropriate modification to account for the fact that $\underline{y}$ is vector valued. Letting

$$I_1 = \int_{t_0}^{t_2} F\,dt; \quad I_2 = \int_{t_2}^{t_1} F\,dt$$

the first variation in I_1 is obtained from the first order terms of

$$\Delta I_1 = \int\limits_{t_0+\delta t_0}^{\bar{t_2}+\delta t_2} F(t,\underline{y}+\underline{\delta y},y'+\underline{\delta y'})dt - \int\limits_{t_0}^{\bar{t_2}} F(t,\underline{y},\underline{y'})\ dt$$

which may be rewritten as

$$\Delta I_1 = \int\limits_{t_0+\delta t_0}^{t_0} F(t,\underline{y}+\underline{\delta y},\underline{y'}+\underline{\delta y'})dt + \int\limits_{\bar{t_2}}^{\bar{t_2}+\delta t_2} F(t,\underline{y}+\underline{\delta y},\underline{y'}+\underline{\delta y'})\ dt$$

$$+ \int\limits_{t_0}^{\bar{t_2}} \left\{ F(t,\underline{y}+\underline{\delta y},\underline{y'}+\underline{\delta y'}) - F(t,\underline{y},\underline{y'}) \right\} dt$$

Using the mean value theorem for the first two integrals, and expanding the first term in the third integral in a Taylor's series around the minimizing curve y, and retaining only the first order terms results in

$$\delta I_1 = [-F]\Big|_{t_0}\ \delta t_0 + [F]\Big|_{t_2_-}\ \delta t_2 + \int\limits_{t_0}^{\bar{t_2}} \left\{ \left(\frac{\partial F}{\partial \underline{y}}\right)^T \underline{\delta y} \right.$$

$$+ \left(\frac{\partial F}{\partial \underline{y'}}\right)^T \underline{\delta y'} \Bigg\}\ dt$$

Noting that

$$\left(\frac{\partial F}{\partial \underline{y'}}\right)^T \underline{\delta y'} = \frac{d}{dt}\left[\left(\frac{\partial F}{\partial \underline{y'}}\right)^T \underline{\delta y}\right] - \left(\frac{d}{dt}\ \frac{\partial F}{\partial \underline{y'}}\right)^T \underline{\delta y}$$

and carrying out the integral of the first term gives

$$\delta I_1 = [-F]\Big|_{t_0}\ \delta t + [F]\Big|_{t_2^-}\ \delta t_2 + \left[\left(\frac{\partial F}{\partial \underline{y'}}\right)^T \underline{\delta y}\right]_{t_0}^{\bar{t_2}}$$

$$+ \int_{t_0}^{\bar{t_2}} \left[\left(\frac{\partial F}{\partial \underline{y}} \right)^T - \left(\frac{d}{dt} \frac{\partial F}{\partial \underline{y}'} \right)^T \right] \delta \underline{y} \, dt$$

but to first order

$$\delta \underline{y} \bigg|_{t_i} = \delta \underline{y}_i - \underline{y}'(t_i) \, \delta t_i$$

so that

$$\delta I_1 = - \left[F - \left(\frac{\partial F}{\partial \underline{y}'} \right)^T \underline{y}' \right] \bigg|_{t_0} \delta t_0 - \left(\frac{\partial F}{\partial \underline{y}'} \bigg|_{t_0} \right)^T \delta \underline{y}_0$$

$$+ \left[F - \left(\frac{\partial F}{\partial \underline{y}'} \right)^T \underline{y}' \right] \bigg|_{\bar{t_2}} \delta t_2 + \left(\frac{\partial F}{\partial \underline{y}'} \bigg|_{\bar{t_2}} \right)^T \delta \underline{y}_2$$

$$+ \int_{t_0}^{\bar{t_2}} \left\{ \left(\frac{\partial F}{\partial \underline{y}} \right)^T - \left(\frac{d}{dt} \frac{\partial F}{\partial \underline{y}'} \right)^T \right\} \delta \underline{y} \, dt$$

A similar development may be made for δI_2, and using these in Equation C.21, the total variation becomes

$$\delta G_1 = \int_{t_0}^{t_1} \left[\left(\frac{\partial F}{\partial \underline{y}'} \right)^T - \left(\frac{d}{dt} \frac{\partial F}{\partial \underline{y}'} \right)^T \right] \delta \underline{y} \, dt + d\Phi + \underline{e}^T \, d\underline{\psi}$$

$$+ \left[\left(F - \left(\frac{\partial F}{\partial \underline{y}'} \right)^T \underline{y}' \right) \delta t + \left(\frac{\partial F}{\partial \underline{y}'} \right)^T \delta \underline{y} \right]_{t_0}^{t_1}$$

$$+ \left[\frac{\partial F}{\partial \underline{y}'} \bigg|_{\bar{t_2}} - \frac{\partial F}{\partial \underline{y}'} \bigg|_{t_2^+} \right]^T \delta \underline{y}_2$$

$$+ \left\{ \left[F - \left(\frac{\partial F}{\partial \underline{y}'} \right)^T \underline{y}' \right]_{\bar{t_2}} - \left[F - \left(\frac{\partial F}{\partial \underline{y}'} \right)^T \underline{y}' \right]_{t_2^+} \right\} \delta t_2 \quad \text{(C.22)}$$

If δG_1 is now set to zero, the following necessary conditions on the minimizing curve are identified:

Euler-Lagrange Equations

$$\left(\frac{\partial F}{\partial \underline{y}}\right) - \frac{d}{dt}\left(\frac{\partial F}{\partial \underline{y}'}\right) = 0 \tag{C.23}$$

Weierstrass-Erdmann Corner Conditions (at Discontinuities in y')

$$\left[F - \left(\frac{\partial F}{\partial \underline{y}'}\right)^T \underline{y}' \right]\Bigg|_{t_i^-} = \left[F - \left(\frac{\partial F}{\partial \underline{y}'}\right)^T \underline{y}' \right]\Bigg|_{t_i^+} \tag{C.24}$$

$$\left(\frac{\partial F}{\partial \underline{y}'}\right)\Bigg|_{t_i^-} = \left(\frac{\partial F}{\partial \underline{y}'}\right)\Bigg|_{t_i^+}$$

Transversality Conditions

$$d\phi + \underline{e}^T d\underline{\psi} + \left[\left[F - \left(\frac{\partial F}{\partial \underline{y}'}\right)^T \underline{y}' \right]\delta t + \left(\frac{\partial F}{\partial \underline{y}'}\right)^T \delta \underline{y}\right]_{t_0}^{t_1} = 0 \tag{C.25}$$

The above three necessary conditions were obtained by consideration of the first weak variation of the augmented functional, G_1. When a strong variation is considered at an arbitrary point along a trajectory satisfying the above conditions, a derivation analogous to the one given in Section C.4 yields a further necessary condition:

Weierstrass Condition (at all points along the curve)

$$F(t,\underline{y},\underline{Y}') - \left(\frac{\partial F}{\partial \underline{Y}'}\right)^T \underline{Y}' \geq F(t,\underline{y},\underline{y}') - \left(\frac{\partial F}{\partial \underline{y}'}\right)^T \underline{y}' \tag{C.26}$$

These four necessary conditions make up the "Multiplier Rule" and must be satisfied in conjunction with the system equations (C.16 and C.17) to obtain the functions $\underline{y},\underline{y}'$, yielding the minimum value of f_o. It should be noted that when F does not contain t explicitly:

$$F - \left(\frac{\partial F}{\partial \underline{y}'}\right)^T \underline{y}' = \text{Constant} \qquad (C.27)$$

8.   Identification of the Optimal Control Problem
       as a Problem of Bolza   .

The optimal control problem is to find the control vector

$$\underline{u} = [u_1 u_2 u_3 \text{---} u_m]^T$$

which minimizes the functional

$$f_o(\underline{x},\underline{y}) = \int_{t_0}^{t_1} L(\underline{x},\underline{y},t)dt + \phi(\underline{x}(t_0),t_0,\underline{x}(t_1),t_1) \qquad (C.28)$$

subject to the differential and algebraic constraint conditions:

$$\frac{d}{dt} x_i = f_i(\underline{x},\underline{u},t) \quad i = 1,2, \text{ --- } L \leq n \qquad (C.29)$$

$$0 = f_i(\underline{x},\underline{u},t) \quad i = L+1, \text{ --- } n \qquad (C.30)$$

$$0 \geq f_i(\underline{x},\underline{u},t) \quad i = n+1, \text{ --- } J, \; j-n \leq m \qquad (C.31)$$

$$\psi_i(\underline{x}(t_0), t_0, \underline{x}(t_1),t_1) = 0 \quad i = 1,2, \text{ ---} r \leq 2n+2 \qquad (C.32)$$

in the state variables

$$\underline{x} = [x_1 x_2 \text{--- } x_n]^T$$

When the inequality constraints are not present, comparison of the resulting set with the equations for the problem of Bolza suggests defining the 2m relations:

$$u_{i-n} = \frac{d}{dt} x_i$$
$$x_i(t_0) = 0 \qquad i = n+1, \; n+2, \text{ --- } n+m \qquad (C.33)$$

This change of variable turns the optimal control problem with no inequality constraints into the problem of Bolza discussed above for the case of an n + m state vector:

$$\underline{y(t)} = [x_1(t) \ x_2(t) \ \text{---} \ x_n(t) \ \bigg| \ u_1(t)dt$$

$$\bigg| u_2(t)dt \ldots \bigg| \ u_m(t)dt]^T$$

having a first time derivative given by:

$$\underline{y'(t)} = [x_1'(t) \ x_2'(t) \ \text{---} \ x_n'(t) \ u_1(t) \ u_2(t) \ \text{---}$$

$$u_m(t)]^T$$

The performance index, Equation C.28, is identified with the Bolza functional of Equation C.15; the end conditions, Equation C.32, with the Bolza end conditions of Equation C.17, and the equality constraints, Equations C.29 and C.30, with the Bolza constraints of Equation C.16.

The use of the Bolza problem necessary conditions, Equations C.23 through C.26, to derive necessary conditions for the optimal control problem, will be illustrated in the next section where inequality constraints are also treated.

9. Control Problem with Inequality Constraints That Are Explicitly Dependent on the Control

In this section the classical calculus of variations is extended to treat the case wherein both inequality and equality constraints exist on the control and state variables. The problem is stated in the previous section as Equations C.28 through C.32. It is assumed that $u_i(t)$ are piecewise, continuous functions of time, and that L and f_i i=1,2,---J possess continuous first and second partial derivatives in all their arguments.

First, the inequality constraints are transformed to equality constraints by the introduction of "slack variables":

$$z_i^2(t) \geq 0 \qquad\qquad i = n+1, \ n+2, \ \text{---} \ J \qquad (C.34)$$

such that Equation C.31 becomes:

$$f_i(\underline{x},\underline{u},t) + z_i^2(t) = 0 \quad i = n+1, \ldots J$$

$$J-n \leq m \tag{C.35}$$

Next, an augmented functional is defined that appends to f_o, the products of Lagrange multipliers times the constraint relations:

$$G_2 = \phi(\underline{x}(t_0),t_0,\underline{x}(t_1),t_1) +$$

$$\sum_{i=1}^{r} e_i \psi_i (\underline{x}(t_0),t_0,\underline{x}(t_1),t_1) \tag{C.36}$$

$$+ \int_{t_0}^{t_1} \left\{ L(\underline{x},\underline{u},t) - \sum_{i=1}^{L} p_i \left[\frac{dx_i}{dt} - f_i \right] \right.$$

$$\left. + \sum_{i=L+1}^{n} p_i f_i - \sum_{i=n+1}^{J} v_i (f_i + z_i^2) \right\} dt$$

(compare with Equation C.18). The integrand of the augmented functional is denoted by F.

$$F = L(\underline{x},\underline{u},t) - \sum_{i=1}^{L} p_i \left[\frac{dx_i}{dt} - f_i \right] + \sum_{i=L+1}^{n} p_i f_i \tag{C.37}$$

$$\sum_{i=n+1}^{J} v_i (f_i + z_i^2)$$

(Compare with Equation C.20.) For conciseness, a function H is defined as:

$$H(\underline{x},\underline{u},\underline{p},t) = L + \sum_{i=1}^{n} p_i f_i \tag{C.38}$$

Here, the Lagrange multiplier vectors:

$$\underline{p}(t) = [p_1(t)\ p_2(t)\ \text{---}\ p_n(t)]^T$$

$$\underline{e} = [e_1 e_2\ \text{---}\ e_r]^T \tag{C.39}$$

$$\underline{v}(t) = [v_{n+1}(t)\ v_{n+2}(t)\ \text{---}\ v_J(t)]$$

have been introduced.

A straightforward repetition of the steps outlined in Section C.7 for the classical problem of Bolza can now be repeated for the optimal control problem. The control variables are thought of as part of the state vector (viz., Equation C.33) and variations taken with respect to them also.

An alternate approach illustrated here is to identify the Bolza problem state vector as

$$\underline{y}(t) = [Z_{n+1}Z_{n+2} \text{ --- } Z_J x_1 x_2 \text{ --- } x_n \int u_1 dt \int u_2 dt \text{ ---}$$
$$\int u_m dt]^T$$

possessing the first-time derivative:

$$\underline{y}'(t) = [Z'_{n+1}Z'_{N+2} \text{ --- } Z'_J \; x'_1 \; x'_2 \text{ --- } x'_n \; u_1 u_2 \text{ --- } u_m]^T$$

and to then make direct use of the Bolza problem necessary conditions, Equations C.23 through C.26. The fact that f_o and f_i i=1,2, ---n are independent of x'_i and that F is independent of $\int u_i dt$ and Z'_i will be seen to simplify the results.

From Equation C.23 and the definitions of $\underline{y}$ and $\underline{y}'$, it is seen that the first J-n Euler-Lagrange equations are

$$\frac{\partial F}{\partial Z_i} - \frac{d}{dt}\frac{\partial f}{\partial Z_i} = 0 \qquad i = n+1, \; n+2, \text{ --- } J$$

or upon expansion of F:

$$2 \, v_i Z_i = 0 \qquad\qquad i = n+1, \; n+2, \text{ --- } J \qquad (C.40)$$

The next L equations are

$$\frac{\partial F}{\partial x_i} - \frac{d}{dt}\frac{\partial F}{\partial x'_i} = 0 \qquad i = 1, \; 2, \text{ --- } L$$

or upon expansion of F:

$$\frac{\partial L}{\partial x_i} + \sum_{j=1}^{n} p_j \frac{\partial f_j}{\partial x_i} - \sum_{j=n+1}^{J} \nu_j \frac{\partial f_j}{\partial x_i} + \frac{d}{dt} p_i = 0$$

However, from the definition of the Hamiltonian, H, this may be written as:

$$\frac{\partial H}{\partial x_i} - \sum_{j=n+1}^{J} \nu_j \frac{\partial f_j}{\partial x_i} = -\frac{d}{dt} p_i \qquad i = 1,2, \text{---} L \qquad (C.41)$$

The next n-L Euler Lagrange equations are

$$\frac{\partial H}{\partial x_i} - \sum_{j=n+1}^{J} \nu_j \frac{\partial f_j}{\partial x_i} = 0 \qquad i = L+1, L+2 \text{---} J \qquad (C.42)$$

since F contains no dependence on x_i', $i = L+1, L+2, \text{---} n$.

Finally, the last m Euler-Lagrange equations are:

$$-\frac{d}{dt} \frac{\partial F}{\partial u_i} = 0 \qquad i = 1, 2, \dots m$$

or

$$\frac{\partial F}{\partial u_i} = C \qquad i = 1, 2, \text{---} m$$

Upon expansion, this becomes:

$$\frac{\partial L}{\partial u_i} + \sum_{j=1}^{n} p_j \frac{\partial f_j}{\partial u_i} - \sum_{j=n+1}^{J} \nu_j \frac{\partial f_j}{\partial u_i} = C \quad i = 1, 2, \text{---} m$$

or, using the definition of H:

$$\frac{\partial H}{\partial u_i} - \sum_{j=n+1}^{J} \nu_j \frac{\partial f_j}{\partial u_i} = C \qquad i = 1, 2, \text{---} m \qquad (C.43)$$

It will be shown below that the value of the constant is zero.

Next, examining the corner conditions, it is seen that from Equation C.24, the second corner condition:

$$\frac{\partial F}{\partial y'}\Big|_{t^-} = \frac{\partial F}{\partial y'}\Big|_{t^+}$$

reduces to

$$P_i\Big|_{t^-} = P_i\Big|_{t^+} \qquad\qquad i = 1, 2, \text{---} L \qquad\qquad (C.44)$$

and

$$\frac{\partial F}{\partial u_i}\Big|_{t^-} = \frac{\partial F}{\partial u_i}\Big|_{t^+} \qquad\qquad i = 1, 2, \text{---} , \qquad\qquad (C.45)$$

since $\frac{\partial F}{\partial y'} = 0$ for all components of y' except x_i' $i = 1, 2,$

--- L, and u_i, $i = 1, 2,$ --- m. For the same reason, the

first corner condition

$$\left[F - \left(\frac{\partial F}{\partial y'}\right)^T y'\right]\Big|_{t^-} = \left[F - \left(\frac{\partial F}{\partial y'}\right)^T y'\right]\Big|_{t^+}$$

reduces to

$$H\Big|_{t^-} = H\Big|_{t^+} \qquad\qquad\qquad\qquad (C.46)$$

where account has been taken of the continuity of $\frac{\partial F}{\partial u_i}$ across

the corner. The corner conditions are seen to reduce to

continuity requirements on the Lagrange multiplier,

$$p_i, \; i = 1, 2, \; \text{---} \; L; \; \text{on H, and on} \; \frac{\partial F}{\partial u_i}, \; i = 1, 2, \; \text{---} \; m.$$

The transversality condition, Equation C.25 becomes:

$$d\Phi + e^T d\psi + \left[H\delta t - \sum_{i=1}^{L} p_i dx_i + \sum_{i=1}^{m} \frac{\partial F}{\partial u_i} du_i\right]\Bigg|_{t_0}^{t_1} = 0$$

Now both Φ and ψ are independent of u_i, $i = 1, 2,$ --- m.

Thus, it is necessary that

$$\left.\frac{\partial F}{\partial u_i}\right|_{t_0} = \left.\frac{\partial F}{\partial u_i}\right|_{t_1} = 0 \qquad i = 1, 2, \text{---} m$$

and that

$$d\underline{\Phi} + \underline{e}^T d\underline{\psi} + \left[H\delta t - \sum_{i=1}^{L} p_i d\underline{x}_i \right]\Bigg|_{t_0}^{t_1} = 0 \qquad (C.47)$$

Since $\dfrac{\partial F}{\partial u_i}$ is required to be zero at the end points,

continuous across corners, and constant between corners, it is clear that it is zero everywhere, and the constant in the Euler-Lagrange equation is zero.

Finally, the Weierstrass condition, Equation C.26, becomes:

$$L(\underline{x},\underline{\hat{u}},t) - \sum_{i=1}^{L} p_i \hat{x}_i' + \sum_{i=1}^{n} p_i f_i(\underline{x},\underline{\hat{u}},t)$$

$$- \sum_{i=n+1}^{J} \nu_i (f_i(\underline{x},\underline{\hat{u}},t) + z_i^2)$$

$$- \left\{ - \sum_{i=1}^{L} p_i \hat{x}_i' + \sum_{i=1}^{m} \frac{\partial F}{\partial u_i} \hat{u}_i \right\}$$

$$\geq L(\underline{x},\underline{u},t) - \sum_{i=1}^{L} p_i x_i'$$

$$+ \sum_{i=1}^{n} p_i f_i(\underline{x},\underline{u},t) - \sum_{i=n+1}^{J} \nu_i (f_i(\underline{x},\underline{u},t) + z_i^2)$$

$$- \left\{ - \sum_{i=1}^{L} p_i x_i' + \sum_{i=1}^{m} \frac{\partial F}{\partial u_i} u_i \right\}$$

where the notation $\hat{x}'$ and $\underline{\hat{u}}$ has been used to denote a first strong variation in $\underline{x}'$ and $\underline{u}$ about their optimal values. This reduces to:

$$H(\underline{x},\hat{\underline{u}},\underline{p},t) \geq H(\underline{x},\underline{u},\underline{p},t) + \sum_{i=1}^{m} \frac{\partial F}{\partial u_i} (\hat{u}_i - u_i)$$

$$+ \sum_{i=n+1}^{J} \nu_i (f_i(\underline{x},\hat{u},t) - f_i(\underline{x},\underline{u},t)$$

Now, along an optimal trajectory $\frac{\partial F}{\partial u_i} = 0$, as shown above.

In addition, f_i, $i = n+1$, $n+2$, --- J is either less than zero or zero. If it is less than zero, then $\nu_i = 0$ from the first Euler equation; if it is equal to zero, then admissible variations, $\hat{u}$, must also maintain $f = 0$. As a result, the Weierstrass condition reduces to the simple expression:

$$H(\underline{x},\hat{\underline{u}},\underline{p},t) \geq H(\underline{x},\underline{u},\underline{p},t) \qquad (C.48)$$

Collecting the above results together, the necessary conditions for optimal control consist of:

The Euler-Lagrange Equations:

$$\nu_i Z_i = 0 \qquad\qquad i = n+1, n+2, --- J \qquad (C.49)$$

$$\frac{\partial H}{\partial x_i} - \sum_{j=n+1}^{J} \nu_j \frac{\partial f_j}{\partial x_i} = -\frac{d}{dt} p_i \qquad (C.50)$$

$$i = 1, 2, --- L$$

$$\frac{\partial H}{\partial x_i} - \sum_{j=n+1}^{J} \nu_j \frac{\partial f_j}{\partial x_i} = 0 \qquad (C.51)$$

$$i = L+1, L+2, --- n$$

$$\frac{\partial H}{\partial u_i} - \sum_{j=n+1}^{J} \nu_j \frac{\partial f_j}{\partial u_i} = 0 \qquad (C.52)$$
$$i = 1, 2, --- m$$

The Corner Conditions:

$H(\underline{x},\underline{u},\underline{p},t)$ is continuous across corners $\qquad (C.53)$

$$p_i(t) \qquad i = 1,2, \ \text{---} \ L \ \text{are continuous}$$
$$\text{across corners} \qquad \qquad (C.54)$$

The Transversality Condition:

$$\underline{d\phi} + \underline{e}^T \ \underline{d\psi} + \left[Hdt - \sum_{i=1}^{L} p_i dx_i \right]\Bigg|_{t_0}^{t_1} = 0 \qquad (C.55)$$

The Weierstrass Condition:

$$H(\underline{x},\hat{\underline{u}},\underline{p},t) \le H(\underline{x},\underline{u},\underline{p},t) \qquad (C.56)$$

The first Euler-Lagrange equation requires the product, $\nu_i Z_i$, to be zero. The value of Z_i must differ from zero when the i-th inequality constraint is satisfied in the inequality sense (see Equation C.35) and must be identically zero when it is satisfied in the equality sense. Hence the value of ν_i is zero when the constraint is satisfied in the inequality sense and nonzero only when the trajectory lies on the constraint surface; i.e., when the constraint is satisfied in the equality sense. The other Euler equations are not affected by $\underline{\nu}$ then, only when the trajectory lies in the interior of the space bounded by the inequality constraints. Indeed, by a review of the steps in the above derivation, it is easily verified that for the simpler optimal control problem having no inequality constraints, the necessary conditions are those given above with ν_i, $i = n+1, n+2, \ \text{---} \ J$ everywhere set to zero.

In the case in which inequality constraints are present but do not depend explicitly on the control; i.e., Equation C.31; and is replaced by

$$\chi_i(\underline{x},t) \le 0 \qquad i = n+1, n+2, \ \text{---} \ J \qquad (C.57)$$

$$j-n \le m$$

the equations presented above are not applicable. Bryson and Dreyfus (13), however, have shown how to extend the classical Bolza problem to treat this case also. Their approach is to put the constraint, Equation C.57, into a form which does

depend explicitly on the control; then append the modified
equations to the augmented functional in the manner illustrated above. Given the new augmented functional, the
necessary conditions for optimality are easily derived by
the procedure illustrated. This case is discussed and the
necessary conditions displayed in Section III.B.3 of the
text.

EXTRAPOLATION LENGTHS
IN
PULSED NEUTRON DIFFUSION MEASUREMENTS

N. G. Sjöstrand
Department of Reactor Physics
Chalmers University of Technology
S-40220 Göteborg, Sweden

I. INTRODUCTION

The pulsed neutron source method has been widely used for the determination of thermal neutron diffusion parameters of various materials. Descriptions of the method and reviews of performed work have been given by several authors (*1, 2, 3, 4*). As can be seen from these works, there are many difficulties both on the experimental side and in the theoretical interpretation. In fact, several discrepancies exist between the results of different experimentalists and also between results from the pulsed source method and from other types of experiments. In later years, many theoretical papers have considerably advanced the understanding of the problems. It is regrettable that they were not available when most of the experiments were performed.

The pulsed source measurements are usually interpreted in terms of a size-dependent parameter, the so-called geometric buckling. The aim of the present article is to discuss a weak point in the pulsed source method; viz., how the buckling is related to the dimensions of the body under investigation. This is usually done via an extrapolation length, a quantity that can be defined in different ways. For large bodies of simple shape, this may not be a serious problem; however, for small systems, the correct assignment of extrapolation lengths is essential for a reliable interpretation of experiments. The presence of edges or corners is one of several sources of uncertainty in this connection. An interesting question is whether the decay constant can be related to a unique buckling, or if a dependence on the shape of the body exists.

II. FUNDAMENTAL REMARKS

A. The Pulsed Neutron Method

In this context, the pulsed neutron source method is used to determine the decay constant, λ, of a population of thermal neutrons in a finite, homogeneous, non-multiplying system, when the number of neutrons decreases exponentially through absorption and leakage (much of the discussion can also be applied to fast neutron systems). Higher modes in space and energy are assumed to be absent (except for a special case treated in section III.D). The possible influence of modes from a continuum is also neglected. The continuum exists when the neutrons can have long flight times due to very low speed or due to infinite dimensions of the system in at least one direction. If there is a continuum, its lower boundary (the "Corngold limit") is at $(v\Sigma_t)_{min}$. Problems related to the continuum of eigenvalues for the decay constant have been discussed recently by Corngold (5) and will not be taken up here.

In the interpretation of pulsed diffusion measurements, it is usually assumed that the decay constant can be expressed as a function of the buckling B^2 in the following way:

$$\lambda = \lambda_o + D_o B^2 - CB^4 + FB^6 + \ldots \tag{2.1}$$

Here, $\lambda_o = \overline{v\Sigma_a}$ is the decay constant for the infinite medium,

D_o is the diffusion constant for neutron density,

C is the diffusion cooling coefficient,

F is a parameter without special denomination.

The B^6 term seems to be the last one which might have experimental significance. The buckling in Equation 2.1 is often identified with the geometric buckling; i.e., the lowest eigenvalue of the Helmholtz equation:

$$\nabla^2 R(r) + B^2 R(r) = 0 \tag{2.2}$$

for the spatial variation of the asymptotic neutron flux. This quantity is assumed to vanish at the extrapolated boundary of the system under consideration. The extrapolated boundary is situated at a certain distance, called the extra-

polation length, outside the physical boundary of the system.
Thus in this context, by extrapolation length is meant the
so-called extrapolated end point, z_o; i.e., that distance
from the surface at which the asymptotic neutron flux inside
the body extrapolates to zero. With these assumptions, the
following well-known formulae for the buckling of simple
bodies are obtained:

Sphere with radius a: $\quad B^2 = \dfrac{\pi^2}{(a + z_o)^2}$ $\hspace{2cm}$ (2.3)

Cylinder with radius a, height h:

$$B^2 = \left(\frac{2.405}{a + z_o}\right)^2 + \frac{\pi^2}{(h + 2z_o)^2} \hspace{2cm} (2.4)$$

Parallelepiped, sides a,b,c:

$$B^2 = \frac{\pi^2}{(a+2z_o)^2} + \frac{\pi^2}{(b+2z_o)^2} + \frac{\pi^2}{(c+2z_o)^2} \hspace{1.5cm} (2.5)$$

Here it is assumed that the extrapolation length is the same
for the different directions. As will be seen in the following, this may not be always a valid assumption.

Diffusion parameters are usually determined in the following way by a pulsed neutron source experiment: the decay
constant is measured for several systems of different sizes.
The geometric buckling for each system is calculated according to formulae (2.3)-(2.5) with the assumption of suitable
extrapolation lengths. By fitting the relation (2.1) to the
data, the different coefficients in this relation are obtained.
Obviously, an uncertainty in the extrapolation length influences the buckling, and hence these coefficients, especially
for the higher terms. Examples of this have been given in
several papers. According to Sjöstrand et al. (6), an increase of the extrapolation length in a polyethylene measurement by 10% gave a decrease in λ_o of 0.4%, an increase in D_o
of 1.0% and a decrease in C of 15%; thus, for a meaningful
interpretation of pulsed diffusion measurements, it is of the
utmost importance that correct extrapolation lengths are
used. The bucklings calculated with these extrapolation
lengths should be such that when they are inserted in Equation
2.1, the correct diffusion parameters are obtained. This

means that a consistent way of using the equations 2.1 - 2.5 and defining the parameters in them must be established.

B.   Connection Between Pulsed and Critical Systems

As is well known (7), there is a close connection between a pulsed moderator system and a critical, stationary system. This is due to the fact that an exponential decay (decay constant λ) in the pulsed system corresponds to an extra absorption of $(-\lambda/v)$ in the critical system; thus, results from calculations on a critical system characterized by c (the number of secondary neutrons per collision) and ℓ (the total mean free path) can be interpreted in terms of pulsed experiments, if the following transformation is made:

$$c = \frac{\Sigma_s}{\Sigma_s + \Sigma_a - \lambda/v} \tag{2.6}$$

$$\ell = \frac{1}{\Sigma_s + \Sigma_a - \lambda/v} = c\ell_s \tag{2.7}$$

Case and Zweifel (8) have pointed out that this should be generally considered only as a *formal* connection between the two types of systems. For example, when the cross-sections in Equations 2.6 and 2.7 vary with energy, there is little probability that c and ℓ in the critical system will show the corresponding variation. However, the connection is very useful in the energy independent case. If the angular distribution of neutrons resulting from collisions is the same in the two systems, then results obtained for a critical system can be applied to the pulsed system and vice versa.

A consequence of the analogy is that in the energy independent case, the spatial neutron distribution is the same in a pulsed and a critical system of the same size and shape; therefore, the extrapolation lengths are also equal in the two cases. However, for pulsed systems, the extrapolation length is preferentially referred to the mean free path for scattering, ℓ_s, whereas, for critical systems, it it is more natural to express the extrapolation length in terms of the total mean free path, ℓ. The identity:

$$z_0/\ell_s = c \cdot z_0/\ell \tag{2.8}$$

will frequently be used in the following discussion to apply results from criticality calculations to pulsed systems.

Most of the pulsed experiments correspond to c values in the interval, $1 < c < 1.3$. The upper limit is equivalent to $B^2 \ell_s^2 = 0.66$ and is reached for spheres of diameter $6.3 \ell_s$, infinite slabs of thickness $2.4 \ell_s$, or infinite cylinders of diameter $4.5 \ell_s$.

III. THE MONOENERGETIC CASE WITH ISOTROPIC SCATTERING

For a realistic treatment of the extrapolation length problem, energy-dependent transport theory must be applied; however, it is useful to first study the monoenergetic case in detail, and in the beginning, to consider isotropic scattering only.

A. The Half-Space Results

In the classical Milne problem, the neutron distribution is studied at the boundary of a homogeneous, non-absorbing half space of an isotropically scattering medium. It has been shown that the asymptotic neutron flux; i.e., the neutron flux more than several mean free paths from the boundary, has a space dependence such that it extrapolates to zero at a distance $z_o = 0.71045 \ell_s$ outside the boundary. In addition, the real flux contains a transient, which is of importance only a few mean-free paths into the medium $(8,9)$. A value with 25 significant digits for the extrapolation length has been given by Cohen (10). The variation of z_o with the number of secondary neutrons per collision, c, has been studied thoroughly. Accurate values in the interval of c between 0 and 1 have been given by Pomraning and Lathrop (11). Using Equation 2.8, it is found that for $0.6 < c < 2$, the expression $z_o = 0.71045 \ell_s$ is correct to within 0.7%. Close to c = 1, the following approximation is valid:

$$z_o/\ell_s = 0.710446 \ [1 + 0.0199 \ (1 - c)^2] \qquad (3.1)$$

(Note that Case and Zweifel (8) have a minus sign instead of

the plus sign here. This misprint appears also in the original reference (*12*).)

B. The One-Dimensional Cases

Regarding bodies with at least one finite dimension,
it should first be noted that the early work of Marshak
(*13*) indicated only a small influence on the extrapolation
length, due to finite slab dimensions. Von Dardel and
Sjöstrand (*14*) mentioned that P_3 calculations showed the
extrapolation distance to be very close to the half-space
value for pulsed thin slabs. Later, it was shown (*7*) that
the monoenergetic P_3 approximation gives the following expression in the isotropic case (if B^4 and higher terms are
neglected):

$$z_o/\ell_s = 0.7051 \ (1 - 0.0256 \ B^2\ell_s^2) \qquad (3.2)$$

In these calculations, the Marshak boundary conditions were
used, and the extrapolated end point was obtained from the
position in which the term corresponding to the asymptotic
neutron flux had its zero. According to Williams (*15*), the
coefficient in front of the buckling should be - 0.0148
instead; however, as will be seen later, this difference is
of minor importance.

In the calculations mentioned so far, the separation of
the flux into one asymptotic and one transient part is relatively straightforward. When it comes to thin slabs, however, the transients at the boundaries influence a large part
of the volume; therefore, a true asymptotic flux may not exist,
and it will not be possible to use this concept in deriving an
extrapolation length. Instead, the following procedure should
be used (a flow diagram is given in Figure 5.2 in connection
with the discussion of the general case). A relation between
decay constant and buckling corresponding to Equation 2.1 is
established by solving the transport equation with the assumption of a space dependence of e^{iBr}. In the monoenergetic
case with isotropic scattering, the exact result (*7*) is:

$$\lambda = v\Sigma_a + v\Sigma_s \left(1 - \frac{B\ell_s}{\tan B\ell_s}\right) \qquad (3.3)$$

The transport equation is then solved numerically for the
system in question, with the boundary condition that no neu-

trons enter from the outside. This gives the decay constant
as a function of the dimensions of the system. In this way,
a connection between buckling and dimensions is established.
The extrapolation length is then readily obtained through the
use of Equations 2.3 to 2.5. It should be observed that with
this definition, the extrapolation length loses the physical
meaning associated with its name, at least for small systems.
The extrapolation length should be regarded primarily as a
correction quantity to be used in formulae 2.3 - 2.5 in order
to obtain the correct buckling for Equation 2.1.

Carlvik (16) performed accurate calculations for slabs
and spheres, starting from the Boltzmann integral equation
and expanding the flux in Legendre polynomials of the coordinate. Isotropic scattering was assumed, and the elements
of the resulting matrix were obtained by recurrence formulae.
The decay constant was given as a function of $\Sigma_s d$, where d
is the thickness of the slab or the diameter of the sphere
(obviously, the value of the absorption cross-section does
not influence the extrapolation length). From Carlvik's
tables, Equation 3.3 and Equations 2.3 and 2.5, the extrapolation lengths plotted in Figure 3.1 are derived. For
both slabs and spheres, the extrapolation length is ℓ_s in the
limit of zero thickness and diameter. This is easily seen
when Equation 3.3 is combined with Equations 2.3 and 2.5,
since these cases correspond to $B\ell_s = \pi/2$ and $B\ell_s = \pi$,
respectively.

Similar results regarding the extrapolation length for
the monoenergetic slab problem with isotropic scattering have
been given by several authors. The results generally agree
with those of Carlvik (16), although it may be difficult to
see this when the data are presented only in the form of a
curve. Variational methods were used by Kladnik (17), Judge
and Daitch (18) and by Goldschmidt (19). Only in the case
of Kladnik is it clear that the values deviate somewhat from
those of Carlvik (16) at small slab thicknesses. It is interesting to note that the conclusions reached by Erdmann and
Shapiro (20) about the good accuracy of Kladnik's results
were largely based on values close to the point where Kladnik's
curve crosses the correct one. Direct numerical solutions
of the transport equation for both slabs and spheres were
obtained by Ghatak and Ahmed (21) and Erdmann and Shapiro
(20). Further, Mockel (22), Dorning (23) and Wood and

Williams (24) presented monoenergetic results as limiting
cases in their energy-dependent calculations.

The analogy between critical and pulsed systems has been
used by Sjöstrand and Dahl (25), who calculated extrapolation
lengths from the very accurate critical dimensions of slabs
and spheres given by Kaper et al. (26). The behavior at
small bucklings is of special interest. From Figure 3.2,
it can be seen that the derivative of the extrapolation
length with respect to the buckling of spheres and slabs
is very close to zero when the buckling approaches zero.
In fact, for $B^2\ell_s^2 \leq 0.1$, the curve can be described by the
equation:

$$z_o/\ell_s = 0.710446 \ (1 + 0.0024 \ B^4\ell_s^4) \qquad (3.4)$$

Thus, at $B^2\ell_s^2 = 0.1$, the fractional deviation from the infinite medium value is only 2.4×10^{-5}. These results illustrate the limited accuracy of the P_3 calculations, which also
gave a $B^2\ell_s^2$ term as shown in Equation 3.2; however, the absence of this term has not been formally proved.

It is interesting to note that by using Equation 3.1
and expressing c in terms of the equivalent buckling through
the use of Equations 2.6 and 2.1, the same expression as that
in Equation 3.4 is obtained for small bucklings, except for
the coefficient in front of the buckling term which is now
0.0022 instead of 0.0024. This indicates a close relationship between the results for a half-space and those for a
slab or sphere in the limit of large dimensions.

Rather little information seems to be available on the
third one-dimensional case, the infinite cylinder. Accurate
extrapolation lengths can, however, be derived from the calculations of Westfall and Metcalf (27), who used the singular
eigenfunction expansion technique to determine the critical
dimensions of homogeneous cylinders. Applying the equiv-
alence relations between critical and pulsed systems as
before, the cylinder curve in Figure 3.2 is obtained. Results in general agreement with these can be derived from
the works of Hembd (28) and Kavenoky (29). It is interesting to note that the extrapolation length rises rather
steeply at small bucklings, in contrast to the behavior for
slabs and spheres. It is therefore not possible to use
Equation 3.4 as an approximation for small bucklings. In-

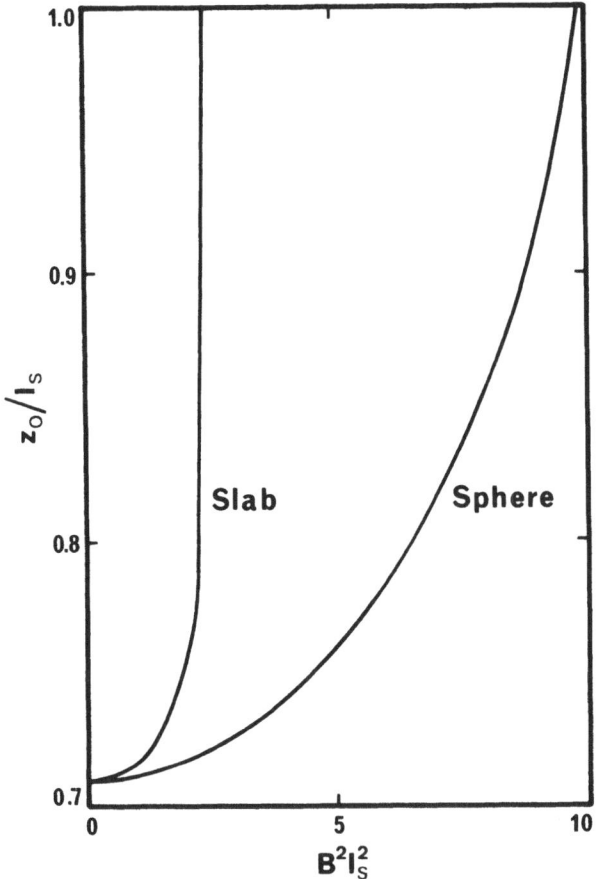

Figure 3.1
Extrapolation Lengths for Slabs and Spheres
As A Function of Buckling

(Derived from the work of Carlvik(16))

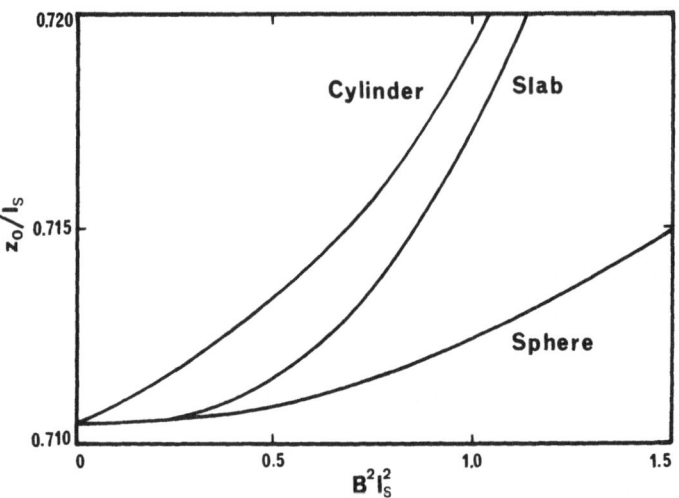

Figure 3.2
Extrapolation Lengths for Slabs, Spheres
and Infinite Cylinders for Small Bucklings
(Based on References (25) and (27))

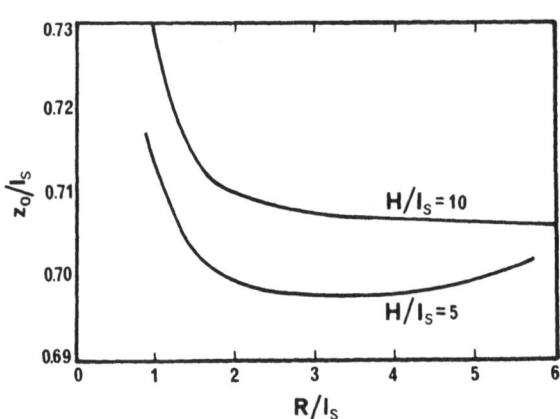

Figure 3.3
Extrapolation Length as a Function of Radius
for Cylinders of Two Different Heights
(Derived from the work of Sahni (30))

stead, the following approximate expression is valid for $B^2 l_s^2 \leq 0.1$:

$$z_o/l_s = 0.71045 \ (1 + 0.0058 \ B^2 l_s^2) \tag{3.5}$$

Judge and Daitch (18) showed with a variational method that the Corngold limit $\lambda^* = v(\Sigma_s + \Sigma_a)$ is reached for a radius of 0.75 l_s. Since the buckling at the Corngold limit is determined by $Bl_s = \pi/2$, it follows from Equation 2.4 that $z_o = 0.78 \ l_s$ for this size. If this is correct, then a curve for the cylinder case in Figure 3.1 should cross the slab curve.

C.   Two- and Three-Dimensional Cases

          The only study of a truly three-dimensional problem is the work of Wood and Williams (24), who investigated the effect of leakage in the y-z directions on the extrapolation length in the x-direction of a parallelepiped. They assumed that the solution to the Boltzmann equation in its integral form (with isotropic scattering) can be written:

$$\phi(x,y,z) = e^{iB_y y \ + \ iB_z z} \ \psi(x) \tag{3.6}$$

They then solved the resulting equation by a variational method. From this, the extrapolation length in the x-direction was obtained as a function of the transverse buckling. Obviously, the extrapolation lengths in the y and z directions must be known, but they are evaluated iteratively invoking symmetry arguments. The results show that an increasing transverse buckling decreases the longitudinal extrapolation length. For a cube, the extrapolation length at $B^2 l_s^2 = 0.1$ is 0.8% lower than 0.7104 l_s. This deviation is larger than for a sphere, slab or infinite cylinder and goes in the opposite direction; however, it is still very small.

          These results of Wood and Williams (24) can be compared to extrapolation lengths derived from the work of Sahni (30). He used an integral transform method to obtain the critical sizes of infinite rectangular prisms and finite cylinders. A difficulty in the interpretation is that the extrapolation

lengths in the two dimensions cannot be determined uniquely
from the criticality parameter, c. It is therefore assumed
here that they are equal. This may seem to be an oversimp-
lification, but it is justified if the extrapolation length
is needed as a correction quantity in experiments. From the
tables of Sahni it is then possible to derive extrapolation
lengths for various combinations of radii and heights (note
that in his Table III, the half-heights are given, not the
full heights as indicated). Through the use of Equation 2.7
and numerical interpolation, the results can be transformed
to dimensions in terms of the scattering mean free path. The
curves shown in Figure 3.3 were derived in this way. The
extrapolation length reaches a value close to or somewhat
above 0.71 ℓ_s at large radii and increases rapidly at small

radii. In contrast to the one-dimensional case, however,
there is a pronounced minimum in between. For the various
curves, the minimum seems to occur not far from the point
where the height is equal to the diameter. For a height of
5 ℓ_s the minimum value is about $z_o = 0.697$ ℓ_s and the corresponding buckling is $B^2 \ell_s^2 = 0.60$.

Figure 3.4 shows the extrapolation length as a function
of buckling for cylinders with height equal to diameter. In
the region presented, the extrapolation length decreases when
the buckling increases. The same figure also gives the extrapolation length for infinite quadratical prisms, derived
from Sahni's (30) results in a similar way. Here, the decrease is stronger for small bucklings, but there is a minimum and a sharp rise after that. For a buckling $B^2 \ell_s^2 = 0.1$,
the extrapolation length for the cylinder is 0.7085 ℓ_s

according to the curve, and for the prism 0.7077 ℓ_s. This
can be compared with the value $z_o = 0.7045$ ℓ_s derived from
the results of Wood and Williams (24) for a cube. It is
probable that the decrease in extrapolation length would be
faster for a cube than for an infinite, quadratic prism, so
the results are in reasonable agreement.

Finite cylinders have been studied also by Horie and
and Nishihara (31), using a variational method. The extrapolation lengths derived from their criticality parameters
show, in general, good agreement with the work of Sahni (30).

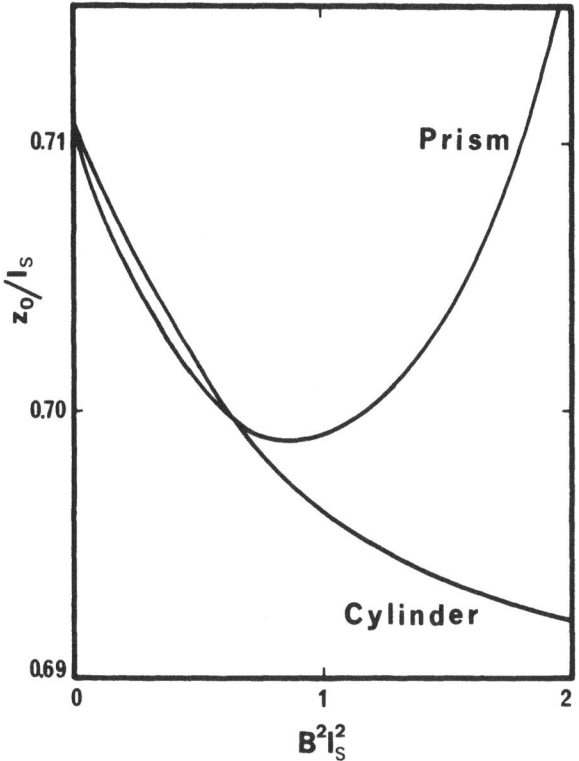

Figure 3.4

Extrapolation Lengths for Cylinders with Height
Equal to Diameter and Infinite Quadratic Prisms

(Derived from the work of Sahni(*30*))

D. The Extrapolation Length of a Higher Harmonic Mode

   As has been shown by Lopez and Beyster (*32*), it is
possible to determine the decay constant of higher harmonic
modes by a Fourier analysis of the time-dependent flux distribution obtained by a movable detector. In such an analysis,
an assumption about the extrapolation length must also be made.
For the monoenergetic case with isotropic scattering, Erdmann
and Shapiro (*20*) have pointed out that the slab solution
(where it exists) for the i-th harmonic mode is identical to
that of the (i - 1)/2-th harmonic mode of the sphere. This

means that the extrapolation length for the first harmonic
mode in a slab can be obtained from the sphere curve in
Figure 3.1 up to the point corresponding to the Corngold limit.
It is, however, more interesting to compare the extrapolation
length for the first harmonic mode in a slab of thickness d
with the extrapolation length of the fundamental mode for
thickness d/2, since they should be equal for large d. In
Figure 3.5, such a comparison is made. The use of an extrapolation length corresponding to the fundamental mode at half
the thickness apparently will be a good approximation for
slabs thicker than about 4 ℓ_s.

E. Conclusion

As a general conclusion for the monoenergetic case
with isotropic scattering, it can be stated that the influ-

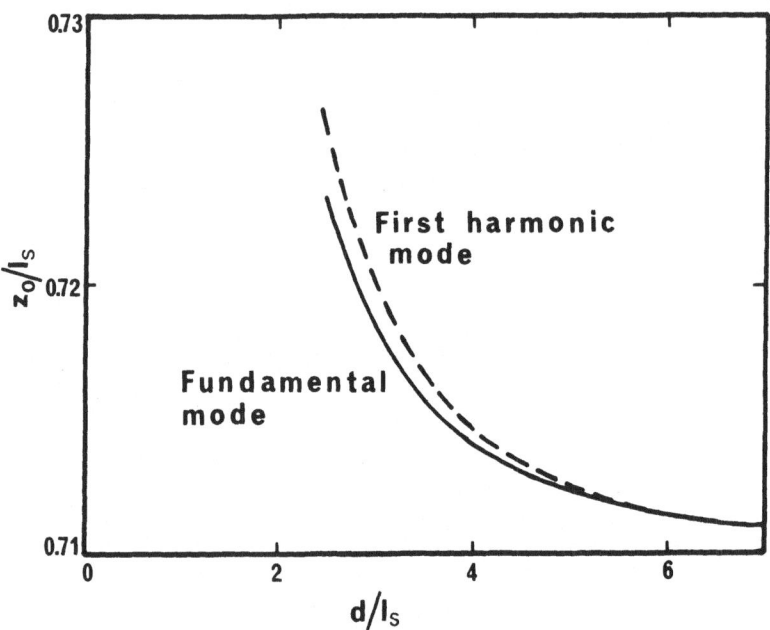

Figure 3.5

Extrapolation Length for the First Harmonic Mode
In a Slab Compared to That of the Fundamental Mode
For a Slab of Half the Thickness

ence of shape and size on the extrapolation length is negligible for most experimental systems.

IV. THE MONOENERGETIC CASE WITH ANISOTROPIC SCATTERING

A. General Remarks

When the scattering is anisotropic, the scattering function is usually expanded in a series of Legendre polynomials:

$$f\ (\Omega' \rightarrow \Omega\ ) = \frac{1}{4\pi} \sum_{n=o}^{\infty} (2n + 1)\ b_n\ P_n\ (\cos\Theta)\ (4.1)$$

where Θ is the scattering angle. We follow here the notation of Davison $(9, \textit{Chapter 17})$ and note that $b_o = 1$ and that b_1 is the average cosine of the scattering angle.

The simplest method for treating anisotropic scattering is to use the transport approximation; i.e., to replace the mean free path for scattering with the transport mean free path $\ell_{tr} = \ell_s/(1 - b_1)$. As shown by Davison (9), this corresponds to setting $b_n \equiv b_1$ for all $n \geq 1$. The transport approximation is generally used when assessing extrapolation lengths in pulsed neutron experiments, but there are few investigations of the validity of this procedure.

B. Half-Space Results

For linear anisotropic scattering (i.e., b_2 and higher order coefficients are equal to zero), the extrapolation length for an infinite, source-free half space close to $c = 1$ is, according to Davison (33):

$$\frac{z_o}{\ell_{tr}} = 0.710446 - \frac{(1 - c)}{(1 - b_1)}\ (0.507828\ b_1$$

$$- 0.155968\ b_1^2)$$

$$+ \frac{(1 - c)^2}{(1 - b_1)^2}\ (0.014120 - 0.704901\ b_1 + 2.207424\ b_1^2$$

$$- 1.630403\ b_1^3 + 0.426819\ b_1^4) \qquad\qquad (4.2)$$

This formula should be accurate to the order of $(1 - c)^3 / (1 - b_1)^3$. Note that in the later work of Davison (9) there is a misprint in the corresponding formula. It can be seen that for anisotropic scattering, there is a stronger dependence on c than for isotropic scattering.

Equation 4.2 can be compared with calculations by Lois and Goldstein (34), who simulated the half-space problem with a slab and a plane source. An analytic method for solving the double spherical harmonic approximation was used as the basis for a computer program. The calculations were performed for values of c ranging from 0.8 to 1.1 and for b_1 up to 0.5. The agreement with the formula is rather good.

Su (35) performed calculations based on the method of Case for solving the transport equation. The results have been published in part in a note by Su and McCormick(36). With b_1 = 2/3, there is excellent agreement with the formula (4.2) in the region of c-values from 0.5 to 1. The difference is always well below the value indicated for the first omitted term. Thielheim and Claussen (37) also based their calculations on the Case method. Numerical results obtained from Thielheim (38) for several b_1 values and a large range of c values show excellent agreement with the above formula within the accuracy stated.

The extrapolation length for an infinite, source-free half-space has also been studied for various degrees of non-linear anisotropy. Already, Mark (39) showed that for pure quadratic anisotropy (i.e., b_1 = 0), the extrapolation length for c = 1 could be written

$$\frac{z_o}{\ell_{tr}} = 0.710 \ (1 + 0.011 \ b_2 + 0.001 \ b_2^2) \qquad (4.3)$$

For the same case, Shure and Natelson (40) give the results shown in Table 4.1.

Except for the last value, there is good agreement with Equation 4.3. Vanmassenhove (41) obtained z_o/ℓ_{tr} = 0.7150 for b_2 = 0.40, which is in reasonable agreement with the results in the table. Van de Hulst (42) used the so-called Henyey-Greenstein scattering function for various degrees of anisotropy. The above-mentioned works by Lois and Goldstein

(34) and by Su (35) have also included various types of non-linear anisotropy. As an example, it can be mentioned that the last author has compared the extrapolation lengths obtained for $b_1 = 2/3$ with those that result when $b_2 = 1/4$ and $b_4 = -1/24$ are included as well. For c values in the range of 0.9 - 1, there is less than 1% difference. The work by Claussen and Thielheim (43) for values of b_2 up to 0.4 corroborates the general conclusions that the influence of quadratic or higher order anisotropy on the extrapolation length is small, close to c = 1.

TABLE 4.1

The Extrapolation Length for Different Degrees of Quadratic Anisotropic Scattering, according to Shure and Natelson (40).

b_2	0.00	0.04	0.10	0.25	0.36
z_o/ℓ_{tr}	0.7104	0.7108	0.7114	0.7125	0.7148

C.   Finite Systems

For linear anisotropic scattering in a slab, Sjöstrand (7) obtained the following expression in the P_3 approximation:

$$z_o/\ell_{tr} = 0.7051 \; [1 - B^2 \ell_{tr}^2$$

$$(0.0256 - 0.2825 \; b_1 + 0.0893 \; b_1^2)]$$

(4.4)

For isotropic scattering, the corresponding formula (Equation 3.2) was not very accurate, so that Equation 4.4 can be expected to give only the order of magnitude of the influence of anisotropy.

Using the same method as mentioned above for isotropic scattering, Carlvik (44) has given some results on critical parameters for slabs and spheres with linearly anisotropic scattering. By using Equations 2.6 and 2.7 as before, they can be interpreted in terms of a pulsed system. Due to the preliminary nature of the work, it could only be said that the slab results agree rather well with Equation 4.2.

The only detailed investigation of the influence of anisotropy on the extrapolation length in finite systems seems to be the work by Lathrop and Leonard (45). Their results are given for critical slab reactors. The authors assumed that the scattering has one isotropic and one anisotropic part. For the latter, elastic hydrogen scattering was adopted; i.e., $b_1 = 2/3$. The ratio of isotropic to anisotropic scattering was varied, so the effective range of b_1 covered was from 0.05 to 0.57. The results are in good agreement with Equation 4.2. In fact, for $1 \leq c \leq 1.3$ and $b_1 \leq 0.3$, the deviations are only 1 per mille or less.

Calculations also were made by Lathrop and Leonard (45) including the next term in the scattering function for hydrogen; i.e., $b_2 = 1/4$. The deviations from the transport approximation value $z_o = 0.7104 \; \ell_{tr}$ are then smaller than when only the b_1 term is included. This is to be expected, since the transport approximation, as stated earlier, implies that $b_1 = b_2$. The difference between the results without and with the b_2 term is less than 3% in the above-mentioned region of interest.

D. Conclusions

In attempting to summarize the situation for the anisotropic scattering case, it should be noticed first that the largest b_1 values occur for hydrogeneous substances, where the average b_1 over a thermal spectrum may amount to 0.25 (7). For b_1 values of this magnitude, it has been shown that Davison's formula, Equation 4.2, is a good approximation for slabs with $1 \leq c \leq 1.3$. Expressing c in terms of the buckling by use of Equations 2.6 and 2.1, the dependence of z_o on the buckling is, in the first approximation:

$$z_o/\ell_{tr} = 0.7104 \; [1 + B^2\ell^2_{tr} \; (0.238 \; b_1 - 0.073 \; b_1^2)] \quad (4.5)$$

Thus, for a slab with $B^2\ell^2_{tr} = 0.2$, the fractional deviation of the extrapolation length from its infinite medium transport approximation value is only 1.1% for $b_1 = 0.25$. With more terms in Davison's formula and in Equation 2.1, a good approximation of the extrapolation length is also expected for slabs of larger bucklings.

Because of the analogy between slabs and spheres (shown, for example, in the isotropic case) the results for slabs probably are also applicable to spheres. There are, however, no results for cylinders and cubes which are more commonly used in experiments. In the isotropic case, a much larger deviation from the infinite medium value was found for cylinders and cubes than for slabs and spheres, and a similar situation may exist here; therefore, no general conclusions can be drawn regarding the validity of the transport approximation for experimental systems, although it is expected to be rather good.

A warning should be added for small systems combined with large anisotropy, where the situation seems to be complicated. As discussed by Sjöstrand (46,47), the buckling evaluated from the anisotropic eigenvalue equation corresponding to Equation 3.3 may be complex. It still would be possible to connect buckling and dimensions through an extrapolation length, but this quantity then also must be complex. This extreme case occurs close to the Corngold limit when b_1 is larger than 0.206, so it is doubtful whether it has any practical significance.

V. THE ENERGY DEPENDENT CASE

A. General

Before entering into the detailed discussion of the extrapolation length problem, some general remarks on the neutron energy spectrum may be motivated. For a more complete treatment, the reader is referred to Williams (48), for example.

First it should be noted that the neutron spectrum in a pulsed moderator will be diffusion cooled, because the leakage probability is larger for neutrons with larger velocity. The smaller the system, the stronger the cooling effect will be. The diffusion cooling influences the decay constant mainly through the diffusion-cooling coefficient C in Equation 2.1. For a relatively large system (say, $B^2 \ell_{tr}^2 < 0.01$) the diffusion-cooled spectrum can be assumed to be the same over the whole volume. If the spectrum is known, the energy-dependent problem can be reduced to an energy-independent one by a suitable averaging of the cross-sections. As a starting point, it is sometimes useful to consider the spectrum as Maxwellian with a lower temperature than the moderator temperature.

The second point to be emphasized is that in reality, there are always energy transients at the boundaries of a system. Their magnitude and importance depend on the size and shape of the system and on the properties of the moderating material. In large systems, they will occupy only a small part of the volume, and in such a case, the assumption of a constant spectrum is valid. In small systems, however, the region affected by transients will be a considerable fraction of the volume. This has been shown by theoretical work by several authors. Williams (48) calculated the average neutron energy across slabs of beryllium and water. For a Be slab 41 cm thick and a water slab 14 cm thick (in both cases $B^2 \ell_{tr}^2 \leq 0.01$) the average neutron energy is constant over about 0.9 of the slab thickness. For thinner slabs (8.6 cm Be and 4 cm H_2O), there is a variation over a large part of the volume. Similar results were obtained by Rönnberg (49) by Monte-Carlo calculations. As seen in Figure 5.1, the average neutron velocity in an H_2O sphere of radius 7 cm ($B^2 \ell_{tr}^2 = 0.03$) is constant to within about 0.5 cm from the surface. Ahmed and Ghatak (50) found from calcula-

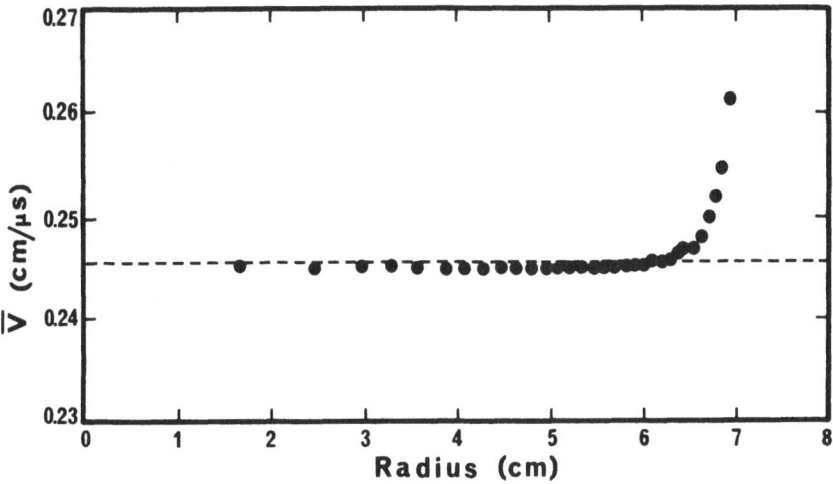

Figure 5.1
Average Neutron Velocity as a Function of Radius
in a Water Sphere of Radius 7 cm
(From Rönnberg (49))

tions on Be slabs of up to 20 cm thickness that the spatial
distribution of neutrons with a particular energy is close
to a cosine form, but that the curves for neutrons with different energies differ very much from each other.

B.   The Definition of the Extrapolation Length

Because of energy transients, it is more difficult
in the energy-dependent case to separate out the asymptotic
part of the neutron flux and from this to derive an extrapolation length. It is therefore essential that the extrapolation length be defined in a way similar to that used in
the monoenergetic case. The flow diagram in Figure 5.2 shows
the appropriate procedure. For a one-dimensional system,
this procedure is straightforward, but for a two- or three-
dimensional body, an approximate technique similar to that
developed by Wood and Williams (24) may be required.

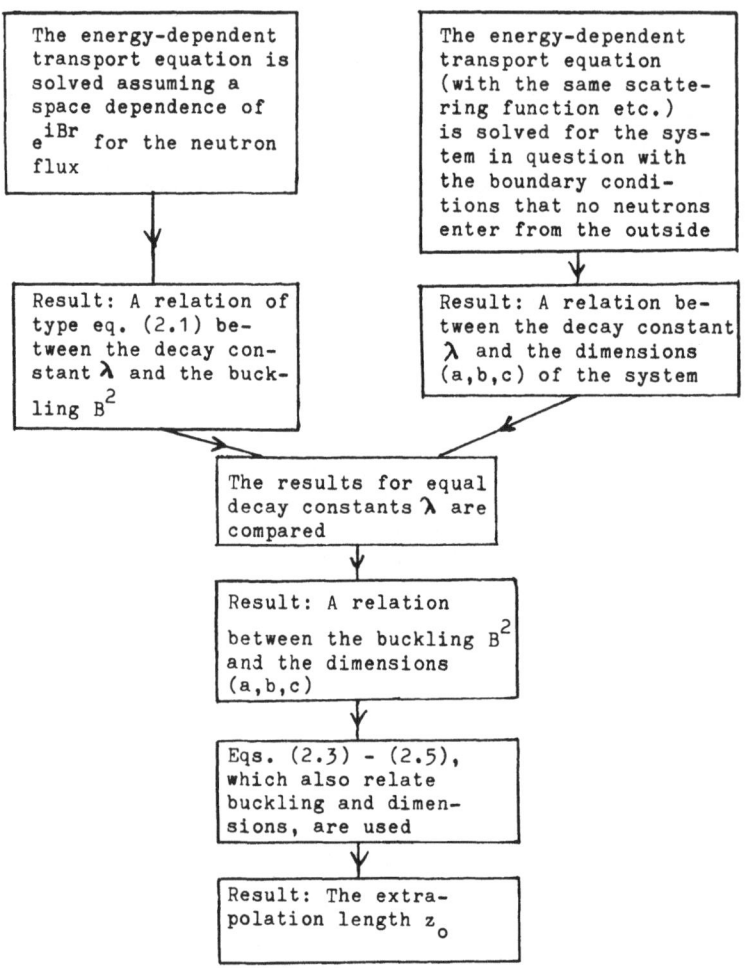

Figure 5.2

Procedure to Obtain Extrapolation Lengths
In the Energy-Dependent Case

The extrapolation length so found must now be regarded
even more strictly as a correction quantity without any simple
connection to the actual behavior of the neutron flux close
to the boundary. This can be illustrated from the above-
mentioned work by Ahmed and Ghatak (50) on beryllium slabs.
For the extrapolation lengths calculated from the cosine
distribution at each energy, there is a variation by more
than a factor of 5 in the energy region of interest. The
representation of the neutron flux with a single function
e^{iBr} is therefore definitely far from the physical reality.

In this connection, it is relevant to mention a different
approach to the extrapolation length problem; namely, the
introduction of an energy-dependent buckling. This has been
done by Beynon (51) and by Ahmed and Mohan (52). The latter
authors refer to the above-mentioned cosine distribution
found by Ahmed and Ghatak (50) and define the buckling through
an energy-dependent extrapolation length, $z_o(E) = 0.71 \, \ell_{tr}(E)$.
In this way, they calculate the decay constants for some systems of water and beryllium and claim to obtain better agreement with experiments. They also find that the decay constant is not a unique function of buckling, but that it depends on the shape of the system. This will be discussed in
Section VII; however, the use of such an energy-dependent
extrapolation length does not make it possible to interpret
experimental results in a unique way. The definitions of
buckling and extrapolation length described earlier do not
suffer from this disadvantage, and are therefore to be pre-
ferred.

To simplify the discussion of the extrapolation length
in the energy-dependent case, it should be noted that this
quantity is influenced by:

1. the size of the system

2. the shape of the system

3. the energy variation of the scattering cross-
   section

4. the detailed properties of the scattering
   function

For monoenergetic neutrons, the influence of the size and
shape of the system was found to be small for experimental
systems. There are, therefore, reasons to believe that the

shape effect also will be small in the energy-dependent case.
The size of the system certainly will have an influence due
to diffusion cooling. The magnitude of this spectrum shift
toward lower energies depends on the scattering function.
If the variation of the scattering cross-section with energy
is large, the influence on the extrapolation length also can
be expected then to be large. The effects of the above factors on the extrapolation length are generally coupled together. In various approximations, however, it may be permissible to separate one or two factors from the others.
This has been verified in calculations of different types;
e.g., in those using variational techniques.

C. The Half-Space Problem

In a fashion similar to the monoenergetic case, it
is useful to consider the half-space problem as a starting
point. The first more general analysis was that of Nelkin
(53), who assumed isotropic scattering and no absorption.
Using a variational method and a Maxwellian as a trial function, he obtained the following formula

$$\frac{z_o}{\overline{\ell}_s} = \frac{3\overline{\ell_s^2}}{8\overline{\ell_s^2}} + \frac{1}{3} \tag{5.1}$$

where the averages are over the Maxwellian. Although this
result is obtained with very simplified assumptions, it gives
an indication of the influence of the energy variation. If
the scattering mean free path is assumed to vary as

$\ell_s(E) = \ell_o(E/E_o)^a$, then

$$\frac{z_o(a)}{\overline{\ell}_s} = \frac{3(1+2a)\,!}{8(1+a)\,!\,(1+a)\,!} + \frac{1}{3} \tag{5.2}$$

This function has its minimum at a = 0; i.e., constant mean
free path, and its value is then 0.7083, which is close to
the exact monoenergetic value; thus, any variation with energy will give a larger extrapolation length. For a 1/v dependent cross-section (a = $\frac{1}{2}$), the result is 0.7577.

Boffi et al. (54) have shown that the same formula (5.1) is also applicable for absorbing media. Kladnik and Kušcer (55) applied the variational method with more accurate trial functions to a monatomic gas of mass 1, and Kladnik (56) extended the results to other masses. The values vary from 0.710 for an infinite mass to 0.735 for mass 1.

Weiss (57) used the heavy-gas model and the P_3 approximation. Neglecting terms of the order of $1/M^2$ and smaller, he obtained the expression:

$$0.7051 \left(1 + \frac{3}{2M} + \frac{2}{3M} \right)$$

for the extrapolation length in terms of the inverse of the bound cross-section Σ_b. Since the average scattering mean-free path is:

$$\ell_s = \frac{1}{\Sigma_b} \left(1 + \frac{3}{2M} \right)$$

and the transport mean-free path is:

$$\ell_{tr} = \ell_s \left(1 + \frac{2}{3M} \right)$$

the result is simply that $z_o = 0.7051 \, \ell_{tr}$. Thus, the energy dependence of the cross-section will only begin to have an influence on the extrapolation length in the next order approximation.

Williams (58) has solved the Milne problem with an essentially exact method using a simple separable scattering kernel. This represents a medium with very strong thermalizing properties, but it is shown that the extrapolation length is very insensitive to the scattering kernel. With the free-gas cross-sections used, the energy variation has the largest influence, the energy transfer effects being about eight parts in 10^4. For hydrogen gas $z_o/\ell_s = 0.7239$. With an infinitely small width of the scattering function, the corresponding value is 0.7246. Using an accurate angular distribution obtained by iteration, Williams found a formula similar to that of Nelkin (5.1):

$$\frac{z_o}{\ell_s} = 0.309401 + 0.401924 \frac{\overline{\ell_s^2}}{\overline{\ell_s^2}} \qquad (5.3)$$

For a constant cross-section, this gives $z_o/\overline{\ell}_s = 0.7113$, and
for a 1/v cross-section, 0.7643. This formula should be
more accurate than Nelkin's, and it is believed to give an
upper boundary to z_o. In a later paper, Williams (59) showed
that the conclusions remain the same, even for more realistic
scattering kernels.

Kladnik (60) used the variational method and three
different scattering kernels for water. With the monatomic
gas model, the extrapolation length obtained was 0.329 cm,
with a Nelkin kernel, 0.339 cm, and with an Egelstaff kernel, 0.305 cm. Expressed in terms of the transport mean free
paths of the different models, the results are 0.723, 0.746
and 0.744, respectively. These values are of the order of
magnitude expected, bearing in mind the strong energy variation of the scattering cross-section for water. It is com-
forting to see that the realistic scattering models of Nelkin
and Egelstaff give values that are in such good agreement.

The main purpose of the work of Arkuszewski (61) was
to study the influence of absorption on the Milne problem.
In the zero absorption limit, his results for a separable
kernel and free gas cross-sections agree very well with those
of Williams (58). Further indications that the extrapolation
length is insensitive to the scattering properties of the
medium have been given by Eisenhauer (62).

D. **Slabs and Spheres**

Gelbard et al. (63) probably were the first to
study the energy-dependent problem in finite systems; however,
their full report has not been available to us. Gelbard and
Davies (64) reported the use of the P_3 approximation and the
Radkowsky scattering kernel for water. Extrapolation lengths
were obtained for slabs and spheres in the buckling range
$0 - 1$ cm^{-2} (the upper limit corresponds to $B^2 \ell_{tr}^2 = 0.2$). The
variation with buckling was found to be the same for both
shapes. The results are shown in Figure 5.3 which has been
taken from Elkert (65). Emon (66) found similar results for
water slabs using the P_7 approximation and four energy groups.

In a later work, Schmidt and Gelbard (67) studied the problem more closely, probably stimulated by the poor agreement
with early experiments. They now used the Nelkin kernel,
which is more realistic than that of Radkowsky, and analyzed
the numerical procedure in detail. Their result for a slab
can be written approximately as

$$z_o/\ell_{tr} = 0.748 - 0.053 \, B^2 \qquad (5.4)$$

where ℓ_{tr} is, here and in the following, the transport mean
free path averaged over the infinite medium (not diffusion-
cooled) spectrum, if nothing else is stated. For a sphere,
the variation with buckling is somewhat less (see Figure 5.3).
For both shapes, the dependence on buckling is smaller than
in the earlier work. Unfortunately, the new results were
published only in an internal report. They have therefore
been overlooked by many experimentalists who, instead, have
used the earlier and less accurate results for the interpretation of their experiments.

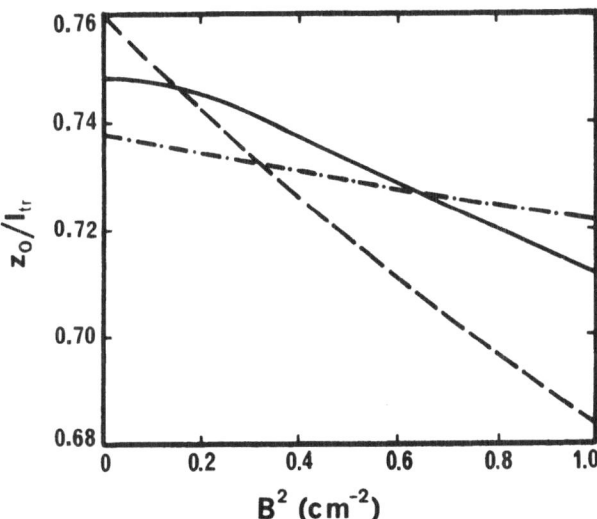

Figure 5.3

Extrapolation Length as a Function of Buckling for Water
Spheres (65). The full line represents the results of
Schmidt and Gelbard (67), the dashed curve that of Gelbard
and Davies (64) and the dot-dashed curve that of Walker et
al. (69).

Williams (*15*) studied the slab problem using the L_n method (which is based on a Laguerre polynomial expansion in energy). One interesting general conclusion for an infinite half-space is that no matter what the variation of the mean-free path with energy may be, the multivelocity effect increases the extrapolation length over the monoenergetic value. This agrees with the conclusions drawn from Equation 5.2. The extrapolation length as a function of buckling was calculated for water and beryllium, assuming suitable values for the thermalization parameter, M_2. For water, there is general agreement with the results of Gelbard and Davies (*64*). The extrapolation length for beryllium increases with buckling, but, as will be seen later, this is probably a consequence of the approximations used.

Vértes (*68*) used a method similar to that of Williams (*15*) and found similar results for a slab of water. The variational method was applied by Kladnik (*17*), who obtained results for various slab thicknesses with the free gas scattering model.

A simple approach was used by Walker et al. (*69*), who calculated the extrapolation length for a half-space according to formula (5.1), using an effective width model for water. They then assumed $z_0(B^2) = 0.7378 \ \bar{\ell}_{tr}(B^2)$; i.e., for each buckling they calculated the diffusion-cooled spectrum and averaged the transport mean free path over the spectrum. As can be seen in Figure 5.3, this procedure also gives reasonable results.

Mockel (*22*) used the invariant embedding method and a separable scattering kernel. He has given curves over the extrapolation length as a function of slab thickness for a $1/v$ kernel and for a cross-section of a free gas of mass 1. It is interesting to note that there is a minimum at a slab thickness of about one mean free path. For larger thicknesses, the curves have the same general behavior as those of Schmidt and Gelbard (*67*).

Slabs and spheres of beryllium have been studied by Ahmed and Ghatak (*50*). They assumed the scattering to be isotropic in the laboratory system and the absorption to be of the $1/v$ type. The scattering kernel used was said to be very satisfactory in calculations of transient spectra in beryllium.

The transport equation in its integral form was solved numerically, but a limiting factor in the calculations was that
only 10 thermal groups could be used. The extrapolation
length was calculated for three slab thicknesses and the
results were given in relation to the transport mean free
path, averaged over the energy spectrum at the center of the
slab. They found that the extrapolation length increases
with increasing buckling. This behavior agrees with that
found by Williams (15), but it is not in accordance with the
more recent work by Wood and Williams (24) which is discussed in Section E.

Dorning (23) used a separable scattering kernel and a
1/v dependent scattering cross-section in calculations of
the extrapolation length for spheres and slabs. His slab
curve agrees with that of Mockel (22). In the sphere curve,
there is also a minimum which appears to be close to the
minimum size for which a discrete decay constant exists. A
realistic scattering kernel for water was used in a later
work by Dorning (70). Here, he obtained extrapolation lengths by comparing the decay constants from very accurate
thirty-group S_n calculations with those of a thirty-group
asymptotic transport theory calculation. Figure 5.4 shows
that the extrapolation distance for spheres agrees very well
with the result of Schmidt and Gelbard (67). It is believed
that Dorning's values are more accurate for spheres with a
radius smaller than about 5 cm, whereas the values of Schmidt
and Gelbard are better above 8 cm radius. In this realistic
curve by Dorning there is also a large decrease in the extrapolation length at small radii, but there is no actual
minimum.

E. Parallelepipeds

As in the monoenergetic case, the only study of a
truly three-dimensional problem is in the work of Wood and
Williams (24) who investigated the extrapolation length in
the x direction of a parallelepiped. In the energy-dependent
case, they made an assumption similar to Equation 3.6:

$$\phi(\mathbf{r},E) = e^{iB_y y + iB_z z} \psi(x,E) \tag{5.5}$$

Using the method described earlier, they first calculated
the extrapolation length in the x direction as a function
of the transverse buckling. Then the extrapolation lengths
in the y and z directions were evaluated iteratively, invoking symmetry arguments. Calculations were made for water,

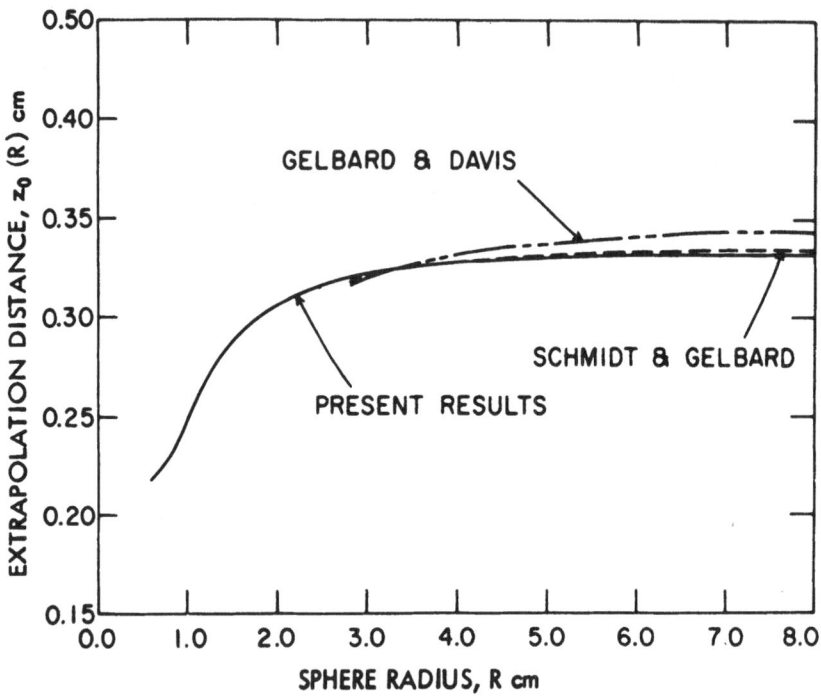

Figure 5.4

Extrapolation Lengths For Water Spheres
(According to Dorning (70))

beryllium, beryllium oxide and graphite. For water, the
effective width model of Egelstaff was used. The inelastic
scattering in the polycrystalline materials was treated by
a multiphonon expansion. For beryllium, the Sinclair phonon
frequency distribution was used, for beryllium oxide the
Debye model, and for graphite a simple phonon distribution
due to Egelstaff. All these scattering kernels described
in detail by Williams (48) are believed to be realistic;
however, the scattering was in all cases assumed to be isotropic. As in the monoenergetic case, it was found that for
all the moderators, the longitudinal extrapolation length decreases when the transverse buckling increases. The variation is small for water but rather large for the crystalline

materials. For water, when comparisons are made at the same total buckling, it is found at $B^2 \ell_{tr}^2 = 0.055$ that z_o / ℓ_{tr} for a slab is 0.712 and for a cube, 0.708. For beryllium, the corresponding values at $B^2 \ell_{tr}^2 = 0.026$ are 0.771 and 0.723; thus, there is some shape dependence of the extrapolation length, which must be taken into account, at least for crystalline materials.

The general behavior of the extrapolation length is, still according to Wood and Williams (24), that it decreases with increasing buckling for all the moderators investigated (see Figure 5.5). For water, this is a natural consequence of the spectrum shift toward lower energies. The situation is more complex for the crystalline materials. Spectrum calculations indicate that increasing buckling means fewer neutrons below the Bragg cutoff energy. Since $\ell_{tr}(E)$ is large in this region, this will tend to decrease the average ℓ_{tr}. As the Bragg cutoff occurs at higher energy for beryllium than for graphite, it might be expected that the variation is smaller for graphite. This is in accordance with observation. The results found by Wood and Williams for the polycrystalline materials are believed to be more accurate than those reported earlier, since the scattering kernels used are more adequate. Wood and Williams finally remark that in spite of some remaining approximations (e.g., those regarding corners), they believe that the theory in its present form gives the extrapolation length to better than 5%.

F. Conclusions

From what has been said in this chapter, it is clear that the calculation of accurate extrapolation lengths is a very difficult problem in the energy-dependent case. The variation of the transport mean-free path, ℓ_{tr}, with energy, is the factor that has the largest influence. The consistent results for water, with its smooth energy variation, is therefore not surprising. The dependence on the shape of the system is so small here that it can be neglected for most experimental systems. It seems reasonable to assume that the results for other hydrogeneous substances such as hydrocarbons will be similar to those for water.

For crystalline materials with sharp Bragg peaks in the scattering cross-sections, the calculations seem to depend

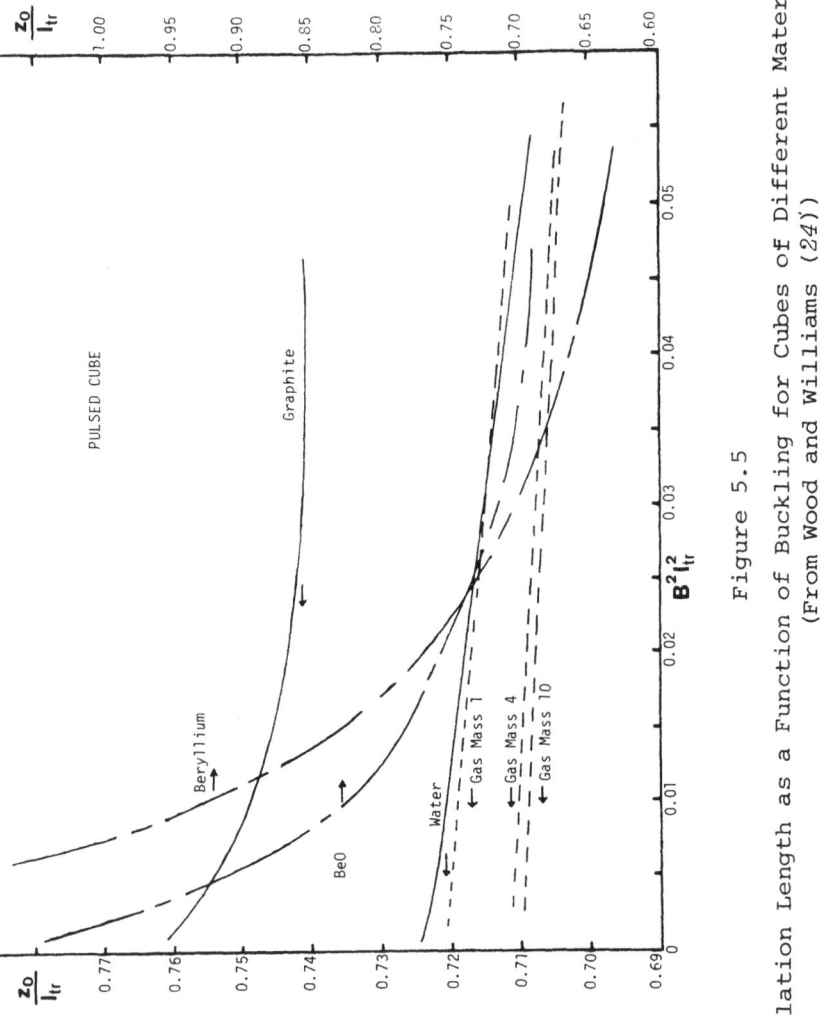

Figure 5.5

Extrapolation Length as a Function of Buckling for Cubes of Different Materials
(From Wood and Williams (24))

very much on the choice of the scattering function and on
the approximations made in the calculations. The work of
Wood and Williams (24) shows that the dependence of the extrapolation length on buckling and shape is strong. It
would be reassuring to have other calculations available to
check the statement of Wood and Williams that the extrapolation lengths can now be calculated to better than 5%.

VI. EXTRAPOLATION LENGTH MEASUREMENTS

A. General

        Because of the importance of correct extrapolation
lengths in the interpretation of pulsed source measurements,
there have been several attempts to measure the extrapolation
length directly in experimental systems. A detector is inserted and its response is studied as a function of time and
position. The spatial distribution of the fundamental mode
is fitted to the type of function expected in the actual case.
From this, the effective dimensions of the system are obtained,
and since its real physical size is known, the extrapolated
end point is readily calculated.

        As discussed in Section V (A), it is the asymptotic neu-
tron flux density that should be extrapolated. The transients
close to the boundary must be avoided. Wood and Williams (71)
have indicated the importance of the transients by calculating the error obtained in the buckling by fitting the true
flux to a cosine distribution in slabs of different materials.
Table 6.1 shows some of their results. For systems that are
not smaller than those given here, it seems possible to correctly assign bucklings by omitting points closer than three
to four mean free paths from the boundary. However, the extrapolation length is more sensitive to the effect of transients, so for this quantity, the error may be one order of
magnitude larger than for the buckling.

        There are also many experimental difficulties as well as
these interpretational ones. For instance, it is of the utmost importance that higher harmonics are absent, that the
detector positioning is correct, that the disturbance caused
by the detector is known and that container wall effects are
eliminated. Perhaps the most **crucial** difficulty is that
caused by room return neutrons. As shown by Williams (72),
such neutrons may have a large influence on the flux in the
outer parts of the system and therefore on the extrapolation
length. At least some of the many observed discrepancies be-

tween theory and experiment must be ascribed to the unusually
difficult experimental problems.

TABLE 6.1

Error in Buckling Obtained
By Fitting the True Flux to a Cosine Distribution
In Slabs of Different Materials
(According to Wood and Williams (71))

Material	ℓ_{tr}	Slab Thickness	Distance, Nearest Points to Boundary	% Error in B^2
	cm	cm	cm	
H_2O	0.44	4	1.11	0.1
Be	1.5	16	4.0	0.4
C	2.5	28	8.2	2.0

B. **Water**

Assuming D_o = 36,000 ± 500 cm^2/s for water at 20°
C, the transport mean-free path is 0.435 ± 0.006 cm. From
the discussion in Section V (D), it is clear that the theoretical results of Schmidt and Gelbard (67) are applicable.
For zero buckling, their value is z_o/ℓ_{tr} = 0.748, which gives
z_o = 0.325 cm with an estimated uncertainty of ± 0.008 cm.
According to Equation 5.4, the extrapolation length will decrease somewhat with increasing buckling; however, with the
relatively large geometries used in the measurements reported
here (the largest B^2 is 0.25 cm^{-2}, corresponding to
0.047 $B^2\ell_{tr}^2$), this can at most decrease the last decimal
by a few units. The calculations by Wood and Williams (24)
give somewhat lower values, between 0.309 and 0.313 cm for
the sizes used in the experiments.

The first reported measurement of extrapolation lengths in any pulsed system is that of Campbell and Stelson (73) in water. They used a cylinder 20 cm in diameter and 20 cm in height ($B^2 = 0.08$ cm^{-2}). With a small LiI crystal detector on a lucite light guide, they obtained an extrapolation length of 0.46 ± 0.05 cm. Similar results were obtained by DeJuren et al. (74), who used a water cylinder with a diameter of 16 cm and a cube of 11 cm side. Corrections were made for the depression effect of the detectors which were either a ^{6}LiI or a boron polyester crystal. The value obtained for the cube was 0.42 cm, and the result for the cylinder was even higher; thus, in both these early experiments, the measured values were much too high. In this connection, the work by Gaerttner et al. (75), which gave 0.56 cm also should be mentioned. The reason for this high value was probably the effect of the reentrant hole used in extracting the neutron beam. Walker et al. (69) performed measurements in cubic boxes with side lengths of about 10 and 18 cm, respectively ($B^2 = 0.25$ and 0.09 cm^{-2}). A small boron or lithium-loaded ZnS (Ag) phosphor on a long perspex light guide was used as a detector. A careful analysis of the harmonic modes was performed, and corrections for the detector disturbance were made. The results were 0.35 ± 0.02 cm for the smaller cube and 0.38 ± 0.04 cm for the larger one. These values are still somewhat high but not very much so.

The most recent measurements of extrapolation lengths in pulsed systems have been performed by Shalev et al. (76) and by Bowen and Scott (77). Shalev et al. used a parallelepiped (11 x 21 x 21 cm) with $B^2 = 0.11$ cm^{-2}. Data were taken at 11 spatial points using a small fission counter. A detailed analysis led to a value of 0.300 ± 0.015 cm at 20° C. In contrast to all earlier results, this is a little low, but it still can be regarded as being in agreement with the theoretical values.

Bowen and Scott (77) studied a cylinder (diameter and height, 18 cm) and a cube (side, 18 cm). In both cases, the buckling was about 0.09 cm^{-2}. The flux mapping was performed using a fission counter. Two independent measurements in the cylinder gave 0.323 ± 0.013 and 0.306 ± 0.030 cm, respectively. Two measurements in the cube gave values of 0.302 ± 0.018 and 0.322 ± 0.012 cm. The agreement with the theoretical value is good. An experiment was also performed with a "flat cylinder" having a diameter of 22 cm and a height

of 10 cm (B^2 = 0.13 cm^{-2}). Surprisingly, the result here was
0.467 ± 0.035 cm. In another series of experiments, the water
moderators were put inside pressure vessels and measurements
were performed at temperatures up to 250°C. For the larger
of the two geometries used (B^2 = 0.05 cm^{-2}) the results agreed
reasonably well with theory, but the accuracy was poorer than
in the room temperature measurements.

C. Dowtherm A

The only other hydrogeneous substance for which extrapolation lengths have been measured is Dowtherm A, which
Walker et al. (69) also used in the above-mentioned measurements. The result was 0.45 ± 0.02 cm for a 10 cm cube
(B^2 = 0.24 cm^{-2}) and 0.48 ± 0.03 cm for an 18 cm cube
(B^2 = 0.085 cm^{-2}). There are no theoretical calculations
available for Dowtherm A, so a comparison is made with the
work on water. From the measurements by Küchle (78) and by
Demanins et al. (79) a value of D_o = 49,000 ± 600 cm^2/s is
taken at 20°C. The transport mean free path is then 0.592
± 0.007 cm. Assuming that the energy variation of the cross-
sections has the same effect as in water, the same half-space
value for z_o/ℓ_{tr} (= 0.748) can be used. Taking some uncertainty into account, a value of z_o = 0.443 ± 0.010 cm at zero
buckling is obtained. For B^2 = 0.24 cm^{-2} (corresponding to
$B^2\ell_{tr}^2$ = 0.084) this should be a little smaller; however, there
appears to be good agreement between theory and experiment.

D. Graphite

The only reported measurement in graphite is that
of Davis et al. (80). The experiment was performed in three
different parallelepipeds with bucklings of 2.4 · 10^{-3},
3.7 · 10^{-3} and 7.1 · 10^{-3} cm^{-2} (the last value corresponding
to $B^2\ell_{tr}^2$ = 0.045). An enriched BF$_3$ counter was used for the
flux traverses, and the influence of the detector disturbance
was taken into account. An average value of 1.825 ± 0.025 cm
was obtained. The diffusion constant was determined to be
D_o = (2.0896 ± 0.0093) · 10^{5}cm^2/s for the graphite, which had
a density of 1.689 g/cm^3. From this, ℓ_{tr} = 2.525 ± 0.011 cm
and z_o/ℓ_{tr} = 0.723 ± 0.012, which is a plausible value from
the theoretical point of view. Wood and Williams (24) get
0.743 for this type graphite and this is not much outside the
error limits.

E. Beryllium Oxide

        Measurements on beryllium oxide have been performed
by Ritchie (81), who used five parallelepipedical structures
in a buckling range of 0.009 to 0.022 cm^{-2} (the largest value
corresponding to $B^2 \ell^2_{tr}$ = 0.058). For the flux distribution
measurements, BF_3 counters were used, and adequate corrections seem to have been applied. The extrapolation length
decreases with increasing buckling, which is in general agreement with the theoretical work by Wood and Williams (24), but
the experimental values all lie above the theoretical ones.
In a later work, Ritchie and Moo (82) have discussed the
reasons for this discrepancy. They state that the experimental extrapolation lengths should not be compared to those
derived from the shape of the neutron flux but from the shape
of the curve corresponding to the reaction rate of the detector. For the 1/v detectors used in the experiments, this
will make a large difference, as seen from Figure 6.1.
Ritchie and Moo (82) used two different data sets for the
scattering kernel of BeO. For the most realistic one (where
an "extinction correction" was introduced to take the deviations from an ideal polycrystalline material into account)
there is good agreement with the experimental values. From
Figure 6.1 it is also seen that the extrapolation lengths obtained with this model and using the shape of the neutron
flux agree rather well with the results of Wood and Williams
(24).

F. Conclusions

        From this chapter, it is clear that there are many
difficulties involved in the experimental determination of
extrapolation lengths and in the interpretation of the results; however, there is in general rather good agreement
between more recent measurements and theoretical predictions.
In some cases, the earlier discrepancies may have been caused
by a not sufficiently accurate description of the complicated
interaction between the neutrons and a crystalline medium.
Generally, however, the reasons for the discrepancies are
most probably on the experimental side.

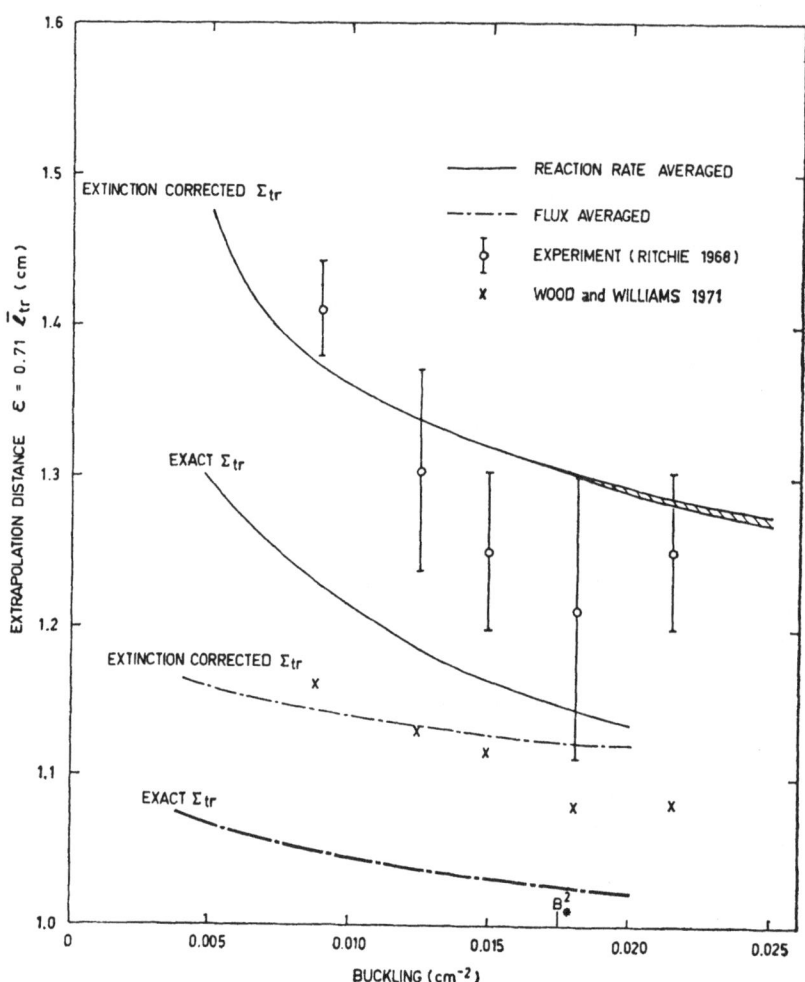

Figure 6.1

Comparison of Calculated and Measured Extrapolation Lengths
In BeO   Assemblies
(From Ritchie and Moo (82))

VII.   EXPERIMENTAL INFORMATION ON SHAPE DEPENDENCE

A. General

As discussed in Section V.B, a "shape effect" on the decay constant can be predicted by the use of energy-dependent buckling; however, the definitions of buckling and extrapolation length adopted in the present paper are such that the decay constant should be a unique function of buckling. It is therefore disturbing that some experimentalists have found different decay constants for systems that have the same buckling but different shapes. This may be caused by experimental errors or by a bad choice of extrapolation lengths, but it is not impossible that the theoretical background might be wrong in some respect. It is therefore important in this connection to study the experimental evidence for and against such a shape effect.

B. Experiments

Hall et al. (83) were the first to find an indication of a shape effect. Using cylindrical water moderators, they obtained the decay constant curves shown in Figure 7.1 for "square" and "flat" systems. At $B^2 = 0.7$ cm^{-2} (corresponding to $B^2 \ell_{tr}^2 = 0.14$) the approximate dimensions of the two shapes were 6 cm diameter, 6 cm height and 8.6 cm diameter, 4 cm height. In the interpretation, various extrapolation lengths were tried, but the figure is given for a constant extrapolation length of 0.33 cm. With this assumption, the diffusion parameters in Table 7.1 were obtained. In spite of the large difference between the two curves, only the diffusion cooling coefficients deviate somewhat outside the errors. Apparently, the experimental uncertainties were rather large. As an example, it may be mentioned that the time analyzer had only nine channels.

A similar tendency for a flat system to have a smaller decay constant than a cubic system has been reported by Beckurts and Wirtz ($2,Page\ 388$). In Figure 7.2, the measurements of Lopez and Beyster (32) on water are compared with those of Küchle (78). The systems of Lopez and Beyster were close to a cubical shape, the smallest size being 7.62 x 7.62 x 5.08 cm (actually, this gives a buckling of about 0.59 cm^{-2}, so results other than those given in the above reference probably have been included in the curve by Beckurts and Wirtz).

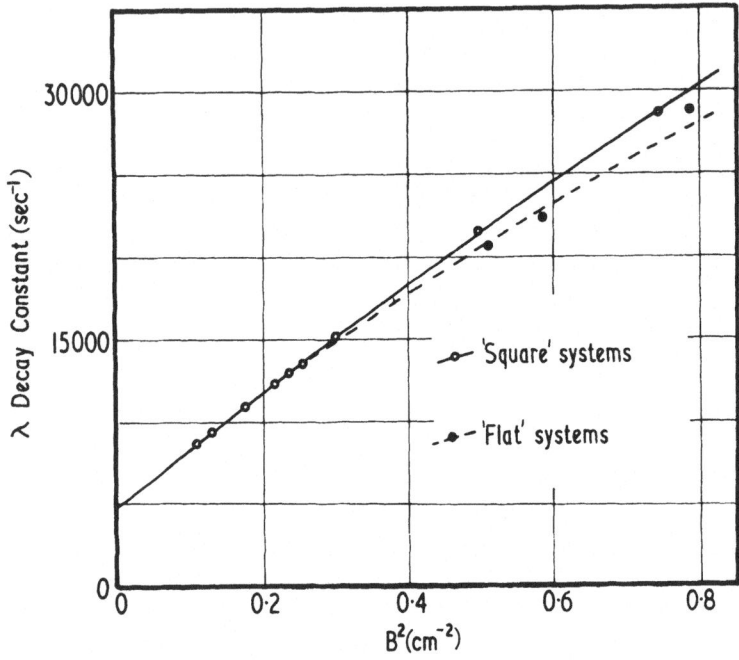

Figure 7.1

Decay Constant as a Function of Buckling
For "Flat"and "Square" Water Cylinders
(According to Hall et al. (83))

TABLE 7.1

Diffusion Parameters for "Flat" and "Square" Cylinders
(Obtained by Hall et al. (83))

	$v \Sigma_a (s^{-1})$	$D_o (cm^2 s^{-1})$	$C (cm^4 s^{-1})$
Flat	4900 ± 150	35630 ± 1170	8050 ± 1670
Square	4880 ± 160	35020 ± 1300	3620 ± 2090

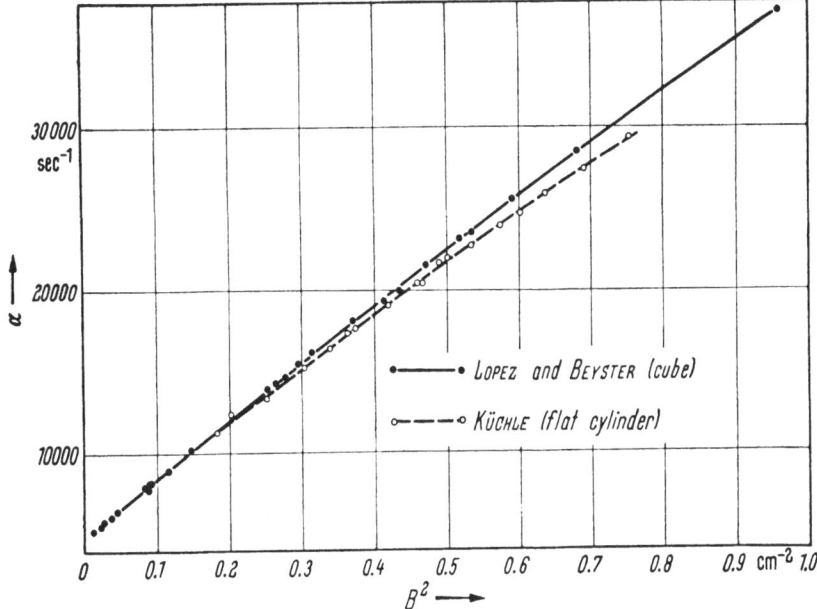

Figure 7.2

Decay Constant as a Function of Buckling
For Cubes (*32*) and Flat Cylinders (*78*) of Water
(From Beckurts and Wirtz (*2*))

Lopez and Beyster interpreted their results for an extrapolation length of 0.32 cm and used also the formulae of
Gelbard et al. (*63*) for infinite water slabs. Küchle used
an extrapolation length of 0.71 ℓ_{tr}, where ℓ_{tr} is averaged
over a Maxwellian at the temperature of the medium; however,
the discrepancy between the two curves is too large to be
caused by such a small difference in the choice of extrapolation length.

Curet et al. (*84*) studied the effect of sample shape in
measurements on paraffin. The various cylinders had equal
buckling, but the height-to-diameter ratios were 0.5, 1.0
and 1.5. The results can be seen in Figure 7.3. A constant
extrapolation length was used, but it is stated that the
extrapolation length corrections applied to the smaller
systems account for only a small fraction of the shape effect obtained. Unfortunately, the experimental details are

not available. It should be observed that in Curet's work
and in those discussed below, the curve for the flat systems
is above that for the square systems, in contradiction to
the results mentioned previously.

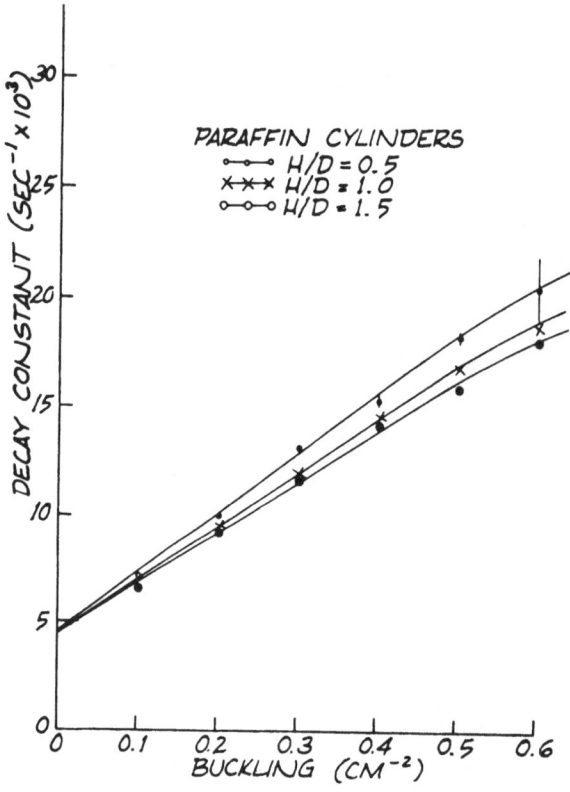

Figure 7.3

Decay Constant as a Function of Buckling
For Paraffin Cylinders of Various Height-to-Diameter Ratios
(According to Curet et al. (84))

Curves similar to those of Curet et al. were obtained by
Freed (85), as can be seen from Figure 7.4. The extrapolation distances calculated by Gelbard and Davies (64) were
used. For a buckling of 0.5 cm^{-2}, the measured decay constants were 22,771, 22,451, and 22,045 s^{-1} for the three
shapes, respectively, so the maximum deviation was 3.2%.
According to Freed, the values for the flat and tall systems
could be made to coincide with the symmetric system if extrapolation lengths of 0.294 and 0.377 cm were used instead of
the 0.338 cm actually employed. From Chapter V, it is clear
that such large changes in the extrapolation length due to the
shape of the system have no support in the presently available
theories.

The most recent work devoted to the shape dependence is
that of Nielsen (86). He performed his experiments at the
same laboratory as Freed and used, partly, the same equipment. One set of data from his experiments on water is given
in Figure 7.5. There is no indication of a shape effect here,
nor is any found in his other experiments. Nielsen also used
polyethylene cylinders and obtained similar results with them.
In the interpretation, he did not use an extrapolation length
but set the incoming neutron current (in the diffusion theory
approximation) equal to zero. This corresponds to an extrapolation length of $\frac{2}{3} \ell_{tr}$ for large systems, and includes a
slight buckling dependence. The effect of the aluminum walls
was not taken into account in the water measurements. Both
these circumstances will change slightly the obtained diffusion parameters; however, they cannot change the general conclusions drawn from the results. Nielsen claims that the
weighting factors used by Freed (85) in the least-squares
fitting routine caused a large spread in the decay constant
values when a point-stripping procedure was used. According
to Nielsen, this fact, together with Freed's method of selecting the "best" decay constant, might have resulted in the
apparent shape dependence. However, it seems to us that the
weighting factors were chosen according to well-established
principles. It is more important that no discussion is given
on how the weighting procedure was performed in the presence
of a background.

C. Conclusions

The above-described experiments are rather conflicting and give no clear evidence of a shape effect on the decay constant. In our opinion, the observed indications of

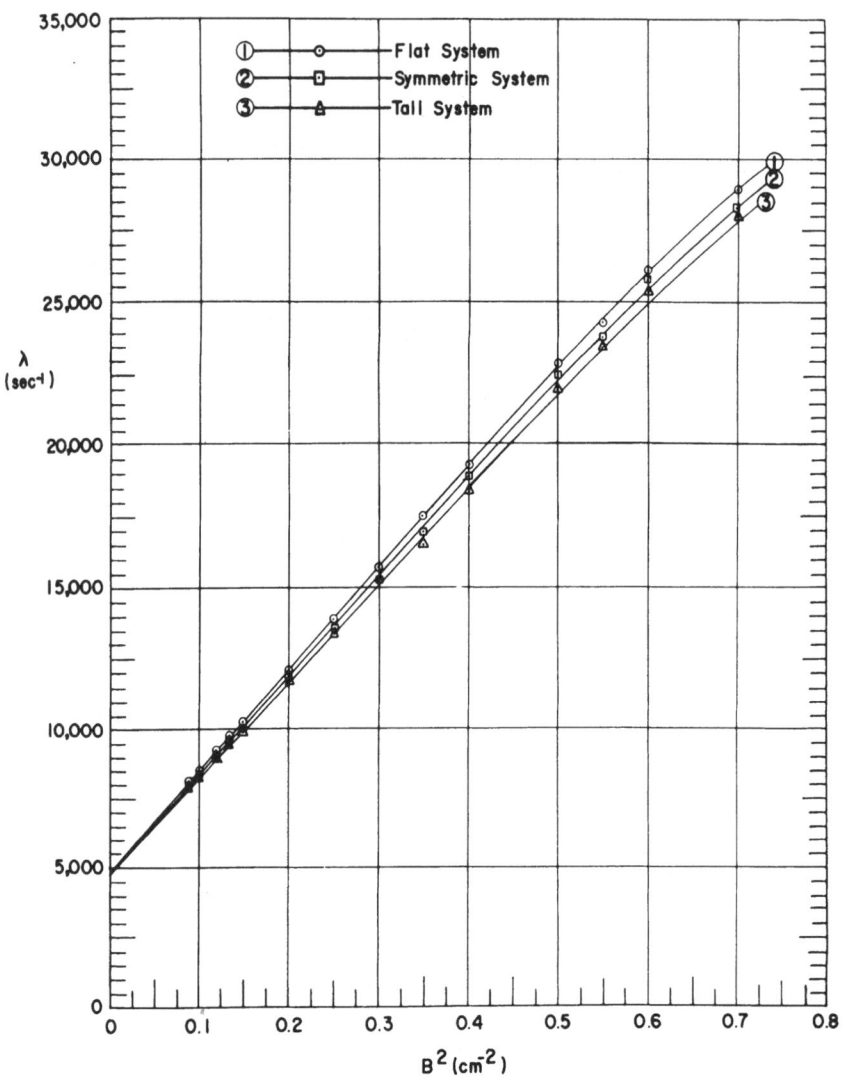

Figure 7.4

Decay Constant as a Function of Buckling
For Cylindrical Water Systems
(According to Freed (85))
Flat systems have a ratio of height to radius between 1 and
1.5; symmetric systems between 1.5 and 2.5; tall systems,
height to radius ratio larger than 2.5.

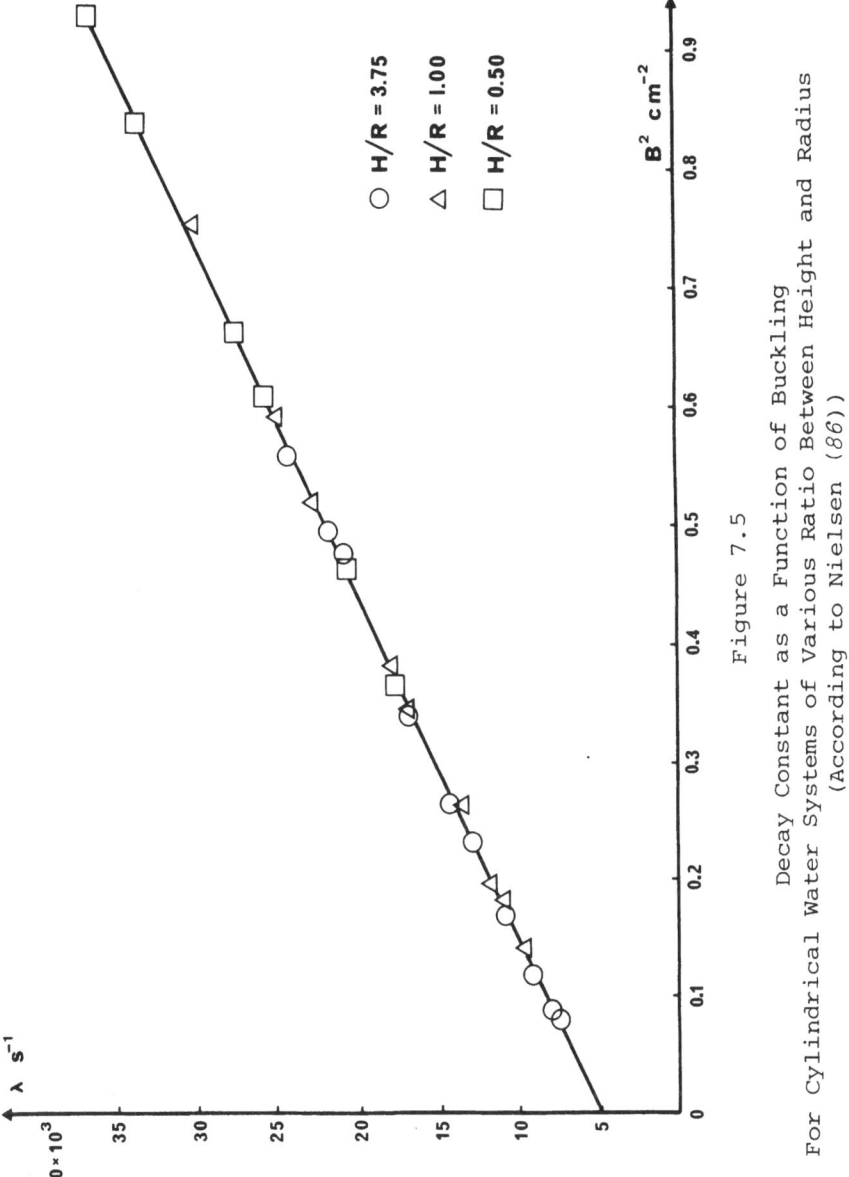

Figure 7.5

Decay Constant as a Function of Buckling

For Cylindrical Water Systems of Various Ratio Between Height and Radius
(According to Nielsen (86))

such an effect are caused by imperfections in the collection
and evaluation of experimental data. A fact that must be observed in this connection is that several other investigators
besides those mentioned here have used differently-shaped
moderator systems but have found no indication of a shape
effect. It is therefore believed that with the definition
of extrapolation length adopted here (see Figure 5.2), the
decay constant will be a unique function of buckling.

VIII. DISCUSSION AND CONCLUSIONS

Buckling and extrapolation length are quantities that
can be defined in different ways. When the aim is to interpret pulsed neutron experiments in homogeneous media, the
definitions should be made in such a way that the parameters
derived are unambiguous. The procedure adopted here, and
presented in Figure 5.2 fulfills this purpose. A consequence is that the extrapolation length loses the meaning
associated with its name, and it should be, therefore, regarded mainly as a correction quantity. In spite of this,
meaningful extrapolation length measurements can be performed, provided that the systems used are not too small and
that great care is used in the interpretation. Generally,
there is rather good agreement between recent measurements
and the theoretical predictions.

The extrapolation length is found to depend on the buckling of the system and on the variation of the scattering
cross-section with energy. The detailed scattering properties of the material and the shape of the system have to be
taken into account to only a smaller extent. If this is
done properly, the decay constant should be a unique function
of the buckling. The uncertainties in the calculated extrapolation lengths are believed to be small (about ± 5% or less)
and cannot explain the rather strong dependence of the decay
constant on the shape of the system which has been observed
in certain experiments. A review of the experiments shows
that the evidence for such a shape dependence is conflicting
and not very strong. It is believed, therefore, that the
shape effects observed in the decay constant are caused by
experimental difficulties.

The diagram in Figure 8.1 may serve as a guide for the
estimation of extrapolation lengths. Where no useful refer-

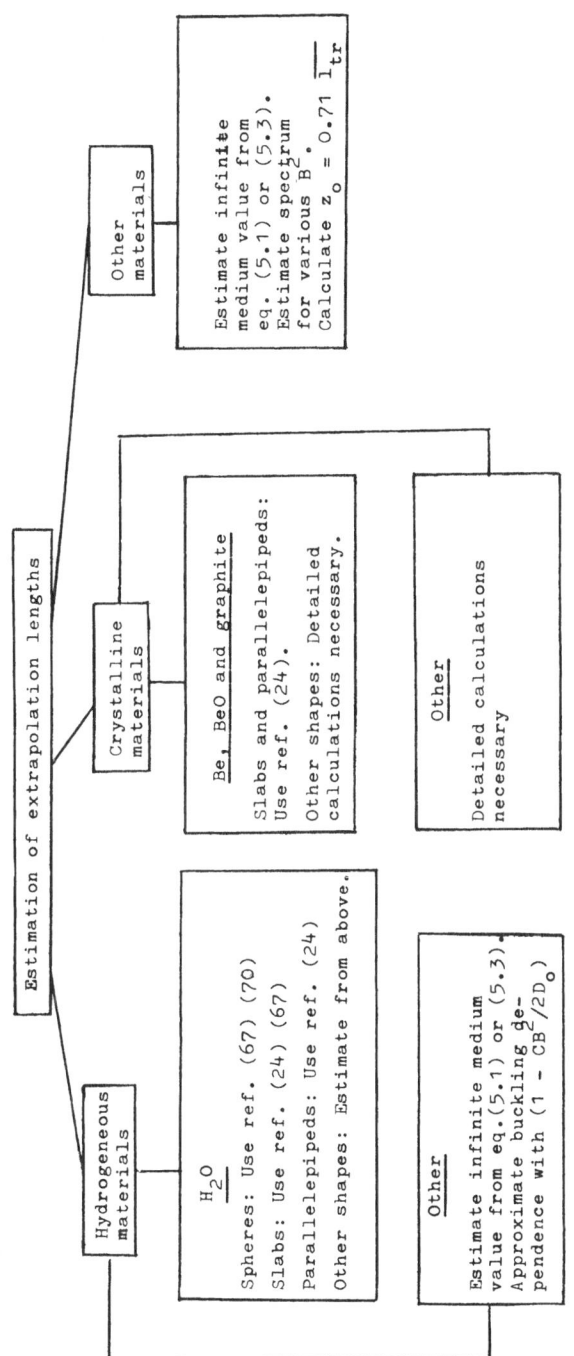

Figure 8.1

Directions for Estimation of Extrapolation Lengths

ences are available, the starting point is the transport mean free path obtained from measurement, as $l_{tr} = 3D_o / \bar{v}$. The simple buckling dependence suggested for hydrogenous media is based on the fact that the scattering cross-section of hydrogen in the thermal region varies approximately as the inverse neutron velocity. The variation of l_{tr} (averaged over the diffusion-cooled spectrum) can be approximated, therefore, by half the variation of the effective diffusion constant with buckling, $D = D_o - CB^2$. A similar procedure may be applicable to other materials where the scattering cross-section varies smoothly with energy; however, in Figure 8.1, the simple approach of Walker et al. (69) is suggested instead. It is believed that these estimates will give the extrapolation length to better than 10% for $B^2 \ell_{tr}^2 \leq 0.1$.

Only homogeneous systems have been treated in this review. The results should also apply to media with heterogeneities that are small compared to the mean free path. The assignment of meaningful extrapolation lengths to systems with large heterogeneities seems to be a very difficult, if not impossible task.

Even though there generally is rather good agreement between theory and experiment, there are some points that require further study:

1.   The presence of edges and corners lead to uncertainties that should be removed.

2.   Very little work seems to have been done on the effect of anisotropic scattering on extrapolation lengths in finite systems.

3.   The reasons for some of the deviations between theory and experiment may be that the scattering process in crystalline materials is not treated with sufficient accuracy.

NOTES ADDED IN PROOF

Section III D (p. 205)

Kerner et al. (87) have given some results for the extrapolation length of higher harmonic modes in slabs; however, they did not use the same definition as here. From the recent and detailed calculations by Hartman (88) it is possible to derive extrapolation lengths of several harmonic modes in slabs and spheres. The results are similar to those shown in Figure 3.5.

Section IV C (p. 210)

Recently, very detailed calculations have been done by Sanchez (89) for linearly anisotropic scattering in infinite cylinders. The extrapolation lengths derived from this work show the same tendency as in the isotropic case, i.e. they deviate from half-space value more than those for spheres and slabs. As expected, this means that Equation 4.5 is not a good approximation for cylinders.

IX. REFERENCES

1. von Dardel, G., and Sjöstrand, N.G. "Diffusion Measurements with Pulsed Neutron Sources," Progress in Nuclear Energy, Ser. I, Vol. 2, PP 183-221, Pergamon Press, London, 1958.

2. Beckurts, K.H., and Wirtz, K., Neutron Physics, Springer-Verlag, Berlin, 1964.

3. Beckurts, K.H., "A Review of Pulsed Neutron Experiments in Nonmultiplying Media, Proceedings IAEA Conference, Pulsed Neutron Research, Karlsruhe, Vol. I, PP 3-33,1965.

4. Cokinos, D.,"The Physics of Pulsed Neutrons," Advances in Nuclear Science and Technology, Vol. 3, PP 1-142, 1966.

5. Corngold, N., "Quasi-Exponential Decay of Neutron Fields," Advances in Nuclear Science and Technology, Vol. 8, PP 1-46, 1975.

6. Sjöstrand, N.G., Mednis, J., and Nilsson, T., "Geometric Buckling Measurements Using the Pulsed Neutron Source Method,"Arkiv Fysik 15, PP 471-482, 1959.

7. Sjöstrand, N.G., "On the Theory Underlying Diffusion Measurements with Pulsed Neutron Sources," Arkiv Fysik 15, PP 147-158, 1959.

8. Case, K.M., and Zweifel, P.F., Linear Transport Theory, Addison-Wesley, Reading, Massachusetts, 1967.

9.   Davison, B., Neutron Transport Theory, Clarendon Press,
     Oxford, 1957.

10. Cohen, E.R., "A Numerical Calculation of the Milne
     Problem Extrapolation Length," Nuclear Science Engin-
     eering, 42, PP 231-232, 1970.

11. Pomraning, G.C., and Lathrop, K.D., "The Extrapolated
     End Point for the Milne Problem," Nuclear Science
     Engineering 29, PP 305-308, 1967.

12. Case, K.M., de Hoffman, F., and Placzek, G., Introduction
     to the Theory of Neutron Diffusion, Los Alamos Scien-
     tific Laboratory, Los Alamos, 1953.

13. Marshak, R.E., "The Milne Problem for a Large Plane Slab
     With Constant Source and Anisotropic Scattering,"
     Physical Review 72, PP 47-50, 1947.

14. von Dardel, G., and Sjöstrand, N.G.,"Diffusion Parameters
     of Thermal Neutrons in Water," Physical Review 96, PP
     1245-1249, 1954.

15. Williams, M.M.R., "Space-Energy Separability in Pulsed
     Neutron Systems," Journal Nuclear Energy 17, PP 55-66,
     1963.

16. Carlvik, I., "Monoenergetic Critical Parameters and
     Decay Constants for Small Homogeneous Spheres and Thin
     Homogeneous Slabs," Nuclear Science Engineering 31,
     PP 295-303, 1968.

17. Kladnik, R., "A Variational Approach to the Time-
     Dependent Slab," Nukleonik 6, PP 147-153, 1964.

18. Judge, F.D., and Daitch, P.B., "Time-Dependent Flux in
     Small Assemblies," Nuclear Science Engineering 20,
     PP 428-435, 1964.

19. Goldschmidt, P., "The Extrapolation Length for Mono-
     energetic Neutrons in Slabs. Variational Calculations,"
     Nukleonik 10, PP 332-334, 1967.

20. Erdmann, R.C., and Shapiro, J.L., "The Extrapolation
     Distance for Monoenergetic Neutrons. An Exact Calcul-
     ation," Nukleonik 9, PP 302-303, 1967.

21. Ghatak, A.K., and Ahmed, F., "Asymptotic Solutions in Transport Theory. Part I - One Velocity with Isotropic Scattering," Journal Nuclear Energy 20, PP 939-951, 1966.

22. Mockel, A., "Invariant Imbedding and Polyenergetic Neutron Transport Theory. Part II - Numerical Results," Nuclear Science Engineering 29, PP. 51-57, 1967.

23. Dorning, J.,"Thermal-Neutron Decay in Small Systems," Nuclear Science Engineering 33, PP 65-80, 1968.

24. Wood, J., and Williams, M.M.R., "Extrapolation Distances in Thermal Neutron Systems. I. The Pulsed Source Experiment," Journal Nuclear Energy 25, PP 101-112, 1971.

25. Sjöstrand, N.G., and Dahl, E.B., "Accurate Extrapolation Lengths for Pulsed Monoenergetic Neutrons in Slabs and Spheres," Nuclear Science Engineering 57, PP 84-86,1975.

26. Kaper, H.G., Lindeman, A.J., and Leaf, G.K., "Benchmark Values for the Slab and Sphere Criticality Problem in One-Group Neutron Transport Theory," Nuclear Science Engineering 54, PP 94-99, 1974.

27. Westfall, R.M., Metcalf, D.R., "Singular Eigenfunction Solution of the Monoenergetic Neutron Transport Equation for Finite Radially Reflected Critical Cylinders," Nuclear Science Engineering 52, PP 1-11, 1973.

28. Hembd, H., "The Integral Transform Method for Neutron Transport Problems," Nuclear Science Engineering 40, PP 224-238, 1970.

29. Kavenoky, A., "The C_N Method in Cylindrical Geometry and One-Velocity Theory," Proceedings Conference on Computational Methods in Nuclear Engineering, CONF-750413, Vol. II, PP III-67 to III-83, 1975.

30. Sahni, D.C., "The Integral Transform Method for Two- and Three-Dimensional Neutron-Transport Problems, Proceedings IAEA Seminar Numerical Reactor Calculations, Wien, PP 197-209, 1972.

31.   Horie, J., and Nishihara, H., "Numerical Solution
       to Critical Problem of Finite Cylindrical Reactors
       by Variational Method," Journal Nuclear Science
       Technology, 11, PP 359-368, 1974.

32.   Lopez, W.M., and Beyster, J.R., "Measurement of
       Neutron Diffusion Parameters in Water by the Pulsed
       Neutron Method," Nuclear Science Engineering 12,
       PP 190-202, 1962.

33.   Davison, B., Milne Problem in a Multiplying Medium
       with a Linearly Anisotropic Scattering, National
       Research Council of Canada, Report CRT-358, 1946.

34.   Lois, L., and Goldstein, H., "Extrapolation Distance
       for Anisotropic Scattering," Nuclear Science Engineer-
       ing 29, PP 148-149, 1967.

35.   Su, S.F., Extrapolation Distance Calculations for
       Transport with Anisotropic Scattering, Thesis,
       University of Washington, 1970.

36.   Su, S.F., and McCormick, N.J., "Extrapolation Distance
       Calculations,"Journal Nuclear Energy 25, PP 657-658,
       1971.

37.   Thielheim, K.O., and Claussen, K., "Comments on Half-
       Space Problems with Linear Anisotropy of Scattering,"
       Kernenergie 16, PP 321-328, 1973.

38.   Thielheim, K.O., Private Communication, 1974.

39.   Mark, J.C., Milne's Problem for Anisotropic Scattering,
       Montreal Laboratory Report MT-26, 1943.

40.   Shure, F., and Natelson, M., "Anisotropic Scattering
       in Half-Space Transport Problems," Annals of Physics
       26, PP 274-291, 1964.

41.   Vanmassenhove, F.R., "Exact Analytical Solution and
       Numerical Treatment of the Milne Problem with Absor-
       ption and Anisotropic Scattering," Physica 42, PP
       179-204, 1969.

42.   van de Hulst, H.C., "Asymptotic Fitting, A Method for Solving Anisotropic Transfer Problems in Thick Layers," Journal Computational Physics 3, PP 291-306, 1968.

43.   Claussen, K., and Thielheim, K.O., "Comments on Half-Space Problems with Quadratic Anisotropy of Scattering," Kernenergie 17, PP 253-258, 1974.

44.   Carlvik, I., Critical Parameters for Homogeneous Slabs and Spheres in the Constant Cross-Section Approximation and with Linearly Anisotropic Scattering, AB Atomenergi Report RFR-228, 1963.

45.   Lathrop, K.D., and Leonard, A., "Comparisons of Exact and S_N Solutions of the Monoenergetic Critical Equation with Anisotropic Scattering," Nuclear Science Engineering 22, PP 115-118, 1965.

46.   Sjöstrand, N.G., "Eigenvalues in Time-Dependent Mono-energetic Neutron Systems with Anisotropic Scattering," Journal Nuclear Science Technology 12, PP 256-257, 1975.

47.   Sjöstrand, N.G., "Complex Eigenvalues of Monoenergetic Neutron Transport Equation with Anisotropic Scattering," Journal Nuclear Science Technology 13, PP 81-84, 1976.

48.   Williams, M.M.R., The Slowing Down and Thermalization of Neutrons, North-Holland Publishing Company, Amsterdam, 1966.

49.   Rönnberg, G., A Monte Carlo Investigation of Quasi-Exponential Decay of Pulsed Neutron Fields in Water, Department of Reactor Physics, Chalmers University of Technology, Report CTH-RF-25, 1973.

50.   Ahmed, F., and Ghatak, A.K., "On the Space-Energy Separability of the Fundamental Mode of the Neutron Transport Operator with Isotropic Scattering," Nuclear Science Engineering 33, PP 106-118, 1968.

51.   Beynon, T.D., "On the Concept of Energy-Dependent Buckling in Multi-group Diffusion Theory," Journal Nuclear Energy, 25, PP 503-511, 1971.

52.   Ahmed, F., and Mohan, R., "Effect of Shape of the
      Assembly on the Decay Constant in Pulsed-Neutron
      Experiments," Nuclear Science Engineering 51, PP
      335-337, 1973.

53.   Nelkin, M., "Milne's Problem for a Velocity-Dependent
      Mean Free Path," Nuclear Science Engineering 7,
      PP 552-553, 1960.

54.   Boffi, V.C., Molinari, V.G., and Parks, D.E., "On the
      Decay of Neutrons in a Subcritical Assembly," Journal
      Nuclear Energy 16, PP 395-403, 1962.

55.   Kladnik, R., and Kuščer, I., "Milne's Problem for
      Thermal Neutrons in a Nonabsorbing Medium," Nuclear
      Science Engineering 11, PP 116-120, 1961.

56.   Kladnik, R.,"Milne's Problem for Thermal Neutrons,"
      Proceedings Brookhaven Conference on Neutron Thermal-
      ization, BNL 719, Vol. IV, PP 1211-1231, 1962.

57.   Weiss, Z., "The $P_N L_J$ Approximation of Neutron Thermal-
      ization in Heavy Gas Moderator," Nukleonika 6,
      PP 703-716, 1961.

58.   Williams, M.M.R., "The Energy-Dependent Milne Problem
      with a Simple Scattering Kernel," Nuclear Science
      Engineering 18, PP 260-270, 1964.

59.   Williams, M.M.R., "A Solution of the Thermal Neutron
      Milne Problem by Perturbation Theory," Nuclear Science
      Engineering 19, PP 353-358, 1964.

60.   Kladnik, R., "The Asymptotic Angular-Dependent Leakage
      Spectrum of Thermal Neutrons," Nuclear Science Engin-
      eering 23, PP 291-298, 1965.

61.   Arkuszewski, J., "Milne Problem for Thermal Neutrons
      with Absorption," Nuclear Science Engineering 27,
      PP 104-119, 1967.

62.   Eisenhauer, C., "Some Results on the Energy-Dependent
      Milne Problem for Thermal Neutrons and Light Gases,"
      Nuclear Science Engineering 19, PP 95-101, 1964.

63.   Gelbard, E., Davies, J., and Pearson, J., Space-
      Energy Separability in the Thermal Neutron Group,
      Westinghouse Electric Corporation Report WAPD-T-1065,
      1959 (also, Trans. American Nuclear Soc. 2:2, Page 95,
      1959).

64.   Gelbard, E., and Davies, J., "The Behavior of Extra-
      polation Distances in Die-Away Experiments," Nuclear
      Science Engineering 13, PP 237-244, 1962.

65.   Elkert, J.,"Determination of the Diffusion Parameters
      for Thermal Neutrons in Water Using the Pulsed Method
      and Spherical Geometries,"Department of Reactor Physics,
      Chalmers University of Technology Report CTH-RF-12,
      1967 (also, Nukleonik 11, PP 159-162, 1968).

66.   Emon, D., A Study of Extrapolation Lengths in Thin
      Slabs in High Order P_N Approximations, Thesis,
      Rensselaer Polytechnic Institute, 1965.

67.   Schmidt, E., and Gelbard, E.M., Improved P_3 Vacuum
      Boundary Conditions and Their Use in Pulse Calcul-
      ations, Westinghouse Electric Corporation Report
      WAPD-T-1788, 1965.

68.   Vértes, P., "Some Problems Concerning the Theory of
      Pulsed Neutron Experiments," Nuclear Science Engin-
      eering 16, PP 363-368, 1963.

69.   Walker, J., Brown, J.B.C., and Wood, J., "Extrapolation
      Distances for Pulsed Neutron Experiments," Proceedings
      IAEA Conference Pulsed Neutron Research, Karlsruhe,
      Vol. I, PP 49-63, 1965.

70.   Dorning, J., "Extrapolation Distances and Diffusion
      Parameters via Pulsed-Neutron Analysis," Nuclear
      Science Engineering 41, PP 22-28, 1970.

71.   Wood, J., and Williams, M.M.R., "The Validity of the
      Buckling Concept and the Importance of Spatial Tran-
      sients in the Pulsed Neutron Experiment," Journal
      Nuclear Energy 21, PP 113-130, 1967.

72.   Williams, M.M.R., "The Influence of Specular and
       Diffuse Reflection on the Extrapolation Distance,"
       Atomkernenergie 25, PP 19-23, 1975.

73.   Campbell, E.C., and Stelson, P.H., Measurement of
       Thermal-Neutron Relaxation Times in Subcritical
       Systems, Oak Ridge National Laboratory Report ORNL-2204,
       PP 34-36, 1957.

74.   DeJuren, J.A., Stooksberry, R., and Carrol, E.E.,
       "Measurements of Extrapolation Lengths in Pulsed Water
       Systems," Proceedings Brookhaven Conference on Neutron
       Thermalization, BNL 719, Vol. III, PP 895-899, 1962.

75.   Gaerttner, E.R., Fullwood, R.R., Menzel, J.H., and
       Toth, G.P., "Measurements of the Spatially-Dependent
       Asymptotic Spectra in Water and Polyethylene,"
       Proceedings IAEA Conference Pulsed Neutron Research,
       Karlsruhe, Vol. I, PP 501-515, 1965.

76.   Shalev, S., Shani, G., Fishelson, Z, and Ronen, Y.,
       "Measurement of the Thermal-Neutron Extrapolation
       Length in a Pulsed Water System," Nuclear Science
       Engineering 35, PP 259-266, 1969.

77.   Bowen, R.A., and Scott, M.C., "Neutron Diffusion
       Measurements in Water from 18°C to 280°C," British
       Journal Applied Physics (J. Phys. D), Ser. 2, Vol. 2,
       PP 401-411, 1969.

78.   Küchle, M., "Messung der Temperaturabhängigkeit der
       Neutronen-Diffusion in Wasser und Diphyl mit der
       Impulsmetode," Nukleonik 2, PP 131-139, 1960.

79.   Demanins, F., Rado, V., and Vinci, F., Determinazione
       dei Parametri di Diffusione dei Mezzi Moderanti con il
       Metodo Della Sorgente di Neutroni Pulsata, Dowtherm A,
       Comitato Nazionale Energia Nucleare Report RT/FI(63)22,
       1963.

80.   Davis, S.K., DeJuren, J.A., and Reier, M., "Pulsed
       Decay and Extrapolation-Length Measurements in Graphite,"
       Nuclear Science Engineering 23, PP 74-81, 1965.

81. Ritchie, A.I.M., "The Measurement of Extrapolation
     Distances in Pulsed BeO Assemblies," Journal Nuclear
     Energy 22, PP 717-734, 1968.

82. Ritchie, A.I.M., and Moo, S.P., "Definition and Use
     of Buckling," Journal Nuclear Science Technology 11,
     PP 535-544, 1974.

83. Hall, R.S., Scott, S.A., and Walker, J., "Neutron
     Diffusion in Small Volumes of Water: Pulsed Source
     Measurements," Proceedings Physical Society 79,
     PP 257-263, 1962.

84. Curet, H.D., Hwu, Y.P., and Robeson, A., "Effect of
     Sample Shape on Decay-Constant Measurements in Paraf-
     fin," Transactions American Nuclear Society, 7, PP 72-
     73, 1964.

85. Freed, D., "The Effect of Sample Geometry on Decay
     Constant Measurements in Water Using the Pulsed Neu-
     tron Technique," Thesis, University of New Mexico,
     1965, (also, Transactions American Nuclear Society, 8,
     PP 433-434, 1965).

86. Nielsen, K., "Data Evaluation and Shape Dependence in
     the Pulsed Neutron Technique," Thesis, University of
     New Mexico, 1968.

87. Kerner, I.O., Kiesewetter, H., and von Weber, S.,
     "Numerische Resultate zu den Eigenwerten und Eigen-
     funktionen der Neutronentransportgleichung für eine
     Platte," Kernenergie, 10, PP 299-306, 1967.

88. Hartman, J., "A Method to Solve the Integral Transport
     Equation Employing a Spatial Legendre Expansion,"
     Report EUR 5339e, 1975.

89. Sanchez, R., "Généralisation de la methode d'Asaoka
     pour le traitement d'une loi de choc linéairement
     anisotrope: données de référence en géometrie cylin-
     drique," Report CEA-N-1831, 1975.

ACKNOWLEDGEMENTS

The author wishes to express his sincere thanks to the many persons who have contributed to the present review through comments and criticism and in other ways. Discussions with Professor M.M.R. Williams were especially helpful. Dr. John McDonald kindly put his linguistic expertise at the author's disposal.

The authors and copyright owners of references 70, 24, 81, 83, 2, and 84 are acknowledged for permission to reproduce Figures 5.4, 5.5, 6.1, 7.1, 7.2, and 7.3.

THERMODYNAMIC DEVELOPMENTS

R. V. Hesketh

Berkeley Nuclear Laboratories
Berkeley, Gloucestershire, United Kingdom

Lux et omnes soni vim transportant

I. INTRODUCTION

To take a step forward, one must first make sure one is
pointing in the right direction. In this paper, I shall first
of all turn around while standing on the spot, and then take
one or two tentative steps. I shall by no means exhaust the
steps that can be taken. To some, my turning around may appear
a retrograde step and they will shake their heads accordingly.
This turning around on the spot has occupied me for six years
now; I feel it timely to commit the essence to paper.

I make no apology for the simplicity of words or equations
that I use. It seems better to me to begin with comprehen-
sible ideas and with quantities which show the correct asymptotic behavior, and to develop these **slowly,** rather than to
attempt the maximum sophistication *ab initio*. Too early an
attempt can lead **precipitously** into sophistry. Examples are
given in the course of this paper.

Need we distinguish thermodynamics by the adjective
"irreversible"? If we examine "equilibrium thermodynamics",
we find it consists of quantities that are forever unobservable. It represents an **asymptotic condition** to which experiments <u>tend</u> but never reach. In <u>The Logic of Modern Physics</u>,
Bridgman writes "the equilibrium concept of temperature is
strictly never exactly applicable ... (in) a great many different sorts of experiment, by methods of *asymptotic approximation* (my italics) ... we establish the existence of various
sorts of physical constants such as constants of **absorption**
and emission and reflection and scattering and fluorescence

and thermal conductivity" (1). To this list we may add other recognized constants, electrical conductivity, ordinary diffusion coefficient, thermoelectric power, etc. (2). Discussion soon reveals that there is no dividing line (3); all physical parameters are the product of non-equilibrium experiments. To observe even the mass of a body, we must accelerate it. To observe the charge on the electron, we must apply a (non-equilibrium) electric field. To measure the Johnson noise of a resistor, we must take power from it. Consider any experiment with which you are familiar, and ask at what point its thermodynamic equilibrium is incomplete. I can discover no "equilibrium experiment". Hence, my first proposal, that the limitations of irreversible thermodynamics, whatever they may be, apply to every experiment ever done. Thermodynamics, if it deals with observables, is *ipso facto* irreversible, and the adjective is redundant.

The belief that there are two distinguishable classes of experiment, equilibrium experiment and non-equilibrium experiment, is widespread, both in the literature and in the standard texts; e.g., the Preface to (4).

Equilibrium remains a useful myth, useful because it is linked to the observable world by the philosophic device of the mathematical limit. It usefully provides parameters for the linear regions of physics. An example is Ohm's Law (3). In an Ohmic conductor, the conductivity has the same assignable numerical value in zero field as it has in a nonzero field. We may generalize to other linear systems. A pertinent example is thermoelectricity (see Section VI).

II.   LINEAR THERMODYNAMICS

Figure 1 shows a graph of y, an arbitrary, cuspless, continuous function of x. In general, an arbitrary function does not pass through the origin, but makes some intercept, y_0, on the y axis. Leaving the y scale unaltered, we may expand the x scale, obtaining the shallower graph of Figure 2. By a sufficiently large expansion, we may obtain a graph of negligible slope; that is to say, if y attains the value, y_1, at the edge of the diagram, then

$$|y_1 - y_0| << y_0. \tag{1}$$

On this inequality rest the thermodynamics of Thomson (5),
Onsager (6), De Groot (2), Prigogine (4) and Zubarev (7).
Within this group the words "pseudo-thermostatic" and "linear
irreversible" are sometimes used as distinguishing labels, but
the words are synonymous, the assumption of all these authors
is the same, that y_0 is nonzero, and that x is restricted to
a range of values in which inequality (1) holds.

The phrase "the linear region of thermodynamics" does
not specify a permissible range of x values other than by
inequality (1). To obtain numerical values, we must call
upon physics, as distinct from formal mathematics. The permissible range of x may be large or small (8), and does not
necessarily correlate with current usage of the word "irreversible".

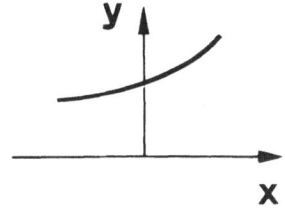

Figure 1

An Arbitrary Continuous Function of y, of x

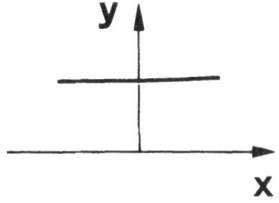

Figure 2

The Same Function as in Figure 1, but with the x Scale
Expanded.

One might perhaps distinguish the flow of electricity through a copper wire as one of the more common examples of linear thermodynamics. Correspondingly, the flow of electricity through a diode is one of the more common examples of nonlinear thermodynamics. Both are irreversible. The flow of electricity is accompanied by Joule heating. Entropy is generated.

III. NONLINEAR THERMODYNAMICS

There is no mathematical restriction to prevent the function y passing through the origin; $y = a\,x^n$ does so. A function that passes through the origin can never satisfy inequality (1), no matter how much we (finitely) expand the x axis. For such a function there is no linear region.

Do we encounter such functions in thermodynamics? A function of odd power is excluded by the requirement that the physical system be stable. The fluxes of physics are proportional to xy (in the present notation) and thus to x^{n+1}. If n is odd, the flux is independent of the sign of x, and drives only in one direction. The thermodynamic fluctuations ever present in all systems thus produce a nonzero flux, even in the absence of external perturbations, and the system collapses to a new stable state.

Even powers of n allow the system to be stable; the flux is an odd power of the perturbation, x, and the rate of entropy generation, x^2y, is an even power, as we require. To the best of my belief, such physical systems have a vanishingly small probability of existence, because we need stochastic processes that give n = 2, 4, etc., *precisely* . The response of a system for which n = 1.99 may be regarded as a superposition of two responses, for one of which n = 1.0 and for the other of which n = 2.0. The presence of the odd value makes the system unstable.

We may therefore exclude the function $y = a\,x^n$ from present consideration. Nonlinearity in physical systems occurs only by the failure of Expression (1), at values of x such that

$$\left| y_1 - y_0 \right| \sim y_0. \tag{2}$$

In such a system, the flux (of electricity, of matter, of heat) is still proportional to xy, but no longer simply to x. A common example is the diode.

IV. THE CONTINUITY OF y

I have introduced y as an arbitrary but continuous mathematical function, and then gradually identified it with the intrinsic parameters - the observable parameters - of physical systems. I have in mind such parameters as conductivity, specific heat, viscosity, mass, electric charge.

Is it proper to treat such observables as continuous functions of the perturbations which may be applied to a physical system? To this question the profession at large does not return a unanimous answer $(3,9)$. The spectrum extends from "obviously yes" to "emphatically no". In thermodynamics, especially (3), the proposal of continuity can be permissibly regarded as a new development.

Consider an everyday, nonequilibrium system with a linear response, a copper wire carrying an electric current. It is a thermodynamic system: $x = (-\nabla V)/T$ and $y = \sigma T$. If we allow that the response is linear, then mathematics alone leads us to expect continuity in σ. We may write

$$\sigma_{\mu\nu} = \text{Lim}_{(\nabla V) \to 0} \left(\underset{\sim}{J}_{\mu} \Big/ (-\nabla V)_{\nu} \right). \tag{3}$$

Mathematically, this limit is well-defined, even in anisotropic materials. Alternatively, we may consider the physics of the conduction electrons; e.g., $(10,11)$. Again, we expect continuity. We may write

$$\sigma = n_{o} \, e^2 \, \tau/m^{*}. \tag{4}$$

$$= n_{o} \, e^2 \, \lambda/v_{F}m^{*}, \tag{5}$$

Do the quantities on the righthand side of Equations (4) and (5) have assignable numerical values, in equilibrium? Our concept of metals allows that in the unperturbed state, electrons travel with a Fermi velocity and collide with phonons. Thus, λ and τ have numerical values, in equilibrium. Thirdly, we might consider the thermodynamics of the conduction

electrons. Fluctuations give rise to Johnson noise. We may
write:

$$\sigma = 4kT \quad \Delta\nu/\overline{V^2}. \tag{6}$$

Equation (6) applies to conditions as close to equilibrium
as we may ever get. If we say that there is a discontinuity
in σ, our justification must be that we are free to postulate
it, not that we can observe it. As emphasized in Section I,
thermodynamics deals with observables. Hence, the postulate
of a discontinuity in σ lies outside the present terms of
reference.

I choose Ohmic conductivity because of its everyday familiarity, and because it highlights the issue of continuity
(3). Without further discussion, I shall assume that thermal
conductivity is likewise continuous, although the contrary is
often vigorously stated. I shall draw upon the continuity of
other intrinsic parameters, enthalpy, thermal energy, monopole electric charge, in subsequent sections.

It remains to ask if there are any physical parameters
which are *discontinuous* as the perturbation of the system
changes. A colleague suggests the magnetization of an Ising
ferromagnet at zero temperature:

$$M = M_0 \, H/\left|H\right|. \tag{7}$$

My own view is that the restriction "at zero temperature"
prevents *M* being classified as an observable. At all observable temperatures, *M*, in real systems, is continuous.

The question of continuity is central to thermodynamics.
The concept of continuity runs (continuously, I hope) through
the length of this paper.

In Figure 1 we have a curve that obeys two conditions of
continuity: the curve itself is continuous, and it is cuspless; *y* is continuous and *dy/dx* is continuous. To the best
of my belief, the intrinisc parameters of real physical systems (and I distinguish the observable physical parameters
from the mathematical models which may be used to represent
them) obey both conditions.

V.   CONTINUITY IN THE LINEAR REGION

Many physicists use linear perturbation theory. Tacitly, implicitly, they assume *continuity* between the unperturbed state and the perturbed state. If in Figure 2 we take y to be continuous, we do no more than our colleagues. Explicitly, they assume a linear response. If in Figure 2 we write

$$y = y_0, \tag{8}$$

we again do no more than our colleagues. Only in so far as a physical system has a nonlinear response is the difference between y and y_0 significant. (Experiment can test the validity of Equation (8).)

If we follow the mathematical texts (*2, 4, 7*) and *assume* a linear response, then

$$y \equiv y_0. \tag{9}$$

In a system that is linear by *definition*, the intrinsic physical parameters have but one assignable value; Equation (9) holds. To paraphrase: the phenomenological coefficients of linear irreversible thermodynamics are *all, without exception*, equilibrium parameters.

There is the obvious corollary: no *new* intrinsic parameters come into existence (i.e., take nonzero numerical values) as a physical system goes from equilibrium to disequilibrium.

The formal mathematical statement, Equation (9), bears upon physical theories of solids and gases; e.g., (*12,13*).

VI.   EXAMPLES OF EQUILIBRIUM PARAMETERS

A.   The Screened Monopole Charge

In a metal, one may have a solute ion whose valence differs from that of the solvent ions. All such impurity ions are screened by the conduction electrons. The screening length is a few atomic spacings and the overall screening is complete (*11, 14, 15*); i.e., there is no residual monopole charge over a macroscopic volume (a macroscopic

volume is one measurable by *statistical* techniques, by looking
at an assembly of many solute atoms and taking an average by
classical techniques). The completeness of the screening is
specified by the Friedel sum rule; e.g., (*11*). If we choose
a macroscopic volume of small physical size; e.g., if we embed
our assembly of solute ions in a very thin film, the screening remains complete.

The completeness of the screening in equilibrium is
commonly agreed.

The monopole charge of the solute is an intrinsic physical parameter of the thermodynamic system. In the linear
region of disequilibrium, therefore, by Equation (9), the
screening remains complete, and the monopole charge zero.
This is a general result (*16*). It bears on the present con-
troversy (*13, 17, 18, 19, 20, 21*) concerning the magnitude
of the "direct force" in electromigration. The present
recipe is simple: the direct force is always identically
zero.

B.   The Enthalpy of a Solute in a Solid

A defect in a solid (e.g., a lattice vacancy) has an
enthalpy of formation, w_f. It appears in the chemical potential and in the overall free energy of the system. It
is clearly an equilibrium quantity. On the other hand, the
enthalpy of migration, w_m, does not appear explicitly in the
chemical potential, nor in the overall free energy. Does
this mean that the conclusion of Section V, that no new physical parameters come into existence in disequilibrium, is
false? I think not; w_m is the value which the local thermal
energy must attain before the defect can jump from one site
to another. At classical temperatures, w_m is a fluctuation
in $3kT$, and this already appears in the equilibrium functions.

If we go to disequilibrium in a *linear system*, w_f retains
its equilibrium value. Yet in thermomigration in solids, it
is presently the custom to replace it by zero (or even $-w_f$);
e.g., (*22,23*). Equally, it is supposed that w_m ceases to be
a fluctuation in the thermal energy and becomes an energy
which the defect carries with it.

Inspection shows that the error associated with supposing
the defect to carry (to "transport") an enthalpy w_m, or

$(w_m - w_f)$, is to take the temperature to be *discontinuous in real space*. The danger was foreseen by Shockley (*24*). Here is a further aspect of continuity, continuity of temperature throughout real space. This concept occurs again in Section XI.

C. The Electronic Entropy of a Solute in a Solid

Suppose again that we have a lattice vacancy in a metal. To create it we remove an uncharged entity, both the ion and the conduction electrons that screen it. The conduction electrons that we remove have a thermal energy, γT^2. Since we remove this energy, the vacancy has a vibrational electronic entropy of formation (c.f., Equation (10) of (*8*)):

$$s_{v,e} \;=\; \gamma T. \tag{10}$$

Employing the thermodynamic relation:

$$\left(\frac{\delta w_f}{\delta T} \right)_p \;=\; T \left(\frac{\delta s_v}{\delta T} \right)_p \tag{11}$$

we obtain (or return to) the conclusion that the formation enthalpy, w_f, contains a contribution γT .

In metals, self-diffusion experiments, thermomigration, electromigration and vacancy concentration experiments are carried out at temperatures well above the Debye temperature (lithium comes nearest to being an exception). Accordingly, the lattice specific heat is $3k$ per atom, and accordingly, the phononic vibrational entropy of formation of a lattice vacancy is independent of temperature (*8*). For the lattice (as distinct from the electron gas) the righthand side of Equation (11) is zero. Thus, only the electronic entropy contributes to the temperature dependence of w_f:

$$w_f = w_{f,0} \;+\; \gamma T^2. \tag{12}$$

The suffix 0 **signifies** a quantity independent of temperature.

Analyses of self-diffusion data (*25, 26*) should contain the last term of Equation (12), for it produces a curvature in the Arrhenius plot.

If we draw a tangent to the low temperature region of such a diffusion plot, we may define (on the ordinate scale,

lnD_s) a difference between the curve and the tangent, at the melting point. For silver for example (27):

$$0.1 \leq \ln (D_{s,\ curve}/D_{s,\ tangent}) \leq 0.37 \qquad (13)$$

and $\gamma T/k$ = 0.1   (14)

i.e., the electronic thermal energy accounts for some, perhaps all, of the observed curvature. It becomes evident that we should not attribute *all* the curvature to the existence of di-vacancies.

The same $\gamma T/k$ term appears in the curvature of Arrhenius plots of vacancy concentration.

I have refrained from associating a γT^2 term with w_m (which also appears in the Arrhenius slope for self-diffusion) on the ground that the oscillations in the electron gas are several orders of magnitude more rapid than the oscillations of the lattice, and that therefore, the fluctuations in the electron gas will not contribute appreciably to vacancy migration. If I am correct in this, the curvature of self-diffusion plots should be identical with the curvature of vacancy concentration plots.

In the transition metals, $\gamma T/k$ is large. In cobalt at 1600° K,

$$\gamma T/k = 0.96 \qquad\qquad\qquad (15)$$

i.e., D_s is increased a factor of 2.5 above the straight line value. In niobium, the factor is nine. Such large factors are a significant contribution to the large curvature of Arrhenius plots for body-centered cubic metals (28).

D.   The Electrical Conductivity of a Solute in a Solid

To my colleagues who say that "electrical conductivity is not a thermodynamic coordinate and cannot exist in equilibrium" (9), I must return some answer. I find it easiest to think yet again of a lattice vacancy in a metal. To a first approximation, a lattice vacancy is an empty hole, a hole both in the lattice and in the electron gas (see subsection VI.A); the electrons that screened the ion are gone. I differentiate Ohm's Law and find:

$$\frac{\Delta\sigma}{\sigma} = - \frac{\text{empty volume}}{\text{original volume}} . \qquad (16)$$

If the lattice vacancy perturbs the electron gas to the extent that conduction electrons are scattered not from one lattice site, but from N sites, the resistivity of one atomic percent of lattice vacancies (a customary unit) is:

$$R = N/100 \; \sigma. \qquad (17)$$

In the noble metals, **experiment** (29) gives $N \sim 100$. Two things are evident from this concept:

1. The perturbed volume, and hence, the resistivity, exists in equilibrium.

2. There is no *a priori* reason that the resistivity of a lattice vacancy should be independent of temperature, the relative variation of N and σ is not prescribed; it depends on the particular metal.

The resistivity appears in the equilibrium-free energy, as the volume of a perturbation. The resistivity does not tell us what is inside this volume. A volume is always positive. Accordingly, lattice defects, at least in small concentration, always *increase* the resistivity, in accordance with Equations (16) and (17). No negative residual resistivities are listed (29).

A lattice vacancy empties electrons from the perturbed volume. Conversely, antimony in silver adds electrons to the perturbed volume. The number of conduction electrons can be either decreased or increased; thus, the energy perturbation that appears in the chemical potential of the defect can be either positive or negative. The chemical potential contains an energy that is the electrical **resistivity** volume multiplied by an energy density. Thus, it is possible to have a lattice defect that has a nonzero resistance but which gives a zero contribution to the chemical potential; i.e., the average energy density within the perturbed volume is the same as in the **solvent**; there are wrinkles in the energy density, but their sum is zero. On the other hand, it is *impossible* to have a lattice defect whose residual electrical resistance is zero but which nonetheless has a nonzero electronic entropy.

I find it more difficult to form an image of the electrical conductivity of the plain metal (with no defect). I can offer no more than Equations (4) and (5). The mean free path, λ, is a measure of anharmonicity. It is therefore related to derivatives of w and s, analogously to thermal expansion.

The game of relating equilibrium parameters one to another is entertaining. It also bears on the evaluation of experiment (30).

E.   Thermopower

In linear systems, the thermoelectric coefficients obey the standard rules (2, 4, 7) and are therefore, quite rigorously, equilibrium parameters. The numerical values to be used in linear disequilibrium are precisely the equilibrium values, nothing more.

Thomson, Lord Kelvin, uses equilibrium values (5). Is he wrong to do so? De Groot writes (2), "it is remarkable that the derived relations were completely confirmed by experiments on thermoelectricity". Undergraduate textbooks are critical (31): "although Thomson's derivation was not rigorous, since classical thermodynamics is not directly applicable to this problem (the techniques of *irreversible thermodynamics* are necessary), his answers are correct".

I can do no more than leave the reader with the following questions:

1.   If Thomson's "pseudo-thermostatic" methods give different answers to those given by "linear irreversible thermodynamics", in which *linear* (and in the first instance, for simplicity, *stationary*) subsystems do these differences occur?

2.   How big are these differences?

3.   Is there experimental evidence of their existence?

I take it as axiomatic that the differences are in *experimental observables*, and that they are expressible as *numbers*. It is my belief that anyone who insists (with a senior colleague) that Thomson's formulation is fallacious must be equally prepared to insist that Ohm's Law, as a theory, is fallacious.

F. The Reduced Heat of Transport, in a Gas

I restrict consideration to the simplest of all gas
mixtures, two isotopes of a noble gas. These atoms have no
rotational or vibrational states, nor any chemical interaction with each other. To a first approximation, they be-
have as rigid elastic spheres. In a second approximation,
we assign them a spherically symmetric interatomic potential.

Suppose an isotope of mass $(m + \Delta m)$ is present in very
dilute solution in a gas of isotopic mas m. From an accessible text; e.g., (32), we have what is no more than the definition of an invariant in terms of experimental observables.
For the reduced thermal diffusion factor, we have

$$- \frac{\nabla c}{c} = \alpha_o \frac{\Delta m}{2m} \frac{\nabla T}{T}. \tag{18}$$

For all mixtures, standard texts; e.g., (2), give

$$-\left(\frac{\nabla c}{c}\right)_T = \frac{q^{*'}}{kT} \frac{\nabla T}{T}, \tag{19}$$

and for mixtures in which the enthalpy of solution is zero

$$\left(\frac{\nabla c}{c}\right)_T = \frac{\nabla c}{c}. \tag{20}$$

From Equations 18-20,

$$q^{*'} = kT \frac{\Delta m}{2m} \alpha_o. \tag{21}$$

Experimenters use α_o because it is independent of the temperature gradient (the system has a linear response). Thus,
$q^{*'}$ is an equilibrium parameter.

Notice, that to obtain Equation (21) I have done no more
than put together two perfectly standard notations, that of
kinetic theory (12, 32, 33, 34) and that of linear (irreversible) thermodynamics (2, 4, 7).

Let us put in some numbers. For neon, experiment gives $\alpha_o \simeq 0.5$, at room temperature and above (32); thus, for ^{22}Ne in dilute solution in ^{20}Ne,

$$q^{*'} \simeq kT/40. \tag{22}$$

Where is this thermal energy located? Do equilibrium representations (35) contain it? To this question, I return in Section XI.

VII.   THE NATURE OF x

A.   Electric Force

I introduced x as an independent mathematical variable, and have already spoken of it as a thermodynamic driving force. To expand and clarify, I consider four examples.

An electric force is the product of an electric field and a charge. For the field we all write, without hesitation, $-\nabla V$; thus, for a flux of electric charge, the relationship between the y and x of Figure 2 gives

$$\underset{\sim}{J} = -\sigma \nabla V. \tag{23}$$

For a system with a linear response, Equation (23) is Ohm's Law.

B.   Thermal Force

Fourier's Law is

$$\underset{\sim}{J} = -\sigma \nabla T, \tag{24}$$

where $\underset{\sim}{J}$ is now a flux of thermal energy rather than of electricity, and σ is thermal rather than electrical conductivity. The force is $-\nabla T$.

C.   Configurational Force

I dislike the above subtitle but I cannot think of a better one. I mean the force that drives an isothermal diffusion experiment. From an energy $kT \; ln(c)$, we obtain a force $-kT \; \nabla(c)/c$, and hence, a matter flux

$$\underset{\sim}{J} = - \frac{Dc}{kT} \; kT \; \frac{\nabla c}{c} \tag{25}$$

$$= - D\nabla c. \tag{26}$$

It is sometimes said that this force is not real, in the sense that Newton's forces are real. I think this leads to inconsistency. The jar of wine that is fermenting before me has near its floor a stationary (i.e., unchanging with time) exponential atmosphere of particles that are more dense than the liquid. Though more dense, they are held up by the force $-kT \; \nabla(c)/c$. Shall I say then that they are held up by an unreal force? When one recalls the trouble into which Newton ran in regard to action at a distance, a distinction in the present case between real and unreal does not seem profitable.

D.   Another Thermal Force

A subsystem of impurities (in either solid, liquid or gas) can have a vibrational thermal energy, conventionally written $-s_v T$. In a temperature gradient, this gives rise to a force $s_v \nabla T$ (for simplicity, I take s_v constant). This force gives rise to an impurity flux

$$\underset{\sim}{J} = \frac{Dc}{kT} \; s_v \nabla T. \tag{27}$$

E.   Entropy Generation

All the above forces generate entropy in the prescribed manner $(2, 4, 7)$. We rewrite Equations (23) and (24) in conjugate form:

$$\underset{\sim}{J} = (\sigma T) \; (-\nabla V/T), \tag{23a}$$

and

$$\underset{\sim}{J} = (\sigma T^2) \; (-\nabla T/T^2) \tag{24a}$$

The four forces generate entropy per unit volume at a rate:

$$\dot{s} = (\sigma T) \; (-\nabla V/T)^2 \tag{28}$$

$$= \sigma (\nabla V)^2/T$$

$$\dot{s} = (\sigma T^2)\,(-\nabla T/T^2)^2$$

$$= \sigma(\nabla T)^2/T^2, \tag{29}$$

$$\dot{s} = (Dc/k)\,(-k\nabla c/c)^2$$

$$= Dk\,(\nabla c)^2/c, \tag{30}$$

and

$$\dot{s} = (Dc/k)\,(s_v\nabla T/T)^2$$

$$= Dc(s_v\nabla T)^2/kT^2. \tag{31}$$

The rate of entropy production is always yx^2.

F.   Conservative Forces

The forces I have used as examples are all conservative forces. All are derived from potentials. There is no foundation whatsoever for the vigorous and widespread belief that "conservative forces are not permissible in irreversible situations *because* entropy is generated". Conservative forces generate **entropy**; not "can", but "*do*". In the preceding subsection, I have done no more than set out explicitly equations that are implicit in the standard **texts** (2, 4, 7).

There is a corollary of some interest in thermomigration. The force appearing in subsection VIII. D may be written:

$$\underset{\sim}{F} = -\nabla(-s_vT)$$

$$= s_v\nabla T$$

$$= s_vT\,\nabla T/T$$

$$= -q^{*\prime}\,\nabla T/T, \tag{32}$$

where

$$q^{*\prime} = -s_vT. \tag{33}$$

Equation (32) is a standard form. If we derive it via the gradient operator, we see that $q^{*\prime}$ is an equilibrium parameter, a quantity of thermal energy.

If s_v is temperature dependent (as in an electron gas; see subsection VI. C) we must, perforce, have a temperature-dependent enthalpy, by Equation (11). The force arising from the temperature dependence of s_v is precisely balanced by that arising from the temperature dependence of w, and so Equation (32) remains valid.

If we postulate a force (in a temperature gradient) arising from a *temperature-independent energy*, E_o, we have:

$$
\begin{aligned}
\underset{\sim}{F} &= -\nabla(E_o) \\
&= -\frac{dE_o}{dT}\, \nabla T \\
&= 0.
\end{aligned}
\tag{34}
$$

Equation (34) makes clear that there is something wrong with the postulated "intrinsic heat of transport" (*22, 23, 36*). As remarked in subsection VI. B, this postulate takes temperature to be a discontinuous field.

As I see it, all the forces of linear thermodynamics are conservative forces. This follows from the linearity of the system. If the system is linear, then there is local equilibrium. If there is local equilibrium, then there is a local equation of state. If there is a local equation of state, then the energies depend on present conditions rather than on past history. If the energies depend on present conditions only, not on history nor on the route taken, then we have a set of potentials. If we have potentials, we thereby have conservative forces.

G.   The Massieu Functions

Disequilibrium that involves a temperature gradient seems to be conceptually more difficult than other forms of disequilibrium. I think this is due to an accident of history. We have become familiar with the principle of minimum free energy, at the expense of a familiarity with the more general principle of maximum free entropy. The Massieu functions (*37, 38*) embody the more general principle. Thus, in an open system, say of lattice vacancies in gold, there exists (in a temperature gradient) a concentration gradient of vacancies, a gradient sloping downward from hot to cold, with an Arrhenius slope, $-w_f$, but the only force making vacancies drift along the gradient is $s_v \nabla T$. Since s_v for lattice vacancies in gold is positive, this force makes vacancies drift *up* the concentration gradient. Were we to have a metal for which $s_v = 0$, we should have a steady concentration gradient, with an Arrhenius slope, $-w_f$, along which there was a *zero* drift velocity. This is readily seen by forming the Planck function

$$\Phi = -G/T, \qquad\qquad (35)$$

and maximizing it.

If we close the boundaries of the system, the flux of defects is necessarily zero in the steady state. (Sr^{++} ions in solid NaCl are such a system.) If the flux is zero, the force $s_v \nabla T$ must be balanced out. It is balanced by a perturbation of the concentration gradient. The Arrhenius slope changes from $-w_f$ to

$$-(w_f + s_v T). \qquad\qquad (36)$$

Notice the plus sign within the bracket, as distinct from the minus sign appearing in the chemical potential. Expression (36) contains no distinction between interstitial solutes and substitutional solutes; it applies equally to carbon in iron (*39*) and to Sr^{++} in NaCl (*40*). Of course, for isotopic solutes (*41*), w_f is zero. When it is nonzero, it *always* appears in the measured Arrhenius slope.

Expression (36) is valid for systems in which diffusive drift of the solvent lattice is negligibly small (the diffusive drift may be small either because D_s is small for the solvent lattice or because the s_v of lattice vacancies is negligibly small). When the solvent lattice drifts, and the boundaries of the specimen are in steady motion, a further term appears in Equation (36); the solute concentration gradient is further perturbed by the need to keep up with the moving boundaries. The Arrhenius slope becomes:

$$-(w_f + s_{v,b} T + D_{s,a} s_{v,a} T/D_b). \qquad\qquad (37)$$

The suffixes a and b here denote solvent and solute atoms

respectively. D_s, the self-diffusion coefficient, is distinguished from D, the diffusion coefficient of one atom.

The "open system" of thermodynamics is a concept often
realized in real experiments, but only because these experiments fulfill a physical condition, the condition of having
sources and sinks for the mobile species (e.g., lattice
vacancies) on a scale much finer than the distances used in
measurement. The dislocation network of real solids is typically on a scale of 10^{-6}m, whereas the specimen size is
typically 10^{-2}m; thus, in any sectioning of the specimen,
one may reasonably assume sources and sinks *within* the
section; the interior of the section remains an "open system". This openness is generally subsumed without explicit
statement of the physical condition.

In a thin, solid specimen, especially one with a low
dislocation density, the condition may not be fulfilled;
the specimen may be an open system only at its periphery.
In its interior, free from sources or sinks, the stationary
condition is $\nabla^2 \underset{\sim}{J} = 0$. Because the mobile species is neither
created nor destroyed, in the interior, this reduces to

$$\nabla^2 \underset{\sim}{u} = 0.$$

In this subsection I have summarized problems peculiar
to temperature gradients. The important equation is
Equation (35).

H.   Entropy Production and Entropy Flow

I consider only the simple case of a flow of thermal
energy. I make the customary assumption of local equilibrium (2, 4, 7, 42). Because there is a local equation of
state, we have as an equality, for an infinitesimal volume
element,

$$Ts\Delta\Omega = q\Delta\Omega. \tag{38}$$

In the flow from one point to another, $q\Delta\Omega$ is conserved, but
T, s, q and $\Delta\Omega$ change. Differentiating,

$$\Delta s = - q\frac{\Delta T}{T^2} - \frac{q}{T}\frac{\Delta(\Delta\Omega)}{\Delta\Omega}. \tag{39}$$

If we have a classical solid or gas, whose specific heat is independent of temperature, $\Delta\Omega$ varies inversely as T. Equation (39) then is:

$$\Delta s = 0; \tag{40}$$

i.e., as thermal energy flows from place to place, the specific entropy, the entropy per unit volume, does not change. On the other hand, if we have a quantum solid, for which $\Delta\Omega$ varies inversely as T^4, then

$$\Delta s = 3q \frac{\Delta T}{T^2}; \tag{41}$$

i.e., as thermal energy flows down the energy gradient, the specific entropy is *reduced* (at lower temperatures, fewer modes are excited, the order is greater, the entropy is less).

If, instead of specific entropy, we consider the extensive entropy; i.e.,

$$TS = Ts\Delta\Omega = q\Delta\Omega = Q, \tag{42}$$

then

$$\Delta S = -Q\frac{\Delta T}{T^2}; \tag{43}$$

i.e., the entropy production is positive. Comparison of Equations (40) and (43) shows that as entropy travels down the temperature gradient, it travels with increasing speed. Entropy is being generated, but the specific entropy does not change. The increasing speed is evident from the expression for the flux of entropy:

$$- \sigma \nabla T/T. \tag{44}$$

Frequently, as in a one-dimensional system, $\sigma\nabla T$ is constant while T decreases.

Being dependent on Equation (24), both the production term and the flow term, Equations (43) and (44) are subject to the *geometry* of the energy flow; linear, cylindrical, etc.

Notice that in deriving both the production and flow, I have never had to use other than a conservative force, a

force derivable from a potential: in Equation (24), a potential T; in Equation (24a), a potential $-1/T$; in Equation (44),
a potential $\ln T$.

VIII.   SUMMARY IN REGARD TO x AND y

By considering the concepts of *continuity* and *linearity*,
and by using illustrative examples, I have endeavoured to
show that the intrinsic parameters of linear (irreversible)
thermodynamics are, without exception, equilibrium parameters.

By considering the concept of *local equilibrium*, and
again using illustrative examples, I have endeavoured to
show that the forces of thermodynamics are conservative
forces, in the widely-understood sense of being derivable
from *potential* by the operation of taking the negative gradient.

Linearity and local equilibrium go hand in hand. The
one implies the other:

1. *If* there is linearity, *then* -synonymously-
   intrinsic parameters are independent of gradients
   in the system (as in Figure 2).

2. *If* there is independence of gradients, *then* there
   can be a local equation of state (and by the
   principle of continuity from equilibrium to
   disequilibrium, there is).

3. *If* there is a local equation of state, *then* there
   are potentials and conservative forces.

To the extent that these conslusions are correct, linear
thermodynamics is purely a matter of equilibrium parameters
and conservative forces.

Nonlinear thermodynamics is symptomatic of large perturbations, of the transition from Aristotleian mechanics
to Newtonian mechanics, transition from the Nernst-Einstein
relation, velocity proportional to impressed force, to
Newton's Law, acceleration proportional to impressed force.
Except in junction devices, the nonlinear region is commonly

inaccessible, requiring perturbations between two and thir-
teen orders of magnitude more severe than are possible in a
laboratory (8).

When present, nonlinearity fulfills two conditions:
the flux is still xy, and the rate of entropy generation is
still x^2y. Since there are only two variables (x and y) the
properly conjugate expression of the nonlinearity in any
physical system is *uniquely determined*; thus, *both* x and y
become functions of the simple quantities T, V, c, etc.
It is to be emphasized that the nonlinear x's are real physical forces.

To illustrate that the physical forces become functions
of the simpler gradients, I consider two examples, a symmetric defect in an electron gas, such as an impurity ion,
and secondly, an asymmetric defect, an edge dislocation with
its line and its Burgers vector orthogonal to a flow of thermal energy. The nonlinearities appearing in Equations (28)
and (30) of Reference (8) modify the present equation (31).
For the symmetric defect in the electron gas,

$$\dot{s} = \frac{Dc}{k} \left\{ s_v \left[\frac{\nabla T}{T} + \frac{\lambda^2}{3} \left(\frac{\nabla T}{T} \right)^3 - \frac{2}{3} \lambda^2 \ln T \; \frac{\nabla T}{T} \; \frac{\nabla^2 T}{T^2} \right] \right\}^2. \quad (31a)$$

For the edge dislocation,

$$\dot{s} = \frac{Dc}{k} \left\{ s_v \left[\frac{\nabla T}{T} + \lambda \; \underset{\sim}{i} \left(\frac{\nabla T}{T} \right)^2 - \lambda \; \underset{\sim}{i} \; \frac{\nabla^2 T}{T^2} \right] \right\}^2. \quad (31b)$$

The curly brackets contain the nonlinear forces. (The unit
vector in Equation (31b) points up the gradient when the
modulus value of the thermal capacity perturbation created
by the defect is greatest at the upper side of the gradient,
and conversely. Equation (31b) ignores the symmetrical part
of the perturbation.)

As is evident from these two equations, I use the phrase
"nonlinear" in the specific sense that the local intrinsic
parameters are functions of the simpler *gradients* (8); e.g.,
∇T. I exclude the sense (4) that the intrinsic parameters
may be functions of the simpler potentials; e.g., T.

The appearance in Equations (31a) and (31b) of only

local *equilibrium* parameters, D, c, s_y, λ, T, suggests an overall conclusion: that in both the linear region and the nonlinear region of thermodynamics we need use only equilibrium parameters. The nonlinear region is distinct from the linear region in that it requires more equilibrium parameters. In the two simple examples given here, the nonlinear region requires the mean free path, λ, while the linear region does not.*

Bearing in mind that complete equilibrium is forever unobservable, whereas nonlinear disequilibrium is a general statement, embracing all disequilibrium, we might make a philosophic generalization and say that all that can be observed can be completely described in terms of that which can never be observed.

IX. NUMERICAL VALUES

Formal mathematical statements, e.g., inequalities (*1*) and (*2*), Equation (3), Equation (44), can give no numerical values. Numerical values can be obtained only from experimental physics, sometimes directly, sometimes with the help of physical models. It becomes necessary to know the mass of a gas atom, a phonon spectrum, an electronic band structure, an interatomic potential. The manipulation of these physical quantities is made much easier by the realization that in systems of *linear response*, one need treat only equilibrium physical quantities.

The remainder of this paper is an application of Section VIII.

---

*

> Nonlinearity appears to be more a matter of semantics than of physics. In these examples, the system is non-linear only because we retain $s_y(\nabla T)/T$ as the real, physical force. If we redefine the force, we remove the nonlinearity. Linearity and nonlinearity are arbitrary qualities dependent upon a choice that we are free to make.

X. RADIATION FORCE

Like a ping-pong ball in a fountain, particles of glass can be supported by a laser beam (43). A comet tail is blown away from the sun (44). Very different in magnitude, both effects arise from the radiation force exerted by photons. A photon of energy, $h\nu$ carries momentum, $h\nu/c$. All traveling waves carry momentum (45). If you doubt reference (45), let me paraphrase: I know of no traveling wave that does not carry momentum. A wave on a string carries momentum, both ends fixed (46), or one end fixed (47).* There is no need for anharmonicity. A continuous, monochromatic laser beam is considered harmonic. The sunlight which strikes the comet tail is harmonic. Electromagnetic theory does not require anharmonicity (48). Yet even the best of solid state textbooks; e.g., (49), keep alive the belief that harmonic waves in solids do not carry momentum. Let us follow Equation (7.9) of Reference (49) and sum the crystal momenta over many collision events:

$$\Sigma \hbar \underset{\sim}{k}' \; = \; \Sigma \hbar \underset{\sim}{k} \; + \; \Sigma \hbar \underset{\sim}{q} \; - \; \Sigma \hbar \underset{\sim}{G}. \tag{45}$$

The summing of momentum is invariant with respect to the frame of reference and Equation (45) is quite general. In experiment, we frequently choose a frame of reference in which the crystal lattice is at rest; i.e.,

$$\Sigma \hbar \underset{\sim}{G} \; = \; 0. \tag{46}$$

We then have a classical conservation law:

$$\Sigma \hbar \underset{\sim}{k}' \; = \; \Sigma \hbar \underset{\sim}{k} \; + \; \Sigma \hbar \underset{\sim}{q} \tag{47}$$

---

*
   If one considers the traveling energy of the wave, rather than the sum of the wave energy and the potential energy stored in the weight hanging from the unfixed end of the string, one obtains the standard answer of references (45, 46).

Equation (47) expresses a rule relevant to Mossbauer momenta
(*50*): if we sum *all* quantum events, we recover a classical
answer. In thermomigration and electromigration, we frequently
sum all events. Moreover, our momentum transfers occur in the
same crystal lattice as the Mossbauer effect, so that the rule
is unquestionably applicable. It is a powerful simplification
to be able to consider real momentum, measured classically.

Momentum, in striking an obstacle, produces a force.
Because Equation (47) excludes G, we do not have to restrict
the concept of phonons to solids with lattices, nor even to
solids; any medium in which sound waves will travel is sufficient. Speech is a propagation of phonons; just say "'ting!"
(The momentum carried by phonons in gases is an old experiment (*51*).) Thus, photons *in vacuo et in vitreo* (*48*), and
phonons in whatever medium, can produce a radiation force on
an obstacle. I shall assume, with Vigoureux (*45*), that the
same is true of plasmons.

In mixtures that are chemically inert, the thermal radiation force is the only force that unmixes the components
in a temperature gradient. It is the simple physical cause
that Chapman sought (*52*).

In chemically active mixtures, radiation force causes
a perturbation to the Arrhenius slope, **from $-w_f$ to $-(w_f + s_v T)$**.
As at Equation (36), the force on the obstacle is determined
by *the perturbation in the thermal energy that the obstacle
creates*.

Radiation force may be treated very simply. A force
per unit area is an energy per unit volume. A force is therefore the product of an area and an energy density. A force
must have direction (occasionally, forces to which no direction can be assigned are reported). Thus, the force on an
area b^2, arising from a thermal energy and a temperature
gradient, is

$$F = -b^2 d \nabla \left(\rho \int_0^T C_p \, dT \right), \qquad (48)$$

where d is a distance along the gradient. If we have a homogeneous medium, F, is uniform across the gradient. In an inhomogeneous medium, F is not uniform. If, for example, the density over a distance, d, and an area, b^2 is $(\rho_0 + \Delta\rho)$, this volume element feels a force that is increased by

$$\Delta F = -b^2 \, d \, \nabla \left(\Delta\rho \int_0^T C_p dT \right) . \tag{49}$$

If the density is $(\rho_0 - \Delta\rho)$, the force **is decreased**. Obviously, forces also arise by changes in C_p.

For a defect such as a lattice vacany in a solid, an equivalent calculation can be made (53, 54). Phonons of long wavelength undergo Rayleigh scattering. For phonons of short wavelength, Morse's correction is made. Phonons are summed over a Debye spectrum. For a defect at one lattice site,

$$\Delta F = - \frac{3}{2} k \nabla T . \frac{\Delta\rho}{\rho} \left[\left(\frac{\Delta\rho}{\rho} \right)^2 + \left(\frac{\Delta\eta}{\eta} \right)^2 \right] . \tag{50}$$

For **defects** in an **electron** gas, rather than in a lattice, one has $2\gamma T$ in place of $\frac{3}{2}k$. To obtain the density and stiffness changes in the electron gas or in the lattice, specific models are necessary.

In the absence of other forces, the **inhomogeneity** (whatever it may be) drifts with a velocity (in the linear region of response)

$$\mu = \frac{D}{kT} \, \Delta F \tag{51}$$

$$= \frac{D}{kT} \, b^2 \, d.\nabla \left(\int_0^T \Delta(\rho C_p) dT \right) . \tag{52}$$

Obviously, $b^2 d$ is the volume of the **inhomogeneity** and $b^2 d \, \Delta(\rho C_p)$ is its defect thermal capacity.

XI. THERMAL DIFFUSION IN GASES

A.   The Radiation Force in a Gas

From Equation (21):

$$\underset{\sim}{F} = -\nabla(q^{*'})$$

$$= -\frac{\Delta m}{2m} \alpha_o \, k\nabla T. \tag{53}$$

Experiment (55) shows that α_o reaches a maximum value of ~ 0.5, and approaches zero as $T \to 0$. Noble gases are comparatively simple. One therefore asks, is α_o calculable?

B. α_o For Rigid Elastic Spheres

The starting point, (53), is the criterion later recognized for solids (8), that the perturbation is weak. In a gas, this means that the energy change along one mean free path is much less than the total energy:

$$\lambda \, \nabla T \ll T. \tag{54}$$

A gas atom is struck by other atoms. Therefore, consider an obstacle of volume Ω struck by a stream of classical particles of mass m, velocity $\underset{\sim}{v}$ and number density N_o. For hard sphere collisions, the scattered momentum is isotropic; thus, on average, the incident momentum is neither transmitted nor reflected, but simply absorbed. The average force on the obstacle is

$$\underset{\sim}{F} = \left(N_o \Omega^{2/3} v \right) (m\underset{\sim}{v}). \tag{55}$$

The first bracket is the number per unit time of particles striking the obstacle. Equation (55) may be written:

$$\underset{\sim}{F} = 2\Omega^{2/3} E \underset{\sim}{i}, \tag{56}$$

where E is the energy density in the particle stream. Now consider six such streams, incident on the obstacle, from all three orthogonal directions. If all are of equal magnitude, the obstacle feels no (average) force. If they are unequal, there is a residual force:

$$\Delta \underset{\sim}{F} = -\frac{2}{3} \Omega \nabla E, \tag{57}$$

where E is the total energy density. Numerical factors associated with the shape of Ω disappear. One may introduce the atomic volume and write:

$$\Delta \underset{\sim}{F} = -\frac{2}{3} \frac{\Omega}{\omega} \nabla (\omega E) \tag{58}$$

$$= -\frac{2}{3} \frac{\Omega}{\omega} \nabla (\frac{3}{2} kT). \tag{59}$$

It remains to calculate Ω/ω. The force we seek is that which unmixes the solute from the solvent, the force that pushes the solute through the solvent. If we replace the solute atoms by solvent atoms (a single gas) $\Delta \underset{\sim}{F}$ becomes zero; thus, $\Delta \underset{\sim}{F}$ arises from the replacement of a solvent atom by a solute atom. We may find Ω/ω by considering a two-body collision.

In the collision of two hard spheres, each of diameter b, the longitudinal momentum transfer, averaged over all impact parameters, is:

$$2 \mu \underset{\sim}{v} \pi b^2. \tag{60}$$

Now make the isotopic substitution $m \rightarrow (m + \Delta m)$ in one sphere. The exchange of momentum is now different, first because of the change in reduced mass, μ, and second, because of the change in $\underset{\sim}{v}$, which is a thermal velocity. The change in reduced mass produces a fractional change in momentum transfer:

$$\Delta m/2m. \tag{61}$$

The change in thermal velocity is slightly less simple because it must include α_o. It gives a fractional change in momentum transfer:

$$-\frac{\Delta m}{4m} (1 - \frac{\alpha_o}{3}). \tag{62}$$

One may think equally well of these fractional changes of
momentum transfer as brought about by a change of collision
cross-section. Since the molecules are spherical, this change,
when rotated about an axis in real space, produces a change
of collision volume (i.e., collision volume for momentum
transfer). Now, Equations 55-59 are based on momentum transfer; thus, from expressions (61) and (62),

$$\frac{\Omega}{\omega} = \frac{3}{2} \frac{\Delta m}{4m} (1 + \frac{\alpha_o}{3})$$ (63)

Substituting into Equation (59) and integrating,

$$q^{*\prime} = \frac{3}{8} \frac{\Delta m}{m} (1 + \frac{\alpha_o}{3}) \ kT.$$ (64)

Comparison with Equation (21) gives

$$\alpha_o = 1.$$ (65)

For simple, rigid elastic spheres, this simple value would
seem to have an *a priori* probability comparable to the value
in use; i.e., 105/118 (56).

C.   Less Rigid Spheres

    A softer, interatomic potential gives a negative contri-
bution to Equations (63-64), and makes α_o less than unity.

For simplicity, suppose a repulsive potential of the form:

$$\phi \propto r^{-z}.$$ (66)

The value attained by ϕ is equal to the kinetic energy of the
colliding particles in the center of mass frame; thus, in
Equation (60) we now have a change in b, as well as in μ and
χ, when an isotopic substitution is made. The interatomic
potential contributes to Ω/ω:

$$(\frac{\Omega}{\omega})_z = 3\frac{\Delta b}{b}$$

$$= -\frac{3}{z} (\frac{\Delta \mu}{\mu} + \frac{2\Delta v}{v})$$ (67)

$$= -\frac{\Delta m}{2m} \frac{\alpha_o}{z}$$ (68)

Thus

$$\frac{\Omega}{\omega} = \frac{3\Delta m}{8m} \left[1 + \frac{\alpha_o}{3} \left(1 - \frac{4}{z}\right) \right] , \qquad (69)$$

and

$$\alpha_o = (1 + 4/3z)^{-1} \qquad (70)$$

Equation (70) shows that α_o does not go negative for *any* simple repulsive potential. In particular, it does not go to zero for Maxwellian molecules, for $z = 4$; instead, it gives $\alpha_o = 0.75$. If we choose to associate the experimental result, $\alpha_o \lesssim 0.5$, with an average value of the hardness parameter z, we have $z \lesssim 1.33$; i.e., the averaged interatomic potential is much softer than Maxwellian repulsion. This average softness is consistent with the small impact energies of thermal collisions, a few multiples of 0.01 eV. The steeply repulsive part of the potential is never called into play.

D.   The Chapman - Enskog Theory

It is well known that the Chapman-Enskog theory predicts zero thermal diffusion for Maxwellian molecules (*12*). We may obtain the Chapman-Enskog result by suppressing the term, $2 \Delta\chi/\chi$ in Equation (67). We then have:

$$\left(\frac{\Omega}{\omega}\right)_z = -\frac{3}{z} \frac{\Delta\mu}{\mu} \qquad (71)$$

$$= -\frac{3}{2} \frac{m}{2m}, \qquad (72)$$

giving $\alpha_o = 1 - 4/z, \qquad (73)$

an equation that gives $\alpha_o = 0$ for $z = 4$.

Dimensional analysis, too, gives cause to doubt the Chapman-Enskog result for Maxwell's molecules. Frankel (*57*) has used dimensional analysis to treat this very problem, and the method is accepted as being applicable (*12, 58, 59, 60*). The method has the advantage of clarity, and none of the physics is lost. The starting point is expression (60). The number of collisions per unit time experienced by a chosen atom is proportional to the number density of other atoms, N_o, as at Equation (55), and to the velocity of these atoms

relative to the chosen atom. The rate of transfer of momen-
tom in one direction, a vector, is therefore proportional to

$$\nabla \ (< \mu \ v^2 \ b^3 \ N_o >_{Av}). \tag{74}$$

Directionality is specified by the gradient operator. v^2
is a scalar quantity. $< >_{Av}$ indicates an average over all
impact parameters and over a macroscopic time. From Equation
(66), the force between two colliding atoms is:

$$\underset{\sim}{F} \ = \ F_o \ \underset{\sim}{r}/r^{z+2}. \tag{75}$$

From the reduced mass, from $\underset{\sim}{v}$ and from F_o, we can form only
one quantity that has the dimensions of length; i.e.,
$(F_o/\mu v^2)^{1/z}$. Substituting for b in Equation (74), the force
supporting thermal diffusion is proportional to:

$$F_o^{3/z} \ \nabla \ (< ((\mu v^2)^{\frac{z-1}{z}} \ N_o >_{Av}). \tag{76}$$

The quantity within the bracket is an *even* power of $\underset{\sim}{v}$. The
average of an even power is *nonzero*. (The average of an odd
power is zero, because in the stationary state, both components of the gas mixture as well as the overall body of gas,
are at rest in our frame of reference, the containing vessel).
Frankel obtains *odd* powers of $\underset{\sim}{v}$ by considering the force from
one unidirectional stream of molecules, as at Equation (55),
instead of the imbalance in six streams, as at Equation (57).
Frankel has

$$< \mu \ \underset{\sim}{\chi} \ v \ b^2 >_{Av}, \tag{77}$$

rather than

$$\nabla \ (< \mu \ v^2 \ b^3 \ N_o >_{Av}). \tag{78}$$

The presence of b^2 rather than b^3 gives:

$$F_o^{2/z} \ < \mu \ \underset{\sim}{\chi} \ v^{\frac{z-4}{z}} >_{Av} \tag{79}$$

in place of Equation (76). For $z = 4$, one is left with

$< \mu \chi >_{Av}$, which averages to zero. The Chapman-Enskog theory makes this mistake; it has an area where it should have a volume.

Though Maxwell's molecules are capable of thermal diffusion, its occurrence depends on assumptions *beyond and out with the atomic potential*. To make this plain, consider again simple, rigid elastic spheres. By common consent, their interatomic potential shows the strongest thermal diffusion. For hard spheres, $z \to \infty$, but $F_o^{1/z}$ remains finite (*12*). Expression (76) then simplifies to

$$\nabla (< \mu \, \mathbf{v}^2 \, N_o >_{Av}). \qquad (80)$$

In a gas mixture, the two species are locally in equilibrium, and locally in equilibrium with each other (I make the same assumption as in Section VII. H). In addition, I also assume equipartition (*12, 22, 34. 35, 56, 57, 58, 59, 60, 61, 62, 63*). Given equipartition, we do not need to distinguish between the two atoms in the mixture. Alike, for solvent-solute collisions, and for solute-solute collisions, we have:

$$< \mu \, v^2 >_{Av} \;\; = \;\; 3kT. \qquad (81)$$

Thus, Expression (80) is **proportional to:**

$$\nabla (kT \, N_o). \qquad (82)$$

Because we use Equation (81) to obtain Equation (82), we lose all distinction between solvent and solute molecules, molecules of mass m and $(m + \Delta m)$, respectively. Expression (82) tells us that each molecule feels the same force; that there is no force pushing one set of molecules through the other set of molecules; that there is no force that must be balanced by a concentration gradient. The lack of distinction between solvent and solute molecules, in Equation (82), specifies that the thermal diffusion is zero. For a gas of rigid, elastic, spherical molecules, the equation of state is:

$$p \, (\underline{\underline{V}} - \underline{\underline{b}})) \;\; = \;\; N_o kT, \qquad (83)$$

whence, the force supporting thermal diffusion is proportional to:

$$\nabla (p \, (\underline{\underline{V}} - \underline{\underline{b}})), \qquad (84)$$

Within Equation (84), only the intrinsic parameter p is subject to the gradient operator; thus, Equation (84) is proportional to:

$$(\underline{\underline{V}} - \underline{b})\ \nabla\ p. \tag{85}$$

Mechanical equilibrium requires:

$$\nabla\ p\ =\ 0, \tag{86}$$

so that thermal diffusion is zero. Moreover, it is zero for the most favorable interatomic potential. If we now look back to Equation (82) and expand it, we see that it specifies that **there are** fewer molecules of both species in the hotter region than in the cold, but this is all.

How, then, does the Chapman-Enskog theory predict thermal diffusion? It does it very simply, by not summing all the momentum transferred to a chosen solute molecule. It does not sum the momentum transferred in collisions with other solute molecules (60). It sums only the momentum transferred in solvent-solute collisions ($12, 60$). Also, it uses Equation (77) rather than Equation (78), (60). On both points, Professor Cowling believes the theory to be correct, while I believe it to be incorrect.

I am aware of comparable theories, all of which, either implicitly or explicitly, also take $q^{*\prime} = 0$ ($57, 58, 61, 62$). To the best of my belief, their nonzero answers are *entirely* dependent on the approximations made. This is not as unfortunate as it might at first sight seem. Were there to be **a satisfactory theory of thermal diffusion in gases that succeeded from the premise,** $q^{*\prime} \equiv 0$, **then** *gas mixtures would lie outside the scope of linear thermodynamics.*

E.   The Continuity of Thermal Capacity

If $q^{*\prime}$ in a gas mixture is nonzero in equilibrium, it is desirable to form some mental image of it. Again, I shall go slowly. Consider first a cube of a classical solid in equilibrium. Cut a large hole in the center, Figure 3a. Plot the thermal energy density along the line AB. In the solid there is everywhere an energy density $(3kT + \gamma T^2)/\omega$. In the hole, the energy density is zero. Figure 3b may be regarded as a superposition of Figures 3c and 3d. We may

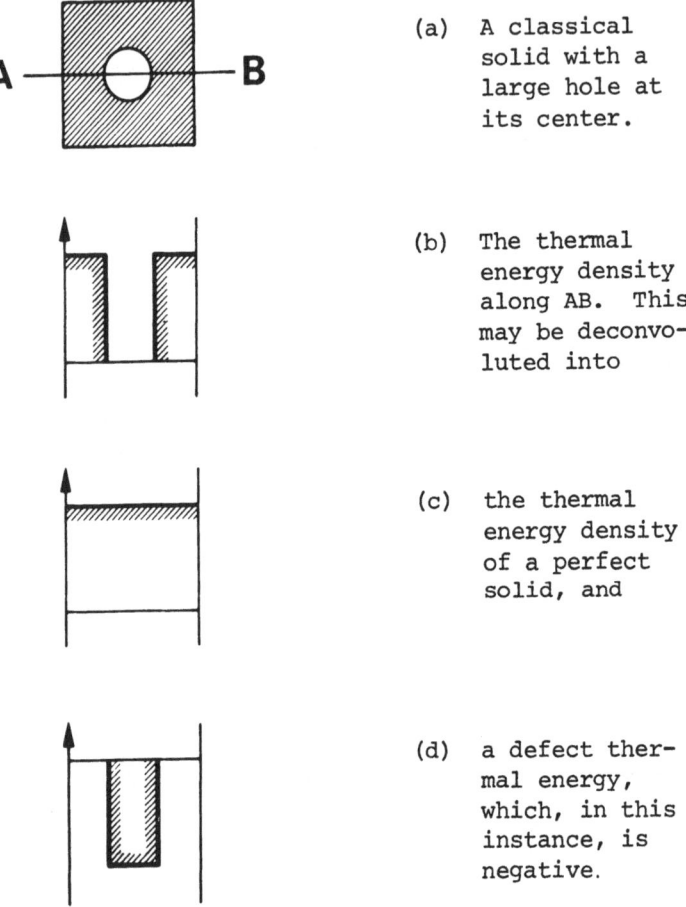

(a) A classical solid with a large hole at its center.

(b) The thermal energy density along AB. This may be deconvoluted into

(c) the thermal energy density of a perfect solid, and

(d) a defect thermal energy, which, in this instance, is negative.

Figure 3

regard the cube as a superposition of a perfect cube and a
defect with a *negative* thermal capacity. The negative thermal capacity per unit volume is - $(3k + \gamma T)/\omega$.

Next, consider one lattice vacancy in an otherwise
perfect cube, Figure 4a. Noble metals justify the simple
picture (*8*, *30*, *54*) that in forming the vacancy, we remove
the kinetic energy of the ion (leaving the thermal potential
energy of the lattice much the same), together with the
thermal energy of the screening electrons. The depth of
the well in Figure 4b is therefore only

$$(\frac{3}{2} kT + \gamma T^2)/\omega.$$

Thermal energy has a relaxation length, a mean free
path for the scattering of the carriers of thermal energy.
Because the relaxation length is nonzero, the sharp corners
of Figure 4d must be rounded off, as in Figure 4e. Like
temperature, *thermal energy density and thermal capacity
are continuous in real space*(in Figures 3 and 4, the temperature along the line AB is as in Figure 5; it is continuous).

Into the lattice vacancy of Figure 4, we might put a
heavy isotope of the solid. It has localized modes, and the
thermal energy density is now as in Figure 4f.

Now return to Figure 3, and put some gas in the hole.
Suppose that in the gas there is a heavy isotopic atom, at
C, Figure 6a. Again, draw the thermal energy density along
the line AB, Figure 6b. Reading from the left: as we move
through the solid and approach the surface, the thermal
capacity falls below the value $(3k + \gamma T)/\omega$ per unit volume
(the temperature is as in Figure 5). We label this perturbation in thermal capacity "surface entropy". In the gas,
the thermal capacity retains some information about the
presence of the surface, until we are several multiples of
λ away from the surface. Thereafter, the energy density
takes the steady value, $3kT/2\underline{V}$ (it is a noble gas). As we
come within a few multiples of λ of the isotope at C, we
encounter a bump (Δm positive) whose integrated thermal
energy is $q^{*'}$.

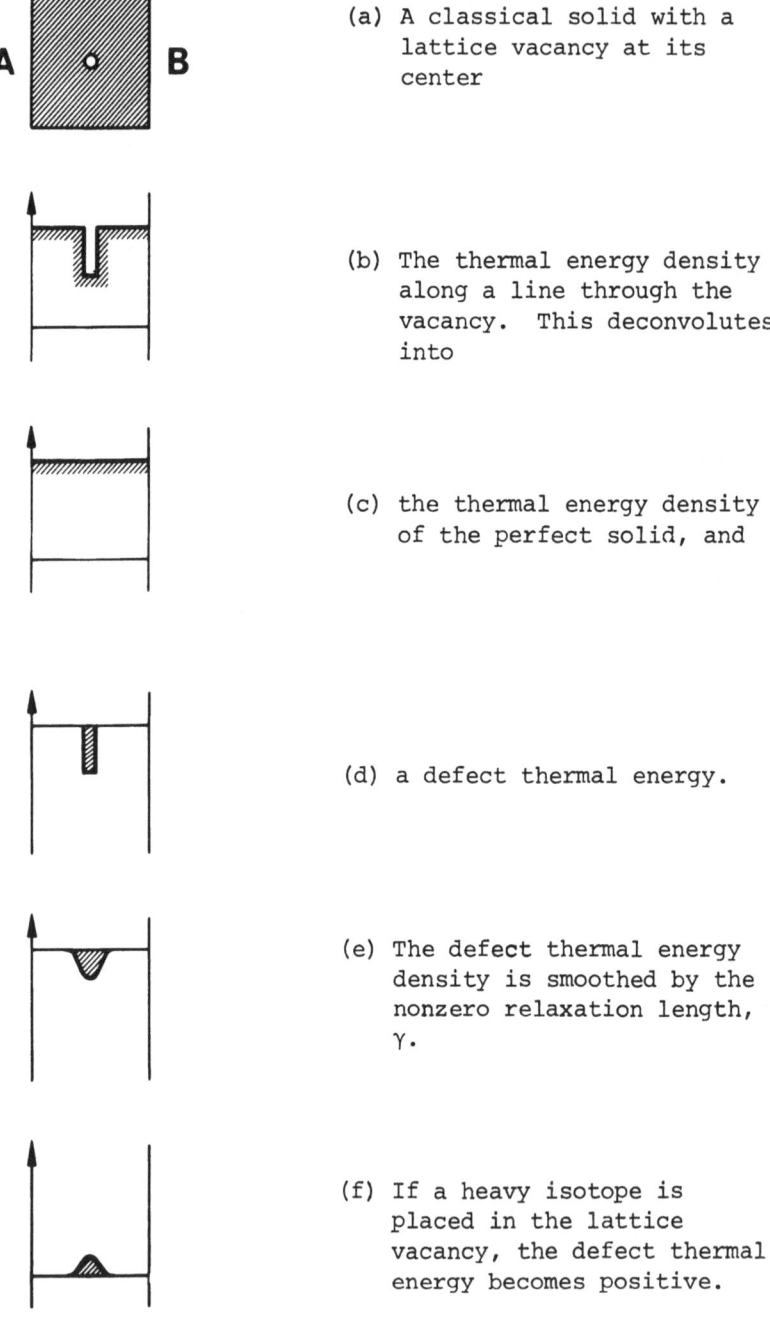

(a) A classical solid with a lattice vacancy at its center

(b) The thermal energy density along a line through the vacancy. This deconvolutes into

(c) the thermal energy density of the perfect solid, and

(d) a defect thermal energy.

(e) The defect thermal energy density is smoothed by the nonzero relaxation length, γ.

(f) If a heavy isotope is placed in the lattice vacancy, the defect thermal energy becomes positive.

Figure 4

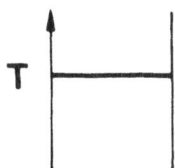

Figure 5
The temperature along the line AB,
in Figures 3, 4, and 6.

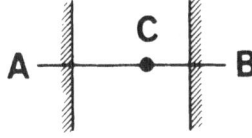

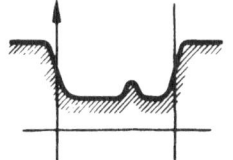

(a) The hole of Figure 3
    with gas inside it.
    At C, there is a heavy
    isotopic molecule.

(b) The thermal energy
    density along the
    line AB.

Figure 6

If my frame of reference is attached to the solid container, Figure 6b is cheating, for I am taking a snapshot rather than integrating over a long time. In a long time, the heavy isotope presently at C wanders over the available space and the bump at C is smeared out (this is why I started with holes in solids). We can escape the accusation of cheating by choosing a frame of reference centered on the molecule at C. There then is always a bump in the thermal energy density, over a volume $\sim \lambda^3$ around the center of our frame of reference.

We are free to choose such a frame of reference because the equations we have developed deal with momentum transfer and not with energy transfer. Momentum transfer is invariant with respect to Galilean transformations, but energy transfer is not.

F. Equipartition in Gases

Does the existence of a nonzero value of $q^{*\prime}$ in equilibrium "cut at the very root of statistical mechanics?" Experimental systems are linear; they are certainly linear for small perturbations; and thermodynamics gives:

$$\alpha_{o, \text{ equilibrium}} = \alpha_{o, \text{ linear disequilibrium}} \quad (87)$$

Clearly, it must be possible to translate between the languages of kinetic theory and linear thermodynamics, *and to retain one and the same conceptual result.*

Of kinetic theory, as presently formulated, we may say:

1. It is inconsistent with thermodynamics. It requires a discontinuity where thermodynamics requires continuity.

2. It does not sum all the momentum transferred to the molecule considered (60).

3. Its collision integrals are dimensionally incorrect. It prefers Equation (77) to Equation (78); hence, its forces lack direction.

A consistent kinetic theory does exist (64). It explicitly uses a volume rather than an area (65) and it re-

quires no discontinuity between equilibrium and disequilibrium (64); i.e., it provides a nonzero $q^{*'}$ in equilibrium.
To the best of my belief, its mathematics is unassailable.
A respected colleague had made a determined attempt (66),
but has reached only Burbury's conclusion (67); i.e., that
there is a lack of molecular chaos which may be described
equally well by either of two schemes, a correlation of
velocities or a correlation of positions. The modern view
is to admit a correlation of positions (66). For example,
a chosen gas atom, in the trough of its interatomic potential, has (on average) a higher number density of neighbors
than elsewhere (than at large distances). Equally, inside
the strongly repulsive part of the potential, the chosen
atom has fewer neighbors than elsewhere - far fewer - it is
difficult to put one atom inside another. Burbury (64, 65)
makes the same point no less explicitly.

I find it quite remarkable that the professional mind
is able to accept *simultaneously* a correlation of molecular
positions, as expressed in the preceding paragraph in terms
of the interatomic potential, and also, an absence of correlation, as expressed in Boltzmann's axiom of molecular
chaos (Sommerfeld, a comparatively recent writer, still calls
this axiom *an assumption* (68)). There seems to be a gestalt
switch at work: we *do* have correlation and we *do not* have
correlation.

A modern standard text (69) considers "a situation of
frequent occurrence", a system of particles having momenta
$\underset{\sim}{p}$ and positions $\underset{\sim}{q}$, in which the energy of the system splits
additively into two independent terms. The splitting is
expressible in either of two equivalent forms; i.e., the
energy splits in *both* forms and we are free to choose either.
In the first form, the first energy term depends upon $\underset{\sim}{p}_i$
alone (and not on $\underset{\sim}{p}_j$, $\underset{\sim}{p}_k$ ...). The second energy term is
independent of $\underset{\sim}{p}_i$. In the second form, the first energy term
depends on $\underset{\sim}{q}_i$ *alone* (and not on $\underset{\sim}{q}_j$, $\underset{\sim}{q}_k$...). The second is
independent of $\underset{\sim}{q}_i$.

The energy terms of Section XI. B do not split in this
simple fashion. Equations (61) and (62) show that $\underset{\sim}{p}_i$ depends
on $\underset{\sim}{p}_j$. Equation (64) gives the energy perturbation arising
from this interdependence.

Reif (69) gives the first energy term (of the first
form) as $\varepsilon_i(\underset{\sim}{p}_i)$, but this is insufficient; for a system of
gas molecules we have the double functional relationship

$$\varepsilon_i \ (\underset{\sim}{p}_i(\underset{\sim}{p}_j \ldots)).$$

Only for a gas of identical molecules does this reduce to
the form given by Reif. It is always true that ε_i is a
function of $\underset{\sim}{p}_i$, but in general, it is *not* true that ε_i
is independent of $\underset{\sim}{p}_j$.

In the second of the equivalent forms, the first energy
term represents the potential energy of the ith particle.
It is from the second form that the failure of the assumption,
in real gases, is perhaps most easily seen. The first energy
term is now dependent on $\underset{\sim}{q}_i$, but independent of $\underset{\sim}{p}_i$. Is it
also independent of $\underset{\sim}{q}_j$, $\underset{\sim}{q}_k$ ...; is it dependent on $\underset{\sim}{q}_i$ alone?
Only if the molecules are far apart. **If they are close to-**
gether, the intermolecular potential energy is nonzero; thus,
in general, the first energy term depends not on $\underset{\sim}{q}_i$ *alone*,
but on the relation of $\underset{\sim}{q}_j$, $\underset{\sim}{q}_k$ ..., to $\underset{\sim}{q}_i$. In general, there
is again a double functional relationship; in place of $\underset{\sim}{q}_i$,
one must write $\underset{\sim}{q}_i \ (\underset{\sim}{q}_j \ldots)$.

For particles that collide, the assumption on which the
equipartition **theorem** is based fails, in the general case;
it is valid only for the restricted case of a system of
identical particles. If the assumption fails, does the
proof which follows it fail? There seems a determination
(which I do not understand) to answer "no". I can discover,
neither by correspondence nor from the literature, any proof
that this assumption is unnecessary. Reif (69) calls his
assumption a condition. He uses italics.

Burbury (64) took the view that the assumption is adequate for *point* atoms but fails for *atoms of nonzero volume*.
I think Burbury is correct. It is worthy of note that Maxwell (70) was explicit in discussing "a system of material
points". The phrase "material point" had at that time a
strict meaning, still found in footnotes (71); it means
(71, 72) a particle of well-defined mass, but of negligible

volume and hence, of infinite density. In modern notation
we might call it a delta function in the density. Part of
the concept has survived into kinetic theory, but part has
been abandoned, and hence, one has an illogical scheme. In
regard to thermal diffusion, it is held that the volume of
a molecule is negligible, but that the shadow of the volume,
the cross-sectional area, is significant (60). To me, this
seems a *non sequitur* of the first magnitude.

In 1900, Burbury's work had no practical application;
the physics of the time could manage without it (thermal
diffusion was unknown) in much the same way as the small
scale navigations of the ancient and mediaeval world could
manage without Copernican astronomy (73).

It is often held that experimental work has vindicated
the Chapman-Enskog theory. A theory must do two things; it
must agree with experiment, and it must be consistent, both
internally, in its axioms, and externally, in regard to
neighboring branches of science (in this case, thermodynamics).
The Chapman-Enskog theory fails both tests of consistency.
In regard to experiment, both Burbury and Chapman-Enskog give
terms in kT. The difference between them amounts to that
between volume and area, between b^3 and b^2. It is always
possible to fit either theory to an *arbitrary* interatomic
potential; those "waxen hypotheses" (74). The crucial test
is comparison with a different class of *experimental data*,
for example, the comparison of thermal diffusion data with
viscosity or thermal conductivity data. Here, the Chapman-
Enskog theory gives an agreement with experiment which is,
"in general, quite poor" (75, also 76).

G.   The Bump in Figure 6

Equation (65) derives from simple, two-body collisions
between hard, classical spheres. It assumes a classical
velocity distribution, a Maxwell-Boltzmann distribution, but
for the solute atom, it scales this according to Equation
(21). At Equation (63), the change in momentum transfer is
expressed as an equivalent change in volume.

We may compare Equation (59) with the general form,
Equation (32), and obtain quite simply:

$$q^{*\prime} = \frac{\Omega}{\omega} kT, \qquad\qquad (88)$$

which is to say, the increase in momentum transfer gives
rise to an increase in translational thermal energy, in the
heavier molecule.

We may form an image: the chosen molecule receives
additional momentum in a collision with a solvent molecule,
and this it retains over a brief interval of time before
passing on the momentum to another solvent molecule. In
laboratory coordinates, the chosen molecule has, on average,
(if all these brief times are summed) a higher energy.

H. Consistency

The concept of a localized change in thermal energy
density reconciles kinetic theory with thermodynamics. The
localization is over a volume specified by the mean free
path.

The concept applies equally to gases, solids, liquids
and plasmas.

I. A localized *Collective* Property

Thermal energy is always a collective property, a distributed property (*77*). Loosely, we assign it to one atom
because that atom is always at its center, as at C, in
Figure 6.

J. Chemically Active Gases

The preceding subsections have all considered *inert*
gases (and model elastic molecules). However, in general,
we have not only radiation force, but also, chemical activity
between the two components of a mixture. The stationary
Arrhenius slope is then specified by Equation (36). If s_v
is negative and w_f is nonzero (e.g., a heavy isotope in a
chemically active gas), the thermal diffusion *changes sign*
when

$$w_f = -s_v T \tag{89}$$

This sign reversal is observed; e.g., (*78*).

K. Solids in Gases; e.g., Dusts in Reactor Gas Circuits

Figure 6 is equally applicable to an isotopic atom and
to a dust particle. Jeans (*35*) applies the principle of

equipartition to the molecules of a gas and to a pendulum swinging in a gas. The concept of $q^{*'}$ covers a similar range of magnitude.

For a dust particle, and using a hard-sphere potential, we again obtain Equation (63) simply from the change of mass. In addition, the real physical volume of a dust particle is greater than that of a gas molecule. The fractional increase in real volume is

$$\frac{\Delta m}{m} \frac{\rho_{gas}}{\rho_{dust}} , \qquad (90)$$

where ρ_{gas} and ρ_{dust} are densities of the gas molecule and of the dust. For carbonaceous dusts in a reactor circuit, the ratio of these densities will be, roughly, one. For metal dusts, the ratio will be, roughly, 1.0 to 0.1. Summing Equations (63) and (90),

$$\frac{\Omega}{\omega} = \frac{\Delta m}{2m} + \frac{\Delta m}{m} \frac{\rho_{gas}}{\rho_{dust}} \qquad (91)$$

$$= \frac{3}{2} \frac{\Delta m}{m} \text{ to } \frac{1}{2} \frac{\Delta m}{m} . \qquad (92)$$

Taking the upper value, the force pushing the dust through gas is, from Equation (88),

$$\Delta \underset{\sim}{F} = - \frac{3}{2} \frac{\Delta m}{m} kT \frac{\nabla T}{T} . \qquad (93)$$

This equation neglects thermal conduction in the dust particle. If thermal conduction is appreciable, Equation (93) also contains, as a factor,

$$\frac{1 - \sigma_{dust}/\sigma_{gas}}{1 + \sigma_{dust}/\sigma_{gas}} , \qquad (94)$$

where the thermal conductivities, σ, are the ordinary macroscopic values (79). The diffusional drift velocity of the dust particle is given by the Nernst-Einstein relation:

$$\underset{\sim}{u} = \frac{D_{dust}}{kT} \Delta \underset{\sim}{F} \qquad (95)$$

If we write the diffusion coefficients inversely as the square roots of the masses, then:

$$\mu = \frac{D_{gas}}{kT} \sqrt{\frac{m_{gas}}{m_{dust}}} \, \Delta F \tag{96}$$

$$= - D_{gas} \frac{3}{2} \frac{\Delta m}{m} \sqrt{\frac{m}{m+\Delta m}} \frac{\nabla T}{T} . \tag{97}$$

If $\Delta m \gg m$,

$$\mu = - \frac{3}{2} D_{gas} \sqrt{\frac{\Delta m}{m}} \frac{\nabla T}{T} . \tag{98}$$

Since the mass of the dust particle goes as the cube of its radius, Equation (93) suggests a force varying as the cube of the radius. Waldmann and Schmitt (80) follow Chapman and hence, give a force depending on the square of the radius. They give a friction term which I presume to correspond to Equation (95), and two expressions for the thermal conductivity factor which differ from Equation (94).

XII.   THERMOMIGRATION AND ELECTROMIGRATION IN SOLIDS

A.   Linear Systems

In the linear region of response, thermomigration and electromigration have no greater limitations (due to irreversibility) than has Ohm's Law. Their intrinsic parameters are simple equilibrium quantities. This can be put to experimental test (30, 81). The simplicity of isotopic defects makes their behavior a crucial test (30).

B.   Thermomigration

In thermomigration, momentum is not carried by bodily movement of the lattice, nor by bodily movement of the electron gas, but by oscillations within the lattice and the gas, by quasi-particles, phonons and plasmons. These quasi-particles are carriers of thermal energy. An impurity species is also a carrier of thermal energy.

These three groups of carriers collide, in collisions that are only slightly perturbed from random. It is the random element that ensures that phonon-plasmon collisions produce a diffusive drift of thermal energy down the energy gradient. The same random element no less surely produces a downward diffusive drift of the thermal energy carried by the impurity species. There is no ambiguity about the sign of thermomigration; it is unequivocally determined by the second law.

Insofar as momentum is carried by quasi-particles, whose interactions are not continuous, it is incident on an obstacle from all directions, and Equation (57) is appropriate. There could be higher order terms which I haven't considered, arising from phonon drag, but at thermomigration temperatures, they are likely to be negligible. The more important thing is that the first order term, Equation (57), contains a characteristic *volume*.

In gases, the temperature dependence of thermomigration can be explored over a significant range, especially toward low temperatures. Is it **mere coincidence** that in the noble gases the temperature dependence of α_Q (*32*, *55*) resembles that of the rotational specific heat of a diatomic molecule? When two gas atoms collide, the line joining their centers rotates, while they are in collision. In the same brief interval that the two atoms look like a diatomic molecule, the whole of thermomigration is determined.

In **solids**, diffusion is too slow to permit experimentation to comparable temperatures. Had we but the aeons, we should expect quantum behavior. In Equation (50) we should write the Debye specific heat in place of $3k$.

C.   Electromigration

In electromigration, on the other hand, **momentum is** carried by bodily movement of the electron gas. The electrons interact continuously, through their Coulomb repulsion, and move cooperatively. In an external electric field, they carry their drift momentum as a single stream. Because of this cooperative behavior, Equation (56), rather than Equation (57) is appropriate; thus, there is a radical difference between thermomigration and electromigration, best shown by the anisotropy of electromigration in hexagonally close-packed single crystals (*82*, *83*).

D.   Anisotropic Electromigration

The species responsible for single crystal electromigration is the lattice vacancy. What cross-sectional area
does it subtend to the electron stream? In particular, how
does the area seen along the [0001] direction in an h.c.p.
crystal compare with the area seen along the $<10\bar{1}0>$ directions?
What is the shape of the lattice vacancy? By Babinet's principle, the vacancy has the same anisotropy as the ion plus
electrons which have been removed to create it (84); hence,
the perturbation in the electron gas, at the lattice vacancy,
has the shape of the Fermi surface, but drawn in real space
rather than in k space. Thus, the lens of k space, in zinc
and in cadmium, becomes a prolate figure in real space, and
the outermost part of the electron scattering surface. Seen
along the hexagonal axis, the lattice vacancy has an area
proportional to:

$$< k^{-2}(0001) >_{Av};   \tag{99}$$

i.e., k^{-2} is averaged in the basal plane. Seen along $[10\bar{1}0]$
the lattice vacancy has an area proportional to:

$$< k^{-2}(1\bar{2}10) >_{Av}.   \tag{100}$$

Other things being equal, the forces of electromigration for
these two single crystal directions stand in the ratio of
Equation (99) to Equation (100). In zinc and cadmium, other
things are very nearly equal; the momentum of an electron
stream along [0001] is only 6% less than for a direction in
the basal plane (83), for a given voltage gradient. If we
assume isotropy in the basal plane, the ratio of Equation
(99) to Equation (100) is:

$$k_{<0001>}/k_{<10\bar{1}0>}.   \tag{101}$$

An experimental value of this ratio is 0.38 (82); thus, for
the same voltage gradient in the two directions, we expect
electromigration forces in the ratio

$$F_{<0001>}/F_{<10\bar{1}0>} = 0.36   \tag{102}$$

The experimental value (82, 83) is between 0.4 and 0.5. This
larger experimental value is consistent with an electron wind
at a slight angle to [0001]. For an angle of $7°$ (82), Equation (102) also gives 0.4.

In zinc and in cadmium, the lattice vacancy is shaped rather like a Rugby ball and feels least force from the electron wind when traveling point first.

E.   Holes and Electrons

The one driving force of electromigration is the momentum of the electron stream. Being scattered by a species within a solid (or liquid) it gives rise to a radiation force, to the "electron wind". It may also act indirectly, by way of phonon drag. The indirect action is usually less than the direct action.

Since the monopole charge on a species is identically zero (see Section VI. A) no net electric charge is carried down the energy gradient, as a species migrates (contrast this with thermomigration in which thermal energy *is* carried down the gradient). The product of the monopole charge and a potential difference through which a species may fall is therefore always zero. The corresponding energy and the corresponding force, the "direct" force, are therefore equally, and obviously, zero.

Textbooks (*86, 87*) are unequivocal in saying that though electrons and positive holes move in opposite directions, *they carry momentum in the one direction.*

In regard to the possibility that holes and electrons carry momentum in *opposite* directions (*88*), it is possible to see one more aspect of continuity. If we start with an electron conductor having a nearly empty band, and gradually (continuously) fill this band, $m^*/|m^*|$ flips discontinuously from +1 to -1 as we approach the top of the band. Suppose we have an impurity, a scattering center, in such a conductor. Does the sign of the electron wind flip as we go through the inflection in the E, k curve? The sign of $\Sigma \hbar k$ in Equation (47) does not flip, and the cross-sectional area for scattering does not flip; cross-sections are always positive. The experimental data on tin, zinc, cadmium, lead, and gamma uranium, all of which, by the sign of the Hall coefficient, could be classed as hole conductors, show no evidence of a reversal of momentum (*83, 88, 89, 90*).

F.   A Classical Force

In discussing Mossbauer momentum, Lipkin points out that if we sum *all* quantum events, we obtain a purely classi-

cal answer (50). Methods other than Mossbauer spectroscopy
support this conclusion; e.g., (91).

In electromigration, all microscopic drift velocities
are summed, into the macroscopic drift velocity, the experimental observable. By virtue of the Nernst-Einstein relation,
all the microscopic momentum transfers are also summed.
Electromigration experiments thus measure a purely classical
quantity. For this reason, we may expect similar behavior
from electron and from hole conductors.

G.   Pure Metals

Lattice planes may be identified by some inert marker
which does not conduct electricity as, in thermomigration,
they are identified by a marker that does not conduct heat
(79). In the frame of reference of such markers, the atom
flow is in the direction of the classical electron stream;
i.e., toward the anode.

H.   Impurities in Metals

In a binary system, the solute can have only a positive
scattering cross-section for electrons; it can have only a
positive electrical resistivity. The sense of the momentum
transfer is therefore still toward the anode, and in the
frame of reference of the crystal lattice, the solute, if
mobile, drifts in this direction.

In the frame of reference of the specimen ends, the
laboratory frame of reference (79), the direction of motion
of the solute is determined by the difference of the solute
and solvent drift velocities. This difference may be of
either sign. Obviously, it is important to define the frame
of reference in which the motion of the solute is described.

I.   Effective Charge

The very existence of the phenomenon of anisotropic
electromigration indicates the inadequacy of the concept of
effective charge. Electric charge is a scalar quantity. It
is meaningless to speak of an effective charge *along the
hexagonal axis* in h.c.p. metals, and the concept fails utterly
when one postulates a *different* effective charge along an *a*
axis.

XIII.   EXPERIMENTAL TESTS

A.   Gases

As remarked at the end of Subsection XI. F, only tests
between *different classes of experiment* are significant. The
poor agreement found in the intercomparison of thermal diffusion data on the one hand and viscosity and conductivity
on the other (*72, 76*) is considerably more significant than
the agreement found in comparisons with model potentials,
vide (74). It seems a reasonable inference that the b^3 of
Equation (78) might be worth trying, in preference to the
b^2 of Equation (77).

To the best of my belief, the present recipe for α_o is
not at variance with experiment.

B.   Solids

In ionic crystals, the entropy of lattice vacancies
measured by two different means (*92, 93*) are in agreement
(*81*).

In metals, the thermomigration and electromigration of
isotopes can be compared with self-diffusion data (*30, 41,
53, 94, 95*). There appear to be two crucial tests; the *sign*
of isotopic thermomigration, and the *existence* of anisotropic
thermomigration. Upon the former, present experimental data
are concordant with the ideas presented here (*30*). To the
best of my belief, no experiments have yet been done upon
the latter. In the absence of phonon drag, anisotropic thermomigration should not exist. No matter how anisotropic the
shape of a lattice vacancy, or of an impurity ion, the heat
of transport should be a scalar quantity.

The realization that $-w_f$ appears in the stationary
Arrhenius slope for instertitial impurities removes the
anomaly with which insterstitials have been associated.

XIV.   CONCLUSION

The whole of physics is irreversible.   If observables could speak, they would doubtless declare "je n'ai pas tout mon equilibre, donc j' existe."

The linear regions of physics are conceptually no more difficult than is Ohm's Law: all we need are *equilibrium parameters* and *conservative* forces.   Rigorously, these are all we have.

When the free entropy of an open system has been maximized, radiation forces may still cause a nonzero diffusive drift of matter.   Radiation forces can arise from photons, plasmons or from any other carrier of thermal energy.   As a stone in water sinks and a bubble rises, the sign of these radiation forces is always such as to transport energy *down* the energy gradient.   If all quantum events are summed, radiation forces are classical.

The kinetic theory of gases in disequilibrium lies within the scope of linear thermodynamics, and the two topics are here reconciled.

The concepts I have discussed bear upon the "doing" of physics.   Especially in solids, they suggest the intercomparison of different classes of experiment, often classes that have been thought to be unrelated to each other.   They also suggest comparatively simple mathematical recipes.

REFERENCES

1.   Bridgman, P.W., The Logic of Modern Physics, Page 123,
      Collier-Macmillan, New York, 1960.

2.   De Groot, S.R., Thermodynamics of Irreversible Proces-
      ses, North-Holland, Amsterdam, 1951.

3.   Hesketh, R.V., "Ohm's Law and Thermodynamic Equilibrium,"
      CEGB Report RD/B/N3561, 1976.

4.   Prigogine, I., Introduction to Thermodynamics of
      Irreversible Processes, Wiley, New York, 1961.

5.   Thomson, W., "On a Mechanical Theory of Thermo-
      Electric Currents", Proceedings, Royal Society Edinburgh,
      3, PP 91-98, 1854; "On the Dynamical Theory of Heat.
      Part VI.   Thermo-Electric Currents," Transactions,
      Royal Society Edinburgh, 22, Part I, PP 123-171, 1854.

6.   Onsager, L., "Reciprocal Relations in Irreversible
      Processes. I.", Physical Review, 37, PP 405-426, 1931;
      "Reciprocal Relations in Irreversible Processes.   II.,"
      Physical Review, 38, PP 2265-2279, 1931.

7.   Zubarev, D.N., Non-Equilibrium Statistical Thermody-
      namics, Consultants Bureau, New York, 1974.

8.   Hesketh, R.V., "How Big is the Linear Region of
      Irreversible Thermodynamics?" Thin Solid Films, 28,
      PP 375-387, 1975.

9.   Letters of Colleagues to R. V. Hesketh, 1975.

10.   Kittel, C., Introduction to Solid State Physics,
      Wiley, New York, 1953.

11.   Ziman, J.M., Principles of the Theory of Solids,
      University Press, Cambridge, 1964.

12.   Chapman, S., and Cowling, T.G., The Mathematical
      Theory of Nonuniform Gases, University Press,
      Cambridge, 1970.

13.   Huntington, H.B., "Effect of Driving Forces on
       Atom Motion," Thin Solid Films, 25, PP 265-280, 1975.

14.   Peierls, R.E., Quantum Theory of Solids, Clarendon
       Press, Oxford, 1955.

15.   Kittel, C., Quantum Theory of Solids, Wiley, New York,
       1963.

16.   Hesketh, R.V., "Atom Transport Down Thermal and
       Electric Gradients in Solids," CEGB Report RD/B/N1468,
       1969; "The Direct Force in Electromigration," CEGB
       Report RD/B/N3560, 1975.

17.   Das, A.K., and Peierls, R.E., "The Force on a Moving
       Charge in an Electron Gas," Journal of Physics C., 6,
       PP 2811-2821, 1973.

18.   Das, A.K., and Peierls, R.E., "The Force in Electromi-
       gration," Journal of Physics C., 8, PP 3348-3352, 1975.

19.   Kumar, P., and Sorbello, R.S., "Linear Response Theory
       of the Driving Forces for Electromigration," Thin
       Solid Films, 25, PP 25-35, 1975.

20.   Landauer, R., "The Das-Peierls Electromigration Theo-
       rem," Journal of Physics C., 8, PP L389-392, 1975.

21.   Sorbello, R.S., "Theory of Electromigration in Metals,"
       Comments on Solid State Physics, B6, PP 117-122, 1975.

22.   Shewmon, P.G., Diffusion in Solids, McGraw-Hill,
       New York, 1963.

23.   Huntington, H.B., "Driving Forces for Thermal Mass
       Transport," Journal of the Physics and Chemistry of
       Solids, 29, PP 1641-1651, 1968.

24.   Shockley, W.,"Some Predicted Effects of Temperature
       Gradients on Diffusion in Crystals," Physical Review,
       93, PP 345-346, 1954.

25.   Seeger, A., "Curved Arrhenius Plots in Self-Diffusion,"
       Comments on Solid State Physics, 4, PP 18-27, 1971.

26. Burton, J.J., "Analysis of Silver Self-Diffusion Data," Philosophical Magazine, 29, PP 121-133, 1974.

27. Rothman, S.J., Peterson, N.L., and Robinson, J.T., "Isotope Effect for Self-Diffusion in Single Crystals of Silver," physica status solidi, 39, PP 635-645, 1973.

28. American Society for Metals, Diffusion in Body-Centered Cubic Metals, ASM, Metals Park, Ohio, 1965.

29. Gerritsen, A.N., Encyclopedia of Physics, (S. Flügge, Editor), Vol. 19, PP 137-226, Springer, Berlin, 1956.

30. Hesketh, R.V., "A Heavy Isotope in a Solid Drifts *Down* a Thermal Energy Gradient, Journal de Physique, Vol. 37, PP 183-188, March, 1976.

31. Rose, R.M., Shephard, L.A., and Wulff, J., Electronic Properties, Wiley, New York, 1966.

32. Vasaru, G., "Thermal Diffusion in Isotopic Gaseous Mixtures," Fortschritte der Physik, 15, PP 1-111, 1967.

33. Hirschfelder, J.O., Curtiss, C.F., and Bird, R.B., Molecular Theory of Gases and Liquids, Wiley, New York, 1954.

34. Grew, K.E., and Ibbs, T.L., Thermal Diffusion in Gases, University Press, Cambridge, 1952.

35. Jeans, J.H., The Dynamical Theory of Gases, University Press, Cambridge, 1925.

36. Crolet, J.L., "Experimental and Theoretical Study of the 'Intrinsic Contribution' to the Heat of Transport," Zeitschrift für Naturforschung, 26a, PP. 907-914, 1971.

37. Callen, H.B., Thermodynamics, Wiley, New York, 1960.

38. Planck, M., Treatise on Thermodynamics, Dover, New York, 1945.

39. Shewmon, P.G., "Thermal Diffusion of Carbon in α and γ Iron", Acta Metallurgica, 8, PP 605-611, 1960.

40.   Allnatt, A.R., and Chadwick, A.V., "Thermal Diffusion
      of Strontium Ions in Sodium Chloride," Transactions
      of the Faraday Society, 63, PP 1929-1942, 1967.

41.   Thernquist, P., and Lodding, A., "Isotope Transport
      Along a Temperature Gradient in Li Metal," Zeitschrift
      für Naturforschung, 22a, PP 837-839, 1967.

42.   Glansdorff, P., and Prigogine, I., Thermodynamic
      Theory of Structure, Stability and Fluctuations,
      Wiley-Interscience, London, 1971.

43.   Ashkin, A., "The Pressure of Laser Light," Scientific
      American, 226, No. 2, PP 63-71, 1972.

44.   Poynting, J.H., "Radiation Pressure," Philosophical
      Magazine, 9, PP 393-406, 1905.

45.   Vigoureux, P., "Radiation Pressure," Contemporary
      Physics, 7, PP 440-446, 1966.

46.   Morse, P.M., and Feshbach, H., Methods of Theoretical
      Physics, Part I, PP 302-306, McGraw-Hill, New York,1953.

47.   Pippard, A.B., and Saxton, W.O. (Editors), Cavendish
      Problems in Classical Physics, Problem 51 (c),
      University Press, Cambridge, Second Edition, 1971.

48.   Burt, M.G., and Peierls, R.E., "The Momentum of a Light
      Wave in a Refracting Medium," Proceedings of the Royal
      Society, A333, PP 149-156, 1973.

49.   Hall, H.E., Solid State Physics, Wiley, London, 1974.

50.   Lipkin, H.J., Quantum Mechanics, New Approaches to
      Selected Topics, North-Holland, Amsterdam, 1973.

51.   Wood, A.B., Acoustics, Blackie, London, 1940.

52.   Chapman, S., "Thermal Diffusion of Rare Gas Constituents in
      Gas Mixtures," Philosophical Magazine, 7, PP 1-16, 1929.

53.   Hesketh, R.V., Diffusion Processes (J. N. Sherwood et
      al. Editors), Vol. 1, PP 231-273, Gordon and Breach,
      London, 1959.

54. Hesketh, R.V., _Atomic Transport in Solids and Liquids_ (A. Lodding and T. Lagerwall, Editors), Verlag der Zeitschrift für Naturforschung, PP 23-35, Tübingen,1971.

55. Paul, R., Howard, A.J., and Watson, W.W., "Isotopic Thermal-Diffusion Factor for Xenon," _Journal of Chemical Physics_, 43, PP 1890-1894, 1965.

56. Jones, R.C., and Furry, W.H., "The Separation of Isotopes by Thermal Diffusion," _Reviews of Modern Physics_, 18, PP 151-224, 1946.

57. Frankel, S.P., "Elementary Derivation of Thermal Diffusion," _Physical Review_, 57, Page 661, 1940.

58. Cowling, T.G., "Approximate Theories of Thermal Diffusion," _Journal of Physics A_, 3, PP 774-782, 1970.

59. Letters to R. V. Hesketh, 1970-1971.

60. Cowling, T.G., Private Communication, 1970-1972.

61. Furth, R., "An Elementary Theory of Thermal Diffusion," _Proceedings of the Royal Society_, A179, PP 461-469, 1942.

62. Monchick, L., and Mason, E.A., "Free-Flight Theory of Gas Mixtures," _Physics of Fluids_, 10, PP 1377-1390, 1967.

63. Waldmann, L., _Encyclopedia of Physics_, Vol. 12 (S. Flügge, Editor), PP 295-514, Springer, Berlin, 1958.

64. Burbury, S.H., _The Kinetic Theory of Gases_, University Press, Cambridge, 1899.

65. Burbury, S.H., "Boltzmann's Law of Distribution $\epsilon^{-2h\chi}$, and van der Waals' Theorem," _Philosophical Magazine_, 2, PP 403-417, 1901.

66. Wyllie, G.A.P., Private Communication, 1973.

67. Burbury, S.H., "On the General Theory of Stationary Motion in an Infinite System of Molecules," _Proceedings of the London Mathematical Society_, 29, PP 225-248, 1898.

68. Sommerfeld, A., Thermodynamics and Statistical
     Mechanics, Academic Press, New York, 1956.

69. Reif, F., Fundamentals of Statical and Thermal Physics,
     Section 7.5, McGraw-Hill, New York, 1965.

70. Maxwell, J.C., On Boltzmann's Theorem on the Average
     Distribution of Energy in a System of Material Points;
     Collected Scientific Papers (W.D. Niven, Editor),
     Vol. 2, PP 713-741, Reprinted by Dover, New York, 1965.

71. Landau, L.D., and Lifshitz, E.M., Mechanics, Page 1,
     Pergamon Press, Oxford, 1960.

72. Maxwell, J.C., Atom; Collected Scientific Papers
     (W.D. Niven, Editor), Vol. 2, PP 445-484, Reprinted by
     Dover, New York, 1965.

73. Kuhn, T.S., The Structure of Scientific Revolutions,
     University Press, Chicago, Second Edition, 1970.

74. Maxwell, J.C., An Essay on the Mathematical Principles
     of Physics, etc. (Review), Collected Scientific Papers
     (W.D. Niven, Editor), Vol. 2, Page 339, Reprinted by
     Dover, New York, 1965.

75. Saxena, S.C., and Raman, S. "Theory and Performance of
     Thermal Diffusion Column," Reviews of Modern Physics,
     34, PP 252-266, 1962.

76. Tokuda, T., Ando, Y., and Fukui, K., "Thermal Diffusion
     of Argon Isotopes," Journal of Applied Physics, 41,
     PP 2854-2859, 1970.

77. Landau, L.D., and Lifshitz, E.M., Statistical Physics,
     Chapter 2, Pergamon Press, London, 1958.

78. Mathur, B.P., and Watson, W.W., "Thermal Diffusion in
     Isotopic $^{16}O_2$-$^{18}O_2$," Journal of Chemical Physics, 51,
     PP 2210-2214, 1969.

79. Biersack, J., and Diez, W., "Motion of Markers and
     Bubbles in Solids by Self-Diffusion in a Temperature
     Gradient," physica status solidi, 27, PP 139-144,
     1968.

80. Waldman, L., and Schmitt, K.H., Aerosol Science (C. N. Davies, Editor), PP 137-162, Academic Press, New York, 1966.

81. Hesketh, R.V., Discussion, Supplément au Journal de Physique, 34, C9-19, 1973.

82. Routbort, J.L., "Electromigration in Zinc Single Crystals," Physical Review, 176, PP 796-803, 1968.

83. Huntington, H.B., Alexander, W.B., Feit, M.D., and Routbort, J.L., "Atomic Transport in Solids and Liquids," (A. Lodding and T. Lagerwall, Editors), Verlag der Zeitschrift für Naturforschung, PP 91-96, Tübingen, 1971.

84. Ziman, J.M., The Physics of Metals, I. Electrons, (J.M. Ziman, Editor), PP 250-282, University Press, Cambridge, 1969.

85. Schwarz, H., "Die Bestimmung der Fermifläche von Kadmium aus der Orientierungsabhängigkeit der Periode der Sondheimer-Oszillationen," physica status solidi, 39, PP 507-514, 1970.

86. Smith, R.A., Wave Mechanics of Crystalline Solids, Chapter 8, Equations (106, 107), Chapman and Hall, London, 1961.

87. Smith, A.C., Janak, J.F., and Adler, R.B., Electronic Conduction in Solids, McGraw-Hill, New York, 1967.

88. Huntington, H.B., "Current Basic Problem in Electromigration in Metals," Transactions, Metals Society AIME, 245, PP 2571-2579, 1969.

89. Ho, P.S., "Electromigration and Soret Effect in Cobalt," Journal of the Physics and Chemistry of Solids, 27, PP 1331-1338, 1966.

90. D'Amico, J.F., and Huntington, H.B., "Electromigration and Thermomigration in Gamma-Uranium," Journal of the Physics and Chemistry of Solids, 30, PP 2607-2621, 1969.

91. Brown, S., and Barnett, S.J., "Carriers of Electricity
     in Metals Exhibiting Positive Hall Effects," **Physical
     Review**, 87, PP 601-607, 1952.

92. Beniere, F., Beniere, M., and Chemla, M., "Conduc-
     tibilitie, Nombres de Transport et Autodiffusion des
     ions dans Differents Monocristaux de Chlorure de
     Sodium," **Journal of the Physics and Chemistry of
     Solids**, 31, PP 1205-1220, 1970.

93. Lowe, I., and Blackburn, D.A., "Measurement of the
     Heat of Transport of K^+ ions in Potassium Chloride,"
     Supplement au Journal de Physique, 34, C9-191-197, 1973.

94. Thernquist, P., and Lodding, A., "Electrotransport of
     Lattice Defects in Lithium Metal," **Zeitschrift für
     Naturforschung**, 23a, PP 627-628, 1968.

95. Lodding, A., Mundy, J.N., and Ott, A., "Isotope Inter-
     Diffusion and Self-Diffusion in Solid Lithium Metal,"
     physica status solidi, 38, PP 559-569, 1970.

ACKNOWLEDGEMENTS

        Professor T. G. Cowling has written to me patiently and
at length. Sir Alan Cottrell's wisdom has shielded me.
Professor F. C. Frank has kept me to the point. Professor
A. P. French and I have hit a similar wavelength. Sir Brian
Pippard's friendly scorn has sharpened my thinking. Professor
W. W. Watson has ben a kind and vigorous mentor. Dr. G. A. P.
Wyllie has put Burbury to the test. All these I thank. This
paper is published by permission of the Central Electricity
Generating Board.

NOMENCLATURE

a	constant of proportionality
a, b	suffixes denoting solvent and solute, respectively
b	atomic diameter
b	excluded volume in a van der Waals gas
c	concentration of an impurity species (solute ion, lattice vacancy, gaseous isotope)
$\underset{\sim}{c}$	velocity of light
C_p	specific heat at constant pressure
d	distance along a gradient
D	diffusion coefficient of one defect (solute ion, lattice vacancy, etc.)
D_s	self-diffusion coefficient
e	electronic charge
E	energy
F_o	force constant in potential
$\underset{\sim}{F}$	force
G	Gibbs free energy
$\underset{\sim}{G}$	reciprocal lattice vector
h	Planck's constant
$\hbar$	$h/2\pi$
$\underset{\sim}{H}$	magnetic field
$\underset{\sim}{i}$	unit vector
i, j	suffixes denoting particles
$\underset{\sim}{J}$	flux
k	Boltzmann's constant
$\underset{\sim}{k}\ \underset{\sim}{k}'$	wave vectors of incident and scattered crystal momentum
m	mass of a gas atom

m^* effective mass of conduction electron

M magnetization

n an index

n_o number density of conduction electrons

N number of atom sites

N_o number density of gas atoms

P gas pressure

$\underset{\sim}{p}$ momentum

$\underset{\sim}{q}$ position coordinate of $\underset{\sim}{p}$; also, at Equations (45-47) a wave vector

q intensive thermal energy

$q^{*\prime}$ reduced thermal energy of transport

Q extensive thermal energy

r radial distance in spherical symmetry

R residual resistivity of one atomic percent of impurities

s specific (intensive) entropy

s_v specific vibrational entropy

$s_{v,e}$ specific electronic vibrational entropy

S extensive entropy

t time, though since I have been concerned principally with stationary states, it does not explicitly appear

T temperature

$\underset{\sim}{u}$ drift velocity

$\underset{\sim}{v}$ mutual velocity of gas atoms

$\underset{\sim}{v}_F$ Fermi velocity

V voltage

$\overline{V^2}$ mean square of noise voltage

$\underline{V}$	gas volume
$(\underline{V} - \underline{b})$	free volume in a van der Waals gas
w_f	enthalpy of formation of a defect structure (a lattice vacancy, the solution of an impurity ion in a crystal, of a second species in a gas)
w_m	enthalpy of migration of a defect
x	an independent variable, and later, a thermodynamic force
y	a dependent variable, and later, an intrinsic parameter of a physical system. y is conjugate to x in regard to entropy production
z	the "hardness parameter" of a repulsive inter-atomic potential
α_o	reduced thermal diffusion factor
γT	the electronic specific heat of an atom in a solid
ε_i	**kinetic energy of the ith particle**
η	stiffness constant for wave propagation
λ	mean free path of a particle in a diffusive system
μ	reduced mass; i.e., $m_1\, m_2/m_1 + m_2$
μ, ν	direction indices
ν	frequency
$\Delta\,\nu$	bandwidth
ρ	mass per unit volume
σ	conductivity, electrical and thermal
τ	relaxation time for scattering in a diffusive situation
ϕ	a repulsive potential
Φ	the Planck function
ω	atomic volume
Ω	volume of an obstacle
$\Delta\Omega$	an infinitesimal volume element

KINETICS OF NUCLEAR SYSTEM

Solution Methods for the Space-Time
Dependent Neutron Diffusion Equation

W. Werner

Laboratorium für Reaktorregelung
und Anlagensicherung Garching
Lehrstuhl für Reaktordynamik und Reaktorsicherheit
Technische Universität München

I. INTRODUCTION

Within the large variety of methods of analysis for transient phenomena in nuclear reactors, this article is limited
to the discussion of computational methods for the analysis
of operational transients, off-normal transients, and hypothetical accidents in power reactors. Though any realistic
analysis of such transients requires a coupled treatment of
thermodynamics, fluid dynamics and neutronic phenomena, all
details of the modeling and computations of thermodynamic
and fluid dynamic quantities are omitted in the following
presentation.

Any spatially non-uniform perturbation of the physical
state of a reactor causes spatially and spectrally non-
uniform transient adjustment of its neutron density distribution.

Disregarding feedback effects, two phenomena necessitate
the consideration of spatial effects: delayed neutron holdback, and the influence of the time derivative of the neutron
flux, where the latter is only of importance in super-prompt
critical transients.

The sometimes significant effects of these phenomena
have been demonstrated in a number of publications; e.g.,
(1-4). The inclusion of the feedback effects, which affect

313

the neutron population in a non-linear fashion, adds another
class of phenomena which necessitates space-time dependent
calculations (5,6).

Though it has been known for some time that space dependent calculations of reactor transients were required in
order to accurately predict neutron flux and power, it has
been only in the last few years that substantial efforts and
progress in the development of computational capabilities
for carrying out such calculations has been reported in the
literature. Some of the reasons for this growing interest
in space dependent calculations are:

1. With increasing size of reactors, the spatial flux
tilting due to local perturbations, and the delayed neutron
holdback effect, become more pronounced.

2. In order to reduce conservatism in modeling of transients, and to be able to set operational safety limits as
realistic as possible more accurate transient calculations
are required.

3. Through the availability of the current generation
of large scientific computers (IBM 370/195 and CDC 7600),
genuine three dimensional solution techniques for the few
group neutron diffusion equation have become feasible, if
advanced numerical techniques are employed.

II. FORMULATION OF PROBLEM

In all following discussions, it is assumed that few-
group neutron diffusion theory is a sufficiently accurate
model for the phenomena to be investigated. Criteria for
this assumption are amply discussed in the literature; e.g.,
(7-9).

Then, the mathematical problem to be solved is an initial-
boundary value problem for the set

$$(v^{-1} \frac{\partial}{\partial t} - L)\phi - \Gamma c = 0, \quad (\frac{\partial}{\partial t} + \Lambda)c - B\phi = 0 \qquad (2.1)$$

of partial and ordinary differential equations.

Herein,

$$\phi(\vec{x},t) = (\phi_1(\vec{x},t),\ldots,\phi_m(x,t))^T, \text{ and}$$

$$c(\vec{x},t) = (c_1(\vec{x},t),\ldots, c_g(x,t))^T, \text{ resp.,}$$

are the vectors of (m) prompt groups, and (g) delayed precursor groups.

Furthermore,

$$L = \nabla \cdot D\nabla - A, \text{ with } D = \text{diag } (D_i),$$

$$(A)_{ik} = \delta_{ik}\Sigma_i^{rem} - \chi_i\nu(1-\beta)\Sigma_k^f - \Sigma_{ik}^s,$$

$$i,k = 1,\ldots,m$$

Since L is an elliptic differential operator of second order the subsystem of partial differential equations in (2.1) is of parabolic type.

The nuclear cross-sections D, Σ^f, Σ^{rem}, and Σ^s are piecewise continuous functions of $\vec{x}$,t and υ, where υ is a vector of thermo- and fluid-dynamic state variables. The interdependence of $\Psi = (\phi,c)^T$ and υ is governed by boundary conditions between the variables of Equations 2.1 and the set of equations describing thermo- and fluid-dynamic properties. The functions $D(\vec{x},t,\upsilon),\ldots,\Sigma^s(\vec{x},t,\upsilon)$ describe the "feedback" of thermo- and fluid-dynamic quantities on Ψ. Finally,

$$(\Gamma)_{ik} = \chi_{ik}\lambda_k, \quad \Lambda = \text{diag } (\lambda_k),$$

$$(B)_{ik} = \upsilon_i\Sigma_k^f, \quad i = 1,\ldots,m, \quad k = 1,\ldots,g.$$

λ_k and β_k, k = 1,\ldots,g, are constants.

Initial conditions (2.2):

In a closed domain (G) in the hyperplane $t = t_o$, $m + g$
functions $\psi_i(\vec{x}, t_o) = \psi_i(\vec{x})$, $i = 1, \ldots, m + g$, are prescribed.

Continuity conditions (2.3):

The interior of (G) may contain surfaces $F_j(\vec{x}) = 0$,
$j = 1, \ldots, J$, across which the nuclear cross-sections D, Σ^f,
Σ^{rem}, Σ^s are discontinuous.

If $M = \{\vec{x} : F_j(\vec{x}) = 0, j = 1, \ldots, J\}$ then

a.   on G/M : $\psi(\vec{x}, t) \varepsilon C^2$.

b.   on M : if $M_j = \{\vec{x} : F_j(x) = 0\}$, $M_j \varepsilon M$, and

$\vec{n} \perp (F_j = 0)$, then

$$\lim_{(\vec{x}\varepsilon F_j(\vec{x}) > 0) \to (\vec{x}\varepsilon M_j)} D(\vec{x})\frac{\partial \psi}{\partial n} = \lim_{(\vec{x}\varepsilon F_j(\vec{x}) < 0) \to (\vec{x}\varepsilon M_j)} D(\vec{x})\frac{\partial \psi}{\partial n}$$

Boundary conditions (2.4):

On the boundary (R) of (G) conditions $\alpha\psi + \beta \frac{\partial \psi}{\partial n} = 0$
have to be satisfied.

In the halfspace $t \geq t_o$ functions $\psi(\vec{x}, t)$ are sought
which satisfy Equation $\overline{2.1}$ and the side conditions 2.3 and
2.4, and for which $\lim_{t \to t_o} (\psi(\vec{x}, t)) = \bar{\psi}(\vec{x})$ holds.

The integers ,m, and ,g, resp., which are the numbers
of prompt neutron groups, and delayed precursor groups,
resp., result from a discretisation in energy space. It is
commonly agreed that for most transient problems in thermal
reactors it is sufficient to use 2 prompt groups and 6 delayed precursor groups. Very rarely it is necessary to use
3 or 4 groups of prompt neutrons.

For Fast Breeder Reactors, at least 6, but often more
than 15 groups of prompt neutrons are needed, together with
6 groups of delayed precursors.

With regard to feedback it is assumed that the relevant
thermo-fluid-dynamic state variables can be computed separately at any desired instant of time $t \geq t_o$.

Since it is obviously impossible to find explicit solutions of this problem for any realistic reactor, numerical
methods must be employed for its solution.

III. SPATIAL APPROXIMATIONS

Common feature of all computational methods amenable to
a feasible implementation into two- and three-dimensional
kinetic codes is a discretisation of the space variables,
$\vec{x}$: G, which is the interior and the natural or artificial
boundary of the reactor, is subdivided into discrete regions,

$$G_j, \ j = 1,\ldots,N, \ \cup G_j = G, \ G_j \cap G_\ell = 0, \ j \neq \ell.$$

In order to simplify the notation, the following derivations are carried out for a rectangular cartesian coordinate system (x,y,z). Then $x = x_i$, $i = 1,\ldots,il$; $y = y_k$,
$k = 1,\ldots,kl$; $z = z_\ell$, $\ell = 1,\ldots,\ell l$ defines the "gridlines"
of the spatial grid; their intersections $x_{ik\ell} = (x_i,y_k,z_\ell)$
are the gridpoints, x_j, of the spatial grid. In this case
the regions G_j are cubes with centers x_j. On the regions
G_j gridfunctions $\Psi_j(\vec{x})$ are defined. For $x \epsilon G_j$, $\Psi_j(\vec{x})$ is
considered an approximation to the true solution $\Psi(x)$.

By defining a neighborhood of G_j through the set of
indices I_j, a difference analogue to the differential operator $\nabla \cdot D \nabla \phi^j$ can be constructed, which can be expressed in
terms of gridfunctions $\Psi_i(\vec{x})$, $i \epsilon I_j$, or of parameters of such
gridfunctions.

Thus, the system of ordinary and partial differential
equations for $m + g$ functions of $\vec{x}$ and t is converted into
a system of ordinary differential equations for $(m + g) \cdot N$
functions of t. To each region G_j there correspond $m + g$
equations of this system. In the right hand side of each of
the first m equations with index j, there appear the gridfunctions $\Psi_i(\vec{x})$, $i \epsilon I_j$, or parameters thereof. Since the
system is very large for the 2- or 3-dimensional case, stor-
age requirements prohibit anything else but the use of one-
step methods for its numerical integration. If the system

is not linear, it is converted, at least locally within the
intervals Δt of the time steps, into a linear one. Thus, all
methods to be discussed in the following lead to the problem
of solving a system of linear equations.

$$A^{n+1}\psi^{n+1} = A^n \psi^n \qquad (3.0)$$

where $\psi = (\psi_1,\ldots,\psi_N)^T$. Instead of the gridfunctions also
parameters of the gridfunctions can appear. Formally, the
various spatial approximations which are discussed in this
chapter, differ with regard to

a. set I_j, which influences the structure of the
    matrices A

b. gridfunctions $\psi_j(\vec{x})$, which influences the value of
    the entries of A.

The various time integration methods to be discussed in
chapter 4 differ with regard to the solution method for
system 3.0.

A. Finite Difference Methods

In the classical approach, Finite Difference (FD)
Methods are constructed by defining gridfunctions
$\psi_j(\vec{x}) = \text{const} = \psi_j = \psi_j(\vec{x}_j)$, $j = 1,\ldots,N$.

The differential operators in 2.1 are replaced by difference operators as follows: In order to keep the computational burden of solving system 3.0 as low as possible,
it is desirable to keep matrix A as sparse as possible. The
most efficient FD-methods are obtained by choosing for I_j
the set of indices corresponding to G_j and its six neighbor
regions adjacent to the surfaces of G_j. This leads to the
well known 7 point formula, which is in use in most FD-codes
and which replaces $\nabla \cdot D\nabla \phi$ by

$$D_{ikl}(\phi_{i+1,k,\ell} + \phi_{i-1,k,\ell} + \phi_{i,k+1,\ell} + \phi_{i,k-1,\ell}$$
$$+ \phi_{i,k,\ell+1} + \phi_{i,k,\ell-1} - 6\phi_{ik\ell})/h^2.$$

(Case of equal mesh size for all coordinates and all adjacent
cubes and equal diffusion constant for all adjacent cubes.)

If the discrete equivalent L_D of the differential operator L is evaluated only at time layer n (or even further backwards) then the matrix A^{n+1} in Equation 3.0 becomes diagonal or block-diagonal with $m \times m$ blocks. If values in time layer n and such values in time layer $n+1$, which are already known (with regard to the direction of progressing through the grid) are used to evaluate L_D, then matrix A^{n+1} becomes lower triangular. In both cases, the solution of system 3.0 can be computed explicitly from already calculated values, therefore, such methods are called "explicit" methods. If A^{n+1} possesses $m \times m$ block entries to both sides of the diagonal $m \times m$ blocks, then the solution at time layer $n+1$ must be calculated by solving a linear system of $N(m+g)$ unknowns, where the essential computational work is caused by the $N \cdot m$ unknowns corresponding to the prompt neutrons. The method is then called "implicit."

Common to all FD-methods using the 7-point approximation of the differential operator is the well known block tridiagonal structure, with $m \times m$ blocks, of the matrix A^n, or of A^n and A^{n+1}. The diagonal blocks $(A)_{j,j}$ describe the intergroup coupling within region G_j, their entries depend on nuclear cross-sections + terms inversely proportional to the square of the spatial mesh size, h. The off-diagonal blocks $(A)_{j,j+1}$ and $(A)_{j+1,j}$ describe the spatial coupling between adjacent regions G_j, G_{j+1}; their entries depend on $1/h^2$. Obviously, the spatial coupling between neighboring regions (for group i) gets lost if $D_i/h^2 << \Sigma_i^{rem} \simeq D_i/L_i^2$ (L_i : diffusion length of group i), in other words, if $h >> L_i$.

The block tridiagonal structure, into which Equation 3.0 is brought by FD-methods is very favorable for its numerical treatment and consequently leads to short computation times per unknown.

In principle, FD-methods would permit the calculation of solutions of 3.0 to any desired degree of accuracy, if the spatial mesh size is chosen "fine enough;" a clarification of this loose terminology comes from the above discussion of coupling properties; h should not be much greater than $\min(L_i)$. For thermal systems, this means that h should not exceed several cm. This condition clearly shows the limitations of FD-methods, as applied to present day thermal reactors: for a large PWR, about $6 \cdot 10^5$ regions G_j would be needed for h = 5 cm, and about $5 \cdot 10^6$ regions for h = 2.5 cm.

Obviously, this would exclude an economical computation of transients in such reactors.

In fast systems, and with present day demonstration reactors, the situation is not as severe.

The discussed upper bounds for the mesh size in FD-methods can obviously not be removed by using higher order approximations for the diffusion-operator. This would introduce extra off diagonal entries in the matrix, but all these entries would be proportional to D/h^2. Thus, an improved spatial coupling for larger mesh size cannot be reached on the basis of FD-methods.

B. Coarse Mesh Methods

For methods to become economically feasible for 3d-problems, it is necessary to greatly reduce the number of unknowns of the problem. This can only be achieved through a considerable increase of mesh size. For large LWR's it is desirable that mesh sizes equal to fuel assembly size can be used in the xy-plane, and about the same size in the axial direction. The solutions obtained with such mesh size should be accurate within 2 - 3% of the true solution of the problem. A higher accuracy seems unnecessary in view of the assumptions entering the derivation of the basic equation 2.1 and of uncertainties of physics parameters. However, convergence to the true solution with decreasing mesh size must be guaranteed.

Basis of the following derivation of a broad class of coarse mesh methods is the concept of "generalized" solutions of a variational principle associated with a differential equation $Lu - r = 0$ (or $u_t + L_u - r = 0$, or $Lu - \lambda u = 0$), which can be shortly described as follows:

Let U_L, W_L be Hilbert-spaces with scalar products: $(\cdot,\cdot)_{U_L}$, $(\cdot,\cdot)_{W_L}$ and $L:D(L) \to W_L$, linear, continuous, $D(L) \subseteq U_L$, $Lu-r=0$.

The associated variational principle considers another pair of Hilbert spaces U,W; $(\cdot,\cdot)_U$, $(\cdot,\cdot)_W$. With $B \varepsilon L(U,W)$, $R \varepsilon L(W)$, an element $u_o \varepsilon U$, satisfying $B(u_o,w)-R(w) = 0$, $\nabla w \varepsilon W$, is called "generalized" solution of $Lu-r=0$, if $D(L) \subseteq U$, $W \subseteq W_L$, dense, and $B(u_o,w) = (Lu,w)_{W_L}$, $R(w) = (r,w)_{W_L}$.

A powerful and flexible method to obtain approximate "generalized" solutions is the Weighted Residual (WR) (also called Galerkin-Petrov) method (*10*).

With a pre-Hilbert space $U' \supseteq U$, $(\cdot,\cdot)_{U'}/_{U \times U} = (\cdot,\cdot)_U$, and $B' \in L(U',W)$, $B'/_{U \times W} = B$, let $P^N \in U'$ and $W^N \in W$ be spaces of "basis functions" and "weight functions" and $\dim(P^N) = \dim(W) = N$. Then, the generalized solution P_{oj} furnished by the WR-method is defined by: $P_{oj} \in P^N$, such that $B'(p_{oj},w_j) = R(w_j)$, $\forall w_j \in W^N$.

Returning to the notations of Equation 2.1, but omitting delayed precursors, and with $(p,w) = \int_{Gx|t_o,t_1|} pwdVdt$, we are led to find P_{oj} such that

$$\left(\frac{\partial p_{oj}}{\partial t}, w_j\right) - (Lp_{oj}, w_j) = 0, \quad \forall w_j \in W^N \tag{3.1}$$

For 3d-problems, computer storage space limitations practically preclude anything else but an approximation of the time derivatives in 3.1 by

$$(p_j^{n+1}, w_j) - (p_j^n, w_j) - \Delta t_n (\alpha_n (Lp_j^{n+1}, w_j)$$

$$+ (1-\alpha_n)(Lp_j^n, w_j)) = 0, \quad \forall w_j \in W^N \tag{3.2}$$

where $\Delta t_n = t_{n+1} - t_n$, and α_n is a weight factor.

By taking bases $\sigma_{j1}(\vec{x}), \ldots, \sigma_{jN}(\vec{x})$ of P^N, resp. $\tau_{j1}(\vec{x}), \ldots, \tau_{jN}(\vec{x})$ of W^N and expressing $p_j^\ell = \sum_{i=1}^{N} \gamma_i^\ell \sigma_{ji}$,

$\ell = n, n+1$, γ_i^n, $i=1, \ldots, N$ being known, a discrete solution of 3.2 can be obtained by computing γ_i^{n+1}, $i=1, \ldots, N$, as solution of a matrix equation

$$\sum_{\mu=1}^{N} B_{i\mu} \gamma_{\mu}^{n+1} = r_i, \quad i=1,\dots,N, \qquad (3.3)$$

with $\quad B_{i\mu} = (\sigma_\mu, \tau_i) - \Delta t_n \alpha_n (L\sigma_\mu, \tau_i).$

If the weight functions are defined as

$$w_{jk}(\vec{x}) = \begin{cases} \neq 0 \text{ on } G_k \\ = 0 \text{ elsewhere} \end{cases}, \text{ then the } B_{i\mu} \neq 0 \text{ only for } i = \mu.$$

In this case the coupling between the unknowns in regions G_i, G_k $i \neq k$ has to be described by appropriate physical boundary (interface) conditions.

Common feature of all Weighted Residual Methods is that the matrix entries describing the coupling between adjacent regions contain not only terms proportional to D/h^2, as in the FD-matrices, but also terms proportional to fission- and absorption cross-sections. Thus, spatial coupling is not lost as h gets large, which means that these methods are potentially suitable for LWR calculations with regions G_j identical with fuel assemblies in the xy-plane and a height of 20-30 cm.

1. Symmetric Weighted Residual Method

    (Finite Element Method)

    The best known Weighted Residual Method is the classical Finite Element (FE) Method (11-13). As basis functions polynomials of order q are used, and $\tau_{ij}(\vec{x}) = \sigma_{ij}(\vec{x})$, $i=1$, ...,N. As regions G_i, $i=1,\dots,N$ triangles or rectangles are mostly taken in 2d, and pyramids or prisms in 3d.

    To be specific, in region G_i, $p_j^i(\vec{x}) = \sum_{k=1}^{K} \gamma_{ik} \sigma_{jk}(\vec{x})$,

where $\sigma_{jk}(\vec{x})$ are polynomials such that

$\sigma_{jk}(\vec{x}) \neq 0$ On the starshaped region around gridpoint k, which is bounded by straight lines (2d) resp., planes (3d) connecting a set I_k of neighboring gridpoints.

$\sigma_{jk}(\vec{x}) = 0$ elsewhere.

The set I_k depends on the degree of the polymial $\sigma_{jk}(\vec{x})$. The number of its elements strongly influences the amount of labor associated with the solution of the matrix Equation 3.3.

For example with one group of prompt neutrons in 2d geometry and a triangular mesh, each line of the matrix B of 3.3 has 7 non zero entries for linear polynomials and 13 non zero entries for quadratic polynomials. For a rectangular mesh there are 9 non zero entries both for linear and quadratic polynomials.

In 3d geometry and a mesh consisting of prisms with triangular cross-sections, each line of B has 21 non zero entries for linear polynomials and 41 non zero entries for quadratic polynomials. For a rectangular mesh 27 entries are non zero for both linear and quadratic polynomials.

In contrast to these figures, FD-matrices have 5, resp. 7 non zero entries in 2d, resp. 3 d in a rectangular mesh.

In general, the number of entries in B increases with increasing degree of the polynomials.

Convergence properties of FE-methods have been explored very intensively (14-16). Error estimates are of the type

$$||p_j - u_o||_{w_2^2(G)} < K \inf_{p_j \in P^N} ||p_j - u_o||_{w_2^2(G)}, \text{ which focuses atten-}$$

tion to the approximation theory problem of how

$$\inf ||p_j - u_o||_{w_2^2(G)} \text{ depends on smoothness properties of}$$

u_o and on the choice of the subshape P^N. If the functions

$\sigma_{ji}(\vec{x}), i=1,\ldots,N$, are taken from the space U of solutions u_o (often called "energy-space") i.e., if their differentiability properties are such that the approximate solution $p_j^{n+1}(\vec{x})$ of Equation 3.2 have the same differentiability properties as the solution u_o, then sharp error estimates of the type

$$||p_j - u_o||_{w_2^2(g)} = O(h^r), \ r \geq q \ \text{ can be derived.}$$

In the 1d-case, functions $\sigma_{ji}(\vec{x})$ can easily be selected which have the desired differentiability properties and simultaneously lead to a narrow bandwidth of the matrix B; in 2d-and 3d-problems, however, the desired smoothness properties can only be obtained at the cost of a great number of entries in matrix B, which makes the solution of 3.3 so expensive, that, in practice, the advantage of large mesh size becomes offset by the expensive solution of the resulting equation. Therefore, all efficient FE-codes for multidimensional problems use "nonconforming" basis functions $\sigma_{ji}(x) \notin U$. This reduces the number of entries in B, but it makes convergence increasingly uncertain with decreasing smoothness properties. For this situation, the "patch" test has been devised, which in effect, is often not much more than a test for consistency. Among the WR-methods considered in section B, FE-methods are most severely affected by non-conforming basis functions, since nonconformity reappears in the weight functions which are identical with the basis functions.

For the solution of the FE-form of Equation 3.3, which has a positive definite symmetric matrix B, direct Choleski technique can be used or iterative methods like relaxation or conjugate gradient techniques. In 3d, the direct solution technique is probably too costly; among the iterative solutions, conjugate gradient methods look attractive in view of the relatively small number (several 1000) unknowns.

Though FE-methods have been in use for many years in areas like stress analysis or fluid flow problems (mostly 2d-problems which permit an easy direct solution), experience with applications to neutron flux calculation is very limited and it is not yet possible to form a firm opinion on the most efficient solution technique for Equation 3.3.

The superiority of FE-methods over other methods in its traditional fields of application mainly results from its flexibility in choosing the regions G_i, which permits one to cover the domain D in which the solution has to be calculated, by a fine mesh in regions where this is desired, and by a coarse mesh in other regions. In other methods, which are based on the use of a regular grid, this is not possible, since the fine mesh used in some region is projected parallel to the grid lines into portions of the domain G, where a fine mesh may not be required. Thus, FE-methods require considerably fewer unknowns in such applications. This advantage is of more theoretical value in neutron flux calculations, especially of LWR's, with their regular fuel assembly structure.

2.   Unsymmetric Weighted Residual Method

Weighted Residual Methods permit the construction of computational schemes, which are more efficient than FE-methods for cases which imply a regular grid structure.

The unsymmetric WR-Methods to be discussed in this section use different spaces P^N and W^N. The first method of this type, which has been implemented into a 3d neutron kinetics code for LWR-calculations (17-19) uses weight functions

$$w_i(\vec{x}) = \{ \begin{array}{l} 1 \quad \vec{x} \in G_i \\ 0 \quad \vec{x} \in G_i \end{array}$$

and polynomials of degree 2 or 3 for the basis functions, which are defined as follows (for shortness of notation, the derivation is made for a rectangular grid):

a.   the solution $\phi(\vec{x})$ is represented inside G_i, $i-1,\ldots,N$, explicitly by a polynomial $P_i^k(\vec{x})$ of degree k

b.   support of $P_i^k(\vec{x})$ is G_i

c.   discrete points of support of $P_i^k(\vec{x})$ are the center-point, $\vec{x}_i$ of G_i, and the 6 midpoints $\vec{x}_i \pm \vec{h}_i/2$, of the 6 surfaces of G_i. At these points $P_i^k(\vec{x})$ has to satisfy: $P_i^k(\vec{x}_i) = \phi(\vec{x}_i)$, $P_i^k(\vec{x}_i \pm \vec{h}_i/2) = \phi(\vec{x}_i \pm \vec{h}_i/2)$.

If these conditions do not suffice to define the polynomial $P_i^k(\vec{x})$ (as is certainly the case for $k \geq 3$) then the surplus coefficients are determined by a local (within G_i) variational principle or by local Galerkin weighting.

Thus, Equation 3.2 takes on the form

$$\int_{G_i} P_i^k(\vec{x})\,dV\Big|^{n+1} -\int_{G_i} P_i^k(\vec{x})\,dV\Big|^n -\Delta t_n\,(\alpha_n \int_{G_i} L P_i^k(\vec{x})\,dV\Big|^{n+1}$$

$$+(1-\alpha_n)\int_{G_i} L P_i^k(x)\,dV^n) = 0,\ i-1,\dots,N. \qquad (3.4)$$

Here, we are faced with the already mentioned case that matrix B of Equation 3.3 has non zero entries only in its diagonal blocks. Thus, no spatial coupling between regions G_i, G_k, $i \neq k$ is described by Equation 3.3. Therefore, spatial coupling between regions G_i, G_{i+1}, separated by surface $S_{i+1/2}$, must be obtained from the exploitation of the continuity condition $P_i^k(x_i + h_i/2) = P_{i+1}^k(x_{i+1} - h_{i+1}/2)$.

and

$$D_i(x_i + h_i/2)\frac{\partial}{\partial x_j}P^k(x_i + h_i/2) =$$

$$D_{i+1}(x_{i+1} - h_{i+1}/2)\frac{\partial}{\partial x}P_i^k(x_{i+1} - h_{i+1}/2) \qquad (3.5)$$

x being the coordinate direction vertical to $S_{i+1/2}$.

This permits one to express the "surface" values

$\Phi_{i-1/2}^n,\ \Phi_{i+1/2}^n,\ \Phi_{i+3/2}^n,\dots$ along a grid line by center

point values $\Phi_{i-1}^n, \Phi_i^n, \Phi_{i+1}^n,\dots$ along the same grid line, and compute them numerically by solution of a linear matrix equation

$$\bar{A}\Phi_{1/2} = \bar{D}\Phi', \quad \Phi_{1/2} = (..,\Phi_{i-1/2}, \Phi_{i+1/2},..)^T, \quad \Phi' =$$

$$(.., \Phi_{i-1}, \Phi_{i+1},..)^T.$$

In view of the restricted set of approximating poly-nomials $p_i^k(\vec{x})$, the linear equation associated with interface $S_{i+1/2}$ at $x_{i+1/2}$ contains the center point values Φ_i, Φ_{i+1}, and the surface values $\Phi_{i-1/2}$, $\Phi_{i+1/2}$, $\Phi_{i+3/2}$ for each group flux. Thus, in the most general case, the equations are block tridiagonal, with block size equal to the number of prompt groups. Upon solution of $\bar{A}\Phi_{1/2} = \bar{D}\Phi'$, continuity Equation 3.5 associated with $S_{i+1/2}$ can be solved for

$$\Phi_{i+1/2} = a_i (\frac{\Phi_{i-1/2}}{\Phi_i}) \cdot \Phi_i + a_{i+1} (\frac{\Phi_{i+3/2}}{\Phi_{i+1}}) \cdot \Phi_{i+1}$$

expressing $\Phi_{i+1/2}$ explicitly in terms of Φ_i and Φ_{i+1}, by using the numerically known ratios $\Phi_{i-1/2}/\Phi_i$ and Φ_{i+3}/Φ_{i+1} for the determination of the coupling coefficients a_i, a_{i+1}. For advancing the solution from time-layer n to n+1, the values of a_i and a_{i+1} at $t = t_n$ are taken.

Thus, the relation between flux values at the center point and at the midpoints of the 6 surfaces of a box is transformed into a relation between flux values at the center point of the box, and flux values at the center points of its 6 adjacent boxes.

Thus, the matrix corresponding to the discrete form of Equation 3.4 has a structure very similar to FD matrices, the only difference being that all entries in the off-diagonal blocks are, in general, non zero, while in FD matrices, only the diagonal entries in the off-diagonal block are non zero. Thus, for the solution of the discrete form of Equation 3.4 use can be made of the powerful methods developed for the solution of FD matrices, and short computation times per unknown can be expected. Convergence properties of the described method are studied in (20). It turns out that, due to the unsymmetry between basis functions and weight functions, the method is not as sensitive to nonconformity of basis functions as are FE-methods. For the described (nonconforming) quadratic polynomials, error estimates $||P_j^2 - u_o||_{L_2(G)} \leq ch|u_o|_{W_3(G)}$ can be derived,

which is better than what can be done in the corresponding case for FE-methods. Due to nonconformity this estimate can very likely not be improved with higher order polynomials. The outlined method and a very similar one (21) to be discussed in Section D are the only methods of the described type, which have yet been implemented in a multidimensional neutron kinetics code and from which experience is available. Theoretically, higher order weight function promise better accuracy, at least for conforming basis functions. For the used nonconforming functions, the situation is yet unclear, and it should be one of the topics of future investigations to find out whether further improvement of efficiency can be obtained through more elaborate weight functions.

C. Synthesis Methods

Basis of the Weighted Residual Methods discussed in the two foregoing sections is Equation 3.3 which is an approximation to the basic variational Equation 3.1, in the sense that it uses restricted function spaces, relative to the function spaces of problem 3.1. Other computational techniques aimed at a manageable number of unknowns, are derived directly from the variational problem 3.1, or, even more generally, from the corresponding variational problem 3.1* of the P_1 equation.

$$(v^{-1} \frac{\partial}{\partial t} + A) \ \phi + \nabla \cdot \vec{j} - \Gamma c = 0$$

$$\nabla \phi + 3A' \vec{j} = 0 \tag{3.5}$$

$$(\frac{\partial}{\partial t} + \Lambda) c - B\phi = 0$$

where the notation of Equation 2.1 is used and where $\vec{j} = (j_x, j_y, j_z)$ are the currents which are not specified by Fick's law, but appear as independent functions. The most successful application of the resulting generalized variational principle has been to Space Synthesis Techniques

(22-26) which have been in use for more than a decade. The
method is specially tailored to geometrical and physical
properties of nuclear reactors. Its essential idea is to
make use of the distinct role of the axial (z) direction,
and to express the three-dimensional neutron flux as a
linear combination of two-dimensional flux shapes $\Phi_i(x,y,z) =$

$\sum_{\ell=1}^{L} \Phi_{i\ell}(x,y)\gamma_{i\ell}(z)$, i=1,...,M. In addition, expressions of

the same type are introduced for the 3 components of the current vector appearing in the variational principle 3.1*.

For the weight functions W_j, expansions $\Phi_i^*(x,y,z) =$

$\sum_{\ell=1}^{L} \Phi_{i\ell}^*(x,y)\gamma_{i\ell}^*(z)$ in terms of the adjoints of the 2d solu-

tions $\Phi_{i\ell}(x,y)$ are used, and corresponding expressions for
current vector. Again, like in all foregoing discussion,
the time dependence is approximated by a suitable finite
difference representation. The expansion functions $\gamma_{i\ell}(z)$,
$\gamma_{i\ell}(z)$,...,i=1,...,M are determined by making the lefthand
side F of the P_1-equation analogue of Equation 3.2 stationary with regard to variations of $\gamma_{i\ell}^*(z)$,...,i=1,...,M,
i.e., by solving the set $\partial F/\partial\gamma_{i\ell}^*(z) = 0$,...i=1,...M of 4LM
equations. Since the functions $\gamma_{i\ell}(z)$,..., are assumed to be
continuous functions of z, these equations are ordinary
differential equations in z. Thus, the problem of finding
a three-dimensional solution of Equation 2.1 is reduced to
the problem of finding the solutions of 2L two-dimensional
problems (for fluxes and current vectors and their adjoints)
which is the major computational work, and of solving a set
of 4LM coupled ordinary differential equations. One of the
major drawbacks of the method is that no rigorous error
bound can be established. It is not even known if an
increase of the number L of two-dimensional shape functions
improves the accuracy of the result, which even renders impossible the usual empirical means of controlling the cor-
rectness of results. Also, it requires some intuition and
experience with the method to select the proper two- dimensional expansion functions.

Some simplifications to the outlined general methods can
be made, which lead to sizable savings in computational labor
without appreciably affecting the accuracy of results. The

first one is to start from the variational principle 3.1 for
the diffusion equation by assuming Fick's law for the currents, which is perfectly valid for the problems under consideration in this article. Thus, the number of unknowns
can be reduced by a factor 4. Another (independent) reduction by a factor 2 can be gained by using fluxes $\Phi_{i\ell}(x,y)$
instead of their adjoint $\Phi_{i\ell}^*(x,y)$ as weight functions. Again,
the effects of this simplification on the computed results
are small, relative to the general uncertainties of the
method. Another reduction of the number of unknowns is
brought about by observation, that, at a given elevation z,
and with a given set of L shape functions $\Phi_{i\ell}(x,y)$, $i=1,\ldots L$,
the computed solution will strongly depend on shape func-
tions corresponding to radial planes close to z, while the
importance of other shape functions will decrease with increasing distance from z. This leads to the "Discontinuous
Space Synthesis" (27), which at different elevations z, uses
different sets of shape functions; obviously this produces
discontinuities of the computed solution at such locations,
where the set of shape functions is exchanged.

Another, more general variant of synthesis methods is
"Multichannel Synthesis" (28). It partitions the radial
planes into subregions G_j, $j=1,\ldots,N$, with each G_i having
its own shape function $\Phi_{i\ell}(x,y)$ and its own z-dependent
coefficient $\gamma_{i\ell j}$, $j=1,\ldots,N$, $\ell=1,\ldots,L$. As before, equations
coupling all coefficients $\gamma_{i\ell j}$ together are obtained from
the underlying variational principle. On one hand, the
number of unknowns in the problem is increased by a factor N;
on the other hand, the number L of planes where shape functions are needed can be reduced.

Very detailed discussions of the various Synthesis Techniques can be found in (9) and (29).

D. Nodal Methods

Basis for the derivation of nodal methods are the P_1
Equations 3.5 but with the diffusion theory approximation
for the currents. Integrating this set of equations over
region G_i, $i=1,\ldots,M$, assumed to be a parallelepiped, yields
a set

$$v^{-1} \int_{G_i} \frac{\partial \phi}{\partial t} \, dV + \int_{G_i} A\phi \, dV + \int_{S_i} \nabla \cdot \vec{j} \, dS - \int_{G_i} \Gamma c \, dV = 0$$

$$\int_{S_{ik}} \vec{j} \, dS + \int_{S_{ik}} D\nabla\phi \, dS = 0, \quad k=1,\ldots,6 \qquad (3.6)$$

$$\int_{G_i} \frac{\partial}{\partial t} \, dV + \int_{G_i} \Lambda c \, dV - \int_{G_i} B\phi \, dV = 0, \quad i-1,\ldots,N$$

where S_i denotes the surface of G_i and S_{ik} its six individual faces. The final variables of interest are $\int_{G_i} \phi \, dV$; however, currents also appear as auxiliary variables. A first distinction of the various nodal methods in use is made by the selection of these auxiliary variables.

Formulation with net currents:

Let $\vec{J_\Omega}$ be the neutron current through a plane with normal $\vec{\Omega}$. Then

$$\int_{S_i} \nabla \cdot \vec{j} \, dS = (J_{ix+} - J_{ix-}) + (J_{iy+} - J_{iy-}) + (J_{iz+} - J_{iz-}),$$

$$\int_{S_{ik}} \vec{j} \, dS = J_{ix_k}, \quad k = 1,\ldots,6, \quad i = 1,\ldots,M$$

Formulation with partial currents:

Let the net current $\vec{J_\Omega}$ be composed of a partial current $\vec{J_\Omega^+}$ in positive normal direction, and $\vec{J_\Omega^-}$ in negative normal direction, i.e., $\vec{J_\Omega} = \vec{J_\Omega^+} - \vec{J_\Omega^-}$; then we have

$$\int_{S_i} \nabla \cdot \vec{j} \, dS = (J^+_{ix+} - J^-_{ix+} - J^+_{ix-} + J^-_{ix-})$$

$$+ (J^+_{iy+} - J^-_{iy+} - J^+_{iy-} + J^-_{iy-}) + (J^+_{iz+} - J^-_{iz+} - J^+_{iz-} + J^-_{iz-})$$

$$\int_{S_{ik}} \vec{j} \, dS = (J^+_{ix+} - J^-_{ix+}), \quad k = 1, \ldots, 6, \quad i = 1, \ldots, M.$$

In diffusion theory, net current, partial currents, and neutron flux are coupled by the relation

$$J^+_{\vec{\Omega}} = \frac{1}{4}\phi + \frac{1}{2}J_{\vec{\Omega}}, \quad J^-_{\vec{\Omega}} = \frac{1}{4}\phi - \frac{1}{2}J_{\vec{\Omega}},$$

or $$\phi = 2(J^+_{\vec{\Omega}} + J^-_{\vec{\Omega}}) \tag{3.7}$$

What is needed to make a useful computational scheme out of Equations 3.6 are coupling relations between the average flux $\phi_i = \int \phi \, dV$ and the currents. Again, formulations with net currents and partial currents have to be distinguished.

Coupling by net currents:

$$J_{ix+} = a_{i,i+1}(\Phi_{i+1} - \Phi_i),$$

$$J_{ix-} = a_{i-1,i}(\Phi_i - \Phi_{i-1})$$

for 3 neighboring "nodes" G_{i-1}, G_i, G_{i+1}, (in x-direction). On the other hand, difference approximation of the second equation in 3.6 yield

$$J_{ix+} = \frac{2D_i D_{i+1}}{\Delta x_i (D_i + D_{i+1})} \, (\Phi_{i+1} - \Phi_i) ,$$

$$J_{ix-} = \frac{2D_i D_{i-1}}{\Delta x_i (D_i + D_{i-1})} \, (\Phi_i - \Phi_{i-1}) ,$$

Φ_{i-1}, Φ_i, and Φ_{i+1} being the centerpoint flux values in nodes G_{i-1}, G_i, and G_{i+1}.

By taking the centerpoint flux values for average flux values, $a_{i,i+1} = 2D_i D_{i+1}/(\Delta x_i (D_i + D_{i+1}))$ is obtained for the coupling coefficients. Obviously, this approximation is a poor one for the desired node sizes. Therefore, "effective" diffusion constants D' are introduced (30), which are adjusted empirically by comparison with reference solution. In another method (31,32) the parameter of adjustment is the factor $\alpha_i = \Phi_i / \overline{\Phi}_i$.

Coupling by partial currents:

Partial currents directed outward of G_i, are taken proportional to the average flux $\overline{\Phi}_i$, i.e.;

$$J_{ix+}^{+} = \alpha_{i,i+1} \overline{\Phi}_i , \quad J_{ix-}^{-} = \alpha_{i-1,i} \overline{\Phi}_i ,$$

which means that 2 coupling coefficients are needed for each internal interface. They can be obtained (33) through

$$\alpha_{i,i+1} = (\frac{1}{4} \int_{G_i} \phi dV + \frac{1}{2} \int_{S_i} D\nabla \phi dS) / \int_{G_i} \phi dV$$

$$\alpha_{i,i-1} = (\frac{1}{4} \int_{G_i} \phi dV - \frac{1}{2} \int_{S_i} D\nabla \phi dS) / \int_{G_i} \phi dV,$$

where the right hand sides have to be evaluated from some reference solution. Another method that can be considered as a "coupling by partial currents" method uses the production rate $F_i = \nu \int_{G_i} \Sigma_i^f \phi_i dV$ as basic variable (34), and partial currents are assumed to be proportional to F_i, i.e.;

$$J^+_{ix+} = W_{i+1,i}F_i \quad , \quad J^-_{ix-} = W_{i-1,i}F_i \quad , \text{ resp.}$$

$$J^-_{ix+} = W_{i,i+1}F_{i+1} \quad , \quad J^+_{ix-} = W_{i,i-1}F_{i-1}$$

The observation that the coefficient $W_{j,k}$ $i \neq j$, can be interpreted as the transition probability of a neutron generated in G_i, crossing over to G_j serves as starting point for the determination of the W_{ij}'s.

It is a common shortcoming of all nodal methods discussed so far that they require empirical adjustment of parameters on the basis of reference solutions. A set of coupling co-efficients obtained by adjusting to a specific reference solution (which may be very costly to compute) will only be valid for cases which are "similar" to the reference case. Any firm base for defining what "similar" actually means is missing.

Therefore, more modern nodal methods are designed to compute coupling coefficients automatically from the instantaneous neutron flux and currents. A recent, very efficient method (21) of this type is defined as follows: Let $P^k_i(x)$ be the one-dimensional analogue of the polynomials $P^k_i(\vec{x})$ introduced in Section B.2. This polynomial contains flux values Φ_i, $\Phi_{i+1/2}$, $\Phi_{i-1/2}$ as parameters. For a slab along x with cross sectional area S, the quantities $S\Phi_{i\pm1/2}$ are identical to the nodal quantities $\int \Phi dS$. The center-S_{ik} point flux Φ_i can be expressed in terms of nodal variables by solving a numerical quadrature formula

$$\bar{\Phi}_i = S \int_{-h_i/2}^{+h_i/2} P^k_i(x)\,dx = S(\gamma_{i-1/2}\Phi_{i-1/2} + \gamma_i\Phi_i + \gamma_{i+1/2}\Phi_{i+1/2}) \text{ for}$$

Φ_i and substituting the obtained expression into $P^k_i(x)$. Thus, for $x \varepsilon G_i$, a polynomial approximation $S \cdot P^k_i(x)$ is obtained for $S \cdot \phi(x)$, which contains the nodal quantities $S \cdot \phi_{i+1/2}$ and $\overline{\phi}_i$, as parameters, that makes the method much closer to Unsymmetric Weighted Residual Methods than to classical Nodal Methods. Upon substitution of $S \cdot P_i^k(x)$ into the set of nodal equations 3.6, and carrying out the integration with respect to x, a relation containing the surface fluxes $S \cdot \phi_{i \pm 1/2}$ and the average flux $\overline{\phi}_i$ is obtained for each node G_i, i=1,....M. In contrast to the method discussed in Section B.2 (where the auxiliary variables $\phi_{i \pm 1/2}$ are eliminated by exploiting continuity conditions, to yield coupling of the centerpoint fluxes ϕ_i), the surface fluxes are expressed by partial currents through 3.7, which, together with the second equation of 3.6 yields a coupled set of equations for average fluxes $\overline{\phi}_i$ and partial currents. Thus, the system to be solved contains twice as many unknowns as the system of the method of Section B.2.

In sequel, the remaining coordinates are treated in the same way. In order to make the decomposition into a series of one-dimensional calculations meaningful, transversal leakage is taken into account in the treatment of the individual coordinate directions.

If it is possible to rigorously prove that the solution of this sequence of one-dimensional problems converges toward the solution of the original two- or three-dimensional problem then it should be a relatively easy matter to carry over the error bounds given in (20) for the method of Section B.2.

E. General Remarks on Spatial Approximation

All the methods discussed in Sections B-D are aimed at a cheaper way to compute the multidimensional neutron flux distribution in a nuclear reactor than can be provided by FD methods. Historically, Synthesis Methods have been the first step away from FD methods. Their conceptual advantage over Coarse Mesh Methods and Nodal Methods is,

that the very detailed structure in radial planes of nuclear
reactors can be modeled just as well as in FD-methods, and
that large, so to speak homogenized zones appear on the axial
direction - in agreement with the actual physical situation.
The number of variables in the problem is reduced solely by
using relatively few radial planes. In contrast to this,
the other methods described reduce the number of unknowns
by enlarging the grid size more or less uniformly in all
coordinate directions. Since the desired mesh size by far
exceeds the dimensions of the fine structure in radial
planes, the success of these methods depends on how well the
nuclear cross-sections describing the detailed structure can
be represented by equivalent homogenized cross-sections.
This problem is a very difficult one in the multidimensional
case, and theoretically satisfying prescriptions for solving
it are not available (see discussion in (9)). Even if this
problem could be solved exactly, the low order polynomials
coming out as solutions of Coarse Mesh Methods could not be
directly compared with the flux shape obtained from a fine
mesh calculation with all details in nuclear cross-sections.
However, for a problem with cross-sections already being
homogeneous in regions as large as the meshes used by a
Coarse Mesh Method, the latter can provide 3d-solutions just
as accurate as a FD-method, but to this end will require
significantly (some orders of magnitude) fewer unknowns.
As long as conforming basis functions and weight functions
are used, the reduction of computing time is not nearly as
great. However, by relaxing on conformity requirements,
computing time reductions almost proportional to the reduction of the number of unknowns can be achieved without appreciably affecting important quantities of interest like average neutron flux or power per fuel assembly. It should be
pointed out, however, that WR-methods do not rely on the
nuclear cross-sections to be homogenized over the regions G_j.
If a region G_j is composed of subregions with different ab-
sorption-, scattering-, and fission cross-sections and continuously varying diffusion cross-sections, this can be
easily accounted for when evaluating the necessary volume
integrals. This permits, e.g., an accurate modeling of rod
motions. If the diffusion cross-sections change discon-
tinuously, the arithmetic would become very involved, since
higher order polynomials would be required. If the diffusion
cross-sections differ only little, what is mostly the case
within the active core, then smoothing prescriptions are an
acceptable way to circumvent this difficulty. In (35), numerical studies of this problem are presented.

IV. DIRECT TIME INTEGRATION METHODS

The foregoing section was devoted to the discussion of various types of spatial approximations, which, finally, all lead to an initial value problem for a coupled system

$$\frac{d}{dt} \psi = H\psi \qquad (4.1)$$

of ordinary differential equations. The time constants involved in this system differ very widely, which makes the problem to be solved a "stiff" one for practically all applications under consideration in this article.

An initial value problem is called "stiff" in $|t_o, t_1|$ if the eigenvalues (negative inverses of time-constants) $(\lambda_1, \dots, \lambda_K)$, $\lambda_i < 0$, $i=1,\dots,K$, $|\lambda_1| < |\lambda_2| < \cdots < |\lambda_K|$ of its associated Jacobian J_H are widely separated and if $|\lambda_K| \gg ||H\psi||_{L_\infty}$, $t_o < t < t_1$. The numerical treatment of stiff problems is rather unpleasant, since the large time-constants, which dominate the behavior of the solution, require large integration intervals Δt, while the small time-constants, which are insignificant for the behavior of the solution enforce

    a. either inadequately small time steps in all explicit numerical techniques in order to satisfy their inherent stability conditions $|\lambda_K| \cdot \Delta t < c$, c being a method specific constant ranging between 2 and 6, and Δt being the time-step size.

    b. or require the use of time consuming implicit techniques.

Numerous special methods are discussed in the literature, e.g.; (36-38) which are designed to handle stiff problems

more efficiently. Practically all of these methods require
the solution to be stored for a greater number of time steps,
which is not feasible for realistic three-dimensional reactor
configurations. Splitting techniques, e.g. (39) are very
efficient if the stiffness is caused by relatively few of the
individual equations of system 4.1, in which case only these
equations are treated implicitly. Since in Equation 4.1 the
stiffness is caused by the prompt neutron groups, which are
responsible for the bulk of computational burden, practically
nothing can be gained by such splitting techniques. Thus,
all feasible direct methods lead to a discretisation

$$A^{n+1}\psi^{n+1} - A^n\psi^n = 0 \qquad (4.2)$$

of Equation 4.1, with a uniform treatment of all prompt
group fluxes.

The discussions in the following four subsections will
necessitate to make distinction with regard to the structure
of the matrix A^{n+1}. If it is of (cartesian) FD-structure,
i.e.; the solution vector Φ_i, corresponding to region G_i,
$i=1,\ldots,N$ is related to its 4 (in 2d), resp. 6 (in 3d) neigh-
bor point values by a linear relation $\alpha_i\Phi_i + \sum_{j\ I_d} \alpha_j\Phi_j$, I_d
being the set of neighbors, then special
solution techniques can be applied which are more efficient
than the techniques necessary for the case of more general
structure.

In all direct methods the propagation in time of the
approximate solution is performed locally; in each time step
the components $\psi_1^{n+1},\ldots,\psi_N^{n+1}$ of the solution vector ψ^{n+1} are
computed newly for each region $G_1,\ldots,G_N$. This large amount
of computational work is justified only if the spatial neutron
flux distribution changes appreciable within one time step.

The use of direct methods would certainly be uneconomical if the spatial neutron flux distribution does not change within time-steps, since in that case, a lot of computational work is spent to perpetually calculate identical or almost identical spatial neutron flux distributions. Later in this section it will be shown how recent direct time integration methods avoid this deficiency of older methods.

The most straightforward method of solution for Equation 4.1 would be to use the forward difference Euler formula, in which case the matrix A^{n+1} becomes diagonal; spatial coupling would be taken care for only in the matrix A^n. However, since the stability condition of this and all similar explicit methods (40) would untolerably limit the time-step size for most applications, such methods will not be considered here. For FD-methods, a Du-Fort-Frankel type discretisation has proved to be efficient for the analysis of fast excursions (41). Unfortunately, the method cannot be generalized to the treatment of Coarse Mesh equations. Therefore, it will also be omitted from the discussions in this section.

A. Alternating Direction Explicit (ADE) Method

A method designed for FD-structured matrices which leads to simple arithmetic and program organization is the Alternating Direction Explicit Method (ADE-Method) (42,43). It has been successfully implemented into a production code (44). It is based on a multidimensional generalization of Saulev's (45) method.

For shortness of notation a spatially uniform mesh size, h, will be assumed in the following, and Φ_i, Φ_{i+1}, Φ_{i-1}, Φ_{k+1}, Φ_{k-1}, $\Phi_{\ell+1}$, $\Phi_{\ell-1}$ are to denote the values of the discrete solution for region G_i, its left and right, front and rear, upper and lower neighbor, respectively.

Then, one step of the ADE-method computes Φ_i^{n+1} by solving

$$\phi_i^{n+1} - \phi_i^n = \nu\Delta t(\gamma_{i-1}^{n+1}\phi_{i-1}^{n+1}) - (\gamma_i^n\phi_i^n + \gamma_i^{n+1}\phi_i^{n+1})$$

$$+ \gamma_{i+1}^n\phi_{i+1}^n + \gamma_{k-1}^{n+1}\phi_{k-1}^{n+1} - (\gamma_k^n\phi_i^n + \gamma_k^{n+1}\phi_i^{n+1}) + \gamma_{k+1}^n\phi_{k+1}^n$$

$$+ \gamma_{\ell-1}^{n+1}\phi_{\ell-1}^{n+1} - (\gamma_\ell^n\phi_i^n + \gamma_\ell^{n+1}\phi_i^{n+1}) + \gamma_{\ell+1}^n\phi_{\ell+1}^n)/h^2 \text{ for } \phi_i^{n+1},$$

where it is assumed that the spatial mesh is swept in positive x,y,z direction. In successive steps, which together with the first step make up one cycle, the directions of sweep through the mesh are permuted in such a way, that after completion of a cycle, the influence of a perturbation in region G_j, j=1,...,N has been spread out on $\cup G_j$ = G. The superscript of the coefficients γ referring to time layers n and n+1 are to indicate the inclusion of known cross-section changes, e.g., by rod motions.

The method is unconditionally stable, and its temporal truncation error is of order (Δt^2).

Since both time-layers, n and n+1, are treated symmetrically, the basic method does not allow for a weight factor by which it can be given the characteristic of a backward difference method (that this is desirable for large time steps is shown in Section E).

Numerical studies have shown that the basic ADE-method is rather inefficient for realistic reactor problems. The relative increment per time-step $\varepsilon = \max_j \{(\phi_j^{n+1} - \phi_j^n)/\phi_j^n\}$ has to be kept unacceptably small in order to obtain accurate results. However, efficiency can be greatly improved by the use of "frequency transformations", which will be discussed in Section E.

B. Alternating Direction Implicit (ADI) Method

One step further toward the final goal of a fully implicit solution technique for FD-structured matrices in Equation 4.2 is made by the use of Alternating Direction Implicit (ADI) Methods (46,47). The essence of the various types of ADI-methods is to decompose the discrete form L_D

of the differential operator L into its $\bar{x}$, $\bar{y}$, and $\bar{z}$ components L_{DX}, L_{DY}, and L_{DZ}. The time-step from n to n+1 is carried out in 3 steps, wherein the first step treats L_{DX} implicitly, and L_{DY}, L_{DZ} explicitly by forward difference Euler formula, i.e.; if $\phi^{n+1/3}$ is to denote the result of the first step,

$$\phi_i^{n+1/3} - \phi_i^n = v\Delta t (\alpha(\gamma_{i-1}^{n+1/3}\phi_{i-1}^{n+1/3} + \gamma_{i+1}^{n+1/3}\phi_{i+1}^{n+1/3} - 2\gamma_i^{n+1/3}\phi_i^{n+1/3})$$

$$+ (1-\alpha)(\gamma_{i-1}^n\phi_{i-1}^n + \gamma_{i+1}^n\phi_{i+1}^n - 2\gamma_i^n\phi_i^n) + \gamma_{k-1}^n\phi_{k-1}^n + \gamma_{k+1}^n\phi_{k+1}^n - 2\gamma_k^n\phi_i^n$$

$$+ \gamma_{\ell-1}^n\phi_{\ell-1}^n + \gamma_{\ell+1}^n\phi_{\ell+1}^n - 2\gamma_\ell^n\phi_i^n), \quad 1/2 < \alpha < 1.$$

For $\alpha = 1/2$ the temporal truncation error is of order (Δt^2) and for $\alpha > 1/2$ it is of order (Δt). For the determination of α, see the discussion in Section E.

The second step treats L_{DY} implicitly, and L_{DX}, L_{DZ} explicitly, and the third step treats L_{DZ} implicitly, and L_{DX}, L_{DY} explicitly. By itself, each one of the steps is only conditionally stable, due to the presence of the explicitly treated components. In reactor applications, each of the steps is effectively unstable. Stability is reached by a judicious combination of the three individual steps.

This decomposition of a three-dimensional implicit problem into a sequence of 3 one-dimensional implicit problems is attractive, since it promises improved accuracy (over ADE-methods), but requires only minor additional computational work due to the simple block tridiagonal structure of the one-dimensional matrices representing L_{DX}, L_{DY} and L_{DZ}. For all ADI-Methods, the proof that a particular combination of the 3 steps is stable (48) relies on constancy in space of the operators L_{DX}, L_{DY}, and L_{DZ}, and there is indication that this requirement is not merely technical but essential. This is also supported by numerical studies, which show ADI-methods to be successful for simple model problems in reactor physics which involve only minor heterogeneities, but exhibit an increasing tendency toward instability for more complex configurations.

In usual FD applications, constancy of the difference operators is violated through varying nuclear cross-sections and variable mesh size, but this violation is rather mild, so that, effectively, ADI methods can be applied without severe stability difficulties to FD equations (49), but like ADE-methods, they require the use of frequency transformation to become efficient.

However, in FD applications with the great number of variables necessitating the use of external storage devices, a great inconvenience is encountered: the successive treatment of the 3 coordinate directions in the 3 steps of cycle requires transport between internal and external storage devices in varying order, causing very complex programming and long program execution times.

In applications to FD-structured Weighted Residual Methods like (17), this problem does not come up, since no external storage is required, but the stability problem becomes severe. It has been shown in Section 3.B, that such methods possess an advancement matrix the entries of which depend on the solution itself. Thus, the condition of constancy or at least near-constancy of the difference operators, which is essential for the successful use of ADI methods, may be strongly violated, even in the case of constant nuclear cross-sections and spatial mesh width throughout the reactor. Numerical studies of typical LWR application have shown that instabilities eventually occur if the time-step exceeds some tenth of a millisecond. Thus, for most applications of Weighted Residual Methods of this type, classical ADI methods would be extremely inefficient.

C.   Combined ADE-ADI Method

Both the stability properties of the ADE method and the potential of ADI methods to use "optimal" weight factors α are combined in the "almost implicit" ADE-ADI method (51). It advances the solution from time layer n to n+1 by

$$\Phi_i^{n+1} - \Phi_i^n = \nu\Delta t \, (\alpha \, (\gamma_{i-1}^{n+1}\Phi_{i-1}^{n+1} + \gamma_{i+1}^{n+1}\Phi_{i+1}^{n+1} - 2\gamma_i^{n+1}\Phi_i^{n+1})$$

$$+ \, (1-\alpha) \, (\gamma_{i-1}^n\Phi_{i-1}^n + \gamma_{i+1}^n\Phi_{i+1}^n - 2\gamma_i^n\Phi_i^n)$$

$$+ \, \gamma_{k-1}^{n+1}\Phi_{k-1}^{n+1} - 2\,(\alpha\gamma_i^{n+1}\Phi_i^{n+1} + (1-\alpha)\,\gamma_k^n\Phi_i^n) + \gamma_{k+1}^n\Phi_{k+1}^n$$

$$+ \, \gamma_{\ell-1}^{n+1}\Phi_{\ell-1}^{n+1} - 2\,(\alpha\gamma_i^{n+1}\Phi_i^{n+1} + (1-\alpha)\,\gamma_\ell^n\Phi_i^n) + \gamma_{\ell+1}^n\Phi_{\ell+1}^n))$$

$$1/2 < \alpha < 1.$$

(4.3)

for $\alpha = 1/2$, the temporal truncation error is of order (Δt^2), and for $\alpha > 1/2$ it is of order (Δt). In 4.3 the operator L_{DX} is treated implicitly, and the operators L_{DY} and L_{DZ} are treated explicitly, similarly to the ADE prescription, the difference being the use of the weight factor with the center point value. The single step described by this equation is already unconditionally stable. A cyclic interchange of implicitly and explicitly treated coordinate directions is made only for reasons of accuracy.

Equation 4.3 differs from the fully implicit equation to be discussed in the next section only with regard to the terms with the spatial indices k-1, k+1, ℓ-1, and ℓ+1, resp. In the fully implicit case, these terms are weighted sums of quantities in time layers n and n+1, like all other terms. For all problems, where the desired time step size $\Delta t >$ some milliseconds, α is effectively 1, in which case Equation 4.3 differs from the fully implicit formula only with regard to the terms with indices k+1 and ℓ+1, and the relative error committed by using 4.3 instead of the fully implicit equation can be shown to be proportional to $\Phi_t\Delta t/(\Phi \cdot h^2)$, which is small due to the large value of h, and the fairly small value of $\Phi_t\Delta t/\Phi$, representing the relative increment. The appearance of $\Phi_t\Delta t/\Phi$ already indicates that the efficiency of this basic form of the ADE-ADI-method can be greatly increased by frequency transformations.

D. Fully Implicit Method

The methods discussed in the last three sections are

aimed at circumventing the costly procedure to solve Equation 4.2 by a fully implicit method. In combination with techniques to be discussed in Section E some alternating direction methods are indeed superior to fully impicit methods for a large class of few group problems. However, for a greater number of prompt groups, e.q.; in fast systems, alternating direction methods may be less efficient than fully implicit methods.

In view of the large number of variables involved, a direct solution of Equation 4.2 by elimination would certainly be very inefficient, if feasible at all. Therefore, all implicit solution methods make use of iterative techniques. In contrast to alternating direction methods, there is no limitation to FD-structured matrices. Common feature of all iterative techniques is to obtain the solution of a linear system of equations $Au = b$ by generating an infinite sequence of vectors $u^{(0)} \to u^{(1)} \to u^{(2)} \to \ldots$, which converge toward the solution $u = A^{-1}b$. The various types of iterative techniques can all be described by

$$Bu^{(i+1)} = (B-A)u^{(i)} + b \qquad (4.4)$$

or solved for $u^{(i+1)}$

$u^{(i+1)} = (I-B^{-1}A)u^{(i)} + B^{-1}b$, B being an arbitrary non-singular matrix, with rank (B) = rank (A). The method converges to $u = A^{-1}b$, if the spectral radius $\rho(I-B^{-1}) < 1$. Desirable properties of B are:

a. "easy" solution of Equation 4.4 for $u^{(i+1)}$

b. small spectral radius of $I-B^{-1}A$.

Let $A = D - E - F$ be the decomposition of A into its diagonal D, lower triangular -E, and upper triangular -F. Then, Jacobi iteration is obtained by choosing $B = D$, Gauss-Seidel iteration by choosing $B = D - E$. Generally, more efficient than these two techniques are relaxation methods, which are characterized by choosing $B(\omega) = D(I-\omega D^{-1}E)/\omega$, with the relaxation parameter ω, which is determined in such a way, that the spectral radius of $I-B(\omega)^{-1}A$ is made as small as possible. Numerous prescriptions for the optimal deter-

mination of the parameter ω are discussed in the literature
e.g.; (51,52).

A powerful method to further accelerate convergence,
which can always be applied, if the linear equations to be
solved result from local operators, is coarse mesh rebalancing (53,54). Convergence in the mesh system (x_i, y_j, z_k)
is speeded by employing information derived from a coarser
mesh system (X_ℓ, Y_m, Z_n), which is relatively cheap to obtain.
Let $u^{(j)}$ denote an approximation to u at some stage of the
calculation. Coarse mesh rebalancing improves $u^{(j)}$ by the
following procedure:

$v(x,y,z) = \Sigma\Sigma\Sigma c_{\ell mn} \Delta_{\ell mn}(x,y,z)$ is an expansion in terms of
simple functions - mostly pyramid functions - defined for
the coarse grid, with $\Delta_{\ell mn}(x,y,z) = 0$ everywhere, except
on the coarse mesh region (ℓ, m, n). The coefficients $c_{\ell mn}$
are obtained upon insertion of $vu^{(j)}$ into the variational
rebalancing equation (55) corresponding to the original
differential Equation 2.1. Once the expansion coefficients
are known, $u^{(j)}$ is improved through multiplication by the
rebalancing function v. In the iteration step to follow,
$vu^{(j)}$ serves to yield $u^{(j+1)}$. Such combinations of relaxation and coarse mesh rebalancing techniques prove to be
very efficient and are used in several production codes
(56,21). The iterative solution techniques discussed so
far theoretically require infinitely many iterations to
obtain the correct solution of Au = b. Convergence depends
on properties of A and B. For FD-structured matrices convergence criterions are well explored. Usually, convergence is guaranteed by the sufficient diagonal dominance
condition. For matrices as they occur with the FE-method,
the situation is not as clear, since diagonal dominance
cannot be assured, but a quite different iterative technique can be used which is applicable to positive definite
matrices (like FE matrices), and which yields - under the
hypothetical assumption of exact arithmetic without round-
off errors - the correct solution u after m = rank (A) steps.
This so-called conjugate gradient (CG) method (57) is based

on the observation that u minimizes the functional

$$F(z) = \frac{1}{2}(Az-b)^T A^{-1}(Az-b) = \frac{1}{2}z^T Az - b^T z + \frac{1}{2}b^T A^{-1}b.$$ This

suggests the use of a steepest descent technique for making F stationary.

Experience with three-dimensional FE-codes for reactor kinetics is very limited; therefore, no firm opinion can yet be formed on the most efficient solution techniques for FE equations.

A principal advantage of fully implicit methods over competing alternating direction methods is the fact that the knowledge of iterative approximations to the true solution permits in every iteration step computation of feedback influences on nuclear cross-sections accordingly. Naturally, convergence is slowed down by such a procedure since the matrix A is changed in every iteration step.

E. Improved Efficiency Through Frequency Transformation
    and Spectral Matching

It is a general shortcoming of all direct methods in their basic form, that the amount of computational work for one time-step is independent of the spatial change of neutron flux within the step. Even in the case of an asymptotic time behavior time steps cannot be made appreciably larger than for a case with significant spatial flux changes. This is felt most severely in alternating direction methods with their large temporal truncation errors.

Therefore, soon after the introduction of alternating direction methods to neutron kinetics it has been suggested to improve such methods by "frequency transformation" (42): the form of the differential equation 2.1 indicates that in a small time interval $|t_n, t_{n+1}|$ the solution behaves like $\exp(\omega(t-t_n))$, or a linear combination $\sum_i c_i \exp(\omega_i(t-t_n))$. Then, upon substitution of $\phi(\vec{x},t) = \phi'(\vec{x},t) e^{\omega t}$ into 2.1, the equation

$$v^{-1}(\frac{\partial}{\partial t} - (L-\omega))\phi' = 0 \qquad (4.5)$$

is obtained for $\phi'(\vec{x},t)$, which describes the deviation of

$\phi'(\vec{x},t)$ from the anticipated (local or global) behavior $e^{\omega t}$.
Since the temporal truncation error is always proportional
to time derivatives, the error associated with $\phi'(\vec{x},t)$ is

small relative to the error associated with $\phi(\vec{x},t)$, if
$\phi^n e^{\omega \Delta t}$ agrees well with ϕ^{n+1}.

Older methods use past history for the calculation of ω,
either locally (42) or globally (49), by the prescription
$\omega = \ln(\phi^{n+1}/\phi^n)/\Delta t$. As can be expected, this is rather
dangerous in time intervals in which significant reactivity
changes occur. This difficulty is avoided by using volume
averaged kinetic equations for the calculation of ω (58,59);
each of the equations

$$v_i^{-1} \frac{\partial \phi_i}{\partial t} = L_{ii}\phi_i + \sum_{k \neq i} L_{ik}\phi_k + \chi_i \sum_j \lambda_j c_j \qquad i=1,\ldots,M$$

(4.6)

$$\frac{\partial c_j}{\partial t} = -\lambda_j c_j + \nu\beta_j \sum_i \sum_i f_i \phi_i \qquad j=1,\ldots,G,$$

$$L = \begin{pmatrix} L_{11} \cdots L_{1m} \\ L_{m1} \cdots L_{mm} \end{pmatrix}, \text{ is spatially integrated over subvolumes}$$

or the total volume of the reactor to yield (the bar denotes
integrated quantities)

$$v_i^{-1}\dot{\overline{\phi}}_i = (\overline{L_{ii}\phi_i/\overline{\phi}_i})\cdot\overline{\phi}_i + \sum_{k \neq i} (\overline{L_{ik}\phi_k/\overline{\phi}_k})\cdot\overline{\phi}_k + \chi_i\sum_j \lambda_j \overline{c}_j$$

(4.7)

$$\dot{\overline{c}}_j = -\lambda_j \overline{c}_j + \nu\beta_j \sum_i \sum_i f_i \overline{\phi}_i, \qquad i=1,\ldots,m, \quad j=1,\ldots,G$$

For this stiff system of ordinary differential equations
one time-step of the same size as the one employed in 4.6
is performed, which advances the volume integrated quantities
from t_n to t_{n+1}. The quantities $\frac{1}{\Delta t} \ln(\overline{\Phi}_i^{n+1}/\overline{\Phi}_i^n),\ldots$
then serve as mean periods for the integration of system 4.6.
In calculating the coefficients of system 4.7, known cross-
section changes occurring in $|t_n,t_{n+1}|$ are incorporated in

a linear fashion, i.e.; $\overline{L_{11}\phi_1}/\overline{\phi_1}$ is defined as follows:

for $t \in |t_n, t_{n+1}|$:

$$\frac{\overline{L_{11}\phi_1}}{\overline{\phi_1}}(t) = (\frac{\overline{L_{11}\phi_1}}{\overline{\phi_1}})^n + \frac{t-t_n}{\Delta t}((\frac{\overline{L_{11}\phi_1}^{n+1}}{\overline{\phi_1}^n}) - (\frac{\overline{L_{11}\phi_1}}{\overline{\phi_1}})^n)$$

and analogously for the other coefficients of 4.7.

As a consequence of the just described determination of the dominant time history, direct methods take on some of the good characteristics of quasistatic methods (see Section 5), with the time step size becoming comparable to the time intervals, in which quasistatic methods require a new determination of spatial flux distribution, without the uncertainties of quasistatic methods about when to newly compute shape functions. These considerations apply to all direct methods, whether they are alternating directon methods or fully implicit methods. The greatest gain in efficiency, however, is made for alternating direction methods.

Another means of improving on the temporal truncation error of direct methods is the optimal determination of the weight factor α_n in Equation 4.2 (60), known in the literature as "spectral matching". Its essence is best demonstrated by the simple example of one ordinary differential equation $\dot{y} = -\lambda y$, $y(t=0) = y_0$. Its solution is $y = y_0 \exp(-\lambda t)$. A simple, one-step numerical integration technique for this equation could be $y^{n+1} - y^n = -\lambda\Delta t(\alpha y^{n+1} + (1-\alpha)y^n)$, $\alpha \geq 1/2$, or $y^{n+1} = y^n \frac{1-(1-\alpha)\lambda\Delta t}{1+\alpha\lambda\Delta t}$. By requiring that the numerical formula matches the time history of the true solution, i.e.; $\frac{1-(1-\alpha)\lambda\Delta t}{1+\alpha\lambda\Delta t} = \exp(-\lambda\Delta t)$, the optimal value of α is obtained as

$$\alpha = \frac{\exp(-\lambda\Delta t) - 1 + \lambda\Delta t}{\lambda\Delta t(1-\exp(-\lambda\Delta t))} \qquad (4.8)$$

For $|\lambda\Delta t| \ll 1$, $\alpha \simeq 1/2$, for $|\lambda\Delta t| \to \infty$, α is effectively 1.

This analysis can be expanded to the case of systems differential equations, with distinct eigenvalues $\lambda_1, \ldots, \lambda_N$. In reactor kinetics applications $\lambda_{max} = \max_j(\lambda_j) \approx \max_i(v_i D_i / \Sigma_i^{abs})$ which is very large. For this case, a more elaborate analysis indicates that the weight factors α for the prompt neutron groups should all be equal, $\alpha_1, \ldots, \alpha_m = \alpha_p$ and be calculated from 4.8, with $\lambda = \lambda_{max}$; for time steps Δt in the millisecond range or larger, α_p is effectively 1.

V.   INDIRECT TIME INTEGRATION METHODS

As has been pointed out earlier, direct methods in their basic form are rather uneconomical in cases when the changes of spatial flux distribution are small between time steps, since in that case, a lot of computational work is spent to perpetually calculate almost identical flux distributions.

Indirect methods try to reduce this unnecessary load of computational work by making further approximations to the solution $\Psi(x,t)$, which all amount to the extraction of the dominant time behavior of the solution.

A.   Time-Synthesis Method

The space- and time-dependent neutron flux $\phi(\vec{x},t)$ is expanded in known functions of space $\phi_i'(\vec{x})$ and unknown, time dependent expansion coefficients $a_i(t)$; $\phi(\vec{x},t) = \sum_{i=1}^{I} \phi_i'(\vec{x}) a_i(t)$. Formally this expansion is identical to the expansion made to derive the space synthesis equation. By the formal procedure described for space synthesis, a set of ordinary differential equations is obtained for coefficients $a_i(t)$, (24).

Space-time synthesis methods (61,62) combine space synthesis and time synthesis by considering expansions $\phi(\vec{x},t) = \sum_{i=1}^{I} \phi_i'(x,y) a_i(z,t)$, with known two-dimensional expansion functions and unknown coefficients depending on z and t, which are obtained as solution of a system of partial differential equations.

The difficulties encountered with the application of
time synthesis methods are of the same nature as the difficulties encountered with space synthesis methods. In
order to make the expansion meaningful, the expansion functions must represent approximations to a set of states
through which the reactor will go during the transient under
consideration. This rather vague rule leaves it unclear
what set of expansion functions should be taken. Also, it
is impossible to establish rigorous error bounds for the
solution. Nevertheless, space-time synthesis codes (63)
are a very practical and economical tool for the analysis
of such operational transients, which permit to make meaningful guesses of future states.

B. Quasistatic Method

       Quasistatic methods are based on a direct factorisa-
tion $\phi(\vec{x},t) = \phi'(\vec{x},t)A(t)$ of the neutron flux into an amplitude $A(t)$ describing the time behavior of the overall neutron
flux level, and a space- and time-dependent function $\phi'(\vec{x},t)$
describing the local adjustment of the neutron flux distribution. The amplitude function $A(t)$ is obtained as solution
of the associated point kinetics equation, which accounts
for the global reactivity effects resulting from time varying nuclear cross-sections, and for the global feedback
effects. Since the numerical integration of this equation
is relatively inexpensive, a fine temporal mesh size Δt_a
can be used. The shape function $\phi'(\vec{x},t)$, which is assumed
to be only weakly time dependent, relative to $A(t)$, is calculated on a much coarser temporal mesh, with $\Delta t_s \gg \Delta t_a$.
The various types of quasistatic methods differ with respect
to the calculation of the shape functions.

       The quasistatic method (64) computes $\phi'(\vec{x},t)$ as solution
of $-(L-I\dot{A}/A)\phi' = \Gamma c/A$, which accounts explicitly for the
source of delayed precursors, but ignores $v^{-1}\frac{\partial \phi'}{\partial t}$.

       The improved quasistatic method (65) approximately ac-
counts also for this time derivative term by computing
$\phi'(\vec{x},t)$ from

$$-(L-I\dot{A}/A)\phi'(t_j) = \Gamma c/A - R, \qquad (5.1)$$

where $R = o$ for $t = t_o$, and

$$R = \frac{\Phi'(t_j) - \Phi'(t_{j-1})}{v\Delta t_{sj}}$$

where t_{j-1} and $t_j = t_{j-1} + \Delta t_{sj}$ designate to successive points of the temporal mesh for the calculation of the shape function.

Equation 5.1 is solved by an iterative technique as discussed in Section IV.D. Indeed, the improved quasistatic method is very similar to a fully implicit direct method with frequency transformation and spectral matching with the frequency prediction coming from space averaged kinetic equations (see Section IV.E).

Since the solution obtained with quasistatic methods converges toward the solution of direct methods, if the temporal mesh for the calculation of shape functions is sufficiently fine, there are no principal difficulties encountered in its application; yet, some uncertainty exists with regard to a flexible and reliable determination of the intervals Δt_s after which the shape function has to be newly computed.

It has been found that for a broad class of problems, especially fast reactor problems, where the overall flux changes occur much more rapidly than the local adjustments of neutron flux, quasistatic methods are much more efficient than direct methods in their basic form. However, against direct methods equipped with frequency transformation and spectral matching, the advantage of quasistatic methods is expected to be much smaller.

VI. COMPUTATIONAL ASPECTS

In the foregoing sections, the mathematical and physical principles governing the derivation of the most important types of spatial approximations and of solution methods for the resulting discrete systems of equations have been discussed, with only a qualitative mentioning of computational implications. In the last two sections, a quantification of computational aspects will be given for some of the most important methods, based on published results, privately communicated information, and the author's own experi-

ence. Naturally, this discussion is not complete, but represents only a collection of selected results. All results
given refer to three-dimensional problems.

A. Comparison of Efficiency of Spatial Approximation

The most important figures of merit for comparing
different types of spatial approximations are

a. the amount of computational work per space point,
    relative to the amount of work per space point for
    Finite Difference Methods, which definitely require
    the least computational work per space point. As
    measure for the amount of computational work, the
    number of matrix entries for an interior point is
    taken.

b. the number of space points necessary to reach a pre-
    scribed accuracy. As reasonable basis for this
    comparison it is assumed that the computation is to
    be accurate to within 3% of the true solution. When
    comparing fine mesh to coarse mesh results, nuclear
    cross-sections which are homogeneous within rela-
    tively large regions (fuel assemblies of the model
    reactor) are assumed (compare the discussion in
    Section III.E).

Included in the comparison have been only methods which
do not require an adaptation of parameters to reference solutions.

As a unit of comparison for the computation work per
space point, the FD-method with 7 point approximation of
the difference operator is taken. For an interior point,
each line of the FD-matrix contains m+6 entries (8 entries
for the important 2-group case).

The QUABOX/CUBBOX-code (17) and IQSBOX-code (21), which
are both implementations of an unsymmetric weighted residual
method (USWR) (compare the discussions in Sections III.B.2
and at the end of Section III.D) have 7m entries (14 for the
2-group case) for an interior point, independent of the order
of the polynomials chosen as basis-functions. Increasing the
order of the approximating polynomials merely increases the
complexity of arithmetic necessary for the calculation of
the matrix entries.

A Finite Element method with rectangular elements and quadratic basis functions has 27m entries (54 for the 2-group case) for an interior point. Thus, assuming equal solution methods for the matrix equation, the computing time requirements per space point and time-step for FD-, USWR-, and second order FE method for a case with two prompt groups should be, roughly, like 1:2:8. Indeed, these guesses can be almost verified by evaluation of computed results.

In order to judge the accuracy of methods, numerical computations are necessary. Main source of material for these are two ANS benchmark problems. The first of the two problems (66) is a static problem for a PWR, the second one (67) a time time dependent problem for a BWR. Fuel assembly size is 20 x 20 cm^2, for the PWR problem, and 15 x 15 cm^2 for the BWR-problem. For both problems, nuclear cross sections are homogeneous over fuel assemblies. Solutions of the PWR problem have been calculated with the FD codes VENTURE (68) and PDQ-7 (69), the USWR codes QUABOX/CUBBOX and IQSBOX, and the FE code FEM3D (70). Table 1 presents some results of these calculations.

Though the exact solution of the problem is unknown, inspection of the results obtained with various mesh sizes permits some conclusions:

a. FD-methods need to use a mesh size of at most 2.5 cm in order to reach the desired accuracy.

b. The FE-method with quadratic basis functions reaches the desired accuracy with a mesh size of 10 cm.

c. The USWR-methods with polynomials of order > 3 reach the desired accuracy with a mesh size of 20 cm.

d. Though computing times could not be compared directly, since the problems were run on different computers, the computing time estimates given above indicate that the USWR-methods are the most efficient ones. This is substantiated by the reported cpu-times.

That these results are not coincidental but typical for LWR-calculations has been verified by a number of other comparative calculations.

The advantage of USWR methods over FD methods is greatest in the 2-group case typical for thermal reactor applications.

Table 1
Comparison of Computed Results for PWR Benchmark Problem

Computer Code	Mesh Size x,y,z (cm)	Number of Basic Variables (Quarter Core)	Eigenvalue	Peak/ Average Power
VENTURE	10 x 10 x 20	5,206	1.02913	2.567
	5 x 5 x 10	41,648	1.02864	2.504
	2.5 x 2.5 x 5	333,184	1.02887	2.408
	1.66x1.66x3.33	1,124,496	1.02896	2.378
PDQ-7	5 x 5 x 10	41,648	1.03054	2.039
	2.5 x 2.5 x 5	333,184	1.02933	2.266
FEM3D	10 x 10 x 27	3,562	1.02917	2.298
IQSBOX	20 x 20 x 20	2,622	1.02916	2.348
	10 x 10 x 20	5,206	1.02910	2.341
CUBBOX	20 x 20 x 20	2,622	1.02888	2.387
	10 x 10 x 20	5,206	1.02895	2.340

For fast reactor systems the advantage is not as great, since the ratio of matrix entries (FD versus USWR methods) approaches 1:7 as the number of neutron groups increases, and, on the other hand, FD codes can use a larger spatial mesh due to the greater diffusion length.

For a fast reactor 3d-benchmark problem with 4 groups (71), the QUABOX/CUBBOX code with quadratic polynomials required for a more accurate solution about the same computing time as the space synthesis code KASY (72), where the time required to compute the two-dimensional expansion functions was not counted. In view of the uncertainties of synthesis methods this result must be considered as being very competitive.

Also for geometries other than rectangular cartesian, similar advantages of USWR-methods over FD-methods are reported. The spatial approximation used in the TRIMHEX-code (73) also is an USWR-method with piecewise linear basis functions and piecewise constant weight functions. Computing time reductions by a factor 2.5, relative to FD-calculations with comparable accuracy have been found.

B. Comparison of Efficiency of Time-Integration Methods

Comparing the two competing methods - alternating direction methods on one hand, and fully implicit iterative methods on the other hand - is rather problematical, since there is no a priori knowledge of the number of iteration steps required by an iterative method. All three types of alternating direction methods - ADE, ADI, and ADE-ADI in their basic form are unacceptably inaccurate. However, their accuracy can be greatly increased by frequency transformation. This is demonstrated in Table 2, which contains some selected results computed with the combined ADE-ADI method for the TWIGLE-reactor test case (42), which has been used as reference case in many publications. Shown is the thermal neutron flux in the center of the reactor. Using a time-step of 5 ms, a maximum relative increment of 2.5% occurs. Already this small increment leads to an error of 38% if no frequency prediction is used.

With frequency prediction from the past history of the mean neutron flux, i.e.; $\omega = \ln(\bar{\phi}^n/\bar{\phi}^{n-1})\Delta t$, $\bar{\phi} = \int \Phi dV/V$, the error goes down to 5.8% in that part of the transient where reactor reactivity is added ($0 < t < 0.2$ sec).

Shortly after termination of reactivity insertion, the computed solution greatly overshoots the correct solution, leading to 12% error. Obviously, this is caused by the use of past history for frequency prediction. With frequency prediction from the reactor averaged kinetics equations (58), the error goes down to 0.5%, which is perfectly acceptable. Even with a time-step size of 25 ms, leading to a relative increment of 12.5%, the error stays below 2.5%. The error rates of fully implicit methods generally are of the same size.

Computing time per time step of alternating direction methods and of fully implicit methods depend in a nonlinear

Table 2
Influence of Frequency Prediction on Accuracy

Time Step Δt (ms) Method time sec	1.0 Reference Solution	5.0 No Frequency Prediction	5.0 Frequency Prediction from Past History	5.0 Frequency Prediction from Kinetics Equations	25.0 Frequency Prediction from Kinetics Equations
0.0	$1.076^{10^{12}}$				
0.05	1.202	1.089	1.132	1.206	1.190
0.10	1.387	1.128	1.362	1.392	1.393
0.15	1.647	1.193	1.691	1.654	1.665
0.20	2.036	1.284	2.052	2.047	2.070
0.25	2.142	1.387	2.395	2.143	2.195
0.30	2.152	1.478	2.412	2.152	2.170
0.40	2.169	1.629	2.026	2.169	2.169
0.50	2.187	1.747	2.210	2.187	2.187
Max. rel. Increment	0.5%	2.5%	2.5%	2.5%	12.5%
Max. rel. Error	Ref.	38%	12% (5.8%)	0.5%	2.5%

fashion on the number of prompt neutron groups: up to m = 4 groups, the one dimensional implicit equations of alternating direction methods are most efficiently solved for all groups simultaneously, with the computing time rising proportional to the square of the number of groups. From m > 4, "outer" iterations over the groups are more efficient. Fully implicit methods always perform outer iterations over the groups independent of m. Generally, the number of required outer iterations increases with m. For thermal systems with 2 prompt groups, figures of comparison can be obtained from the BWR kinetics benchmark problem mentioned earlier.

The solution of the problem has been calculated with the CUBBOX code and with the IQSBOX code. As discussed earlier, the two codes use very similar spatial approximations. For

time integration, CUBBOX uses the combined ADE-ADI method
with spectral matching and frequency prediction from reactor
averaged kinetics equations, and IQSBOX uses a fully implicit
method with successive overrelaxation and coarse mesh rebalancing. Both methods require about the same number of time-
steps (600-700) for an accurate calculation of time history.
With a spatial mesh size of 15 cm for both codes IQSBOX
requires about 5 times the computing time of CUBBOX. This
single test case indicates a clear advantage of the ADE-ADI
method over the fully implicit method, and this result can
presumably be substantiated for coarse mesh calculations of
thermal (2 group) systems. For smaller mesh sizes, the efficiency of the ADE-ADI-methods decreases rapidly, since the
error term, which expresses the local difference of the
ADE-ADI operator from the fully implicit operator, increases
proportional to $1/h^2$, h being the spatial mesh size. This
behavior is demonstrated strikingly by inspection of results
computed with various mesh sizes. It seems to be that fully
implicit methods are more efficient if h < 5 cm.

The accuracy of fully implicit methods also goes down
with decreasing spatial mesh size, but not as rapidly.

These few examples already indicate that there may be
no generally best time integration method. In order to
determine integration methods optimally suited for certain
classes of problems, much more experience is needed.

With regard to time integration techniques for FE-methods,
information is very rare. Iterative techniques with successive overrelaxation (70) and conjugate gradient techniques
are investigated (74), but conclusions about their efficiencies cannot yet be made.

In conclusion, some absolute figures will be given, which
serve to illustrate the great progress made by the introduction of WR-methods into reactor analysis.

The BWR-kinetics benchmark problem referenced above is a
quarter core simulation of super prompt-critical transient
with adiabatic heatup and space dependent Doppler-feedback.
The transient starts at cold conditions, goes through first
peak, second peak and is followed to 3.0 secs. With a spatial mesh of 15 x 15 x 30 cm^3, the CUBBOX-code requires
10 min of cpu-time on a IBM 370/195 for the whole transient.

In 3d-quarter core simulations of operational transients with
doubling time of ~30 seconds the neutronic part is computed
by the CUBBOX-code in less than real time on an IBM 370/195.

These figures demonstrate that through the use of modern
methods reliable three dimensional neutron kinetics calculation can be carried out with economically tolerable expense.

REFERENCES

1. Yasinsky, J. B and Henry, A. F., "Some Numerical Experi-
    ments Concerning Space-Time Kinetics Behavior," Nucl.
    Sci. Eng., 22, pp. 171-181, 1965.

2. Yasinsky, J. B., "On the Use of Point Kinetics for the
    Analysis of Rod-Ejection Accidents," Nucl. Sci. Eng.,
    39, pp. 241-256, 1970.

3. Kessler, G., "Space-Dependent Dynamic Behavior of the
    Fast Reactors Using the Time-Discontinuous Synthesis
    Method," Nucl. Sci. Eng., 41, pp. 115-148, 1970.

4. Jackson, J. F., and Kastenberg, W. E., "Space-Time
    Effects in Fast Reactor Dynamics," Nucl. Sci. Eng., 42,
    pp. 278-294, 1970.

5. Salah, S., Rossi, G. E., and Geets, J. M., "Consequences
    of Asymmetric Cold Water Addition to a PWR Core from an
    Inactive Loop," Trans. Am. Nucl. Soc., 14, Page 756, 1971.

6. Salah, S., Rossi, C. E., and Geets, J. M., "Three-Di-
    mensional Kinetic Analysis of an Asymmetric Boron Dilu-
    tion in a PWR Core," Trans. Am. Nucl. Soc., 15, Page 831,
    1972.

7. Lamarsh, J. R., Introduction to Nuclear Reactor Theory,
    Addison-Wesley, Reading, Mass., 1966.

8. Meghreblian, R. V. and Holmes, D. K., Reactor Analysis,
    McGraw-Hill, New York, 1960.

9. Henry, A. F., Nuclear Reactor Analysis, MIT Press,
    Cambridge, Mass., 1975.

10. Mikhlin, S. G., and Smolitsky, K. L., Approximate Me-
     thods for Solution of Differential and Integral Equa-
     tions, Elsevier, New York, 1967.

11. Aziz, A. K., and Babuska, I., The Mathematical Founda-
     tion of the Finite Element Method with Application to
     Partial Differential Equations, Academic Press, New
     York, 1972.

12. Strang, G., and Fix, G. J., An Analysis of the Finite
     Element Method, Prentice-Hall, Englewood Cliffs, New
     Jersey, 1973.

13. Kang, C. M., and Hansen, K. F., "Finite Element Methods
     for Reactor Analysis," Nucl. Sci. Eng., $\underline{51}$, pp. 456-495,
     1973.

14. Babuska, I., "Error-Bounds for Finite Element Methods,"
     Numer. Math, $\underline{16}$, pp. 322-333, 1971.

15. Schultz, M. H., "L^2 Error Bounds for the Rayleigh-Ritz-
     Galerkin Method," SIAM J. Numer. Anal., $\underline{8}$, pp. 737-748,
     1971.

16. Nitsche, J. A.,"Convergence of Nonconforming Methods,"
     in Mathematical Aspects of Finite Elements in Partial
     Differential Equations," C. deBoor, Ed., Academic Press,
     New York, 1974.

17. Birkhofer, A., and Werner, W., "Efficiency of Various
     Methods for the Analysis of Space-Time Kinetics," Proc.
     Conf. Mathematical Models and Computational Techniques
     for Analysis of Nuclear Systems, CONF-730414, Vol. 2,
     pp. IX-31-41, 1973.

18. Birkhofer, A., Langenbuch, S., and Werner, W., "Coarse-
     Mesh Method for Space-Time Kinetics," Trans. Am. Nucl.
     Soc., $\underline{18}$, Page 153, 1974.

19. Langenbuch, S., Maurer, W., and Werner, W., "Simulation
     of Transients with Space-Dependent Feedback by Coarse
     Mesh Flux Expansion Method," MRR 145, Proc. of Joint
     NEACRP/CSNI Specialists' Meeting on New Development in
     Three-Dimensional Neutron Kinetics, pp. 173-188, 1975.

20. Schäfer, A., Über das Konvergenzverhalten eines Galerkin-
     Petrov-Verfahrens Thesis, TU Munchen, 1976.

21. Finneman, H., "A Consistent Nodal Method for the Analy-
     sis of Space-Time Effects in Large LWR's," MRR 145,
     Proc. of Joint NEACRP/CSNI Specialists' Meeting on New
     Developments in Three-Dimensional Neutron Kinetics,
     pp. 145-172, 1975.

22. Selengut, D. S., "Variational Analysis of a Multidimen-
     sional System," Page 89, HW-59126, Hanford Laboratory,
     1959

23. Dougherty, D. E., and Shen, C. N., "The Space-Time Neu-
     tron Kinetics Equations Obtained by the Semidirect
     Variational Method," Nucl. Sci. Eng., 13, pp. 141-152,
     1962.

24. Kaplan, S., Marlowe, O. J., and Bewick, J., "Application
     of Synthesis Techniques to Problems Involving Time-
     Dependence," Nucl. Sci. Eng., 18, pp. 163-176, 1964.

25. Yasinsky, J. B., "The Solution of the Space-Time Neutron
     Group Diffusion Equations by a Time Discontinuous Syn-
     thesis Method," Nucl. Sci. Eng., 29, pp. 381-391, 1967.

26. Stacey, W. M. Jr., "Variational Functionals for Space-
     Time Neutronics," Nucl. Sci. Eng., 30, pp. 448-463, 1967.

27. Yasinsky, J. B., and Kaplan S., "Synthesis of Three-
     Dimensional Flux Shapes Using Discontinuous Sets of
     Trial Functions," Nucl. Sci. Eng., 28, pp. 426-440, 1967.

28. Stacey, W. M. Jr., "A Variational Multichannel Space-
     Time Synthesis Method for Nonseparable Reactor Tran-
     sients," Nucl. Sci. Eng., 34, pp. 45-56. 1968.

29. Stacey, W. M. Jr., Space-Time Nuclear Reactor Kinetics,
     Academic Press, New York, 1969.

30. Yasinsky, J. B., and Henry, A. F., "Some Numerical
     Experiments Concerning Space-Time Reactor Kinetics
     Behavior," Nucl. Sci. Eng., 22, pp. 171-181, 1965.

31. Boresen, S., "A Simplified, Coarse-Mesh, Three-Dimen-
     sional Diffusion Scheme for Calculating the Gross Power
     Distribution in a Boiling Water Reactor," Nucl. Sci.
     Eng., 44, pp. 37-43, 1971.

32. Boresen, S., "Characteristics and Performance of the 3D LWR Simulator PRESTO," Trans. Am. Nucl. Soc., 15, Page 956, 1972.

33. Delp, D. L., Fischer, D. L., Harriman, J. M., and Stedwell, M. J., "FLARE - A Three-Dimensional Boiling Water Reactor Simulator," GEAP-4598, General Electric, 1964.

34. Goldstein, L., Nakache, F., and Veras, A., "Calculation of Fuel-Cycle Burnup and Power Distribution of Dresden-I Reactor with the TRILUX Fuel Management Program," Trans. Am. Nucl. Soc., 10, Page 300, 1967.

35. Deppe, L. O., Hansen, K. F., "Applications of the Finite Element Method to Two-Dimensional Diffusion Problems," Nucl. Sci. Eng., 54, pp. 456-465, 1974.

36. Gear, C. W., Numerical Initial Value Problems in Ordinary Differential Equations, Prentice Hall, Englewood Cliff, New Jersey, 1971.

37. Lapidus, L., and Seinfeld, J. H., Numerical Solution of Ordinary Differential Equations, Academic Press, New York, 1971.

38. Enright, W. H., "Second Derivative Multistep Methods for Stiff Ordinary Differential Equations," SIAM J. Anal., 11, pp. 321-331, 1974.

39. Hofer, E., "A Partially Implicit Method for Large Stiff Systems of ODEs with Only Few Equations Introducing Small Time-Constants," SIAM J. Numer. Anal., 13-5, 1976.

40. Richtmeyer, R. D., and Morton, K. W., Difference Methods for Initial Value Problems, Interscience Publishers, New York, 1967.

41. Birkhofer, A., and Werner, W., "Eine Methode zur Berechnung der raum- und zeit-abhängigen Leistungsverteilung in Kernreaktoren," Atomkernenergie, 15, pp. 97-102, 1970.

42. Wight, A. L., Hansen, K. F., Ferguson, D. R., "Application of Alternating-Direction Implicit Methods to Space-Dependent Kinetics Equations," Nucl. Sci. Eng., 44, pp. 239-251, 1971.

43. Ferguson, D. R., and Hansen, H. F., "Solution of the Space-Dependent Reactor Kinetics Equations in Three Dimensions," Nucl. Sci. Eng., 51, pp. 189-205, 1973.

44. MEKIN: MIT-EPRI Nuclear Reactor Core Kinetics Code, 1975.

45. Gordon, P., "Nonsymmetric Difference Equations," J. Soc. Indust. Appl. Math., 13, pp. 667-673, 1965.

46. Peaceman, D. W., and Rachford, H. H. Jr., "The Numerical Solution of Parabolic and Elliptic Differential Equations," J. Soc. Indust. Appl. Math., 3, pp. 42-65, 1955.

47. Douglas, J., and Gunn, J. E., "A General Formulation of Alternating Direction Methods, Num. Math., 6, pp. 428-453, 1965.

48. Janenko, N. N., "Die Zwischenschrittmethode zur Lösung mehrdimensionaler Probleme der mathematischen Physik," Lecture Notes in Mathematics, 109, Springer-Verlag, Heidelberg, 1969.

49. Birkhofer, A., and Werner, W., "Fully Implicit Matrix Decomposition Method for Space-Time Kinetics," Trans. Am. Nucl. Soc., 15, pp. 789-790, 1972.

50. Langenbuch, S., and Werner, W., "Implicit Matrix Decomposition Scheme for Coarse-Mesh Methods", Trans. Am. Nucl. Soc., 21, Page 224, 1975.

51. Varga, R. S., Matrix Iterative Analysis, Prentice-Hall, Englewood Cliffs, New Jersey, 1962.

52. Wachspress, E. L., Iterative Solution of Elliptic Systems, Prentice-Hall, Englewood Cliffs, New Jsersy, 1966.

53. Nakamura, S., "A Variational Rebalancing Method for Linear Iterative Convergence Schemes for Neutron Diffusion and Transport Equations," Nucl. Sci. Eng., 39, pp. 278-283, 1970.

54. Nakamura, S., "Coarse Mesh Acceleration of Iterative Solution of Neutron Diffusion Equation," Nucl. Sci. Eng., 43, pp. 116-120, 1971.

55. Fröhlich, R., "A Theoretical Foundation of Coarse Mesh Variational Techniques," Report CNM-R-2, Vol. 1, Page 219, (CONF-670501), 1967.

56. Anderson, M. M., Buckner, M. R., Carswell, J. H., Dodds, H. L., Gregory, M. V., Honeck, H. C., Routt, K. R., and Stewart, J. W., "Three-Dimensional Coupled Neutronic and Engineering Calculations of Savannah River Reactors," Proc. Conf. Computational Methods in Nuclear Engineering, CONF-750413, Vol. II, VI, pp. 123-141, 1975.

57. Hestenes, M. R., and Stiefel, E., "Methods of Conjugate Gradients for Solving Linear Systems," Nat. Bur. Standards, J. of Res., 49, pp. 409-436, 1952.

58. Langenbuch, S., and Werner, W., "Eine Methode zur Verbesserung der Zeitintegration in 3d Neutronenkinetik-Rechnungen durch eine Form der Periodenfaktorisierung," Proc. Reaktortagung, 1976.

59. Garland, W. J., Vlachopoulos, J., Harms, A. A., "A Summation-Exponent Analysis for Space-Dependent Reactor Transients," Trans. Am. Nucl. Soc., 18, Page 322, 1974.

60. Devought, J, and Mund, E., "A-Stable Algorithms for Neutron Kinetics," MRR 145, Proc. of the Joint NEACRP/CSNI Specialists' Meeting on New Developments in Three-Dimensional Neutron Kinetics, pp. 21-71, 1975.

61. Yasinsky, J. B., "Combined Space-Time Synthesis with Axially Discontinuous Trial Functions," USAEC Report, WAPD-TM-736, Westinghouse Electric Corp., Bettis Atomic Power Laboratory, 1967.

62. Yasinsky, J. B., "Numerical Studies of Combined Space-Time Synthesis," Nucl. Sci. Eng., 34, pp. 158-168, 1968.

63. Henry, A. F., "Review of Computational Methods for Space-Dependent Kinetics," Dynamics of Nuclear Systems, University of Arizona Press, Tuscon, Ariz., 1972.

64. Ott, K., and Madell, J. T., "Quasistatic Treatment of Spatial Phenomena in Reactor Dynamics," Nucl. Sci. Eng., 26, pp. 563-565, 1966.

65. Ott, K., and Meneley, D. A., "Accuracy of the Quasi-static Treatment of Spatial Reactor Kinetics," Nucl. Sci. Eng., 36, pp. 402-411, 1969.

66. Wagner, M. R., Finnemann, H., Lee, R. R., Meneley, D. A., Michelsen, B., Misfeldt, I., Vondy, D. R., Werner, W., "Multidimensional LWR Benchmark Problems," Trans. Am. Nucl. Soc., 23, 1976.

67. Werner, W., Finnemann, H., Langenbuch, S., "Two- and Three-Dimensional BWR Kinetics Benchmark Problem," Trans. Am. Nucl. Soc., 23, 1976.

68. "Nuclear Reactor Core Analysis Code: VENTURE," Oak Ridge National Lab., 1976.

69. Cadwell, W. R., PDQ-7 Reference Manual, WAPD-TM-678, January 1967.

70. Misfeldt, I., "Solution of the Multigroup Neutron Diffusion Equations by the Finite Element Method," RIS-M-1809, Danish AEC, Research Establishment RIS, Denmark, July 1975.

71. Buckel, G., "Vorschlag für ein Benchmark Problem in xyz- und Dreiecks-z-Geometrie," INR Notiz, 335, 1975.

72. Buckel, G., Approximation der stationären, dreidimensionalen Mehrgruppen-Neutronen-Diffusionsgleichung durch ein Syntheseverfahren mit dem Karlsruher Synthese-Programm KASY, KfK-1349, 1971.

73. Dodds, H. L. Jr., Honeck, H. C., Hostetler, D. E., "Coarse-Mesh-Method for Two-Dimensional Mixed-Lattice Diffusion Theory Calculations," Trans. Am. Nucl. Soc., 21, Page 223, 1975.

74. Schmidt, F. A. R., IKE Stuttgart, FRG, Private Communication.

REVIEW OF EXISTING CODES
FOR
LOSS-OF-COOLANT ACCIDENT ANALYSIS

Stanislav Fabic

Chief, Analysis Development Branch
Water Reactor Safety Research
U. S. Nuclear Regulatory Commission
Washington, D. C. 20555

I.   GENERAL CLASSIFICATION OF LOCA CODES

Within the last two years it has become customary in the
United States to categorize all LOCA codes into two main
groupings: the codes employed in the licensing application
which obey the conservative assumptions prescribed by the NRC
Acceptance Criteria described in Appendix K to 10 CFR 100 (1).
These codes employ the so-called Evaluation Model and are
generally referred to as the EM codes. The WREM package includes most of the EM codes necessary for the safety evaluation of the Water Reactors performed independently by USNRC.
The second grouping consists of the so-called Best Estimate
(BE) codes which employ models that are not conservative, but
instead, represent a realistic representation of the physical
processes. Some of the existing BE codes, such as RELAP-4
Mod 5 (developed by the Aerojet Nuclear Company for USNRC)
are simply structured. Advanced BE codes now under development incorporate the present state of the art in both the
modeling and the numerical analysis treatment. The present
EM codes comprise an assembly of codes run sequentially. Each
member of the sequence is a stand-alone code developed for
some special application. For example, the saturated blowdown codes consider the system transients from the beginning
of blowdown to the time when the discharge flow through the
postulated break ceases. The system nodalization is not fine
enough to permit tracking of the rarefaction waves to the extent required for evaluation of the hydraulic loads. Their

365

main purpose is to calculate the overall system depressuriza-
tion including the transients introduced through injection of
the emergency core coolant (ECC). The heat transfer in the
reactor core and the steam generators (in the case of PWR's)
are included in sufficient detail to account for the coolant's
energy state. On the other hand, such heat transfer calculations are not sufficient for evaluation of the maximum cladding and fuel temperature. The so-called hot channel, or the
fuel behavior codes are employed for this latter purpose.
They are "driven" by the boundary conditions calculated by
the system blowdown and/or the system reflood codes described
herein, and account for the structural deformation of the
fuel and cladding, the fission gas pressure, metal/water
reaction and the decay heat generation. Finally, the thermal
hydraulics of the refill/reflood stage is calculated via reflood codes that emphasize the physical process within the
reactor vessel while the emergency coolant water is being
introduced into the core to quench it.

In the case of a Small Break analysis (for the postulated
breaks that are smaller than about 10% of the double-ended
break in the discharge flow area) the core may be only partially uncovered during the whole blowdown sequence; hence,
the reflood codes are not called for in those instances.
However, the (saturated) blowdown codes for the Small Break
application contain special features necessary to account
for separation of steam and liquid, tracking of the liquid or
mixture levels, and calculation of the heat transfer regimes
in the vicinity of the liquid or mixture levels.

When consequences of a large break in a PWR are analyzed,
the Containment code is also utilized in the calculation sequence to account for the change in the pressure of the environment into which the break fluid discharges. In this
case, however, the containment code is not as detailed as
when the Containment Analysis is performed for the purpose
of establishing the pressure and temperature history; hence,
loads experienced by the Containment walls.

The blowdown-induced loads on the primary coolant system
(reactor internals, supports, piping and other system components) are calculated by separate codes that are capable of
tracking the propagation of rarefaction waves throughout the
system. Such calculations are not performed within the EM
sequence. A typical EM sequence for PWR LOCA analysis is
illustrated in Figure 1.

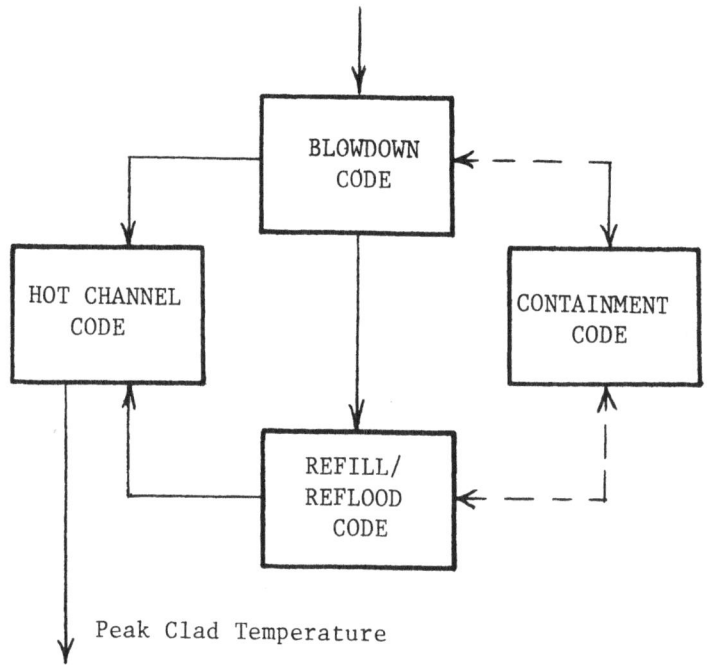

Figure 1

EM Code Sequence

Recent trends in the development of the Best Estimate
(BE) codes have resulted in merging into one large Systems
Code the various functions previously performed via the separate, stand-alone codes. This ensures proper compatibility
and continuity between the various calculational phases.
Furthermore, the same subroutines are being called upon for
evaluation of the equation of state (steam tables), heat
transfer coefficients, thermal properties, etc. It has been
recognized that a general subdivision of the System into
control volumes for which the same conservation equations are
solved, using the same thermo-hydraulic models, is too re-
strictive a procedure. Different system components may require special models to more realistically describe their
behavior, and some may require multidimensional treatment.
For this reason, the Advanced BE Codes now under development

in the United States and elsewhere are being programmed in
a modularized form wherein each system component is represented by its own module. For example, the equations being
solved for the pump module differ from others; the PWR downcomer module may employ a different hydraulic model and in
more dimensions than that for a pipe, etc. The advanced BE
codes will not be utilizing global correlations (such as
FLECHT) which are strongly tied to a specific geometry and
the mode of testing. Local fluid and thermal properties
will be calculated, in more than one dimension where necessary, to allow applicability to the full scale plant accident
analysis.

Code verification process, through which the knowledge
is gained as to how well the important physical processes
have been modeled, could be carried out only with the BE
codes. The latter are also useful for planning experiments
and data interpretation. One of the important uses of a
Best Estimate code is to evaluate the degree of conservatism
employed in the licensing (EM) calculations.

II.   DESCRIPTION OF TWO-PHASE FLOW GOVERNING EQUATIONS

A.   Modeling of Two-Phase Flow

The physics of two-phase flow is extremely complex. It
features an irregular and variable geometric pattern of the
two coexisting phases with sharp discontinuities in the fluid
properties at the interfaces. There are very few flow situations in which the phases separate sufficiently to allow
some tractable definition of the location, the shape and the
extent of the interface. In the majority of cases of interest, the flow topology is far too difficult to allow the use
of the so-called local, instantaneous formulation of the
governing equations. Even in the bubbly flow regime it is
difficult to define the bubble population; the vapor bubbles
differ in size, are not uniformly distributed, and their shape
becomes distorted due to the induced mass and other flow
effects. For these reasons, most of the two-phase flow
analysis performed in the past resorted to drastic simplifications in which the two-phase fluid was viewed as a homogeneous mixture of vapor and liquid, in thermal equilibrium
with each other. The steam tables were utilized to define
the properties of that pseudo fluid and the single phase
governing equations utilized to describe the flow behavior,

in both steady-state and in transients. In some flow regimes
this "homogeneous-equilibrium model" (HEM) gave surprisingly
good results. As our ability to visualize and measure two-
phase flow behavior increased, it became clear that in many
important instances, the two phases are not traveling with
the same velocities (or may even be flowing in the opposite
direction), they are not uniformly distributed, and are not
in thermal equilibrium. As a consequence, a renewed effort
has been underway during the last five years to attempt to
introduce modeling sophistications that could account for
thermal nonequilibrium and nonhomogeneity. Various averaging procedures are being employed which convert the local
instantaneous governing equations written for each phase
to the averaged governing equations for the mixture (2).
The time averaging is attractive because it views the fluid
the way a probe, immersed in the fluid at some fixed location, would view the appearance and disappearance of the two
phases over some short period of time. Whatever the property
this probe measures, the recorded signal over the time period
could be averaged and useful information obtained providing
the chosen time interval is (a) larger than the transport
time of the single bubbles or droplets within the probe domain; and (b) the time interval is short in comparison with
the time constant of the flow dynamics of interest. Once
the local governing equations have been suitably time averaged, one can then proceed to employ them directly or to
average them over the control volume, or area, or the length
of interest, depending on the desired degree of freedom chosen
for the geometric representation of the flow field. In the
so-called lumped parameter approach, the equations are averaged over fairly large control volumes of "lumps", while in
the three-dimensional applications in which the computational
cells are small, the local, time-averaged equations are utilized. Analyses applied to one-dimensional (pipe) geometries
employ the time-averaged governing equations which are also
averaged over the flow cross-section area.

    Most of the licensing type (EM) calculations are being
performed in the lumped parameter mode using the homogeneous-
equilibrium (HEM) model, since they are expected to yield the
bounding results reflecting various conservative assumptions
in the modeling of the physical phenomena. The best estimate
(BE) calculations are much more demanding. The on-going effort
in the NRC-sponsored development of the BE analytical tools
(advanced LOCA codes) being performed at various National

Laboratories reflects a state of the art in both the modeling
of physics and in the numerical solution techniques.

In their most complex (two-fluid) formulation, the governing equations that describe the laws of conservation of
mass, momentum and energy (often referred to as the "conservation" or "field" equations) must be supplemented by the
macroscopic constitutive equations and by the equations that
describe the interfacial transfer conditions for mass, momentum and energy. The macroscopic constitutive laws include
the thermal and the caloric equations of state, the viscous
stress, τ, the conduction heat transfer, q, the interfacial
transfer terms for mass, Γ, momentum, M, and energy, E, between the phases, and others (such as the turbulent fluxes
and surface tension effects when such are considered).
Finally, the closure of the system of the equation demands
the specification of the initial and the external boundary
conditions. Tables 1 and 2 list the practical two-phase
flow models in the ascending degree of complexity. It can
be seen that the least complex formulation contains the
largest number of restrictions and the smallest number of
the governing and the constitutive equations. The restriction on V implies three possibilities: (a) both phases travel
with the same velocity and in the same direction (the homogeneous model); (b) both phases travel in the same direction
and the ratio of the vapor and the liquid velocity (slip)
is defined via a constitutive equation for slip; and (c) the
phases may flow in concurrent or in counter-current manner,
but the difference between the vapor and the liquid velocities, or the difference between the center of mass of the
mixture and the vapor velocity, or the difference between
the center of volume velocity of the mixture and the vapor
velocity, are specified via a constitutive equation.

In multidimensional formulations there will be one conservation of momentum equation for each spatial direction.
In addition, the number of the constitutive equations for
the wall shear, wall heat transfer, M, and E will again be
increased to reflect the increased degrees of freedom. The
same holds for the expression for the relative velocity,
V_r, or the diffusional velocity, V_G-V_m, or the "drift",
$V_G - V_j$. The slip formulation is not employed in multidimensional formulations. The restriction on h_G and/or h_L
implies that $h_G = (h_G(p))_{sat}$ and/or $h_L = (h_L(p))_{sat}$.

TABLE 1

PRACTICAL TWO-PHASE FLOW MODELS (ASSUMING $P_G = P_L = P$)

Model Designation	Restrictions No.	Restrictions Imposed on	Mass.	Mom.	Ener.	No. of Interface Transf. Eqs.	External Constitutive Eqs. No.	External Constitutive Eqs. Type
1V1T	3	$V_G=V_L$, h_L, h_G	1	1	1	0	2	$\bar\tau,\ \bar q$
1VS1T	3	V_G/V_L, h_L, h_G	1	1	1	0	3	$\bar\tau,\ \bar q,\ V_G/V_L = \text{slip}$
1VD1T	3	V_G-V_L, h_L, h_G	1	1	1	0	3	$\bar\tau,\ \bar q,\ (V_G-V_L)$ or (V_G-V_m) or (V_G-V_j)
$1VT_K^{sat}$	2	$V_G=V_L$, $\left\{ h_L \right.$	1	1	2	1	3	$\bar\tau,\ \bar q,\ q_G,\ E$
$1VST_K^{sat}$	2	V_G/V_L, $\left. \text{or} \right.$	1	1	2	1	4	$\bar\tau,\ \bar q,\ q_G,\ V_G/V_L,\ E$
$1VDT_K^{sat}$	2	V_G-V_L, $\left\} h_G \right.$	2	1	1	1	4	$\bar\tau,\ \bar q,\ \Gamma,\ V_r$ (or V_{Gm}, or V_{Gj})
2V1T		h_L, h_G	1	2	1	1	4	$\tau_L,\ \tau_G,\ \bar q,\ M$
1V2T	2	$V_G=V_L$	2	1	2	2	5	$\bar\tau,\ q_L,\ q_G,\ \Gamma,\ E$
1VD2T	2	V_G-V_L	2	1	2	2	6	$\bar\tau,\ q_L,\ q_G,\ \Gamma,\ E,\ V_r$ (or V_{Gm})
$2VT_K^{sat}$	1	h_L or h_G	2	2	1	2	5	$\tau_L,\ \tau_G,\ \bar q,\ \Gamma,\ M$
$2VT_K^{sat}$		h_L or h_G	1	2	2	2	6	$\tau_L,\ \tau_G,\ \bar q,\ q_K,\ M,\ E$
2V2T	0	None	2	2	2	3	7	$\tau_L,\ \tau_G,\ q_L,\ q_G,\ \Gamma,\ M,\ E$

TABLE 2

PRACTICAL TWO-PHASE FLOW MODELS DESIGNATIONS

Designation*	Characteristics	Final, verifiable results
1V1T	(HEM) Homogeneous, Equilibrium	P, $\bar{v}$, α (α = void fraction)
1VS1T	Slip, Equilibrium	P, $\bar{v}$, α
1VD1T	Drift, Equilibrium	P, $\bar{v}$, α
1VT$_K$T$_{sat}$	Homogenous, Partial Non-equilibrium	P, $\bar{v}$, α, T_L or T_G
1VST$_K$T$_{sat}$	Slip, Partial Non-equilibrium	P, $\bar{v}$, α, T_L or T_G
1VDT$_K$T$_{sat}$	(DF) Drift Flux, Partial Non-equilibrium	P, $\bar{v}$, α, T_L or T_G
2V1T	Two Fluid, Equilibrium	P, V_L, V_G, α
1V2T	Homogeneous, Full Non-equilibrium	P, $\bar{v}$, T_L, T_G, α
1VD2T	Drift, Full Non-equilibrium	P, $\bar{v}$, T_L, T_G, α
2VT$_K$T$_{sat}$	Two Fluid, Partial Non-equilibrium	P, V_L, V_K, α, T_L or T_G
2V2T	(TF) Two Fluid, Full Non-equilibrium	P, V_L, V_K, α, T_L, T_G

* 1V1T = one velocity, one temperature model; 2V2T = two velocities, two temperatures model (one for each phase). $T_K T_{sat}$ = one phase assumed at saturation, temperature of the other (K) phase computed. The letters S or D following 1V indicate that either a slip ratio or a vapor drift is used, respectively, in formulation of field equations.

When more than one conservation equation is indicated in Table 1, for either mass, or momentum, or energy, various possibilities exist: either one of these is represented by the particular conservation for the mixture, or one conservation equation is written separately for each phase, or the difference between two phasic conservation equations is employed.

In the United States, the existing codes fall in the 1V1T or 1VS1T or 1VD1T category. The advanced codes now under development for NRC belong to the (DF) and the (TF) category. Examples of the conservation equations for the HEM, DF, and the TF models will be given in sections II.B, II.C and II.D, below, in the vectorial form, indicating the simplifying assumptions. The subscript m will denote the mixture property.

B.   Homogeneous Equilibrium Model

Conservations of mass, momentum and total energy give:

$$\frac{\partial \rho_m}{\partial t} + \vec{\nabla} \cdot (\rho_m \vec{v}_m) = 0 \tag{1}$$

$$\frac{\partial \rho_m \vec{v}_m}{\partial t} + \vec{\nabla} \cdot (\rho_m \vec{v}_m \cdot \vec{v}_m) = -\vec{\nabla}p + \vec{\nabla} \cdot \tau_m + \rho_m \vec{g} \tag{2}$$

$$\frac{\partial \rho_m (e_m + \frac{v_m^2}{2})}{\partial t} + \vec{\nabla} \cdot (\rho_m (e_m + \frac{v_m^2}{2}))$$

$$= - \vec{\nabla}\vec{q}_m + \vec{\nabla} \cdot (\overline{\overline{\P}}_m \cdot \vec{v}_m) + \rho_m \vec{g} \vec{v}_m + Q_m \tag{3}$$

where the surface stress tensor $\overline{\overline{\P}}_m = -(p_m \overline{\overline{I}} - \overline{\overline{\tau}}_m)$, $\overline{\overline{\tau}}_m$ is the viscous stress tensor, $\vec{q}_m$ is the heat flux, Q_m is the body heating, and e_m is the internal energy.

The above balance equations need to be supplemented by the equations of state, the subsidiary equations describing mixture properties, and the constitutive equations for $\overline{\overline{\tau}}_m$, $\vec{q}_m$, and Q_m.

Idaho National Engineering Laboratory (INEL) code SCORE-EVET comes closest to this generalized, three-dimensional (cartesian) definition. As the name suggests, it is a (core) component code. Various simplifying assumptions are made for the constitutive equations (τ, q), mainly based on test data from one-dimensional geometries with steady-state, two-phase flow.

The Pacific Northwest Laboratory COBRA-4 code is also a core component code. It considers the flow as having a predominately axial direction. All the "lateral" flow is lumped into one lateral momentum equation. It is being cast presently in the drift flux format.

INEL's RELAP-4 code is the existing Systems LOCA code. It has a lumped parameter structure in which the spatial effects are integrated over the control volume for the conservation of mass and energy. For example, the mass balance becomes, utilizing the Green's theorem:

$$\int \frac{\partial \rho}{\partial t} \, dV + \oint \vec{n} \cdot (\rho \vec{v}) \, dS = 0 \qquad (4)$$

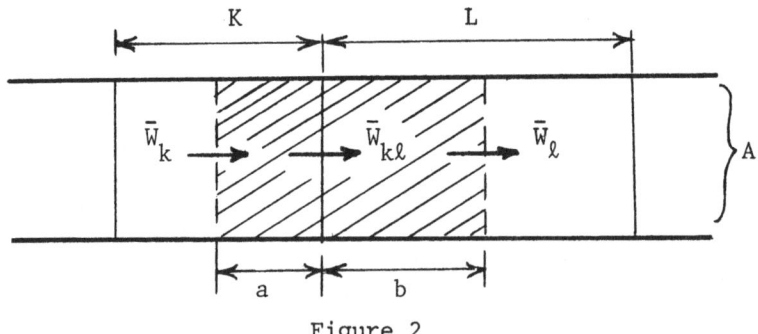

Figure 2

Staggered Control Volumes
The shaded area represents the control volume
over which the fluid momentum is balanced

with $\rho\mathbf{v} = M_j$ (with ρ uniform within V) and $-\vec{n}_i(S\rho v)_i = \overline{W}_{ij} =$ inflow from side "i" into control volume j, the mass balance reduces to:

$$\frac{dM_j}{dt} = \sum_i \overline{W}_{ij} \qquad (5)$$

The momentum equation is centered at the "junction" between two adjacent control volumes. Consider two adjacent control volumes labeled k and ℓ, with equal flow areas, A, in the direction of flow, illustrated in Figure 2.

Integrating the momentum balance over the new control volume, shown shaded, results in:

$$\int_V \frac{\partial\rho\vec{v}}{\partial t} \, dV + \oint_S \vec{n} \cdot (\rho\vec{v}\vec{v} + p - \overline{\overline{\tau}})dS - \int_V \rho\vec{g} \, dV = 0 \quad (6)$$

It is assumed that $(L/A)(dW_{k\ell}/dt)$ represents the rate of change of momentum everywhere in the shaded control volume. With this nomenclature and dividing through by A, the momentum equation, in lumped parameter form, becomes:

$$\frac{1}{2A}(L_K + L_\ell)\frac{d\overline{W}_{k\ell}}{dt} = (p_k - p_\ell) - (\tau_a + \tau_b) -$$

$$(1/2)(\rho_k L_k + \rho_\ell L_\ell)g\cos\theta + (\overline{W}_k v_k - \overline{W}_\ell v_\ell)/A \qquad (7)$$

where $\overline{W}_k$ and $\overline{W}_\ell$ are taken as the arithmetic means of the adjacent two-boundary junctions flows. Note that $\overline{W}_{k\ell}$, the inflow into the control volume ℓ, appears in the balance equations for mass and energy for that control volume.

The term $(L_k + L_\ell)/(2A)$ is denoted as the geometric inertia, I. For the more complex cases in which the flow areas of the adjacent control volumes are not equal, the geometric inertia, I, becomes $(L_k/A_k + L_\ell/A_\ell)/2$ and additional terms appear on the righthand side of the momentum equation. Various assumptions are made in modeling these terms, which need to be verified. The problem becomes even more difficult

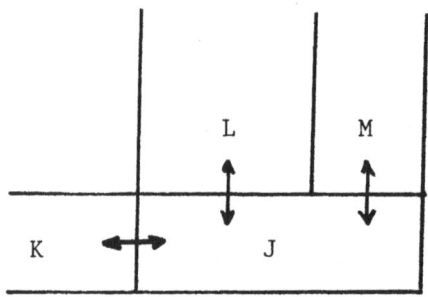

Figure 3

Lumped Parameter Simplification
Of Multidimensional Flow Regimes

when a given control volume, J, connects into more than two
control volumes, exemplified in Figure 3.

The double-ended arrows indicate junctions. It is clear
that it is now quite difficult to define the control volume
boundaries for the momentum balance at each junction. This
obviously calls for a multidimensional flow pattern that
cannot be accurately modeled with the lumped parameter method.
When such cases are encountered, the momentum flux terms are
deleted and code user defines the geometric inertias to his
own choice. The frictional and the form losses are treated
as in the single phase flow, with a two-phase (Baroczy) multiplier. The bubble rise model is employed to achieve phase
separation in any input-defined control volume. When several of
such volumes are connected in a vertical stack, each control
volume ends up with two separate regions. In order to overcome this deficiency, a drift flux term was recently introduced into the energy equation (only). Its performance has
been spotty, mainly because of the difficulties encountered
in specifying the fluid properties at junctions.

C.   Drift Flux (Diffusion) Model

The advantage of the drift flux model (DF) over HEM is
two-fold: (a) the effects of the difference in the steam
and the liquid velocities on the *mixture* dynamics is considered, and (b) thermal nonequilibrium effects can be considered in a simplified manner. The latter is achieved by
considering the dispersed phase to be at saturation while the

continuous phase is in thermal nonequilibrium.

The basic concept is to consider the mixture as a whole, rather than each phase separately. This reduces the total number of field and constitutive equations required in formulation, in comparison with the two-fluid model. It is generally accepted that the diffusion model is appropriate to the mixture where dynamics of two components are closely coupled. Even for the mixtures that are weakly coupled locally, the DF model is adequate where the relatively large axial dimension of the system gives sufficient interaction times. The field equations are as follows:

Mixture continuity:

$$\frac{\partial \rho_m}{\partial t} + \vec{\nabla} \cdot (\rho_m \vec{v}_m) = 0 \tag{8}$$

Diffusion equation for the dispersed phase (2):

$$\frac{\partial \rho_2 \alpha_2}{\partial t} + \vec{\nabla} \cdot (\alpha_2 \rho_2 \vec{v}_m) = \Gamma_2 - \vec{\nabla} \cdot (\alpha_2 \rho_2 \vec{V}_{2m}) \tag{9}$$

Mixture momentum:

$$\frac{\partial \rho_m \vec{v}_m}{\partial t} + \vec{\nabla} \cdot (\rho_m \vec{v}_m \vec{v}_m) = -\vec{\nabla}p + \vec{\nabla} \cdot (\bar{\bar{\tau}} + \bar{\bar{\tau}}^T - \Sigma_k \alpha_k \rho_k \vec{V}_{km} \vec{V}_{km}) \tag{10}$$

Thermal energy (in terms of enthalpy):

$$\frac{\partial \rho_m i_m}{\partial t} + \vec{\nabla} \cdot (\rho_m i_m \vec{v}_m) = -\vec{\nabla} \cdot (\bar{q} + q^T)$$

$$-\vec{\nabla} \cdot (\Sigma_k \alpha_k \rho_k i_k \vec{V}_{km}) + Dp/Dt + \phi_m^\mu \tag{11}$$

For derivation and explanation of the nomenclature, the reader should consult Reference 3. The important point to emphasize is that the above balance laws govern the *macroscopic mixture field*. They were obtained by the time-averaging of the local, instantaneous balances applied to the two-phase systems with interfacial discontinuities.

As shown in Reference 3, these balances must be supplemented
by:

1. the axiom of continuity, $\sum_k \alpha_k = 1$

2. thermal equation of state (for each phase)

3. caloric equation of state (for each phase)

4. definition for ρ_m, i_m

5. **kinematic constitutive** equation for the drift
    velocity $\vec{V}_{2m}$

6. identity of the diffusion velocities:

$$\sum_k \alpha_k \rho_k \vec{V}_{km} = 0 \tag{12}$$

7. mechanical constitutive equations for:

    viscous stress, $\overset{=}{\tau}$

    turbulent stress, $\overset{=T}{\tau}$

8. energetic constitutive equations, for:

    conductive heat flux, $\overline{q}$

    turbulent heat flux, q^T

    dissipation term, ϕ_m^μ

9. constitutive equation for phase change mass
    generation, Γ_2.

10. the interfacial mass transfer law:

$$\sum_k \Gamma_k = 0 \tag{13}$$

It should be noted that in the above formulation the surface
tension effects were ignored and the interface temperature
was set equal to that of the dispersed phase; i.e., $T_{sat}(p)$.

This leads to the implied assumption that $p_G = p_L = p$.
Further simplifications have been introduced in the current
formulations of the NRC's advanced LOCA codes, THOR and TRAC,
insofar as the effects of turbulence are ignored, and in some
cases, the dissipation term, ϕ_m^μ in the energy equation is

deleted. It should be emphasized that the constitutive equations pertain to the macroscopic, time-averaged properties
within a computation cell. Model development experiments

will be utilized for verification of the constitutive equations. Brookhaven's THOR code is one-dimensional (allows a mix of the lumped and the distributed parameter modeling). Los Alamos' TRAC allows mixed dimensions.

D.   Two-Fluid Model

This, the most complex model, considers "everything we know" about the two-phase flow dynamics. It allows separate tracking of the phase velocities and of the phase temperatures, both of which may differ from equilibrium. In view of our basic assumption that $p_L = P_G = p$, the field equations become:

Continuity

$$\frac{\partial}{\partial t}\,(\alpha_k \rho_k) + \vec{\nabla}\cdot(\alpha_k \rho_k \vec{v}_k) = \Gamma_k \tag{14}$$

where $k = 1$ and 2, and Γ_k is the rate of generation of the phase k due to the interfacial mass transfer.

Momentum

$$\frac{\partial}{\partial t}\,(\alpha_k \rho_k \vec{v}_k) + \vec{\nabla}\cdot(\alpha_k \rho_k \vec{v}_k \cdot \vec{v}_k)$$

$$= -\,\vec{\nabla}(\alpha_k p) + \vec{\nabla}\cdot\{\alpha_k\,(\bar{\bar{\tau}}_k + \bar{\bar{\tau}}_k^{\,T})\}$$

$$+ \alpha_k \rho_k \vec{g} + \vec{M}_k \tag{15}$$

The interfacial momentum transfer, M_k, is composed of the following terms:

$$\vec{M}_k = \Gamma_k \vec{v}_{ki} + p_{ki}\vec{\nabla}\alpha_k + \vec{M}_k^{\,d} \tag{16}$$

where M_k^d represents the interfacial **drag on phase, k.** Consequently, the righthand side of the momentum equation becomes:

$$\text{RHS} = -\alpha_k \vec{\nabla}p + \vec{\nabla}\cdot\{\alpha_k\,(\bar{\bar{\tau}}_k + \bar{\bar{\tau}}_k^{\,T})\} + \alpha_k \rho_k \vec{g} + \Gamma_k \vec{v}_{ki} + \vec{M}_k^{\,d} \tag{17}$$

where $\vec{v}_{ki}$ represents the velocity of the interface, and, due to our original assumption, p_{ki} was set equal to p.

Total Energy

$$\frac{\partial}{\partial t} \{\alpha_k \rho_k (e_k + \frac{v_k^2}{2})\} + \vec{\nabla} \cdot \{\alpha_k \rho_k (e_k + \frac{v_k^2}{2}) \vec{v}_k =$$

$$= - \vec{\nabla} \cdot \{\alpha_k (\vec{q}_k + \vec{q}_k^T)\} + \vec{\nabla} \cdot (\alpha_k \overline{\overline{\P}} \cdot \vec{v}_k) + \alpha_k \rho_k \vec{g} \cdot \vec{v}_k + E_k \tag{18}$$

where the interfacial energy transfer term can be approximated by:

$$E_k \simeq \Gamma_k (i_{ki} + \frac{v_k^2}{2}) - p \frac{\partial \alpha_k}{\partial t} + \vec{M}_k^d \cdot \vec{v}_k \tag{19}$$

Often a thermal rather than the total energy balance is utilized. Let $h_k = i_k + (v_k')^2/2$ represent the effective enthalpy which includes the kinetic energy term caused by the turbulent (only) component of the phasic velocity; the latter is often ignored. Then it can be shown (see Reference 3) that

Thermal Energy (in terms of enthalpy) becomes

$$\frac{\partial}{\partial t} (\alpha_k \rho_k h_k) + \vec{\nabla} \cdot (\alpha_k \rho_k h_k \vec{v}_k) = -\vec{\nabla} \cdot \alpha_k (\vec{q}_k + \vec{q}_k^T)$$

$$+ \alpha_k \frac{Dp}{DT} + \alpha_k \overline{\overline{\tau}}_k \vec{\nabla} \vec{v}_k + \Gamma_k h_{ki} + \vec{M}_k^d \cdot (\vec{v}_{ki} - \vec{v}_k) \tag{20}$$

For derivation of the thermal energy based on the internal energy the reader should consult Reference 3.

In all of the above balance equations, the phase density, ρ_k, is averaged only during that portion of the averaging time increment, $[\Delta t]_k$, during which only the phase K is present. The phasic velocities, $\vec{v}_k$, heat fluxes, $\vec{q}_k$, the internal energies, e_k, enthalpies, i_k, and the effective enthalpies, h_k, are all averaged over the total averaging time increment $[\Delta t]_T$.

The above balance equations are supplemented by the interfacial transfer laws:

$$\sum_k \Gamma_k = 0, \quad \sum_k M_k = 0, \quad \sum_k E_k = 0 \tag{21}$$

and various subsidiary equations (such as $\sum_k \alpha_k = 0$), external constitutive equations (for q_k, q_k^T, τ_k, τ_k^T, Γ_k, M_k^d) and the equations of state. The turbulent contributions, denoted by the superscript T, are often ignored. The external constitutive equations may contain partial differentials of the dependent variables. The constitutive equations are dependent on the geometric configuration of the steam water interfaces. This is also true for the drift flux model, although much more drastic simplifications could be tolerated, as for example, the use of the algebraic equations to define some limiting "flow regime maps". For the two-fluid model, however, where all the mass, momentum, and energy transfer terms at the interfaces are very strongly affected by the interface configuration, it may be necessary to define one or more additional transport equations to define the shape and the extent of the interface.

At present, only one code based fully on the two-fluid model is being developed under NRC sponsorship: it is LASL's KACHINA code and its more recent derivatives such as K-FIX and K-TIF. It serves two purposes: (a) to provide a test bed for research in numerical two-phase flow dynamics; and (b) to provide a yardstick for structuring simpler two-fluid descriptions of some system components during some regime of LOCA. LASL's TRAC code (for Systems LOCA) will allow implementation of two-fluid models for those components in which the weak coupling between the phases is anticipated; otherwise, the drift flux model will prevail. Some terms in the above balance equations are ignored in KACHINA and its constitutive equations are, at present, rather rudimentary. Their verification and/or improvement will have to rely solely on the Model Development Experiments that view the basic phenomena, one at a time. The development of the satisfactory two-fluid capability will take a long time, since, apart from the physics-modeling difficulties, extensive problems must be overcome in formulating a feasible and efficient numerical solution method.

E.   Numerical Solution Techniques

   Mathematical models approximate the actual physics problem. Next, a numerical model of the mathematical problem is
needed. The *numerical model* approximates the mathematical
model; however, the error involved in this approximation
can, at least theoretically, be made arbitrarily small.
The numerical solution methods differ greatly in sophistication, depending on the desired degree of accuracy, the
restrictions of the computer storage, and the allowable computer running time. The stability, convergence, and consistency of the numerical models for the solution of the complex, nonlinear balance equations cannot yet be proven on
purely mathematical grounds. Consequently, the numerical
simulation of transient, two-phase flow is an art that requires a great deal of experience gained through experimentation and the use of various heuristic arguments. In such
cases, the numerical analyst must resort to the comparison
of his results with the results of the closed form solutions
of simplified problems (where such can be obtained) and of
course, to the comparisons with a variety of test data.

   Finally, because of the finite word length of digital
computers, the *computer solution* only approximates the solution to the numerical model.

   Since the use of the Method of Characteristics (MOC)
has grown in recent years and since it differs in concept
from other methods, it may be useful to touch upon it first.
The partial differential equations (PDE) are first put in
the canonical form which leads to a set of ordinary differential equations (ODE) which (a) describe the manner in which
the dependent variables change along the characteristic
paths; and (b) describe the characteristic paths themselves.
The set of ODE are then solved by an explicit finite difference method. MOC has found applicability in one-dimensional
modeling of the early blowdown transients during which the
inertial and acoustic effects play a dominant role. Approximate multidimensional solutions could be obtained through
the use of the Equivalent Piping Networks (EPN) (7).

   When only the spatial differentials are discretized, the
initial set of PDE reduces to a set of ODE which can be solved
by either the Runge-Kutta (or equivalent) method (RK), or by
a predictor-corrector method (PC) in an explicit manner. Discretization of differentials in both time and space leads to

a set of algebraic equations. Linearization of PDE's leads to a linear set of algebraic equations that can be expressed in a matrix format and solved implicitly (simultaneously) in a variety of ways. When the original PDE's are first integrated in space, leading to the lumped parameter (LP) geometric representation, the subsequent discretization of the temporal derivatives also leads to a set of algebraic equations which may or may not be linearized. In the former case, the matrix solution method is employed for their implicit solution, while the latter case is solved through a Newton-Raphson iteration scheme. In some cases, the lumped parameter model is solved by the explicit methods mentioned above.

A variety of newer implicit methods have been employed at Los Alamos, starting with ICE (Implicit Continuous-fluid Eulerian), followed by IMF (Implicit Multi-Fluid). The latter is the basis of the KACHINA code. Very recently, the desire to achieve a tighter implicit coupling with the energy balance has led to the FIX (Fully Implicit Exchange) method which is the basis of LASL's K-FIX code. The latter is an improved version of KACHINA, particularly suited for very detailed studies of two-phase transients.

Developers of the advanced best-estimate *systems* LOCA codes are still searching for the optimum solution technique. Such codes are designed to employ a variety of the hydraulic models and a variety of the geometric representations (mixed geometry), to best suit any given system component. Explicit solution techniques are tractable except that they require very small time increments. Implicit techniques, for the whole system, lead to very large, sparse matrices that require large core memories. Very fast computers, with very large core memories (e.g., CRAY-1) **have recently** become operational. Such computers may renew the interest in the explicit solution techniques, many of which are described in Reference 4.

III. PRESSURIZED WATER REACTOR (PWR) CODES

A. Subcooled Decompression

The subcooled decompression period ends after around
100 milliseconds in the case of a full, double-ended break.
During that period, most of the primary system fluid has de-
pressurized down to the local saturation pressure and the
amplitudes of the rarefaction and the compression waves have
decayed down to negligible values. Rarefaction waves origin-
ate at the break and propagate around the system with the
local velocity of sound. Their amplitudes are modified as
the flow area changes are encountered. In addition, when an
area enlargement is encountered, a portion of the primary
rarefaction wave is reflected back as a compression wave.
Since the waves penetrate various regions inside the reactor
vessel at different times, sizable net pressure loads are
generated on the reactor internals, steam generator internals,
as well as the primary system piping. It is important to
analyze these blowdown-induced dynamic loads to assure that
(a) the stresses at various system supports are not exceeded;
and (b) that no significant distortion of the reactor internals takes place that may impair the movement of the control
rods or the supply to the Emergency Core Coolant (ECC).

When the fluid is still in the form of a compressed liquid,
the spatial acceleration terms in the fluid conservation equations are negligible in comparison with the temporal acceleration, and in addition, the energy equation can be ignored,
since insufficient time is available for heat transfer. These
simplifications lead the well-known "wave equation" which can
be solved analytically. In the WHAM code (5), all such calculated forward and reflected waves are tracked in time and
space (one-dimensionally) and their local amplitudes summed
to obtain instantaneous pressure distribution throughout the
system. Special break flow model, based on Burnell's idea
(6), is employed to calculate the magnitudes of the origina-
ting rarefaction waves. Multidimensional effects can be modeled
in an approximate fashion via the "equivalent piping network" (7).
Such fully-subcooled blowdown analyses are applicable during
a very short time period. The do, however, yield upper-bound
values for the blowdown loads, since the appearance of steam
will greatly diminish both the speed and the amplitude

of the decompression waves. The Westinghouse BLODWN-2 code
represents a more realistic calculational model since it
allows for the presence of both the subcooled liquid and the
two-phase mixture. It solves the HEM system of equation via
the one-dimensional method of characteristics (MOC). Equivalent piping network is again utilized for multidimensional
effects. German DAPSY and Japan's DEPCO-MULTI codes have
very similar characteristics. A more recent German code,
DRUFAN, allows for thermal nonequilibrium.

When the blowdown loads are sizable, some reactor internals will experience deflections that cause dynamic variations
in the flow areas seen by the decompression waves. This, in
turn, can result in a very significant change in the local
pressures; hence, loads. In order to account for these
effects (to introduce further realism in the analytic modeling), the description of hydrodynamics is coupled with the
description of structural dynamics. German KRAFT code was a
somewhat crude attempt toward that goal since it utilized a
lumped parameter approach with which a fine wave structure
cannot be tracked.

Los Alamos Scientific Laboratory (LASL) is now in the
process of developing, for USNRC, a multidimensional hydro-
elastic code named SOLA-FLX. This advanced code will represent the state of the art in the ability to calculate both
the blowdown loads and their consequences.

B.   Blowdown Period

During this period, which lasts until the primary system
and the containment pressures have equalized, the code must
be able to calculate a transient flow of the steam-water mixture and to account for heat transfer at all the important
"wetted" boundaries. The latter must include the core fuel
rods and the steam generator tubes. In the best-estimate
analyses, the heat transfer from a variety of "walls" and
internal structures is also accounted for, especially where
the surface-to-volume ratios are large. The local heat transfer regime is not only affected by the local "wall" surface
temperature but also by the local pressure, flow rate, and
the local flow regime (void fraction and void distribution).
The flow distribution and the instantaneous location of the
stagnation point are governed by the resistances offered by

the various flow paths leading to the break, in conjunction
with the work performed on the fluid by the recirculating
pumps. Immediately after the assumed double-ended cold leg
break, the core flow will reverse and rush through the break;
however, as the fluid at the break begins to change phase,
the break flow begins to choke, limiting the discharge rate.
Depending on the number of the intact loops present, the
pumps in these loops may, at that time, overpower the break-
induced flow and cause a return to the positive; i.e., the
upflow of coolant in the core. However, as the coolant void
fraction begins to increase, the pump performance suffers
increased degradation such that the break flow again becomes
a dominant driving force which causes a resumption of a negative (downward) flow in the core. The latter will remain
negative until the onset of the reflooding process. From
the above description, it is clear that in the blowdown
regime, the calculated local flow magnitudes, directions,
and the fluid properties, are greatly affected by the modeling of the break flow (choking), the pump behavior, and by
the resistances for the flow of two-phase mixture. The latter
are usually separated into the frictional and the so-called
"minor" losses. The minor losses comprise the effects of the
sudden flow area changes. Both the frictional and the minor
losses are affected by the presence of voids (steam) and the
so-called two-phase flow multipliers, which are functions of
the local flow regimes, are employed to modify the hydraulic
losses experienced by the pure liquid. Modeling of the coolant pump behavior during LOCA could not be accomplished yet
from "first principles". The so-called homologous curves,
in four quadrants, are obtained from the steady flow measurements with liquid. The measured head and torque flow versus
pump speed are then "degraded" to account for the presence
of steam, for the void fraction above about 5%, and below
about 90%. The degradation multiplier is obtained from experiments. In licensing calculations, assumption is made
that the power supply to the pump motor ceases at the beginning of LOCA. The code then calculates the pump "coastdown"
as effected by the inertia of the rotating parts. Further on,
the "locked rotor" assumption is made to maximize the "steam
binding" effects. In order to account for the uncertainties
in the break flow calculation, the licensing (EM) criteria
demand a series of repetitive LOCA analyses in which both
the break flow areas and the discharge coefficients are dim-

inished in a stepwise manner to explore which combination
results in the largest surface temperature of the fuel cladding. The licensing calculations employ the Moody model for
the break flow while the current best estimate calculations
employ a variety of models, each of which is best suited for
a given flow regime. The advanced BE codes will solve for
the break flow in the same manner as for any interior flow
since their governing equations contain the terms that will
automatically calculate "choking" if and when the right local
fluid conditions are reached.

During the blowdown phase, the analysis must account
for mixing of the injected emergency coolant (a highly subcooled liquid) with superheated steam and for the subsequent
flow of that mixture inside the downcomer. The intended
mission of the injected coolant is to flow downward through
the downcomer, fill up the lower plenum, and start the reflooding process; however, as long as the steam flowing
downward through the core and upward through downcomer, on
its way to the break, has a sufficient momentum, it will
prevent the ECC liquid's downward penetration. This is the
so-called downcomer "flooding" phenomenon. In the licensing
calculations (which employ HEM), a limiting upflowing steam
velocity is specified above which no liquid penetration is
allowed. Should any liquid reach the lower plenum (and/or
fill up the downcomer) during that period, it is artific-
ially removed from those regions at the start of the reflood
calculations. In the best-estimate calculations that employ
hydraulic models that allow counter-current flow of liquid
and steam, the flooding condition is not artifically imposed,
but calculated as a consequence of the calculated hot wall,
condensation, and the interfacial momentum transfer effects.
In the existing BE codes, the limiting values of the counter-
current flow are calculated via correlations of the available
test data obtained with the "separate effects" tests on
downcomer geometries.

The heat addition to or removal from the fluid is calculated in a simpler fashion than in the "hot channel" codes,
mainly because the core is not as finally subdivided or
"meshed", for both the hydraulic and for the heat conduction
calculations. In the licensing calculations, the decay heat
production is increased by 20% above the best estimate value.

Finally, for the Small Break analysis, it is important
to track the water or the mixture level throughout LOCA, and
throughout the system. This is necessary in order to pass
the proper fluid to various flow paths and to account for
the drastic changes in the heat transfer regimes above and
below such liquid or mixture levels. Combinations of the
"bubble rise" and "slip" models are currently employed to
calculate the phase separation.

C.   Reflood Period

During this period, the emphasis is on calculating the
manner in which the injected coolant enters the reactor core
and quenches it. Prior to that time, the fuel cladding temperatures were steadily rising, due to the steam blanketing
of the rods that leads to the poor heat removal capability.
At such high clad surface temperatures, the coolant entering from below cannot wet the surface and thereby quench it,
until that temperature falls below the "sputtering" point.
In the vicinity of that point there is a very steep clad
temperature gradient in the axial direction, when the reflood
rate is low and the axial conduction of heat in the cladding
is chiefly responsible for the propagation of quenching. In
the vicinity of the quench front there is a region of nucleate,
transition, and film boiling in which net vapor is generated.
That vapor escapes the rising liquid level and entrains some
of the liquid in the form of droplets. In the fast reflood
regime (greater than two inches/second) the liquid level
rises faster than the quench front. The region of liquid
(whose bulk may even be subcooled) located above the quench
front is enveloped by a thin layer of vapor across which heat
is removed from the cladding via film boiling. Such a region
may extend a number of inches in length, causing a much less
steep axial gradient in the clad temperature. The total rate
of vapor generation is now larger, causing an increase in the
entrainment of the liquid droplets. The rate of reflood,
i.e., of inflow of the coolant at the core bottom, is governed by the driving force and by the resistance. The former
is caused by the elevation difference, i.e., the hydrostatic
head, between the liquid level in the downcomer and the liquid
level in the core. The resistance is caused mainly by the
pressure force acting on the top of the core liquid level,
which, in turn, is determined by the hydraulic resistance
needed to push the escaping steam and the entrained liquid
droplets, all the way to the break. Evaporation of the entrained droplets, on their way through a steam generator

(which now acts as a heat source) contributes very significantly to the flow resistance. From the above it can be seen that counteracting phenomena take place during reflood: fast reflood rate generates more steam and more entrainment, which generate a higher back pressure, which slows down the reflood rate, which generates less steam and less entrainment, hence a lower back pressure and the cycle repeats itself. Such oscillations, rather than a smooth reflood rate, have indeed been observed in the reflood experiments that featured a gravity feed of water into the downcomer (8). The observed dominant frequency of the reflood oscillation corresponds to the natural frequency of the U-tube manometer in which the two legs of the manometer are formed by the water columns in the downcomer and in the core, respectively. The intermittent surges of the back pressure above the core liquid level drive the manometer oscillations. In the current licensing calculations, the above described *local* phenomena (of heat transfer, vapor generation, and the liquid entrainment) are not considered. Instead, a correlation is utilized for the quench propagation, steam generation and entrainment, as functions of the liquid velocity at the core inlet and the power generation rate; the magnitude of the subsequently calculated backpressure determines the rate of liquid flow at the core inlet. Since such a (FLECHT) correlation was obtained from the forced-feed reflood experiments and the current reflood codes ignore inertial or dynamic effects, the oscillatory reflood behavior is not calculated. In the BE codes now under development, such a global correlation will not be utilized; local effects will, instead, be considered, in conjunction with the dynamic models for the fluid flow in the primary coolant system. In addition, an effort is also being made to consider the mechanism for a partial separation of the entrained droplets in the upper plenum caused by impacting on many control rods and support columns. Those separated droplets that have not been evaporated while in contact with the hot metal will form a liquid pool on top of the upper core plate. The pool will drain into the cooler core channels in which the steam flow is not high enough to prevent a counter-current flow of liquid.

D.   Hot Channel/Fuel Behavior

The purpose of the hot channel or the fuel behavior code is to calculate the thermal, chemical and the mechanical condition of the fuel rod located in the hottest coolant channel. The boundary conditions are specified by the system code.

These may consist of time history of the fluid conditions in
the upper and the lower plenum, or of the flow rate at the
core inlet, or of the local fluid conditions (pressure, temperature, quality or void fraction, mass flux) along the rod
length. The time history and the axial distribution of power generation in the "hot rod" is specified in all cases.
The initial conditions pertaining to the radial temperature
profiles in the fuel pellets and in the cladding, as functions of the axial elevation, of the gap thickness and the
gap gas composition and pressure, etc., depend on the assumed
time of the postulated LOCA with reference to the fuel "life".
While nominally the axial power distribution resembles a
chopped cosine shape, significant skewness of that profile
occurs early as well as late in the life of the fuel rod.
Furthermore, the gas composition, the gap width and the
degree of cracking of the fuel pellets are also dependent
on both the fuel life and on the pre-LOCA operational history. Lately, special fuel behavior codes for the "steady-
state" or pre-LOCA applications have been developed to provide the appropriate initial conditions to the transient
fuel behavior codes. Examples are FRAP-S and FRAP-T, respectively developed by INEL for USNRC. The purpose of a
transient hot channel code is to calculate the peak cladding
temperature, and the degree of the clad oxidation, both of
which are given upper bounds by the USNRC Commission Rules. (1)

    In the process of providing this "end product", the hot
channel code must calculate time histories of the radial and
the axial profiles, in the cladding and in the fuel pellets,
the gap composition and pressure, the clad swelling or ballooning, or the clad collapse, or even the local clad rup-
ture (if such were to occur), axial elongation and bowing of
the fuel rods, clad oxidation and the metal/water chemical
(exothermic) reaction along the exposed clad surfaces where
the needed threshold temperatures have been reached. The
heat transfer coefficients are correlations pertaining to
the convective, nucleate boiling, transition boiling, film
boiling, the dispersed droplets, and the single-phase convective heat transfer regimes. A separate correlation is
employed to define the critical heat flux, CHF, which defines
a boundary between the nucleate and the transition boiling
regimes. A variety of correlations are employed to define
the thermal and the mechanical properties of the cladding
and pellets. While the heat conduction equation is solved
in the transient form, usually in radial direction only (at

a number of axial elevations), the structural deformation is solved in a quasi-steady manner.

E.   Component Codes

The hot channel or the fuel behavior codes are system components codes. Some of such codes have been extended to apply not only to one channel but to the whole fuel bundle or even to the core-wide distribution. COBRA-4 and SCORE are the examples of such codes, although in their present form, they concentrate more on the multidimensional description of the fluid flow and properties rather than on the details of the thermal and structural behavior of individual fuel rods. Eventual coupling of FRAP-T (or of some of the important models in FRAP) with the core-wide hydraulics codes will provide for the desired interaction between the structural deformation of the cladding (ballooning) and the flow distribution. The component codes are extremely useful tools for analyzing the behavior of single system components when the latter are studied experimentally in the separate effects tests. Such experiments allow an in-depth study of the component's behavior when subjected to a variety of the controlled boundary conditions. The Component Code will, in this instance, contain a very detailed geometric and thermo-hydraulic view of that component. Comparisons of its results with test data will constitute an in-depth verification. Sensitivity studies are then performed to ascertain the simplifications one could tolerate in the geometrical as well as in the thermohydraulic description. The simplified version will then become a module of an overall systems code.

IV.   BOILING WATER REACTOR (BWR) CODES

Considering the fact that there is a great deal of similarity in the nature of the thermohydraulic processes in both PWR's and BWR's during a postulated LOCA, it is not surprising that the salient features of all LWR codes are the same. The degree of the initial subcooling in BWR is very slight, and therefore, the subcooled loads are not calculated. On the other hand, initialization of a BWR LOCA analysis is harder, due to boiling of coolant in the core, which calls for careful evaluation of the coolant properties distribution inside the vessel. However, the fuel bundles in BWR are enclosed in individual canisters that prevent lateral mixing and are, therefore, much more suited to the one-dimensional core flow calculations. During the blowdown portion, the behavior of the jet pumps plays an important role that calls for some care in modeling their "off-design" flow condition. The lower plenum flashing provides an important source of the core coolant before ECC becomes effective. The emergency core coolant is supplied via sprays above the upper core support plate. A layer of subcooled liquid is formed over that plate and drained into the core only over those regions where the upflowing steam has insufficient momentum to prevent the establishment of the counter-current flow. In those regions, the rods will first be quenched from the top. The downflowing liquid eventually fills up the lower plenum and starts the bottom reflood process which greatly resembles that of a PWR. The "flooding" correlations, based on largescale test data, are utilized to describe the liquid penetration process through the upper core support plate. Thermal radiation is also considered because the relatively cold canisters provide an important thermal sink for removal of heat from fuel rods before the quenching process becomes effective.

In licensing (EM) calculations, the blowdown-induced thermal-hydraulic transients are subdivided into the "short term" and "long term" codes, dealt with in separate, standalone codes. The "long term" code replaces the refill/reflood code (in Figure 1), its output being the water level, ECCS actuation times, and reflooding times. Both are followed by the "core heatup code" which replaces the "hot channel code" in Figure 1. While the calculation of the gap conductance, and of the critical heat flux are done within the hot channel code in the case of PWR, they are calculated via separate codes in the General Electric Company's BWR analyses, and their output is directed to the core heatup code.

V. LISTING OF LOCA CODES

All systems and components codes, EM and BE versions, for PWR and BWR applications known to the author as of Spring, 1976, are shown in Table 3. It would not be surprising if some of those codes have by now become obsolete. Within the past few years there has been a great proliferation of codes, both in the United States and abroad, commensurate with a great proliferation of knowledge gained, worldwide, from numerous test facilities.

TABLE 3

CODE NAME	ORIGIN		TYPE						APPLICATION				CHARACTERISTICS			REFERENCE	COMMENTS
	COUNTRY	ORGANIZATION	PWR	BWR	LICENSING	BEST ESTIMATE	SYSTEMS	COMPONENT	BLOWDOWN	REFLOOD	HOT CHANNEL	SMALL BREAK	HYDRAULIC MODEL	GEOMETRY	NUMERICS		
RELAP-4 (Mod 3)	USA	INEL, for NRC	X	X	X		X		X		X		1V1T	L.P.	IMPL	9	
RELAP-4 -FLOOD	"	"	X		X		X			X			"	"	"	9	
TOODEE	"	"	X	X	X			X		X	X					9	
MOXY	"	"	X		X			X		X	X	X				9	
RELAP-4 (Mod 5)	"	"	X	X	X	X	X		X	X	X	X	1VD1T	L.P.	IMPL	10	Tracs 2∅ level
RELAP-4 (Mod 6)	"	"	X	X		X	X		X	X	X	X	"	"	"		Being Developed
RELAP-4 (Mod 7)	"	"	X	X	X	X	X		X	X	X	X	"	"	"		"
SCORE-EVET	"	"	X	X		X					X		1V1T	3-D	MAC/ICE	11	Fuel Behavior-
FRAP-S	"	"	X			X		X					NA	1-D		12	Steady State

Code	Source	Equations	Dimensions	Numerical Method	Ref.	Comments
FRAP-T	"	NA	1-D	EXPL.& IMPL.	12	Fuel Behavior—Transient
THOR	BNL for NRC	1VDT$_k$T$_s$	L.P.& 1-D	ICE	14	Being Developed
TRAC	LASL for NRC	1VDT$_k$T$_s$ + 2V2T	L.P., 1-D, 2-D, 3-D	IMF	15	Being Developed as Advanced BE Code
KACHINA	"	2V2T	2-D	IMF	16	2∅, 2-component
SOLA-DF	"	1VDT$_k$T$_s$	1-D, 2-D	ICE	15	
K-TIF	"	2V2T	2-D, 3-D	IMF	15	Two Incompressible Fluids
SOLA-FLX	"	1VDT$_k$T$_s$	1-D, 2-D	ICE	15	Hydro-elastic Code Being Developed
K-FIX	"	2V2T	2-D, 3-D	FIX	15	Being Developed. Replaces KACHINA.
COBRA-4	PNL for NRC	1V1T	Axial, lateral	IMPL	17	Core-wide, or Hot Channel
CRAFT	B&W	1V1T	L.P.	EXPL.	18	
REFLOOD	"	1V1T	L.P.	EXPL.	19	
FOAM	"	1V1T	1-D	EXPL.	20	Calculates Mixtures Level
THETA	"	NA	1-D	EXPL./ IMPL.	21	Hot Channel Code

Code	Ref.	Function	Numerical Method	Model	Nodes	Vendor
FATES	22	System Initialization		EPN	1V1T	CE
CEFLASH-4A	23		IMPL.	L.P.	"	"
CEFLASH-4AS	23	Tracks 2ø Level	"	"	"	"
COMPERC-II	24		QUASI-STEADY STATE	"	"	"
STRIKIN-II	25	Hot Channel Code				"
PARCH	26	Hot Channel Code For Slow Reflood	QUASI-STEADY	1-D	1V1T	"
LAMB	27	Short Term Thermal Hydraulics	EXPL.	L.P.	1V1T	GE
SAFE	27	Long Term T-H	"	"	"	"
CHASTE	27	Core Heat-Up	"	"	"	"
SCAT	27	Calculates Critical Heat Flux	IMPL.	1-D	"	"
SATAN-VI	28		EXPL.	L.P.	1VD1T	W̲
WREFLOOD	29		"	"	"	"

Code	Country	Org.									1V/1T	Dim.	Method	Ref.	Comments
WFLASH	"	"			X	X		X		X	1V1T	L.P.	IMPL.	30	
LOCTA-IV	"	"		X	X	X		X	X	X	N.A.	1-D		31	Early Blowdown Transients
BLODWN-2 & 2-A	"	"	X		X		X	X	X	X	1V1T	1-D EPN	MOC	32	
MULTIFLEX	"	"			X	X		X	X	X	1V1T	1-D EPN	MOC	33	Coupled Structural-Hydraulic Analysis
RODFLOW	CANADA	AECL				X	X	X			1VS1T	1-D	EXPL.	13	For Pressure Tube Reactors
TINA	DENMARK	RISØ	X		X	X	X	X		X	$1VT_2T_s$	AXIAL+ LATER.	IMPL.	34	
DINO	"	"		X		X	X		X	X		1-D		35 (p. 9)	Fuel 2-D
RHC	"	"		X	X		X	X	X	X	2V2T	1-D		36	
DANBLOW	"	"		X	X			X	X	X	3-D	3-D		35 (p. 11)	
DANAIDES	FRANCE	SACLAY				X	X	X	X	X	1V1T	L.P.	EXPL.	37	
CERES	"	"		X			X	X	X	X	"	"	"	35 (p. 73)	
FLIRA-2	"	"	X		X	X	X	X		X		1-D			Tracks Quench Front
CYLSTERE	"	EdF	X	X	X	X	X	X		X	$1VT_kT_s$	1-D EPN		35 (p. 55)	
BERTHA	"	CENG	X		X	X		X		X	$1VT_LT_s$	1-D	MOC	38	Advanced BE Code

Code	Country	Organization												Ref.	Remarks
BRUCH-D	Germany	LRA	X				X				1V1T	L.P.	EXPL.	35 (p.129)	
BRUCH-S	"	"			X	X	X				"	"	"	"	
BLAST-2	"	"	X		X	X	X				1V1T	1-D	MOC	35 (p.131)	With Nucleation Delay
DAPSY	"	"	X			X	X				1V1T	1-D EPN	MOC	"	
DRUFAN	"	"	X		X	X	X				$1VT_L^T{}_s$	1-D EPN	MOC	"	Thermal Non-Equilibrium
NICKY	ITALY	TRC-ISPRA	X		X	X	X				1V1T	1-D EPN	MOC	35 (p.501)	Being Developed
DEPCO-MULTI	JAPAN	JAERI	X	X	X	X	X				1V1T	1-D EPN	MOC	39	Subcooled Decompression
ALARM	"	"	X	X	X	X	X				1V1T	L.P.			
LTTH	"	"	X		X	X			X		"	"			
ASCOT	"	"	X		X	X		X			1V1T	2-D	MOC		
RFLD	"		X		X	X				X	"	L.P.			
NORA	NORWAY		X		X	X	X				1VS2T	1-D EPN	IMPL		Being Developed
RELAP-UK	UK	AEEW	X		X	X	X				1Vs1T	L.P.	IMPL	40	
HUBBLE-BUBBLE	"	AHSB	X		X		X				$1VT_L^T{}_s$	1-D	EXPL	41	Lagrangian Formulation

VI. CODE VERIFICATION,

SENSITIVITY, AND UNCERTAINTY STUDIES

A. Code Verification

In the USNRC sponsored programs, **verification is being**
performed on two levels: (a) the developmental verification,
conducted by the code developers; and (b) the applied verification, conducted by an independent (verification) group.
The developmental verification emphasizes comparisons of
code results with test data from the Separate Effects tests
and the Model Development tests. In contrast to the Integral
System tests (such as LOFT and Semiscale) the Separate Effects
tests examine the behavior of single components, in various
scales. The Model Development tests concentrate on basic
two-phase flow phenomena, aimed at defining various terms
in the constitutive equations. A series of special standard
problems is also being utilized in which the comparisons are
run "blind"; i.e., the code developer does not know the test
data *a priori*. Very few comparisons with Integral Systems
test data are made at this stage in order to verify that
the system dynamics is correctly calculated. The Applied
Verification activity emphasizes Integral Systems test comparisons, with a few Separate Effects tests for single system components.

B. Sensitivity Studies

These are performed by the code developers during the
Model Verification studies, for the purpose of establishing
(a) effects of various modeling assumptions (including geometric nodalization) on some important "output" of every
system component; and (b) effects of the component modeling
assumptions on the System Behavior; in particular, on the
peak clad temperature. Some of these studies may utilize the
same methodology as for code uncertainty studies; however,
the emphasis is on the modeling assumptions rather than on
the uncertainties in input data.

C. Uncertainty Studies

Such studies are performed by the Applied Verification
Group with verified codes. The goal is to establish the uncertainty in the computed peak clad temperature plus some
other effects (e.g., the extent of metal/water reaction,
clad oxidation, hydrogen release), as affected by the uncer-

tainties in some important code input data. These may refer
to the initial conditions, the boundary conditions, and the
coefficients embedded in some of the **important** correlations.
The code sensitivity studies should uncover the important
parameters.

In such studies, either the Response Surface or the
Stratified Monte Carlo techniques are employed in order to
evaluate the mutual interaction of various input uncertain-
ties considered simultaneously. The selected input uncertainties (up to 25) are each assigned their probability
distributions. The output of the uncertainty study yields
a probability distribution of, say, the peak clad temperature,
illustrated in Figure 4.

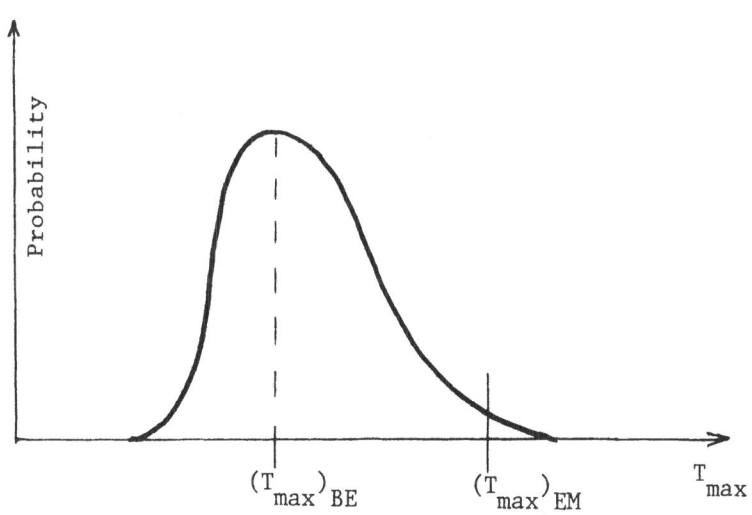

Figure 4

Sample Probability Distribution
For the Peak Clad Temperature

This information must be combined with the information
on the code accuracy to arrive at the total probability distribution. The code accuracy information will be obtained
from the verification process (comparisons with test data)
in which it should be possible to define the code error, say,
due to physics modeling simplifications, or due to nodalization, on some statistical basis. It may also be possible
to obtain the code error information as a function of the
modeling and nodalization simplifications. This will be
useful, since an acceptable code error may be obtained with
simpler models that give significant savings in the computer
running time.

The main purpose of the BE **codes uncertainty studies**
is to supply information on the degree of conservatism embedded in the licensing calculations performed with an EM
code. Various criteria could be employed for obtaining that
information. These are now being evaluated. For example,
one may find the probability that the peak clad temperature
in LWR will exceed the EM value within some confidence band.
Or, the difference between the most probable BE peak clad
temperature and the EM calculated peak clad temperature
could be expressed in either degrees or in the number of
standard deviations. Should such studies determine that the
EM calculations are excessively conservative, it may be
possible to relax the EM criteria in order to arrive at
sufficient conservatism. The BE codes are expected to be
long running and will require very large computer storage.
It is likely, therefore, that the licensing analyses will
continue to rely on EM codes, although more advanced EM
versions may be employed in the future.

VII. REFERENCES

1.   Federal Register, Vol. 39, 10 CFR Part 50, "Acceptance
     Criteria for Emergency Core Cooling System for Light
     Water-Cooled Nuclear Power Plants," No. 3, January 4,1974.

2.   Vernier, P., Delhaye, J.M., "General Two-Phase Flow
     Equations Applied to the Thermodynamics of Boiling
     Nuclear Reactors," STT-CENG (Reprint from EPE-Volume IV,
     No. 1-2, 1968.

3. Ishii, M., _Thermo-Fluid Dynamics Theory of Two-Phase Flow_, Eyrolles, 1975.

4. Roache, P.J., _Computational Fluid Dynamics_, Hermosa Publishers, 1972.

5. Fabic, S., "Computer Program WHAM for Calculation of Pressure, Velocity, and Force Transients in Liquid Filled Piping Networks," _Kaiser_ Engineers Report No. 67-49-R, November, 1967.

6. Burnell, J.G., "Flow of Boiling Water Through Nozzles, Orifices and Pipes," _Engineering_, 164, PP 572-576, 1947.

7. Fabic, S., "Two and Three Dimensional Fluid Transients," _ANS Trans._, 13, No. 1, Page 386, 1970.

8. Waring, J.P., Hochreiter, L.E., "PWR FLECHT-SET Phase B1 Evaluation Report," _WCAP-8583_, Westinghouse Electric Corp., August, 1975.

9. "WREM: Water Reactor Evaluation Model, Revision 1," Division of Technical Review, Nuclear Regulatory Commission, _NUREG-75/056_, May, 1975.

10. Slater, C.E., Katsma, K.R., Fischer, S.R., Rettig, W.H., "Small Break Analysis Models in RELAP-4 Computer Code," _ANS Trans._, Volume 22, PP 264-265, November, 1975.

11. Wnek, W.J., Ramshaw, J.D., Trapp, J.A., Hughes, E.D., and Solbrig, C.W., "Transient Three-Dimensional Thermal-Hydraulic Analysis of Nuclear Reactor Fuel Rod Arrays: General Equations and Numerical Scheme," _ANCR-1207_, November, 1975.

12. Aerojet Nuclear Company, "Quarterly Technical Report on Water Reactor Safety Programs Sponsored by the Nuclear Regulatory Commission's Division of Reactor Safety Research," _ANCR-1262_, November, 1975.

13. Banerjee, S., Hancox, W., Jeffreys, R. and Suladisky,M., "Transient Two-Phase Flow Heat Transfer during Blowdown from Subcooled Conditions with Heat Addition," _AIChE_ Paper No. 24, 15th National Heat Transfer Conference, San Francisco, August, 1975.

14. Brookhaven National Laboratory, "Development of a Computer Code for Thermal Hydraulics of Reactors (THOR)," Quarterly Progress Reports BNL-19978, 50455, 50458.

15. Los Alamos Scientific Laboratory, "Quarterly Reports: Transport Theory, Reactor Theory, and Reactor Safety," LA-5964, 5975, 6029, 6054, 6137, 6155, 6164-PR.

16. Amsden, A.A., Harlow, F.H., Los Alamos Scientific Laboratory, "KACHINA: An Eulerian Computer Program for Multifield Fluid Flows," LA-5680, December, 1974.

17. Wheeler, C.L., Battelle Norwest Laboratory, "COBRA-IV-I: An Interim Version of COBRA for Thermal Hydraulic Analysis of Rod Bundle Nuclear Fuel Elements and Cores," BNWL-1962.

18. Hedrick, R.A., Cudlin, J.J., Holtz, R.C., Babcock & Wilcox Company, "CRAFT-2 Fortran Program for Digital Simulation of a Multinode Reactor Plant during Loss-of-Coolant," BAW-10092, July, 1974.

19. Bingham, B.E., Shieh, K.C., Babcock & Wilcox Company, "REFLOOD - Description of Model for Multinode Core Reflood Analysis," BAW-10093, March, 1974.

20. Dunn, B.M., Morgan, C.D., and Cartin, L.R., Babcock & Wilcox Company, "Multinode Analysis of Core Flooding Line Break for B & W 2568 MW Internal Vent Valve Plants," BAW-10064, § 3.2, April, 1973.

21. Stoudt, R.H., and Heck, K.C., Babcock & Wilcox, "THETA 1-B-Computer Code for Nuclear Reactor Core Thermal Analysis;"B & W Revision to IN-1445 (Idaho Nuclear Corporation, C. J. Hocevar and T. W. Wineinger), BAW-10094, Revision 1, April, 1975.

22. Combustion Engineering, "FATES: Fuel Evaluation Model," CENPD-139-A, July, 1974.

23. Combustion Engineering, CFLASH-4A, A Fortran IV Digital Computer Program for Reactor Blowdown Analysis, CENPD-133-NP, August, 1974.

24. Combustion Engineering, "COMPERC-II: A Program for
    Emergency Refill/Reflood of the Core," CENPD-134-NP,
    August, 1974.

25. Combustion Engineering, "STRIKIN-II, A Cylindrical Geo-
    metry Fuel Rod Heat Transfer Program," CENPD-135-NP,
    August, 1974.

26. Combustion Engineering, "PARCH, A Fortran IV Digital
    Computer Program to Evaluate Pool Boiling, Axial Rod,
    and Coolant Heat-up," CENPD-138-NP, August, 1974.

27. "General Electric Company Analysis Model for Loss-of-
    Coolant Analysis, in Accordance with 10 CFR 50,"
    App. K, Volume 2, NEDO-20566, 76NED2, Class I,
    January, 1976.

28. Bordelon, F.M., Kelly, R.D., Massie, H.W., Jr., Muench,
    R.A., Spencer, A.C., Young, M.Y., Westinghouse Electric
    Corporation, "SATAN-VI Program: Comprehensive Space-
    Time Dependent Analysis of Loss-of-Coolant," WCAP-8306,
    June, 1974.

29. Collier, G., Kelly, R.D., Spencer, A., Waryng, J.P.,
    Westinghouse Electric Corporation, "Calculational Model
    for Core Reflooding after a LOCA (WREFLOOD Code),"
    WCAP-8171, June, 1974.

30. Esposito, V.G., Kesavan, K., Maul, B., Westinghouse
    Electric Corporation, "WFLASH - A Fortran IV Computer
    Program for Simulation of Transients in Multi-Loop PWR,"
    WCAP-8261, Revision 1, July, 1974.

31. Bordelon, F.M., Collier, G., Spencer, A.C., Burman, D.
    L., Ohkubo, M., Yang, J.W., Westinghouse Electric Cor-
    poration, "LOCTA-IV Program: Loss of-Coolant Transient
    Analysis," WCAP-8305, July, 1974.

32. Fabic, S., Westinghouse Electric Corporation, "Descrip-
    tion of the BLOWDOWN-2 Computer Code," WCAP-7918, Rev.
    2, October, 1970 (Reprinted August, 1975).

33. Takeuchi, K., Kowalski, D.J., Esposito, V.J., Bordelon,
    F.M., Westinghouse Electric Corporation, "MULTIFLEX,"
    WCAP-8709, February, 1976.

34. Anderson, P.S., "Status of the Theoretical Foundation
    for Dynamic Subchannel Analysis Including Results for
    One-Dimensional Calculations," NORHAV-D-12, February,
    1975.

35. Commission of the European Communities, European
    Community Light Water Reactor Safety Research Projects,
    Experimental Issue, EUR 5394e, 1975.

36. Anderson, J.G. Munthe, "REMI/HEAT-COOL, A Model for
    Evaluation of Core Heat-up and Emergency Core Spray
    Cooling System Performance for Light Water-Cooled
    Nuclear Power Reactors," RISØ Report No. 296, September,
    1973.

37. Menessier, D., Forge, A., Pedel, A., Lacotte, D.,
    "DANAIDES - Program de Calcul de Transitoire de Decom-
    pression," CEA Sets No. 38, October, 1975.

38. Rousseau, J.C., Boudsocq, G., Marechal, A., Riegel, B.,
    Schall, M., "Code BERTHA - Application aux Experiences
    CANON," CEA-CENG, TT, No. 491, March, 1975.

39. Namatame, K., Kobayashi, K., Japan Atomic Energy Res-
    earch Institute, "Digital Computer Code DEPCO-MULTI
    for Calculating the Subcooled Decompression in PWR LOCA,"
    JAERI-M-5623, February, 1974.

40. Brittain, I., Fayers, F.J., "Some Aspects of Model
    Improvements in RELAP-UK," AEEW-No. 1253, AEE, Winfrith,
    March, 1974.

41. Edwards, A.R., "Conduction Controlled Flashing of a
    Fluid, and the Production of Critical Flow Rates in a
    One-Dimensional System," UKEA Report AHSB(s) R147,
    1968.

CONTROLLED FUSION AND REACTORS OF THE TOKAMAK TYPE

Robert W. Conn

Fusion Technology Program
Nuclear Engineering Department
University of Wisconsin
Madison, Wisconsin, U. S. A.

I.   INTRODUCTION

Research on fusion reactor problems has increased dramatically as the plasma physics of magnetic confinement continues to make substantial progress. As part of this research several studies (1-6) have been completed on the conceptual design of future fusion reactors. The purpose of these studies is to identify the key technological problems associated with fusion reactors and thereby guide future research. A description of one such conceptual design can be used to serve as an introduction to the broad field of fusion technology, particularly if it is combined with other work that generally surveys the technological aspects of most approaches to fusion power (6,7). In this paper, a detailed description of the UWMAK-III conceptual tokamak reactor design (8) is given and it serves to highlight the important technological areas. These areas include (1) plasma engineering problems such as startup, auxiliary plasma heating, plasma burn dynamics at thermonuclear conditions, and impurity control; (2) blanket design and neutronics analysis; (3) tritium breeding, extraction and recycle; (4) materials problems with the blanket structure; (5) thermal hydraulics analysis of the blanket system; (6) superconducting magnet design; (7) induced radioactivity; (8) potential materials requirements for reactor systems; and (9) overall plant design. For completeness, we present first a brief discussion of the basic requirements for fusion grade plasmas. In section II, the basic ideas behind the main magnetic confinement concepts are described. This allows us an opportunity to discuss the basis for the present optimism in the controlled fusion community. After a general description of the basic elements of a fusion re-

405

actor in section III, we turn in section IV to the main focus of the paper, a detailed description of the conceptual tokamak reactor design, UWMAK-III.

The fuel for the first fusion reactors will be the isotopes of hydrogen and helium and will make use of one or more of the following nuclear reactions*:

Deuterium-Tritium (D-T) Cycle:

$$_{1}^{2}D + _{1}^{3}T \rightarrow _{2}^{4}He(3.52 \text{ MeV}) + _{0}^{1}n(14.06 \text{ MeV}) \qquad (1)$$

Deuterium-Deuterium (D-D) Cycle:

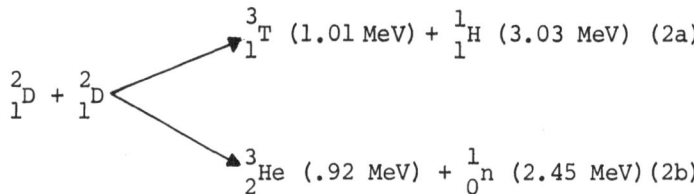

$$_{1}^{3}T \ (1.01 \text{ MeV}) + _{1}^{1}H \ (3.03 \text{ MeV}) \quad (2a)$$

$$_{1}^{2}D + _{1}^{2}D$$

$$_{2}^{3}He \ (.92 \text{ MeV}) + _{0}^{1}n \ (2.45 \text{ MeV}) \ (2b)$$

Deuterium-Helium-3 (D-^{3}He) Cycle:

$$_{1}^{2}D + _{2}^{3}He \rightarrow _{2}^{4}He \ (3.67 \text{ MeV}) + _{1}^{1}H \ (14.67 \text{ MeV}) \qquad (3)$$

Both the D-D and D-T cycles involve the release of an energetic neutron that will not be retained by a confining magnetic field. Fusion reactors based on these cycles will therefore require a system surrounding the reaction chamber (the blanket) that is capable of slowing down and extracting the neutron kinetic energy. A potential advantage of the D-^{3}He cycle is that the reaction products are charged particles which, in addition to offering the potential of direct conversion, will produce neutrons only as a result of concomitant D-D fusion reactions. This neutron production level will be lower than in the D-T cycle (9) and this would alleviate some of the neutron-related problems.

---

*1 mega electron volt (MeV) = 0.16 x 10^{-12} joules (0.16 pJ).

The choice between these potential fuel cycles for the near term becomes clear by examining the cross-sections for the various fusion reactions as shown in Figure 1. The cross-section for the $T(d,n)^4He$ reaction is, at its maximum, a factor of five larger than the cross-section for $^3He(d,p)^4He$ and the cross-section peaks at lower energy.

It is apparent from Figure 1 that the cross-sections for these fusion reactions become large only at high energies, and it is necessary to heat the reacting species to temper-

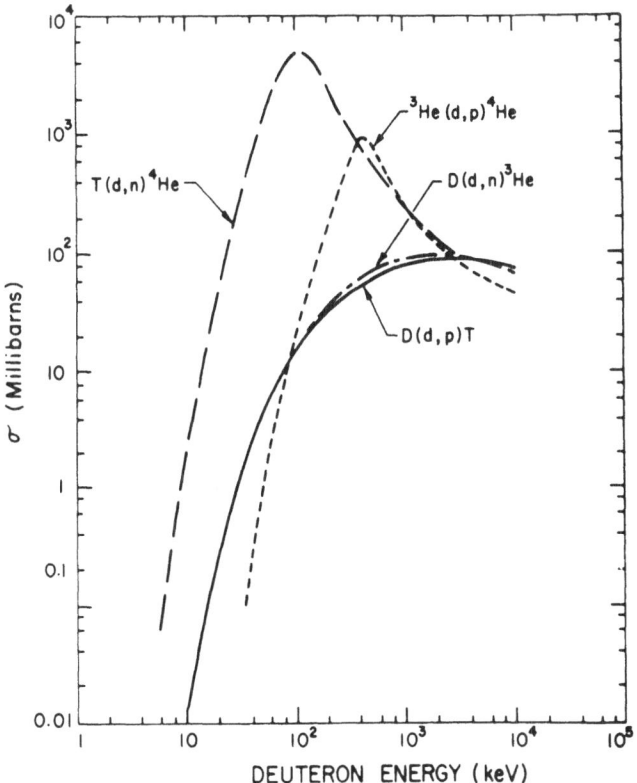

Figure 1

Microscopic Cross-Sections for Various Fusion Reactions

atures where the fusion rate becomes reasonably large. For
the D-T cycle, this means temperatures on the order of 10 keV,
and of course, at such temperatures, the gas will be a fully
ionized plasma.

A nonlinear least-square fit of the T(d,n)^{4}He reaction
cross-section, valid over the entire range from 1 keV to
10 MeV, is:$^{(10)}$

$$\sigma_{D-T}(E) = \frac{A_2/[1 + (A_3E - A_4)^2] + A_5}{E(e^{A_1/\sqrt{E}} - 1)}$$

where E is the energy in eV, σ_{D-T} is in barns (1 barn
$= 10^{-24}$ cm^2) and the constants are:

$$A_1 = 1453$$

$$A_2 = 5.02 \times 10^7$$

$$A_3 = 13.67 \times 10^{-6}$$

$$A_4 = 1.076$$

$$A_5 = 4.08 \times 10^5$$

Similar fits have been developed for the other fuel cycles.$^{(11)}$

Perhaps more relevant is a comparison of the cross-
sections averaged over a Maxwellian distribution of speeds
for both incident particles. This fusion reaction parameter
denoted by <σv> and with units of cm^3/sec is shown in
Figure 2. It is clear that the deuterium-tritium cycle
offers the greatest advantages since the reaction rate parameter is largest and peaks at the lowest temperature, approximately 70 keV. It thus appears certain that the first
fusion reactors will be fueled with deuterium and tritium.

In the longer term, other reaction cycles may prove
feasible and as in the D-^{3}He cycle, can offer certain distinct advantages such as having only charged reaction products. Some of the other fuel cycles involving heavier
elements include ^{6}Li(p, ^{3}He)^{4}He, ^{6}Li(d,t)^{5}Li, ^{7}Li(p,γ)^{8}Be,
and ^{11}B(p,3α).

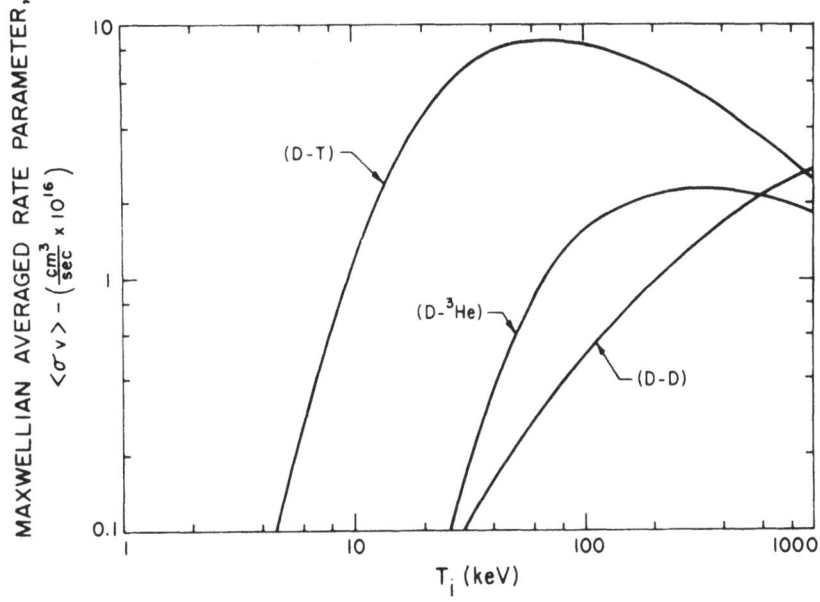

Figure 2

Maxwellian Averaged Fusion Reaction Rates

The basic condition required of a plasma to approximately achieve energy breakeven is typically given as the
Lawson criterion.[12]   While simplified, this criterion provides a general guideline for assessing the status of present
experiments.   A plasma at temperature T composed of electrons
at density n and deuterons and tritons at a density of n/2,
respectively, has a thermal energy of 3 nκT, where κ is
Boltzmann's constant and we have assumed that all species
have the same temperature.   Such a hot plasma will radiate
energy via many processes but for the purpose here, let us
assume that bremsstrahlung radiation is the main loss mechanism.   This loss rate is given for a pure hydrogenic plasma
by:

$$P_x = 4.81 \times 10^{-31} \, n_e^2 \, T^{1/2} \quad (watts/cm^3) \qquad (5)$$

where n_e is the electron density in particles per cm^3 and T_e
is the electron temperature in keV.   Let us assume that the

thermal energy, $3 n \kappa T$, is contained for a characteristic time, τ_E. τ_E is usually referred to as the energy confinement time and accounts for energy losses by all mechanisms except bremsstrahlung. Then one must balance

$$\frac{3 n_e \kappa T}{\tau_E} + P_x$$

against the energy generated by the plasma to maintain an energy equilibrium.

Fusion reactions take place at the rate $(\frac{n}{2})^2 \langle \sigma v \rangle$ and an amount of energy, E_{FUS}, is released per fusion reaction. Assuming that this power plus the bremsstrahlung and plasma thermal energy is available for conversion to electricity at an overall efficiency, η, the Lawson condition arises from the following power balance on a unit volume of plasma:

$$\frac{3 n \kappa T}{\tau_E} + P_x = \eta \left(\frac{n^2 \langle \sigma v \rangle}{4} E_{FUS} + P_x + \frac{3 n \kappa T}{\tau_E} \right). \qquad (6)$$

Solving for $n \tau_E$, using $P_x = C n^2 \sqrt{T}$ gives

$$n \tau_E = \frac{3 kT \ (1-\eta)}{\eta \left[(\frac{\langle \sigma v \rangle}{4}) E_{FUS} + C \sqrt{(T)} \right] - C \sqrt{T}} \qquad (7)$$

which is a function only of the plasma temperature. The curve of $n \tau_E$ versus T for the D-T and D-D cycles is shown in Figure 3. The minimum product of plasma density times the plasma energy containment time required to meet the Lawson criterion is approximately 5×10^{13} sec-cm^{-3}. Importantly, a plasma that meets the Lawson criterion is not necessarily ignited and self-sustaining because the analysis does not include the possible direct heat deposition in the plasma of some or all of the energy of the fusion reaction products. The condition that characterizes an ignited and self-sustaining plasma is thus another important criterion.

The ignition condition is obtained by balancing the power lost from the plasma due to bremsstrahlung and thermal losses in a time, τ_E, against the power deposited in the plasma from the slowing down of fusion reaction products. Letting f denote the fraction of fusion energy deposited in the plasma, this balance is

$$\frac{n^2 <\sigma v>}{4} fE_{FUS} = \frac{3nkT}{\tau_E} + P_x \tag{8}$$

and the ignition condition on $n\tau_E$ becomes

$$n\tau_E = \frac{3kT}{\frac{<\sigma v>}{4} fE_{FUS} + C\sqrt{T}} . \tag{9}$$

For the D-T fuel cycle, fE_{FUS} is 3.52 MeV, the energy of the alpha particle that would be contained in the plasma by a confining magnetic field. This criterion is also shown in Figure 3, and the minimum $n\tau_E$ is now approximately 2×10^{14} sec-cm^{-3}. Thus, it is somewhat more difficult to meet the ignition condition in a D-T mixture than to meet the Lawson criterion.

From these considerations we see that one basic require- ment of fusion grade plasma is an $n\tau_E$ value on the order of 10^{14} sec-cm^{-3}. It is also seen in Figure 3 that this minimum in the $n\tau_E$ curve occurs at approximately 25 keV and that the $n\tau_E$ requirements increase sharply for T < 10 keV.

Another indication of the optimal temperature range is obtained from the expression for the fusion power per unit volume,

$$P_{FUS} = \frac{n^2 <\sigma v>}{4} E_{FUS}. \tag{10}$$

In magnetic confinement systems, an important figure of merit is the ratio of plasma kinetic pressure, p, to the pres- sure exerted by the magnetic field, $B^2/2\mu_0$, where B is the mag- netic field strength and μ_0 is the permeability of vacuum. The ratio of these two quantities is denoted by β

$$\beta = \frac{p}{B^2/2\mu_0} . \tag{11}$$

In many of the magnetic confinement systems important today, such as the tokamak, β is limited and should not exceed some maximum value. If we consider β as fixed and solve equation (11) for n, we can express P_{FUS} as

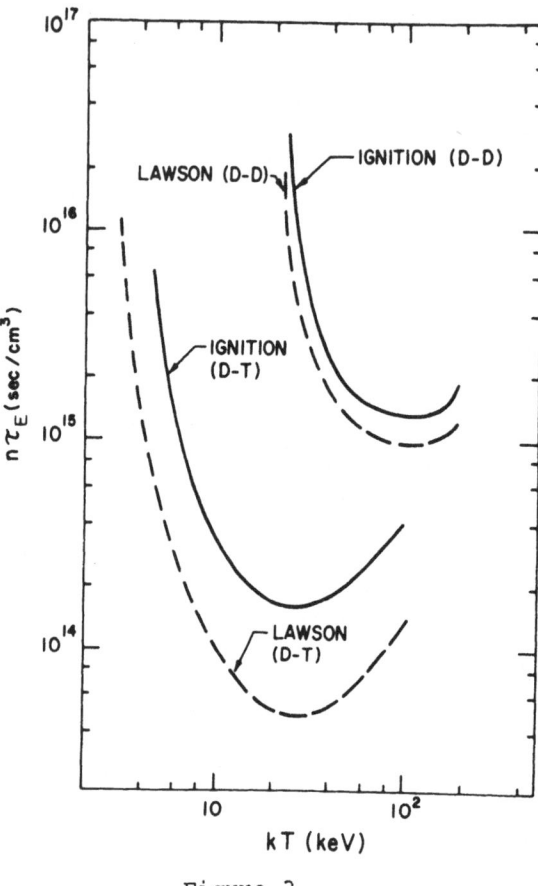

Figure 3

Lawson Criterion for the D-T and D-D Fusion Fuel Cycles

$$P_{FUS} = \frac{\langle\sigma v\rangle}{4(\kappa T)^2} \frac{\beta^2 B^4}{(2\mu_o)^2} E_{FUS}. \qquad (12)$$

Thus, for fixed β and a given magnetic field stength, the power output of a fusion system depends only on the parameter, $\langle\sigma v\rangle/T^2$. This is plotted as a function of temperature in Figure 4. We see again that for a given temperature, the D-T cycle will have the highest power density for a fixed reaction rate and that this reaction parameter is a

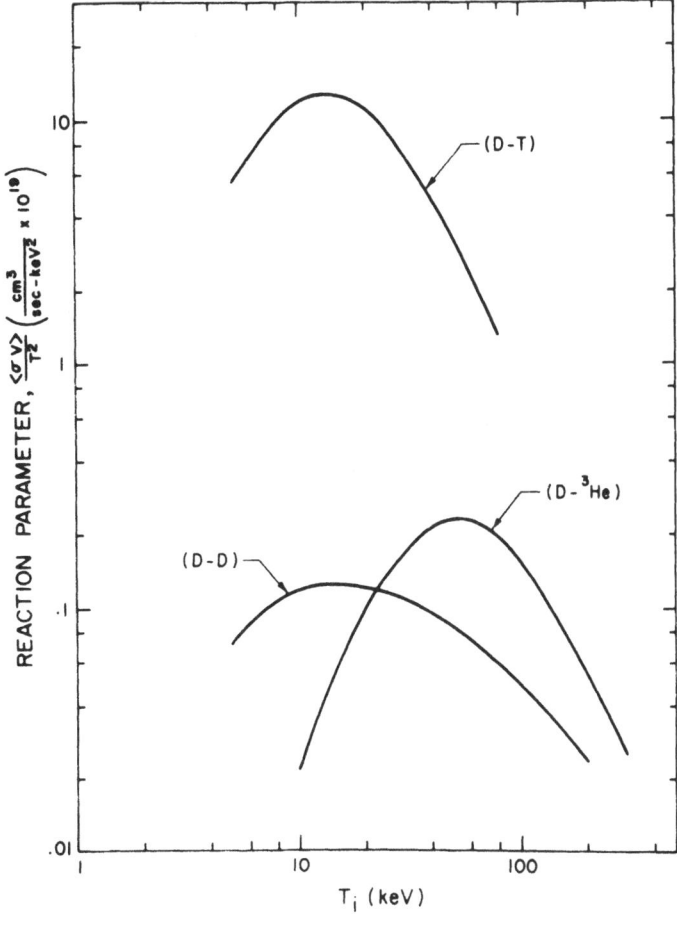

Figure 4

Reaction Parameter as a Function of Temperature

maximum at about 14 keV. The maximum is relatively broad, extending from 10 keV to 20 keV. Thus, the basic tempera- ture requirement is T > 10 keV, and for a β-limited plasma, the optimum is about 14 keV.

A general note of caution is required in considering these basic criteria. While they are certainly useful in generally providing a yardstick against which to measure

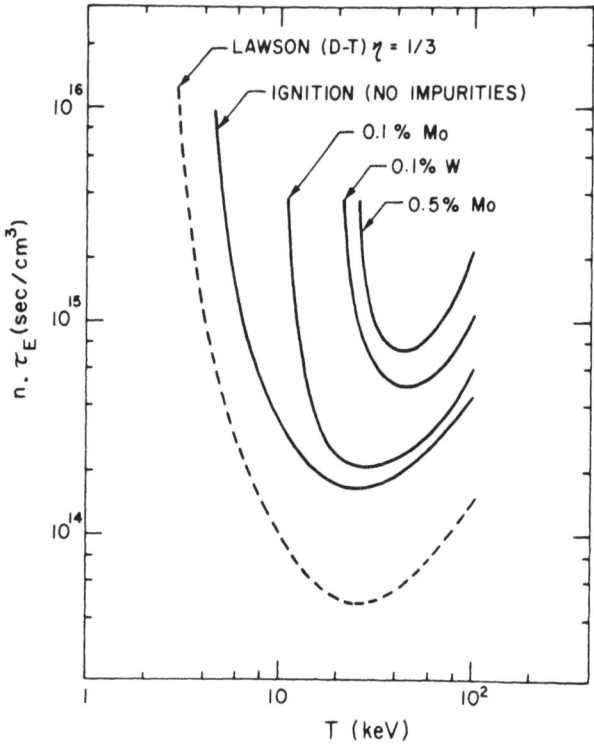

Figure 5

Effect on the Lawson $n\tau_E$ Criteria of the
Explicit Inclusion of Line and Recombination Radiation (13)

our present condition, these criteria can be markedly altered
by a more detailed analysis. In particular, a separate accounting of the effects of impurities has an important impact (13). High Z impurities such as Fe, Mo, or W, will
generally not be fully stripped of all electrons, even at
plasma temperatures of 10 keV. The resulting line and recombination radiation is typically much larger than bremsstrahlung. This, in turn, impacts on the minimum $n\tau_E$ requirement
to meet the Lawson and ignition conditions, as shown in
Figure 5. Thus, while $n\tau_E \simeq 10^{14}$ sec-cm^3 and $T \simeq 10$ keV
indicate general goals, other effects can substantially impact

on these values. Such effects are normally included in more
elaborate computer simulations of reacting plasmas (14) such
as are carried out as part of detailed fusion reactor design
studies (1-5) and in integrated studies comparing the power
balances of several confinement approaches (7).

II. CONFINEMENT OF FUSION GRADE PLASMAS

At the high temperatures required to achieve a significant fusion reaction rate, the gas of hydrogen isotopes is a
fully ionized plasma that cannot be contained by structural
materials. The approach that has dominated the controlled
fusion effort until very recently has been that associated
with magnetic fields. Various magnetic field topologies
have been invented to achieve the goal of stable plasma
confinement and these can be broadly classed as closed or
open systems. To provide a basis for the discussion of
fusion technology problems and conceptual reactor design,
let us review the essential ideas of the three most widely
studied approaches to magnetic confinement, the magnetic
mirror, the theta pinch, and the low β toroidal system
tokamak (for Toroidal Kamera Magnetik or toroidal magnetic
chamber). This will be followed by a brief discussion of
the most recent experimental results to illustrate the basis
for the present optimism in the fusion community. A general
review of the plasma physics of mirrors, pinches, and closed
line systems can be found in the paper of Post.(15)

A. Mirror Systems

We begin with a brief study of mirror systems because
the discussion of particle motion in a magnetic mirror is
also helpful in a description of tokamaks. Open-ended containment of plasmas includes magnetic mirror systems and
linear theta pinches. A magnetic mirror relies on the fact
that a charged particle will exhibit a gyromotion about a
magnetic line of force and that, on approaching regions of
increasing magnetic field, the particle will have its motion
parallel to the field line countered by the opposing force
of the field. If the field is large enough, the parallel
motion can be reflected. The gyromotion arises from the
force,

$$F = q(\underline{E} + \underline{v} \times \underline{B}) + \underline{F}^{ext} \tag{13}$$

which acts on a charged particle in an electromagnetic field. Here, q is the particle charge, $\underline{v}$ is its velocity, and $\underline{E}$ and $\underline{B}$ are the electric and magnetic fields, respectively, and $\underline{F}^{ext}$ is an arbitrary external force. Since the $\underline{E}$ field gives rise only to rectilinear motion, consider $\underline{E} = \overline{0}$, and let us also take $\underline{F}^{ext} = 0$. In a constant magnetic field, the force will act perpendicularly to both $\underline{v}$ and $\underline{B}$. One can then show that the kinetic energy is a constant of the motion and that the remaining particle motion is circular. The frequency of this gyromotion is the gyrofrequency,

$$\omega_c = \left(\frac{qB}{m}\right) \tag{14}$$

which, for electrons, is 2.8 MHz per gauss. The corresponding radius of the gyro orbit is the Larmor radius,

$$\rho_c = \left(\frac{mv_\perp}{qB}\right). \tag{15}$$

where $v_\perp$ is the velocity perpendicular to the magnetic field. As typical values of these characteristic quantities at thermonuclear conditions, we note that for a 50 kG magnetic field, the electron gyrofrequency is 140 GHz, while the ion gyrofrequency is smaller by the ratio of electron to ion mass, m_e/m_i. In the general case where $v_\parallel \neq 0$, such particle motion generates a helix about the magnetic field line m_e/m_i. In the general case where $v_\parallel \neq 0$, where $v_\parallel$ is parallel to the magnetic field, such particle motion generates a helix about the magnetic field line.

In most thermonuclear confinement schemes, the electric and magnetic fields are slowly varying over one gyro period, and to a good approximation, the magnetic flux enclosed by the gyro orbit is a constant. Under these circumstances, the orbit is said to be adiabatic, a condition that can be expressed equivalently as conservation of the magnetic moment, μ, given by

$$\mu = \frac{qv_\perp^2}{2\omega_c} = \frac{W_\perp}{B}. \tag{16}$$

$W_\perp$ is the particle kinetic energy perpendicular to the magnetic field. From conservation of the total kinetic energy, W, and the adiabatic invariance of μ, we find that

$$W = W_\parallel + W_\perp = W_\parallel + \mu B \tag{17}$$

where $W_{//}$ is the kinetic energy parallel to B. The principle
of magnetic mirrors is based on this equation. Since both W
and μ are constants, $W_{//}$ must decrease as the particle moves
into regions of higher field, and if the field is high enough
the particle will be reflected. Such a magnetic field arrangement is shown in Figure 6. While this simple field topology can confine single particles, it cannot, in a stable
way, confine a plasma. The reasons are related directly to
the fact that a plasma is a collection of interacting charged
particles that can act collectively to support wave motions
and exhibit macroscopic motions ordinarily associated with
fluids. In addition, the interaction and scattering of particles leads to changes in their velocity components and
concomitant losses through the ends of magnetic mirrors.

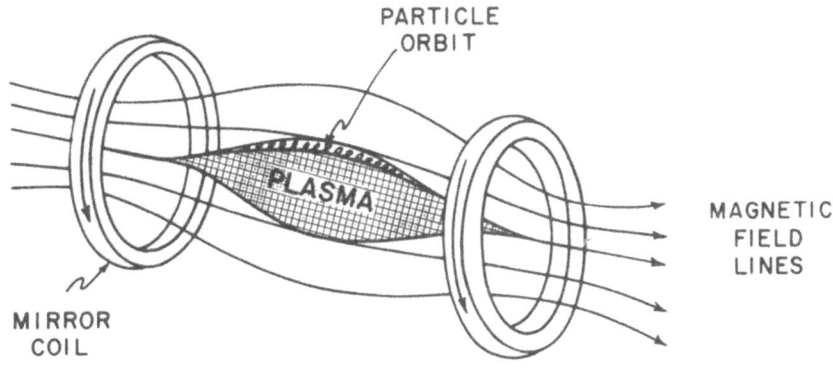

Figure 6

The Simple Magnetic Mirror

Note that not all particles will be trapped in a magnetic
mirror with a finite maximum field. Clearly, a particle with
only energy parallel to the field will not be reflected. Therefore, a mirror trap would lose particles out the ends even if
there were no configuration space diffusion or problems of
stability. Velocity space diffusion due to collisional scattering that changes the pitch angle with respect to the field
is sufficient to cause particle loss.

A simple estimate of the scaling of $n\tau_p$ for mirror systems can be obtained by taking the pinch angle scattering rate to vary as the ion-ion collision frequency, ν_{ii}, for large angle scattering. ν_{ii} is proportional to $n/T^{3/2}$ so taking $\tau_p \sim 1/\nu_{ii}$ implies $n\tau_p \alpha \; T^{3/2}$, which is the observed scaling in stabilized mirrors. The scaling is also predicted to depend on the mirror ratio R, defined as the ratio of maximum to the minimum fields in the magnetic mirror system. More detailed calculations[16] show that

$$n\tau_{pi} = 2 \times 10^{10} \; E_i (keV)^{3/2} \; Log_{10} \; R. \qquad (18)$$

Mirror ratios of 2 are typical in present experiments, and R values between 2 and 10 are generally discussed for mirror power reactors. Note that because losses are due to pitch angle scattering in velocity space, $n\tau_p$ is not dependent on physical dimensions. The preferential leakage of plasma from parts of velocity space leads to an anisotropic particle distribution function that can drive instabilities and cause anomalously fast particle loss.[15,17]

Charged particles drift in plasmas as a result of changes in the effective radius of curvature associated with the particle motion as it gyrates about its main line of motion or its guiding center. The force on the particle is Equation (13). If we decompose $\underline{v}$ into components, $\underline{v}_\perp$ and $\underline{v}_\parallel$, perpendicular and parallel to $\underline{B}$, and do the same for $\underline{F}_{ext}$ and $\underline{E}$, the result of the parallel force is rectilinear motion, which we can neglect. The remaining equation for $v_\perp$ is

$$m \; \underline{\dot{v}} = \underline{F}^{ext} + q(\underline{E}_\perp + \underline{v} \times \underline{B}). \qquad (19)$$

This motion is not purely circular, but contains an additional component, or drift. The drift velocity is given by:

$$\underline{V}_D = \frac{\underline{E} \times \underline{B}}{B^2} + \frac{\underline{F}^{ext} \times \underline{B}}{q \; B^2} \qquad (20)$$

The first term is the $\underline{E} \times \underline{B}$ drift of the guiding center motion and is independent of charge. The second term is due to other forces, either external or arising from the nature of the confining fields. The forces associated with these drifts can drive instabilities when the local radius of curvature points out of the plasma, as it does in the case of the simple magnetic mirror field, rather than into it.

A magnetic mirror field structure that does have field lines always bending away from the plasma would be one in which the magnetic field is smallest at the plasma center and increases in all directions as one moves out. This configuration corresponds to a magnetic well or minimum B topology. Experiments by Ioffe[18] first showed that a coil configuration as shown in Figure 7a, which produces such a magnetic well, confines a plasma that is not subject to the gross MHD instabilities of simple mirrors. Magnetic wells have since been produced by other coil arrangements such as the baseball and Yin-Yang coils shown in Figures 7b and 7c, respectively. Note that the magnetic field lines are always convex to the plasma and the field increases away from the center. In such a minimum B configuration, it is not topologically possible for magnetic flux tubes to interchange, so that flute modes do not occur. These modes can also be suppressed by "line tying," by which we mean connecting the field lines coming out of the mirror throats to a good conductor such as metal plates at the ends of the mirror. Since the field lines are frozen into a good conductor, an interchange of magnetic flux tubes is prevented.

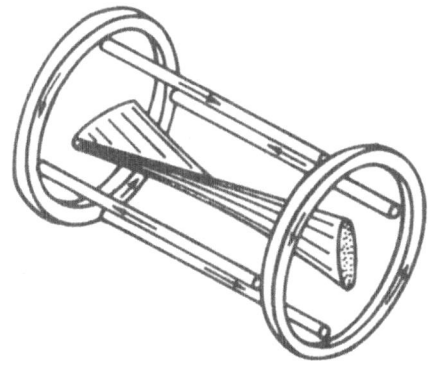

Figure 7a

Minimum B Magnetic Mirror Created using Ioffe Bars

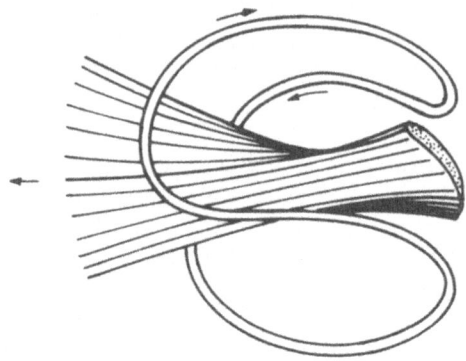

Figure 7b
Minimum B Magnetic Mirror Produced by Coils
Following the Pattern of Seams on a Tennisball

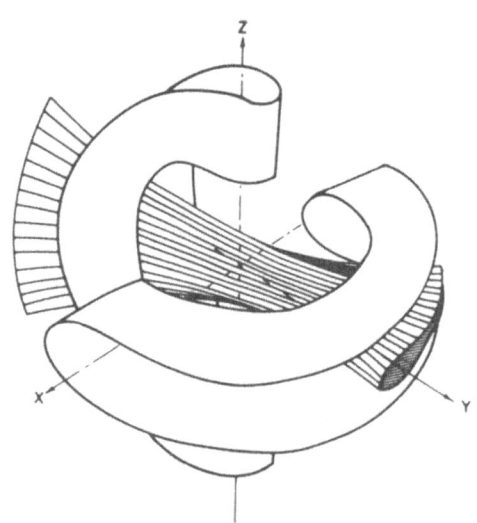

Figure 7c
Minimum B Magnetic Mirror Configuration
Created with Yin-Yang Magnet Shapes

With these advances, the gross MHD instability problems essentially have been overcome in mirrors. The remaining difficulties lie in velocity space microinstabilities[17] and in the practical questions of continuously fueling a mirror plasma to maintain the ion temperature against the end losses. Even without microinstabilities, ion energies of about 200 keV ultimately will be required before classical losses can be reduced enough to produce a favorable $n\tau_E$. In addition, even at reactor temperatures, the end loss problem is still dominant and the high temperatures required can be maintained only by injecting particles into the plasma at or above the desired operating ion temperature. This will most likely be achieved by the injection of neutral beams of particles at energies greater than 100 keV and at equivalent currents in the hundreds at ampere range. Thus, neutral injection and high circulating power are characteristic of mirror systems.

Results from recent experiments on mirror machines have been very encouraging. These results have been obtained on the 2X-IIB machine shown in Figure 8, and operated by the Lawrence Livermore Laboratory.[19] There have been several important results. It had been predicted theoretically that the plasma in this experiment should be unstable to ion cyclotron instabilities[20,21] such that the particle confinement time would be only several tens of microseconds. Initial operation did observe ion cyclotron fluctuations, although the confinement time was somewhat greater than 200 microseconds; however, these microinstabilities have been stabilized by the use of a streaming plasma[22] and by neutral gas fed into the mirror ends.[23] Typical plasma results using the stabilizing streaming plasma are a plasma density of $5 \times 10^{13} \text{cm}^{-3}$, ion temperature of 13 keV, electron temperature of 140 eV, and central β values of 0.4-0.5. With gas feed, the line density has been raised to $1.2 \times 10^{14} \text{cm}^{-3}$ and β values of about 1, without any decrease in the confinement time.

The overall energy confinement time includes many effects, the most important of which is energy loss by fast ion scattering into the mirror loss cone and losses due to electron drag. The typical value for both the density-particle lifetime product and the density-ion cooling time product is $7 \times 10^{10} \text{ cm}^{-3}\text{s}$. This leads to an $n\tau_E$ product of $3.5 \times 10^{10} \text{cm}^{-3}\text{s}$. In the gas feed experiments, $n\tau_E$ has been as high as $4.8 \times 10^{10} \text{ cm}^{-3}\text{s}$.

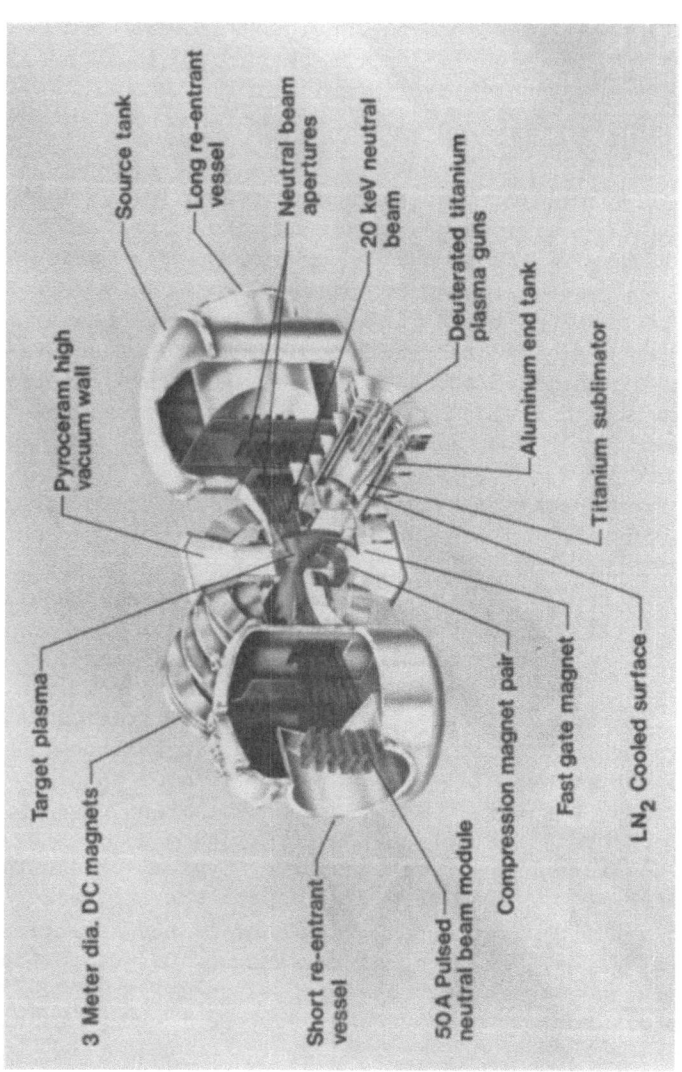

Figure 8

2X-IIB Mirror Experiment
Operated at the Lawrence Livermore Laboratory

Equally important are the scaling laws observed for $n\tau_p$ and $n\tau_E$, since this indicates what we might expect in future machines. It is found that $n\tau_p$ for the hot ions scales as $E_i^{3/2}$, as predicted theoretically. Also, the ion cooling rate by rethermalization onto the electrons produces an $n\tau_E$ for the energy containment in the hot ion component that varies as $T_e^{3/2}$, also predicted by theory. Thus, there appears to be considerable correlation between experimental results and theoretical expectations.

Finally, the 2X-IIB has demonstrated methods for starting up a mirror machine in a DC magnetic field.[24] The method consists of injecting a low density streaming plasma into the magnetic field that forms a sufficiently dense target that the injection and trapping of neutral beams leads to a satisfactory buildup of the density.

B. Theta Pinches

Pinch devices are the second major class of confinement systems and linear pinches are another example of open-ended machines. The simplest case is the Z pinch in which an applied electric field induces an axial current, j_z, to flow in a cylindrical plasma producing a poloidal field, B_θ. The body force, $\underline{j} \times \underline{B}$ points inward and confines the plasma. Such a scheme is illustrative, but as with the simple mirror, is MHD unstable. In particular, it is unstable against modes having the form $A = b(r)e^{i(\omega t - m\theta - kz)}$, where A can be any of the parameters, such as density and magnetic field, which characterize the plasma. An $m = 0$ mode corresponds to the sausage instability in which the discharge necks off at various points along the axis. This can be stabilized by adding a strong longitudinal magnetic field to effectively stiffen the overall field structure. The $m = 1$ mode corresponds to a radial motion of the plasma column and continues to be a mode of great concern.

An analysis of a simple pinch can be based on the single fluid equilibrium equation,

$$\underline{j} \times \underline{B} = -\nabla p \tag{21}$$

and leads to the condition that $\beta = 1$. The addition of a

longitudinal field lowers β and a diffuse toroidal pinch
with strong longitudinal fields will be discussed shortly.
Pinch discharges with β ~ 1 are inherently pulsed devices.
The dynamic Z pinch and the θ pinch are prime examples in
which the experimental time scale is in microseconds. The
simple snow model illustrates the basic concepts. The plasma
is taken to have high conductivity and the j_z current is
assumed to flow in a thin surface layer. The $\underline{j}$ x $\underline{B}$ force
causes an inward radial velocity of the surface layer which
acts like a piston to confine and compress the plasma inside.
The implosion phase is described by a one-fluid momentum
balance equation:

$$\frac{\partial}{\partial t} \underline{U} = \underline{j} \times \underline{B} - \nabla p \tag{22}$$

and Ampere's law:

$$\nabla \times \underline{B} = \mu_o \underline{j}. \tag{23}$$

Typically, ∇p is neglected during the early implosion phase.
The plasma will continue to implode until the increased gas
pressure balances the magnetic pressure.

The θ pinch produces this same effect by a high voltage,
low inductance, fast-rising coil surounding the plasma column,
as shown in Figure 9. The j_θ current in the coils produces a
fast-rising longitudinal magnetic field, B_z. The resulting
transformer action induces surface currents to flow in the
θ direction opposite to the current in the coil, thus excluding the longitudinal field from the plasma interior.
The $\underline{j}$ x $\underline{B}$ force then acts to compress the plasma.

Linear theta pinch experiments at Los Alamos Scientific
Laboratory have achieved densities of 10^{16} to 10^{17} cm^{-3} and
ion temperatures up to 5 keV; however, the confinement time
is typically limited by end losses to about 10 microseconds
(25).

In an effort to overcome the end losses, much work is
continuing on toroidal pinch discharges such as the Scyllac
experiment shown in Figure 10. This torus has a wall minor
radius of 44 mm and a major radius of 4 m, giving an aspect
ratio of about 900. Nevertheless, even in such a large as-

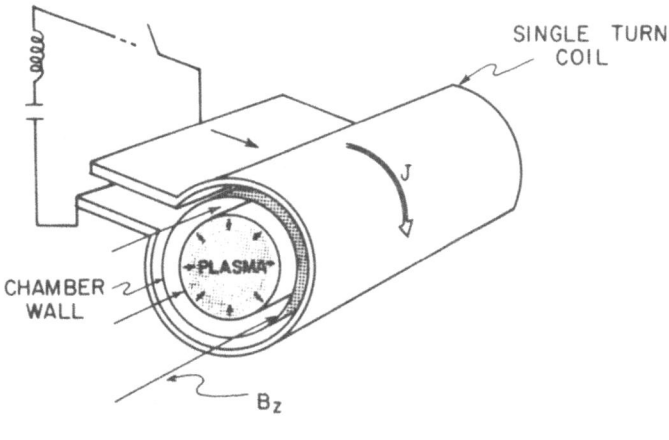

Figure 9

Schematic of Theta Pinch Operation

pect ratio system, a magnetic field gradient will exist that
gives a rise to difficult toroidal equilibrium and stability
problems. The outward force on the toroidal plasma is given
by $\beta\ B_o^2 a^2/(4R)$, where a is the plasma radius and R is the
major radius of the torus. Of particular concern is the
m = 1 kink mode. Theoretical work has indicated the use of an
additional field that varies as $\sin(\ell\theta - hz)$ for ℓ and h as
integers can be used to stabilize the plasma. Scyllac is
designed to produce ℓ = 0 and 1 fields superimposed on
the average field, B_o, which can be as much as 51.5 kG.
When superimposed on the main axial field, an ℓ = 1 field
produces a rippled field line structure. The inner ripples
are larger so that the length of the field lines tend to
be equalized. Experiments on the five-meter sector of
Scyllac with externally imposed ℓ = 1 windings have observed
the m = 1 kink with a low growth rate. No higher m modes
were observed. Experiments on the full Scyllac torus have
been completed (26) with results that were not markedly
improved over those obtained in linear devices or in sectors

Figure 10

Scyllac Toroidal Theta Pinch
At Los Alamos Scientific Laboratory

of the torus.[25] Measurements at 3.5 μs into the discharge indicate that the electron density is $2.5 \times 10^{16} cm^{-3}$, the electron temperature is 600 eV, the ion temperature is 800 eV and β is 0.95. However, the overall confinement time was limited by the onset of m = 1 kink mode to 6-10 μs; thus, the product of $n_e \tau_E$ is about $2 \times 10^{11} cm^{-3} s$.

Stabilization of the plasma is possible with either a strong feedback system or by currents driven in the walls as the plasma moves. Wall stabilization is not possible in Scyllac because the ratio of wall radius to plasma radius is too large. Therefore, feedback control of the m = 1 mode using ℓ = 0 fields is being used; however, the feedback system is not strong enough to stabilize Scyllac at its full design field, and experiments in a derated torus with B_o = 15 kG are proceeding. The confinement time has increased to about 15 microseconds, but at lower values of density.

An optimum cycle for theta pinch operation and one which has received attention in reactor studies is the staged theta pinch.[3] The cycle is shown schematically in Figure 11. The first stage consists of shock or implosion heating of the plasma by a high voltage, low inductance system, with a rise time in about 100 nanoseconds. The magnetic field would rise to tens of kilogauss, imploding a fully ionized plasma with an initial density of about $10^{15} cm^{-3}$. The shock heating will take place in several microseconds, after which the plasma volume will thermalize. The resulting plasma then would be compressed adiabatically by a low-voltage, high-inductance compression coil with a rise time of about a millisecond. In a reactor, this would be followed by a burning phase of about 50 msec, with some direct recovery of energy due to expansion of the plasma against the field. The shock heating coils would produce fields of about 10-20 kG and the adiabatic compression coils would have maximum fields in the 100-200 kG range. In present experiments, a single energy source powers both the shock heating and adiabatic compression coils. In the staged theta pinch, these two coils would be driven by separate sources.

C. Tokamaks

The major closed confinement system is the low β device called tokamak. In contrast to theta pinches, these machines

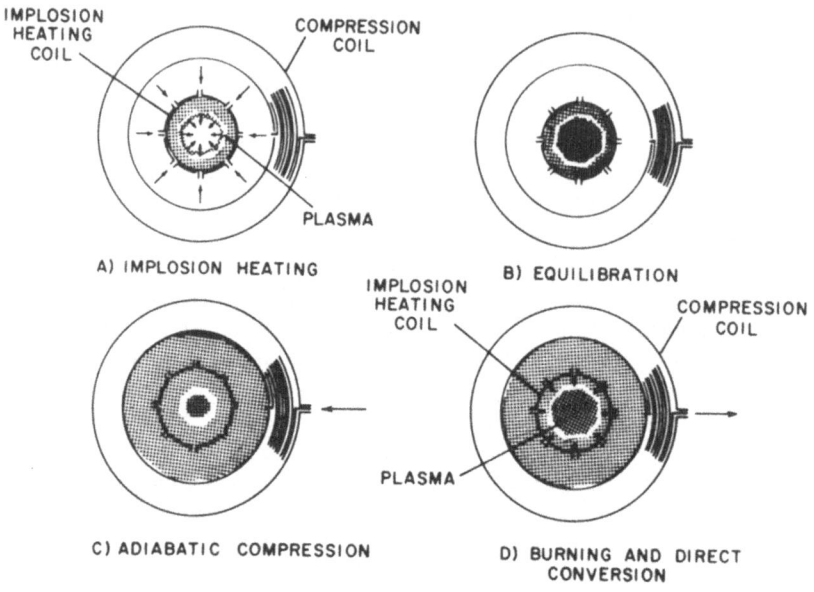

Figure 11

Plasma Cycle for the
Proposed Theta Pinch Reference Reactor Concept

tend to be more like steady state systems. For future reference, the basic coordinates in a toroidal system are shown
in Figure 12, where r is the minor radius, R is the major
radius, R/r is the aspect ratio, ϕ denotes the toroidal angle,
and θ denotes the poloidal angle. A tokamak[27-28] is a
diffuse toroidal Z pinch in which a nested set of flux surfaces is generated by the current that is induced to flow in
the plasma itself. The device is axisymmetric and the plasma
is confined by the poloidal magnetic field, B_θ, generated by
a current carried in the plasma, and by the toroidal field,
B_T. To drive the plasma current, the plasma itself is treated
as a single turn secondary of a transformer. Energizing the

primary side produces a toroidal electric field, E_T, which causes gas breakdown and drives the toroidal plasma current. The transformer may have an iron or an air core and can have either normal or superconducting windings. An air core transformer with superconducting windings allows for the smallest aspect ratio and this type of transformer probably will be used in reactor systems. Present day experiments use copper coils for all windings and most machines have an iron core transformer. A schematic of the basic features of a tokamak is shown in Figure 12, and a drawing of the Princeton Large Torus (PLT) is shown in Figure 13.

Plasmas have a finite electrical resistivity that is independent of density and varies with temperature as $T_e^{-3/2}$. Spitzer [29] has derived the expression:

$$\eta_{sp} = \frac{1}{\sigma_{sp}} = 1.65 \times 10^{-9} \frac{Z_{eff} \ln \Lambda}{T_e^{3/2}} \quad \text{(ohm-m)} \tag{25}$$

where T_e is in keV and Z_{eff} is defined by:

$$Z_{eff} = \frac{\sum_j n_j Z_j^2}{n_e} \tag{26}$$

and the sum on j extends over all ionic species. As a result of this resistivity, plasma heating in tokamaks will occur because of the plasma current, although this ohmic heating is insufficient to ignite a fusion reactor plasma. Nevertheless, this self-heating aspect of tokamaks has been an integral part of their success in reaching high plasma temperatures. Early experiments reported electron temperatures of about 1.5 keV and ion temperatures approaching 1.0 keV with ohmic heating alone. [27,30,31,32] Methods to provide additional plasma heating include adiabatic compression of the plasma in the major radius (the ATC experiment) [33], and the injection of high energy neutral beams. [34,35,36]. Central ion temperatures of 2 keV have been achieved.

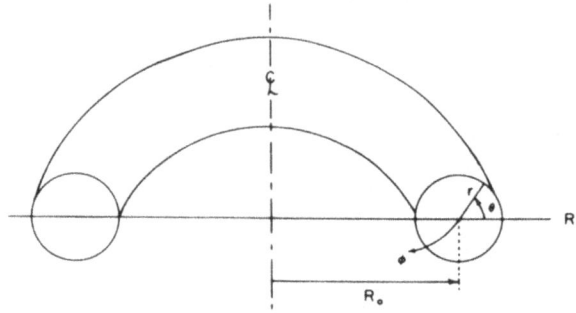

a. BASIC COORDINATES FOR TOKAMAKS

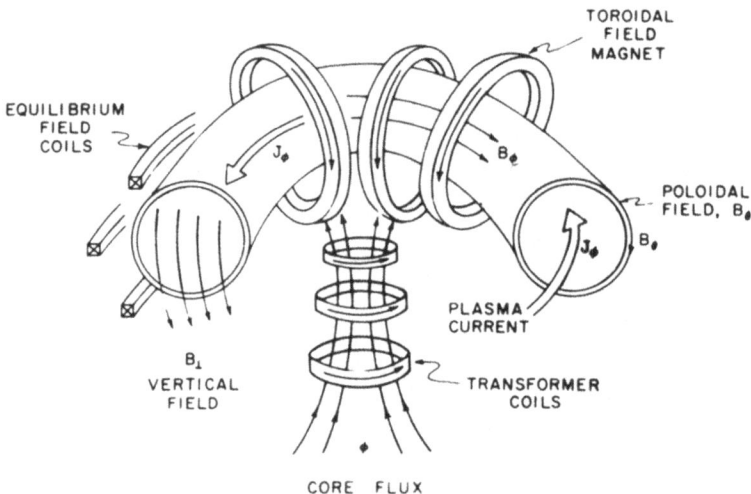

Figure 12

Schematic of the Main Magnetic Fields
and Operating Mode of a Tokamak

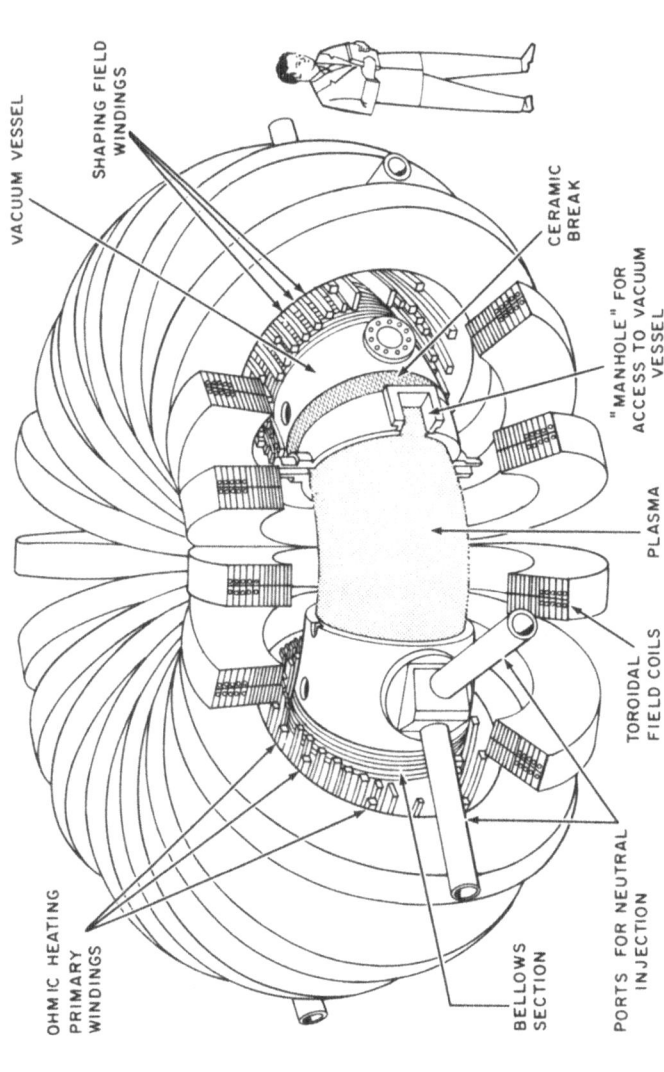

VACUUM VESSEL

SHAPING FIELD
WINDINGS

CERAMIC
BREAK

"MANHOLE" FOR
ACCESS TO VACUUM
VESSEL

OHMIC HEATING
PRIMARY
WINDINGS

BELLOWS
SECTION

PORTS FOR NEUTRAL
INJECTION

TOROIDAL
FIELD COILS   PLASMA

Figure 13

Model of the Princeton Large Torus (PLT), a Research Device, Designed to Test Scaling
Laws of Plasma Confinement. The cutaway view shows the plasma column surrounded by
coils which provide magnetic confining fields. Sponsored by the United States Atomic
Energy Commission, construction of the PLT at the Princeton University Plasma Physics
Laboratory began in June, 1972, and began operation in December, 1975.

The tokamak is inherently a pulsed device, since it requires an externally produced electric field. However, at high temperature and low collision frequency, as in the reactor plasma regime, trapped particle effects have led to the prediction that there will exist a diffusion-driven current (a current proportional to a density gradient) that can provide the required rotational transform and trapping magnetic field, even without an externally-produced electric field (37,38). This is the so-called "bootstrap current" and is predicted to exist with $E_T = 0$ when $\overset{o}{\beta_\theta} \simeq \sqrt{A}$. $\overset{o}{\beta_\theta}$ is the ratio of the peak plasma pressure to poloidal magnetic field pressure at the plasma boundary. Under these circumstances, plasma operation could be truly steady state unless other factors, such as impurity buildup, limit the burn time. If steady state operation is not possible, long burn should be feasible if impurities can be controlled using devices known as diverters (39).

A main characteristic of tokamaks is the use of a strong toroidal magnetic field to suppress MHD instabilities. In particular, a strong toroidal field, B_T, will suppress the helical kink instability if the MHD stability factor,

$$q(r) = \frac{r}{R} \frac{B_T}{B_\theta(r)} \tag{27}$$

is greater than one throughout the plasma and the vacuum region. Further, the plasma current that produced B_θ provides the rotational transform required to suppress the flute instabilities and to counteract the grad B_T drifts. For a current density that decreases with $r (j_\phi(r)$ decreases from the plasma center to the edge), the criterion to suppress local flute perturbations is the Kruskal-Shafranov limit,

$$q(r) > 1. \qquad\qquad o \leq r \leq a \tag{28}$$

where a is the plasma radius. From this condition, we see why tokamaks are low β devices. At $r = a$,

$$q(a) = \frac{1}{A} \frac{B_\phi}{B_\theta(a)}, \text{ and } \beta = \frac{\beta_\theta}{q(a)A}.$$

Since B_θ is the main confining field, we expect β_θ

$$\beta_\theta = \frac{P}{B_\theta^2/2\mu o} \text{ to be about 1.}$$

Since A is typically greater than 3, and q(a) is also about 3, we expect $\beta \lesssim \beta_\theta/9$. Typical values of β in present devices have reached about 0.01, and it is hoped that values of 0.05 can be achieved in reactors with circular cross-sections. Higher β may be possible in noncircular, vertically elongated tokamak plasmas such as in the Doublet devices (40,41) shown in Figure 14. The value of β scales as S^2, where S is the plasma shape factor defined as the ratio of the perimeter in the poloidal direction to the toroidal direction:

$$S = \frac{\oint_\perp ds}{2\pi R}.$$

One hopes to increase the current, and therefore, the β, in noncircular machines, while keeping the same q. Doublet machines at General Atomic Company will investigate these possibilities.(41)

The basic ideas regarding equilibrium of a plasma in a tokamak are relatively straightforward. If the plasma were a straight cylinder, the poloidal magnetic pressure would be equal around the plasma boundary. However, because of the toroidal curvature, the poloidal field is stronger on the inside of the plasma loop than on the outside. Without additional fields, the loop would tend to expand in major radius; therefore, it is necessary for plasma equilibrium to have an additional vertical field produced by external windings, which subtracts from the poloidal field on the inside loop and adds to it on the outside, thus producing an equilibrium configuration (see Figure 12). In many present-day experiments (26,27) eddy currents driven in a conducting shell surrounding the plasma generate the vertical field required to maintain the discharge in equilibrium, but in reactors, it is clear that an externally produced vertical field will be required. Indeed, this vertical field, added to the poloidal field generated by the plasma current, produces a resultant magnetic field topology that leads naturally to a poloidal field diverter.

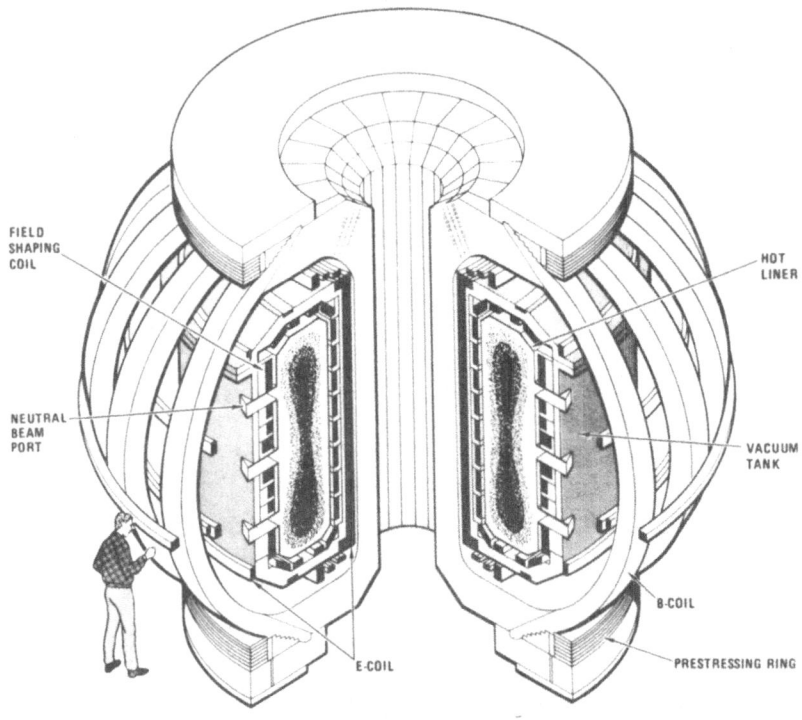

Figure 14

Doublet-III Noncircular Tokamak Experiment
Under Construction at General Atomic Company

The combination of vertical and poloidal field can be
combined to produce a separatrix that becomes a natural plasma boundary. Two such configurations are shown in Figure 15.
Magnetic surfaces inside the separatrix remain closed in the
plasma volume; however, particles on field lines outside the
separatrix are carried away from the plasma surface and can
be collected in regions beyond the main vacuum chamber walls.

This would prevent sputtering by charged particles diffusing
from the plasma of high Z wall atoms. As noted earlier, high
Z impurities in a plasma substantially increase radiation
losses and thereby increase the $n\tau_E$ required for a workable

reactor. The field configuration "b" in Figure 15 is called
a double null system and it lends itself naturally to tokamaks in many ways including vertical elongation of the plasma
cross-section.

The poloidal field in a tokamak generated by the plasma
current provides a rotational transform to the plasma. When
this poloidal field is superimposed on the strong toroidal
field, B_T, the resultant field lines are helical. The ro-

tation generated by the poloidal field compensates for the
outward drift of particles caused by the gradient in the
toroidal field. The toroidal field is stronger on the inside of the torus, varying as 1/R from the inside to the
outside of the loop.

This gradient in B_T can produce a class of particles
that cannot reach the inside of the plasma loop. Since
effects associated with these trapped particles are most
important in the reactor regime of plasma parameters (high
temperature, low collision frequency), we mention briefly
the basic effects of particle trapping on diffusion and
conduction phenomena in tokamaks.

Since the magnetic field is stronger on the inside,
particles with mostly perpendicular velocity will be reflected by this higher field, and thus be mirror trapped.
In particular, if the particle kinetic energy is less than
μB^{max}, the particle will be reflected. Circulating, or untrapped particles have $W > \mu B^{max}$, and these follow orbits
such as in Figure 16a. The fraction of trapped particles
depends on minor radius and an average value is $A^{-1/2}$.

In the absence of drift effects, the trapped particles
follow an arc between the turning points. The gradient in
B_T produces a drift, as discussed earlier, and leads to a
displacement of these arcs such that the resulting orbits,
projected on the (r, θ) plane, have the banana shape shown in

Figure 16b. The width of this banana orbit for electrons
at the midplane is given by:

$$\Delta r = \frac{m \, v_{\parallel}}{e \, B_{\theta}} \quad \frac{1}{\sqrt{A}} \, \rho_{\theta}^{e} \tag{29}$$

where ρ_{θ}^{e} is the electron gyroradius in the poloidal magnet

field. This is important because it means the step size
that enters the diffusion coefficient is the gyroradius in
the *weaker* magnetic field. This leads to an enhanced diffusion in high temperature toroidal plasmas, as compared to
classical diffusion in a cylinder. Nevertheless, the result
yields basically classical scaling laws for particle and
energy containment times, which is one of the most encouraging aspects of the recent tokamak research.

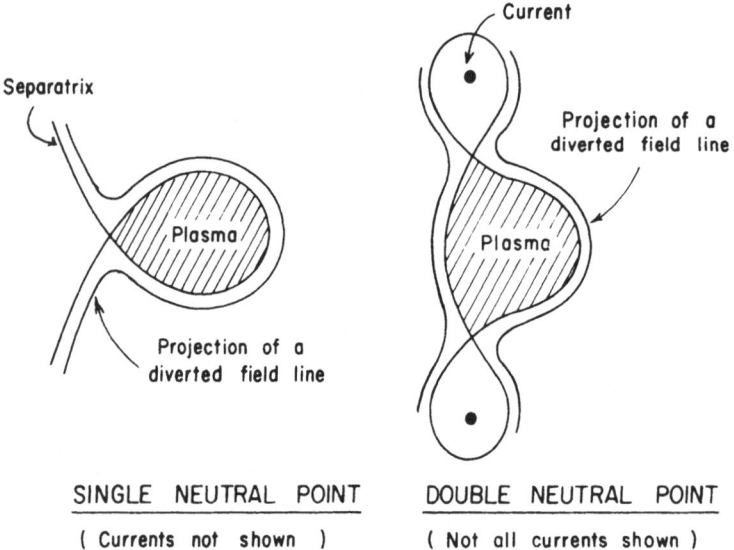

SINGLE NEUTRAL POINT   DOUBLE NEUTRAL POINT

(Currents not shown)   (Not all currents shown)

Figure 15

Characteristic Shapes for a Single and Double Null
Poloidal Diverter on a Tokamak. The plasma is bounded
by a separatrix (magnetic limiter).

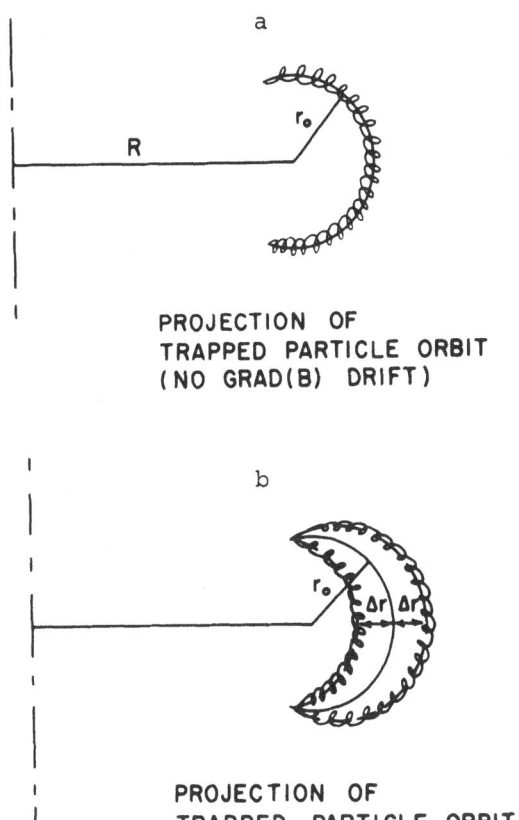

PROJECTION OF
TRAPPED PARTICLE ORBIT
(NO GRAD(B) DRIFT)

PROJECTION OF
TRAPPED PARTICLE ORBIT
(WITH GRAD(B) DRIFT)

Figure 16

Projections of Trapped Particle Orbit in Tokamak

In the high temperature regime where CTR plasmas will operate, the plasma is essentially collisionless. In particular, the electron-ion collision frequency, ν_{ei}, that scales as $T^{-3/2}$, is less than the bounce frequency of oscillation between mirror points given by:

$$\omega_{bounce} = \frac{v_{th}^{e}}{q\,R\,A^{3/2}} .$$
(30)

The diffusion coefficient in this so-called banana regime is[37]

$$D_{\perp}^{e} = 1.12 \ (1 + \frac{T_i}{T_e}) \ q^2 A^{3/2} \nu_{ei} (\rho_{\phi}^{e})^2 \ ; \nu_{ei} < \omega_{bounce} \quad (31)$$

where T_i and T_e are the ion and electron temperatures, and ν_{ei} is the electron-ion collision frequency. Since the classical diffusion coefficient is $\nu_{ei}(\rho_{\phi}^{e})^2$, there is considerable enhancement of the diffusion coefficient due to trapped particle effects. The basic theory governing collision diffusion in low β systems is referred to as neoclassical transport theory.[42]

The heuristic explanation for the form of the neoclassical diffusion coefficient is obtained by recalling that the fraction of particles trapped is in the order of $1/\sqrt{A}$. Also, trapped (untrapped) particles that make a transition to being untrapped (trapped) because of collisions with other particles have a step size for diffusion equal to Δr. The diffusion coefficient will be obtained from an expression like

$$D_{\perp}^{e} = \frac{(\text{step size})^2}{\text{characteristic time}} . \quad (32)$$

The characteristic time is the collision time for a particle to scatter from trapped to untrapped and vice-versa. This collision time is shorter than the $90°$ collision time because it involves smaller scattering angles to make the trapped-untrapped transition; thus, the characteristic time becomes:

$$\tau_{eff} = \frac{\tau_{90°}}{A} \quad (33)$$

Further, the characteristic step size can be written in terms of ρ_{ϕ}^{e} as:

$$\Delta r = \sqrt{A} \ q \ \rho_{\phi}^{e} . \quad (34)$$

Noting that the particle flux, Γ, is proportional to $\frac{dn_t}{dr}$, where n_t is the number of trapped particles, and that

$$n_t \simeq \frac{1}{\sqrt{A}} \, n, \tag{35}$$

we have

$$D_\perp^e \simeq q^2 \, A^{3/2} \nu_{ei} \, (\rho_\phi^e)^2 \tag{36}$$

$$D_\perp^e \simeq q^2 \, A^{3/2} \, D_\perp \, \text{class} \tag{37}$$

A similar analysis can be performed for the thermal conductivity and the results show the importance of including neoclassical effects in the analysis of tokamak reactors.

In the collisional regime, where $\nu_{ei} \gg \omega_b^e$, toroidal effects also add to the classical diffusion coefficient, and Pfirsch and Schlutter (43) have derived the expression:

$$D_\perp \simeq q^2 \, \nu_{ei} \, (\rho_e^e)^2 \, . \tag{38}$$

Connecting this collisional regime with the collisionless, or banana regime is the plateau regime, where D is independent of ν_{ei}.

The scaling of $D_\perp$ with ν_{ei} predicted by neoclassical theory is shown in Figure 17. Hinton and Rosenbluth[44] have given an expression that continuously joins the plateau and banana regimes. In addition, starting with the early work of Sagdeev and Galeev,[45] a reasonably complete neoclassical theory has been developed, including a set of transport equations.[42]

The electrical conductivity, σ, in present machines is anomalously low and has added to the effectiveness of ohmic heating. Neoclassical theory predicts a decrease in σ in the banana regime, since trapped particles do not participate in carrying the main current. A recent study has found[49]

$$\sigma = \sigma_{sp} (1 - 1.95 \, (\tfrac{r}{R})^{1/2} + .95 \, (\tfrac{r}{R})) \, . \tag{41}$$

Unfortunately, this formula is derived assuming $\omega_b^e \gg \nu_{ei}$, which does not hold in present tokamak plasmas. The conductivity in present devices appears anomalously low compared to the Spitzer value by factors of 3 to 6. This appears to be

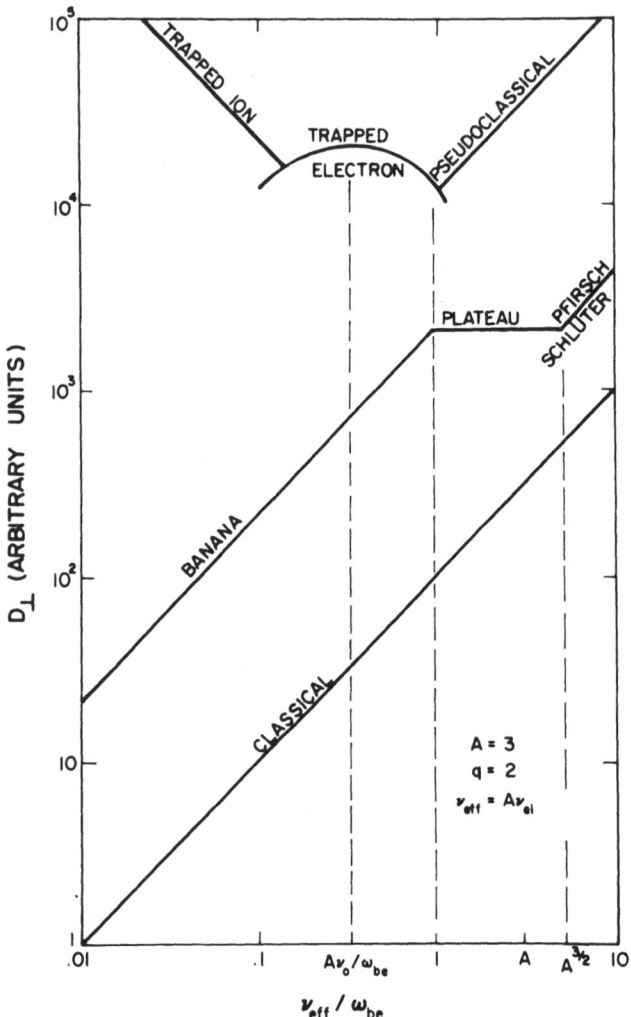

Figure 17

Predictions from Various Theories
For the Diffusion Coefficient as a Function of
Collision Frequency in Tokamaks

due to the presence of impurities, but this problem is not completely understood.

In experiments, there is evidence that the ion thermal conductivity follows neoclassical theory[46] but the electron thermal and electrical conductivities are anomalous. The electron thermal conductivity is higher than predicted by neoclassical theory for present experiments, and follows an empirical pseudoclassical scaling[47,48] where

$$\chi_e^{pc} \sim C_o \; \nu_{ei} \; (\rho_e^{\theta})^2 \tag{39}$$

$$D_{\perp}^{pc} \sim C_1 \; \nu_{ei} \; (\rho_e^{\theta})^2 \tag{40}$$

where χ_e represents the electron thermal conductivity. Typically, C_o is about 10, and C_1 is approximately $1/3 \; C_o$.

A most relevant aspect of tokamak transport for future systems is the prediction that various microinstabilities associated with particle trapping will cause further enhanced diffusion.[50-53] Of particular concern are the trapped electron and trapped ion modes. The anticipated diffusion coefficients expected from these modes are not well known, but estimates, as shown in Figure 17, show what is predicted by theory. As can be seen, the trapped ion mode diffusion coefficient greatly exceeds the neoclassical value of $D_{\perp}$ in the low collision frequency regime relevant to reactors. Interestingly, this may be beneficial, because large reactor systems would have extremely long confinement times (thousands of seconds) based on neoclassical theory.[1] The long confinement time would, in turn, lead to alpha particle buildup and to a reduction in the ion population. A burn would therefore be terminated by the inability to remove the ash from the fusion plasma. Trapped particle instabilities yield estimated diffusion coefficients short enough to prevent this problem, yet long enough to yield acceptable $n\tau_E$ values in large systems.[5,13] This is at odds with the traditional quest in plasma physics for ever-improved particle confinement.

Note that in toroidal devices such as tokamaks, particle loss is by cross-field diffusion in real space rather than

via pitch angle scattering in velocity space as in mirror
machines. The particle containment time is estimated from

$$\tau = \frac{a^2}{\alpha D_\perp} \tag{42}$$

where a is the characteristic length of the system and α is
a constant. Typically, a is the minor radius of the torus
and α is taken as 2. The formula illustrates that τ depends
on system size and that one can expect improvements as we
proceed to build larger devices.

Tokamaks today carry plasma currents ranging from 25kA
to 400 kA, and large devices beginning operation in early
1976 will carry currents approaching 1 MA. We noted earlier
that central electron temperatures approaching 3 keV, and
central ion temperatures of 2.0 keV have been achieved in
several machines.[34-36] The density values range from
$3 \times 10^{13} \text{cm}^{-3}$ to 10^{14}cm^{-3}, with characteristic energy confinement times of 10-20 ms. Recent studies on the Alcator
tokamak[32] have raised the density to $5 \times 10^{14} \text{cm}^{-3}$ with
τ_E = 20 ms, giving a value of $n\tau_E$ equal to 10^{14}cm^{-3}-s. In
addition, the experiments operate for many energy confine-
ment times, typically several hundred milliseconds. A pulse
duration of up to one second should be achieved in the largest present-day experiments. Auxiliary heating of the
plasma by both neutral beam injection and by adiabatic compression has been found to follow classical predictions, and
no instabilities have been observed. Thus, tokamaks have
produced high temperatures, high $n\tau_E$ values, long pulse times,
and scalable heating schemes.

D.   Present Status and Future Directions

A summary of the present state of plasma confinement results can be seen in Figure 18, where we plot the $n\tau_E$ and T_i
values achieved by various experiments. The largest $n\tau_E$
value has been achieved in tokamaks, while the highest ion
temperature has been achieved in mirror machines. The Lawson
criterion is also shown on the figure to give some idea of
the meaning of these experimental achievements.

Based on these results, several large Tokamak experiments are planned for operation around 1980.[54] The Tokamak Fusion Test Reactor will be built by the United States at the Princeton Plasma Physics Laboratory, and will be the first experiment capable of burning D-T. The European Community is planning the Joint European Torus and Japan will build the JT-60. Both experiments are planned for operation, primarily with hydrogen. The Soviet Union is planning to construct the largest device, T-20, scheduled to operate between 1983 and 1985, and will be capable of D-T operation. A summary of the key parameters for each of these experiments is given in Table 1. Future mirror and pinch experiments are also under consideration. The MX machine proposed at Lawrence Livermore

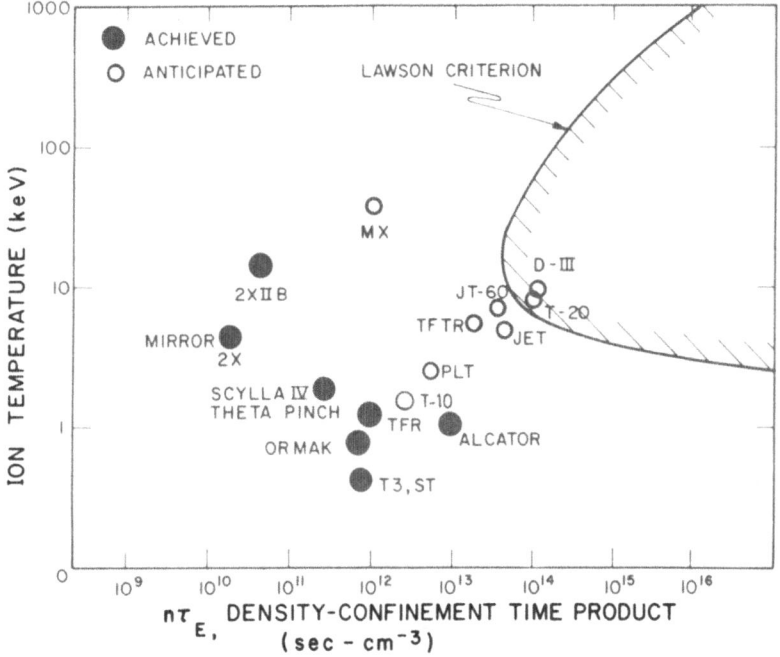

Figure 18

The Lawson Criteria and Values of $n\tau_E$ and T_i Achieved in various experimental devices. The open circles are the anticipated values in experiments now beginning operation or under construction.

Table 1

Characteristic Parameters for the Next Generation
of Large Tokamak Experiments (54)

		Experiment			
Parameter	Units	TFTR[(a)]*	T-20[(b)]*	JT-60[(c)]*	JET[(d)]*
Plasma Current	MA	2.5	6	3.3	4.8
Major Radius	m	2.48	5	3	2.96
Plasma Half-width	m	0.85	2	1	1.25
Plasma Half-height	m	0.85	2	1	2.10
Aspect Ratio		2.9	2.5	3	2.37
Elongation Ratio	-	1	1	1	1.68
Toroidal Magnetic Field (on Axis)	T	5.2	3.5	5	3.4
Max. Toroidal Magnetic Field	T	9.5	7.8	11	6.9
Vacuum Vessel Volume	m^3	64	400	100	190
Vessel Surface Area	m^2	110	400		450
Neutral Injection Energy	keV	60,120	80-160	50-100	80,160
Neutral Injection Power	MW	40	60	10-20	3-25
Neutral Injection Pulse Length	sec	0.1-0.5	2-13	10	0.3-10
Number of TF Coils	-	20	24	24	32
Peak Power for TF Coils	MW	488	1200	350	425
Peak Power for Ohmic Heating Coils	MW	34	600	NA	186
Peak Power for Equilibrium Field Coils	MW	200	420	NA	96
Total Peak Power Including other Sources	MW	700	1680	NA	700

---

*   (a) Tokamak Fusion Rest Reactor (USA)
     (b) Tokamak - 20 (USSR)
     (c) Japanese Tokamak - 60 (Japan)
     (d) Joint European Torus (EEC)

Table 1

Anticipated Performance Characteristics

Mean Ion Temp.	keV	6	7-10	5-10	5
Mean Electron Temperature	keV	6	7-10	5-10	5
Mean Ion Density	cm^{-3}	$4\text{-}8x10^{13}$	$0.5\text{-}5x10^{13}$	$2\text{-}10x10^{13}$	$5x10^{13}$
Poloidal Beta	-	1-2.5	1	1	1
Toroidal Beta	-	0.007-0.015	0.03	0.02	0.03
Energy Confinement Time	sec	0.2	2	0.2-1	1
$n\tau_E$	$cm^{-3}\text{-}sec$	$1.5x10^{13}$	10^{14}	$2\text{-}6x10^{13}$	$5x10^{13}$
D-T Neutron per Pulse		10^{18}	$10^{20}\text{-}10^{21}$	No DT Expts	10^{20}
D-T Pulses during Machine Life		$4x10^{3}$	$1x10^{5}$	0	$1\text{-}5x10^{3}$

Laboratory is the most likely device. Its expected performance parameters are included in Figure 18.

Overall, the prospects for progress in the magnetic confinement approach to fusion are bright. It is therefore auspicious to consider the technological problems posed by fusion reactors.

III. THE BASIC ELEMENTS OF FUSION REACTORS

As fusion power comes closer to reality, one must face the hard technological questions associated with obtaining electrical power from controlled fusion. Over the past six years, a relatively extensive effort has been started to understand and to solve these technological problems. For definiteness, we shall restrict our discussion to the D-T fusion fuel cycle.

The D-T fusion reaction generates 17.58 MeV of which 14.06 MeV, or 80% comes off as a neutron that escapes from the confinement region. Thus, the electrical power we seek can be produced economically only if we recover the neutron kinetic energy in a region surrounding the plasma. This region is the blanket and is hown in the schematic illustration, Figure 19. The neutron slows down by collisions with the structure and coolant in this zone and the heat is removed. The hot coolant then goes through a power cycle to generate electricity. Generally, greater than 95% of the neutron energy is deposited in the blanket zone; however, neutrons and gamma rays from neutron-induced nuclear reactions do leak out, so that a radiation shield is required behind the blanket. The shield also limits the heat input to the magnets that provide the confining magnetic field. This is especially important when the coils are superconducting.

Tritium does not occur naturally, and has a radioactive half life of 12.6 years; therefore, it is necessary to breed tritium using the high energy neutrons and lithium. Natural lithium has two isotopes, ^{6}Li (7.42%) and ^{7}Li (92.58%), and the nuclear reactions leading to tritium production are:

$$^6\text{Li} + \text{n} \rightarrow {}^4\text{He} + \text{T} + 4.86 \text{ MeV}$$

and

$$^7Li + n \rightarrow {}^4He + T + n - 2.87 \text{ MeV.}$$

The first reaction is exothermic, while the second can occur only when the incident neutron energy exceeds 2.87 MeV. Cross-sections for these reactions are shown in Figure 20. Thus, the blanket region in Figure 19 serves a second auxiliary purpose of breeding tritium. In addition, if liquid lithium is used as both coolant and breeder, the bulk of the heat will be generated within the lithium itself, so that the liquid metal acts more like a heat transport medium.

The first set of problems common to all D-T systems is directly related to the 14 MeV neutron reaction product.

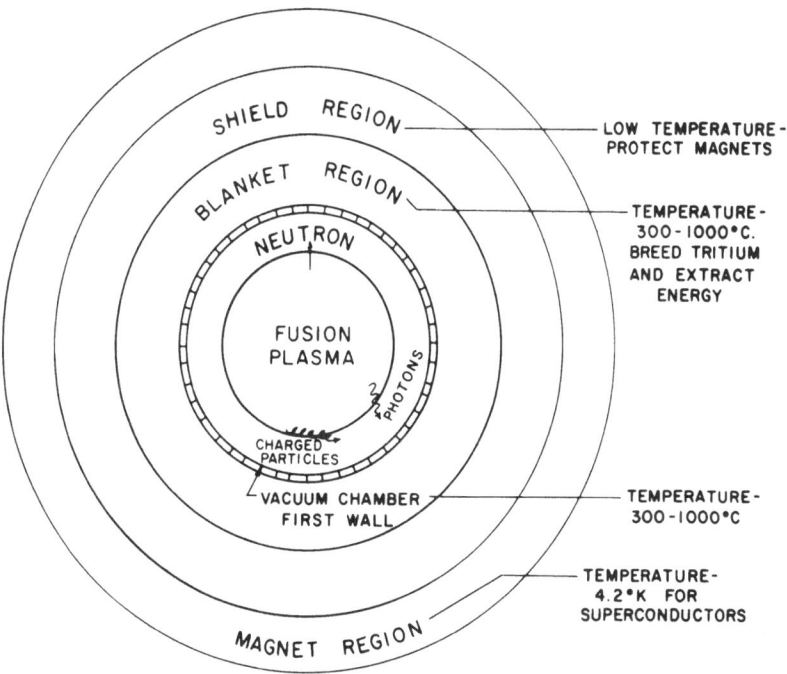

Figure 19

Schematic of a Fusion Reactor
Depicting the Major Systems

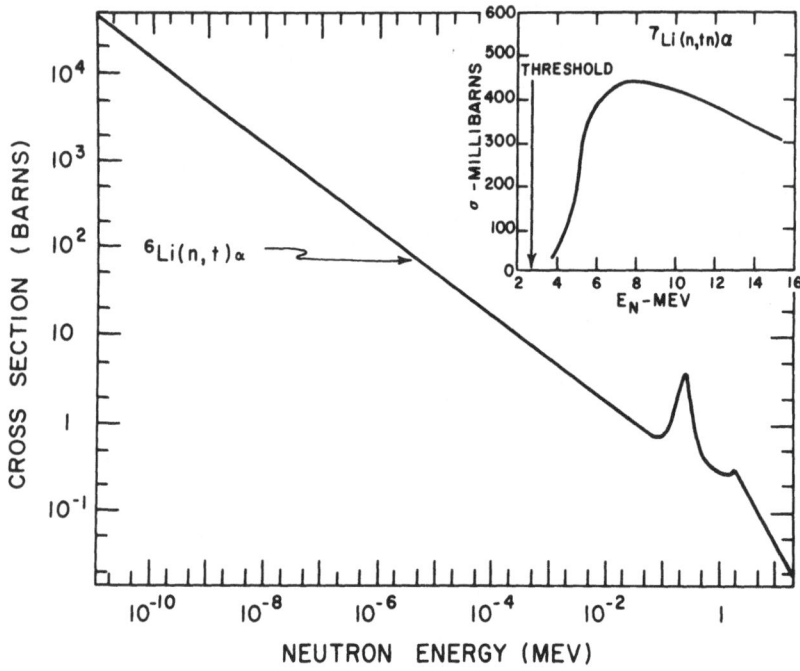

Figure 20

Neutron Reaction Cross-Sections
For the Production of Tritium from Lithium

The plasma chamber constitutes the first material barrier and
is subject to bombardment by 14 MeV neutrons, electromagnetic
radiation, and charged particles diffusing from the plasma.
This first wall is thus subject to high neutron and photon
fluxes, high heating rates, and pressure from the back side
due to a coolant required to keep the first wall at an acceptable temperature. Typical heating rates are roughly 10 to
50 watts/cm^3. The 14 MeV neutron power passing through the
first wall will probably be in the range from 1 - 10 MW/m^2.
A wall loading of 1 MW/m^2 is a 14 MeV neutron current of
4.43×10^{13} n/cm^{2}s. Thus, first wall problems alone include
the bombardment of the first wall by neutron, photon, and
charged particle fluxes, radiation damage induced by the
14 MeV neutrons, and heat transfer and fluid flow problems
related to cooling the first wall.

The sputtering at the first wall due to neutrons and charged particles is an important source of first wall erosion.[55] It also can generate high Z impurities that can return to the plasma with detrimental effects on plasma performance. In addition, in theta pinches, radiation damage as well as thermal cycling on an insulator required on the front wall are important. Clearly, first wall surface and bulk materials problems is of prime importance.

The neutrons escaping the plasma mostly penetrate the first wall and slow down, depositing energy in the surrounding fusion blanket structure. The problems here include heat transfer and fluid flow using gas or liquid metal cooling. There will be continued radiation damage, and one must consider materials compatibility of the coolant and structural materials. Since tritium is required, one must consider tritium breeding, extraction and control, and since neutrons are interacting with the blanket materials, there will be radioactivity and afterheat. The calculation of neutron and photon fluxes in the blanket and shield, and the calculation of nuclear heating and gas production rates are important problems in the neutron transport theory and analysis of fusion systems.

The shield behind the blanket further attenuates the energy flux such that the heat load to the next element, namely the magnets, is acceptably low. Typically, such systems are about one meter thick and attenuate the energy flux by another four orders of magnitude.

For magnetic confinement such as tokamaks and mirrors, superconducting magnets are required. This involves the technology problems of the design and construction of large bore superconducting magnets, reliability of operations, and stress analysis to determine the structural requirements.

This simplified view brings together most of the major features anticipated in future fusion power reactors. In addition to the specific areas mentioned, there are also the many and varied aspects of power systems analysis. The hot coolant, on leaving the blanket, must be coupled via heat exchangers to an electrical power generating system, either via a steam cycle or a direct cycle with gas cooling. Also, we shall require the capability to control tritium flow, to carry out safety analyses associated with fusion systems, to

investigate potential environmental and resource questions,
and to begin to assess the potential economics of fusion
power.

To consider in greater depth the technological problems
of all approaches to fusion is a massive undertaking. Since
the greatest effort worldwide is on tokamak systems, we shall
consider in detail a particular conceptual tokamak power
reactor design. This self-consistent analysis will serve as
a useful guide to tokamak fusion technology.

IV. THE UWMAK-III CONCEPTUAL TOKAMAK REACTOR

A. Overview

The conceptual design of a fusion reactor inevitably
involves extrapolation of our present knowledge, particularly
in the area of plasma confinement scaling. Nevertheless, it
is an important undertaking, for it is the only way to de-
develop a reasonably self-consistent view of fusion reactor
problems and to assess their relative importance. In two
previous studies,[1,5] we developed conceptual fusion reactor designs based on the philosophy that design decisions,
wherever possible, should be based on present technology.
The choice of 316 stainless steel as the structural material
is a primary example. The UWMAK-III[8] conceptual tokamak
reactor design study has had as its principal goal the detailed, self-consistent examination of a tokamak reactor
where design choices are made that represent reasonable
extrapolations of present technology in the context of the
general time frame envisioned for the widespread practical
application of controlled fusion. The prime examples are
the choice of the molybdenum based alloy, TZM (99.4% Mo,
0.5% Ti, 0.08% Zr, 0.01% C) as the structure material, and
the use of aluminum as the stabilizer and structure in the
toroidal field magnets. On the other hand, many of the design features are generic and therefore applicable to near-
term tokamaks as well. Examples here include the unique
blanket design, the MHD analysis and the plasma shape, the
method of impurity control and particle collection using a
diverter, the use of RF heating to bring the plasma to ignition conditions, the tritium extraction and recycling
processes, and the general design approach to module removal.

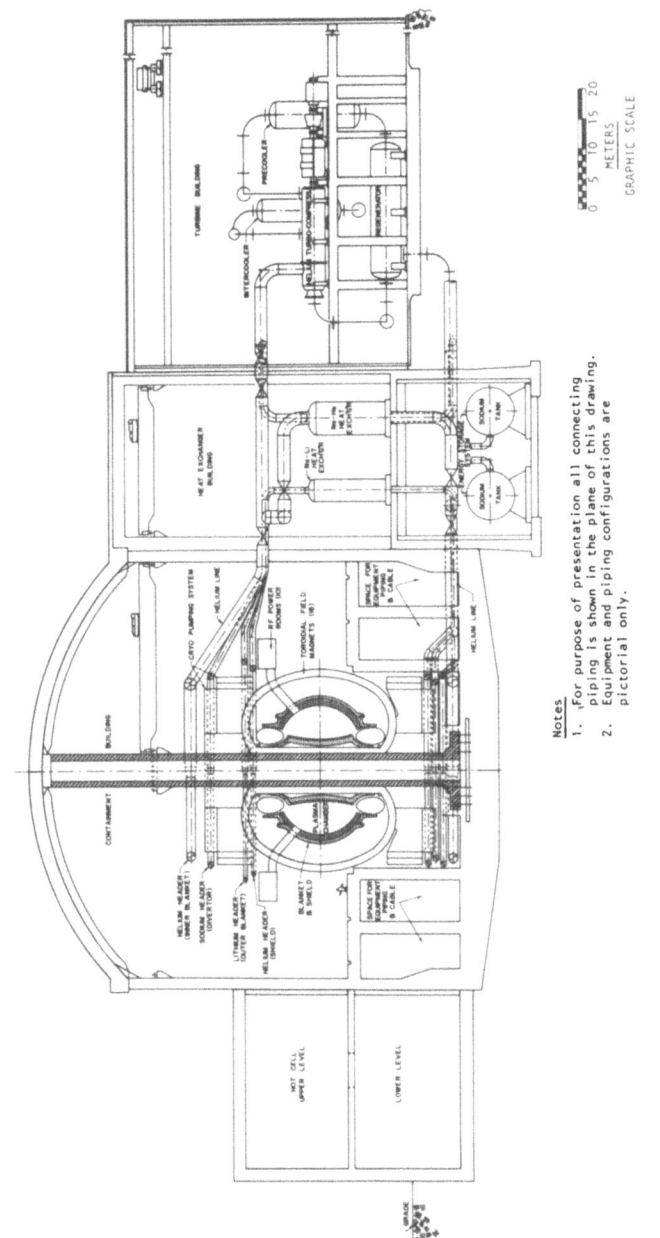

Figure 21

Cross-Section View of UWMAK-III Conceptual Tokamak Power Reactor (8)

Notes

1. For purpose of presentation all connecting piping is shown in the plane of this drawing.
2. Equipment and piping configurations are pictorial only.

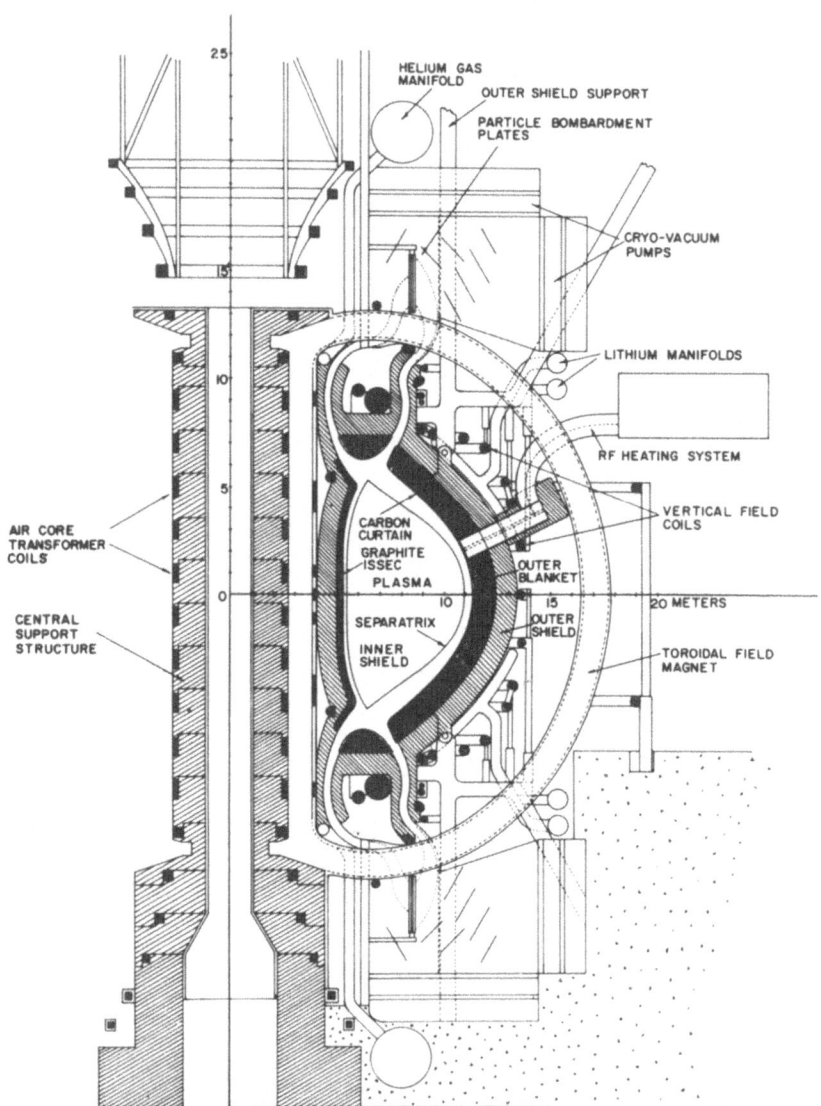

Figure 22

View of Right Half Cross-Section of Nuclear Island
in UWMAK-III

Table 2

MAJOR CHARACTERISTICS AND PARAMETERS OF UWMAK-III

Structural Material	Mo Alloy, TZM
Coolant:	
Inner Blanket	He
Outer Blanket	Lithium
Fuel Cycle	(D-T), Li
No. Toloidal Field Magnets	18
Magnet Superconductor	NbTi
Magnet Structural Material	Al Alloy, 2219
Maximum Magnetic Field	8.75 T
On Axis Magnetic Field	4.05T
Plasma Dimensions:	
Major Radius	8.1m
Half Width	2.7m
Height to Width Ratio	2
Plasma Shape	"Triangular D"
Plasma Current	15.8 Ma
Impurity Control Method	Diverter + Low Z Liner
Plasma Heating Method	RF (Fast Wave - 60 MHz)
Burn Time	1800 sec
Duty Factor	0.947
Breeding Ratio	1.25
Average Neutron Wall Loading	2.5 MW/m^2
Power Cycle:	
Blanket	Closed Cycle He
Diverter	Na-Steam
Power Output During Burn	5,000 MW(th)
Average Electrical Output	1985 MW_e
Net Plant Efficiency	41.9%

An overall view of the UWMAK-III reactor is shown in Figure 21, and a detailed cross-section view of the nuclear island is given in Figure 22. A summary of the major parameters and characteristics of the design is given in Table 2. The reactor is designed to generate 5,000 MW_t during the deuterium-tritium plasma burn, and 4,735 MW_t continuously. The net electrical output is 1,985 MW_e. The design has a

number of unique features, one of which is the noncircular
plasma shape produced by discrete coils outside the blanket
and shield zones. The height to width ratio of the plasma
is approximately 2, and the shape factor, the ratio of the
circumference to the inscribed circle, is 1.6. The plasma
is bounded by a separatrix with two null points symmetrically
located above and below the horizontal midplane that provides
for a poloidal diverter.

The reasons for choosing a noncircular cross-section are
to take advantage of the potential for higher total beta and
therefore to increase the power density in the plasma without
substantially increasing the toroidal magnetic field (40).
The plasma current is 15.8 MA and q near the separatrix is
above 3. This appears reasonable for future machines and
detailed MHD calculations of plasma equilibrium and stability
have been carried out to determine the final plasma shape.

A second major characteristic of UWMAK-III is its substantially reduced size compared with the earlier studies
UWMAK-I (1) and UWMAK-II (5), which have circular plasmas
with a major radius of 13m. A comparison of the plasma shape
and the toroidal field magnet size for UWMAK-II and UWMAK-III
is shown in Figure 23. The size reduction is made possible
by the design decision to increase the 14 MeV neutron wall
loading from ~ 1 MW/m^2 previously, to approximately 2.5 MW/m^2
in UWMAK-III, and by the higher value of power density in the
noncircular plasma. The higher value of neutron wall loading
was, in turn, made possible by the unique blanket design
based on the concept of using graphite as a neutron spectrum
shaper to protect the first structural wall from radiation
damage.(56)

A unique feature of the UWMAK-III is the use of RF heating to raise the plasma to ignition conditions. The purpose
is to assess the potential of RF heating in reactor systems
since previous studies of Tokamak reactors have employed
neutral beam injection heating (1,5). A study has been
made of several supplementary RF heating wave modes including
fast magnetosonic waves at ω_{CD}, $2\omega_{CD}$ (ω_{CD} is the deuteron
ion cyclotron frequency) and the two-ion hybrid resonance,
lower hybrid waves, and low-frequency Alfven waves. Because
of the potential for generating large amounts of RF power,
accessibility of the wave and energy deposition in the plasma

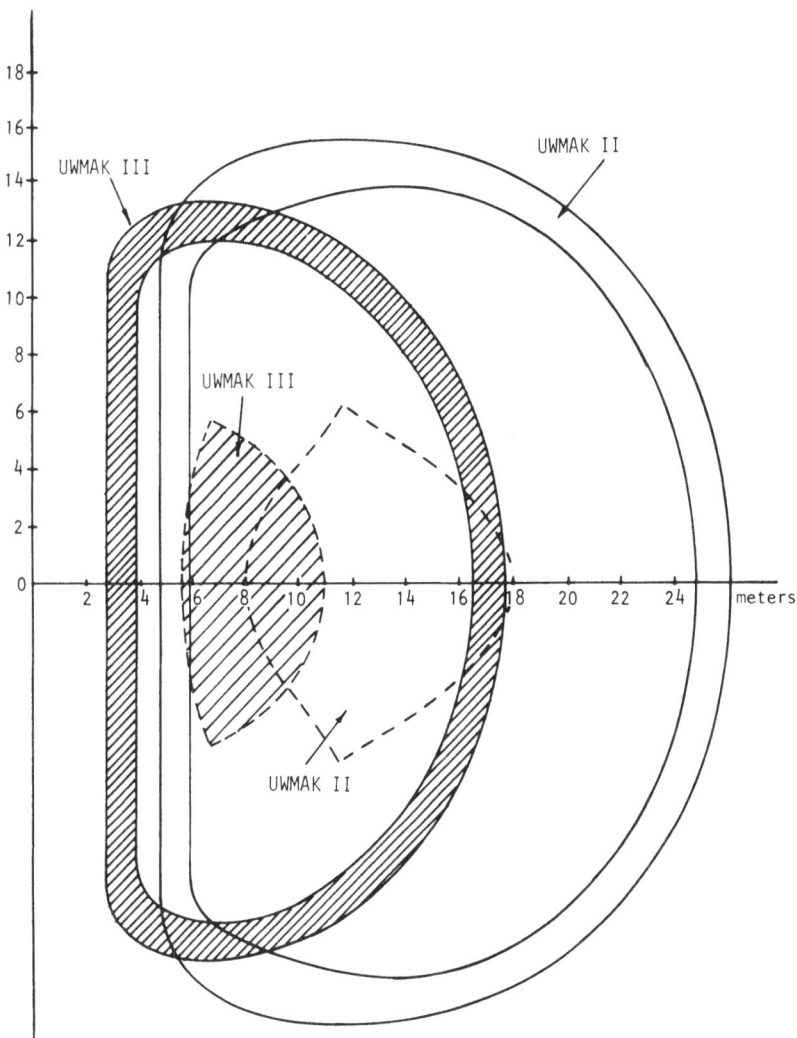

Figure 23

Comparison of the Plasma and Toroidal Magnet Size
In UWMAK-II[5] and UWMAK-III

core, and direct heating of the ions, we have concentrated
our studies on the fast magnetosonic wave.

Another unusual feature in this work is the magnet
design which uses NbTi superconductor with Al as the stabilizer and a high-strength Al alloy as the structural material.
The use of high purity aluminum as the stabilizer allows
one to reduce the amount of stabilizer, compared to the
use of Cu, by a factor of between 2 and 5.   Alternately,
because the resistance of Al is lower than that for Cu, the
heat generated because of I^2R losses is less when the same
amount of material is used. All this makes for a better
stabilizer. In addition, experiments at Wisconsin[57] have
shown that when reinforced as it is in a magnet, high-purity
Al does not develop a high resistivity when stressed beyond
its yield point. The use of an Al alloy provides a structure
of almost equal strength to the use of 316 stainless steel at
about one-third the weight. There will now be a compatible
contraction of both the conductor and the structure on
cooldown.

The blanket and shield design is another unique feature
that can be most readily understood by examining the schematic
illustrated in Figure 24. The inner blanket and shield closest to the torus centerline is *not* designed to breed tritium,
but only to extract heat and protect the superconducting
toroidal field magnet. As such, it really constitutes a
shield that is divided into hot and cold zones. Greater than
99% of the energy incident on the inner shield is deposited
in the hot zone composed of structure (TZM), boron carbide
(B_4C), and helium coolant. The cold inner shield uses 316

stainless steel as the structure. In front of the first wall
of the inner shield is a 25 cm graphite zone that is passive
(not actively cooled) and that serves to soften the neutron
spectrum incident on the first structural wall. This, in
turn, reduces the radiation damage to the structure, and the
inner hot shield is designed for plant life.

The outer blanket is cooled with liquid lithium which
also acts as the breeding material. The overall breeding
ratio is 1.25. The structural material in the outer blanket
is not protected neutronically and therefore will have to be
replaced periodically. The wall life is estimated to be
about two years; however, the outer portion of a torus is
located where space is plentiful, so the difficulties involved
in blanket module replacement are at least minimized.

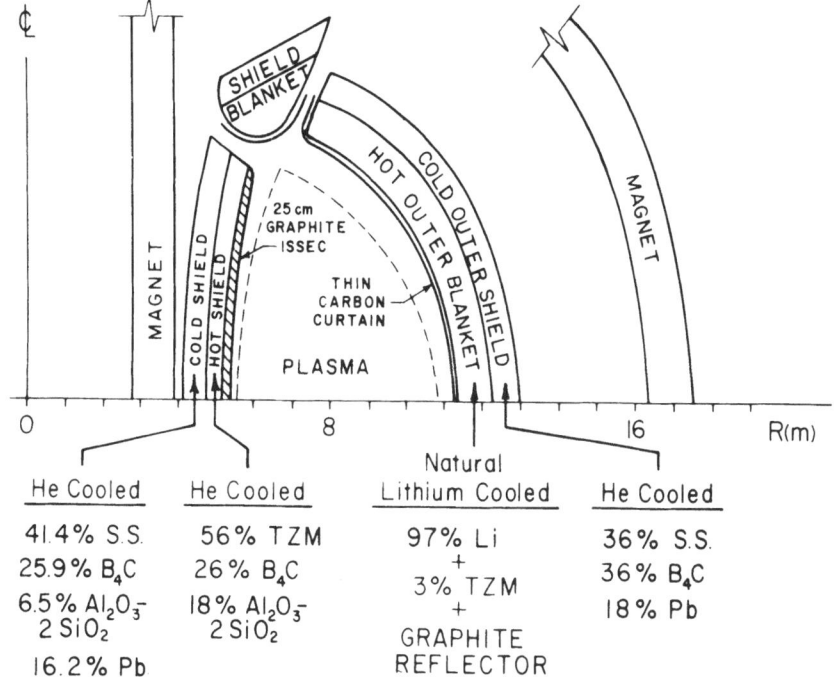

Figure 24

Schematic of the Blanket and Shield Design for UWMAK-III

The use of a refractory metal alloy as the structural material was motivated primarily by the desire to increase the operating temperature of the coolant to increase the overall efficiency of the power cycle. The choice of the molybdenum-based alloy, TZM, over alloys of vanadium or niobium was based on the good high-temperature creep strength and fatigue life of TZM, its compatibility with both helium and liquid lithium, its low permeation to tritium, its good neutronics properties, its relatively low cost, and the good resource position of the United States with respect to Mo. The problem associated with joining TZM is recognized and examined in detail later on.

A final feature of the UWMAK-III design is the use of
a helium power cycle. The coolant from the inner blanket
goes directly to a helium turbine, while the lithium from
the outer blanket goes to an intermediate, Li-Na heat exchanger, and the sodium subsequently goes to an Na-He heat
exchanger from which the He is used to drive the turbines.
Sodium from the intermediate loop plays the dual role of
isolating the tritium-containing-lithium from the power
cycle and providing a working fluid for thermal energy
storage that can be used during the down part of a tokamak
plasma cycle to keep a continuous hot helium flow to the
turbines.

B. Analysis of the Plasma

Noncircular plasma cross-sections offer the potential for
achieving higher values of total β (the ratio of plasma to
magnetic field pressure) compared with circular plasmas of
equal poloidal beta, β_θ, and safety factor, q. The scaling
is:

$$\beta = \beta_\theta \left(\frac{S}{qA}\right)^2$$

where A is the aspect ratio and S is the plasma shape factor.
For UWMAK-III, A equals 3 and S equals 1.6. Of course, the
potential benefits of noncircular tokamaks will be realized
only if an equilibrium plasma can be made that is stable at
values of q and β_θ similar to those in circular machines.
If, for example, q must be increased to maintain stability
in noncircular systems, the desired increase in β will not
come about; therefore, for UWMAK-III, a detailed MHD equilibrium and stability study has been carried out to determine
the proper plasma shape[58]. A detailed magnetic flux plot
is shown in Figure 25. The code used to perform this analysis was developed by the MHD group at Princeton.[59] A
list of the characteristic parameters that go along with
this plasma design is given in Table 3. The plasma has
been satisfactorily tested for stability against rigid displacements, general kink modes and localized interchange
modes. The large triangularity of the plasma shape is the
result of forcing the vacuum vertical field to have good
curvature in the region to the right of the magnetic axis
at R_m = 9.0 m.

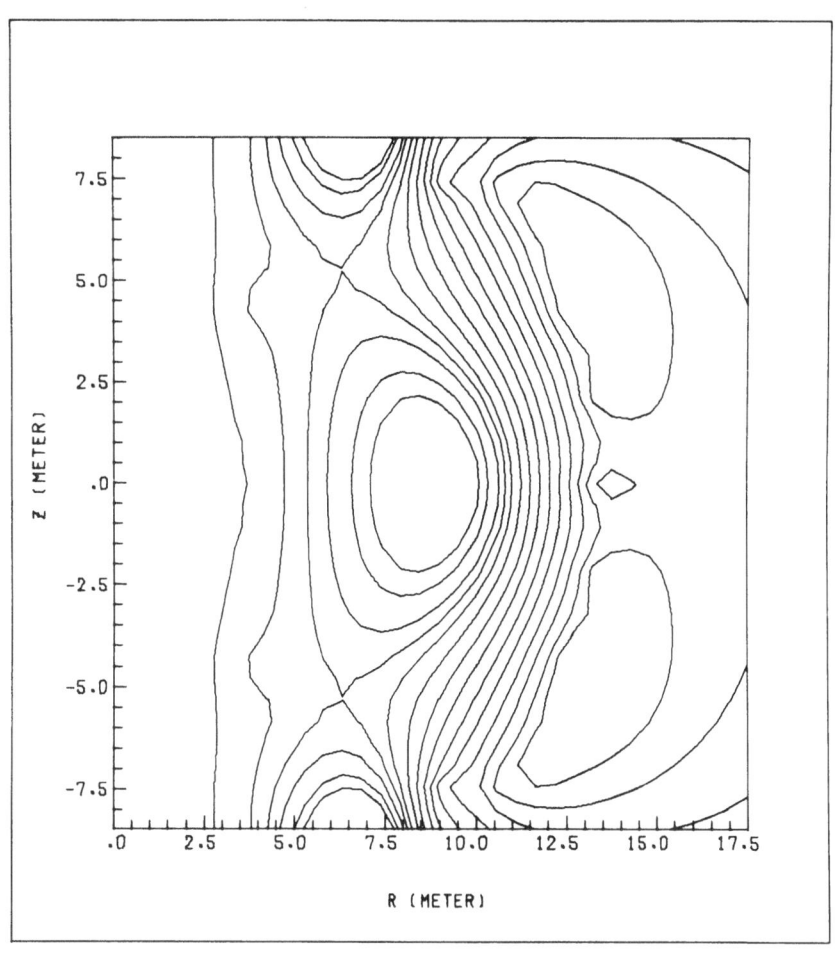

Figure 25

Contour plot of ψ-surfaces from MHD Equilibrium Calculations
for the D-shaped Plasma in UWMAK-III.   The two null points
are symmetric about the plasma midplane at approximately
R equal to 6.4 m.

Table 3

Plasma Parameters of UWMAK-III from MHD Analysis

R_o = 8.1 m I_p = 15.8 MA

B_o = 4.05 T $\overline{\beta}_\theta$ = 2.2

R_{in} = 3.75 m   q(a) $\simeq$ 3.5

B_{max} = 8.75 T q_o $\simeq$ 1.0

R_m = 9.0 m   a = 2.7 m

R_j = 10.0 m   b $\simeq$ 5.4 m

Δ_m = 0.8 m   S $\simeq$ 1.6

Δ_j = 1.2 m   b/a = 2.0

The design of the transformer to drive the plasma current during startup and burn is based on an air core transformer with superconducting windings. The plasma current is assumed to rise linearly in 15 seconds to the design value of 15.8 MA. The vertical field coils, or diverter coils, provide approximately 96% of the necessary startup flux, with the transformer coils providing the remaining amount. A list of all transformer and diverter coils is given in Table 4, together with their current carrying capabilities. The total flux swing during the current rise is 254 V-s.

The flux swing required for the burn phase was determined using a space-dependent transport model of plasma behavior. The complete reactor burn cycle is listed in Table 5. Using a general expression for the plasma conductivity, including trapping and impurity effects, 107 V-s is necessary to drive the plasma current for 1800 s.

Table 4

TRANSFORMER PROGRAMMING
(All Currents in MA)

Coil No.	R(m)	±Z(m)*	Initial Current	Current (End of Startup)	Current (End of Plasma Burn)
T1	2.502	1.0	4.0528	3.5463	-5.1528
T2	2.508	3.0	4.0376	3.5330	-5.1336
T3	2.522	5.0	3.9900	3.4901	-5.0956
T4	2.546	7.0	3.9168	3.4273	-4.9798
T5	2.591	9.0	3.8034	3.3281	-4.8356
T6	2.675	10.997	3.5917	3.1428	-4.5665
T7	2.854	12.989	3.2211	2.8185	-4.0953
T8	3.227	14.950	2.5632	2.2429	-3.2590
T9	3.857	16.844	1.7693	1.5482	-2.2497
T10	4.725	18.643	1.1418	0.9991	-1.4516
T11	5.489	19.932	0.4159	0.3639	-0.5287
T12	19.0	5.0	0.1588	0.1390	-0.2019
T13	9.0	10.5	0.1512	0.1323	-0.1923

Diverter

Coil No.	R	±Z(m)*	I(MA)
D0	6.8	9.0	15.50
D1	3.9	0.0	.40
D2	3.9	1.0	.40
D3	4.0	4.5	2.00
D4	4.6	5.5	-0.60
D5	3.9	9.0	-1.20
D6	5.9	9.5	-2.20
D7	6.8	13.0	-1.30
D8	8.3	11.7	-4.30
D9	8.5	9.9	-3.00
D10	8.8	10.0	-1.53
D11	8.8	9.3	-1.65
D12	8.8	9.9	-1.10
D13	8.9	7.7	-1.60
D14	9.3	7.5	-1.62
D15	10.8	7.3	-1.55
D16	11.8	6.8	-1.60
D17	12.8	5.3	-1.65
D18	13.0	4.3	-1.65
D19	13.5	2.3	-1.52
D20	13.5	0.0	-1.52

* For each coil at location (R,Z), there is a similar coil at (R-Z) for Z = 0.

Table 5

UWMAK-III Reactor Cycle

Time (Sec.)   Phase

0-15   Startup: Plasma and
                                     diverter currents rise
                                     to full value; transformer
                                     currents begin to drop.

15-30   RF Heating: Plasma and
                                     diverter currents at full
                                     value; transformer currents
                                     continue to drop.

30-1830   Burn: Transformer currents
                                     drop to maximum negative
                                     value.

1830-1850   Shutdown: Plasma and diver-
                                     ter currents drop to zero;
                                     transformer currents rise
                                     to provide negative startup
                                     flux.

1850-1890   Pumpout and Recharge: First
                                     20 seconds, hydrogen gas
                                     pumped into cool plasma,
                                     residual gas pumped out.
                                     Final 20 seconds, trans-
                                     former currents reset to
                                     initial values.

1890-1900   Final Pumpout and Refill:
                                     Chamber purged and refilled
                                     with fresh fuel.

The power requirements during the 15s current rise
time are met by a combination of power from the grid and
from an energy storage system. With UWMAK-III, we propose
to take 500 MW off the grid for 15s. The remaining 413 MW
that are required are provided from an inductive energy
storage unit similar in design to that used in UWMAK-II.
The only other large power demand is for the RF heating
system and that follows the current rise phase. As such,
this requirement is not added to the 913 MW peak load.

The plasma reaches a temperature of only 1.8 keV during
the ohmic heating phase; thus, auxiliary heating is required
to ignite the plasma and bring it to its steady state operating condition during the burn period. We have examined
RF heating for this purpose and have chosen to concentrate
on the fast magnetosonic wave at $\omega = 2\ \omega_{CD}$. *(60)* Analysis
shows that the ion cyclotron harmonic absorption term occurs
in a vertical band of finite width, Δ, centered at the plasma
major radius. The finite width is less than the plasma radius
and is due to thermal effects and the 1/R variation of the
toroidal field. The energy deposited in this slab is then
uniformly distributed over the flux surfaces due to the rotational transform. The expression for the ion heating
profile is directly proportional to R/r and is therefore
strongly peaked at the plasma center.

The existence of toroidal eigenmodes *(61)* also has a
strong effect. In smaller devices, these eigenmodes are
well separated. However, we have found that in large reactor
systems like UWMAK-III, the density of toroidal eigenmodes in
frequency domain becomes a near continuum *(60)*. This means
the eigenmodes will always be available to heat the plasma,
but the cumulative effect of all excited modes on the loading
of the launching structure and heating efficiency is a subject
requiring further research.

The theoretical expression for plasma ion heating by the
fast magnetosonic wave is used, together with a global model
of plasma behavior *(13,5)* to analyze the heatup rate. At the
end of the 15 s ohmic heating phase, 40 MW of RF wave heating
at 60 MHz is turned on; Figure 26 shows the ion and electron
temperatures during this period. After 8 s, the plasma
reaches ignition, and if the RF power is left on for 12 s,
the plasma reaches the equilibrium temperature of $T_i \simeq 19$ keV,
$T_e \simeq 23$ keV after 15 s.

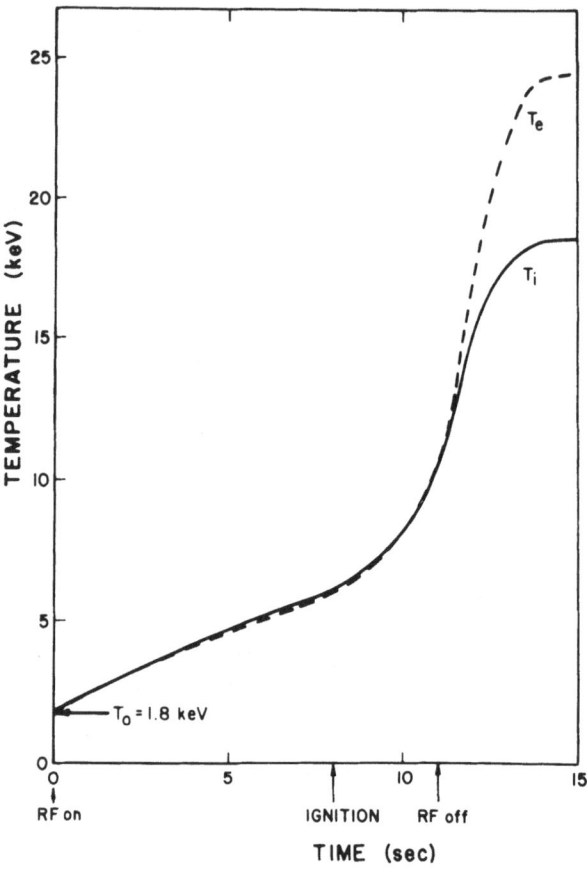

Figure 26

Magnetosonic Wave Heating to Reactor Operating Conditions
P_{RF}= 40 MW

An alternate case was considered in which 200 MW is used.
In this case, ignition occurs after 1.32 and the wave heating
is turned off 0.02 s later. The ions in this case are driven
to about 450 eV higher temperature than the electrons. When
the supplementary heating is extinguished, the alpha heating
is large enough that the plasma continues to heat, and reaches
equilibrium in 15 s. In this fast heatup case, it is important that the ions are heated sufficiently more than the
electrons, so that when the auxiliary heating is turned off,
the initial small drop in ion temperature does not cause the

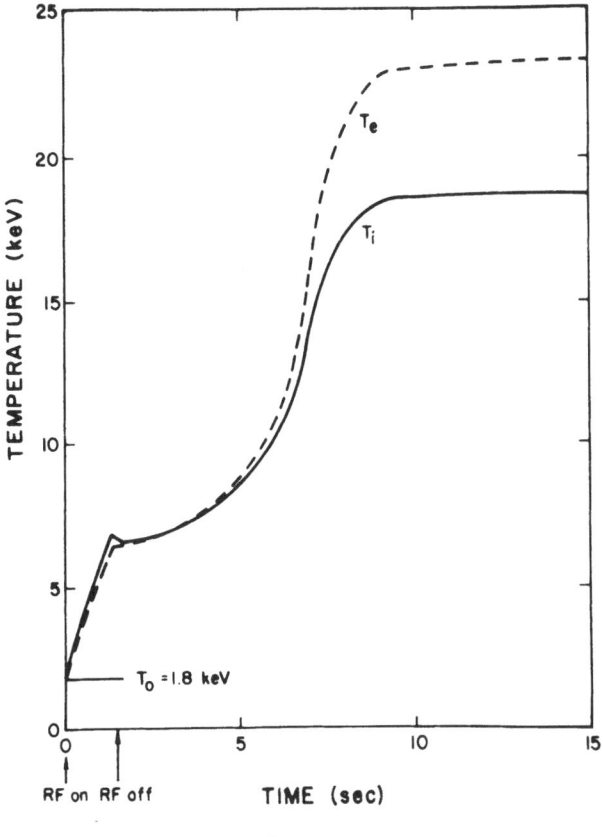

Figure 27

Magnetosonic Wave Heating to Reactor Operating Conditions
P_{RF}=200 MW

plasma to become unignited. This scenario is shown in
Figure 27.

   An important problem with RF heating in addition to
adequate coupling of the waves to the plasma is the design
of a suitable launching structure located on the torus wall.
It must be capable of handling 200 MW of power at 60 MHz for
a startup period of up to 15 seconds. It must have the
property that the excited RF fields have a toroidal magnetic
field and an associated orthogonal electric field with the
orientation shown in Figure 28a. The launching structure must
also be designed for the excitation of modes that lead to
energy deposition near the plasma core.

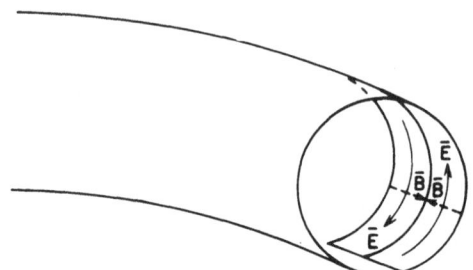

a) IDEALIZED MAGNETOSONIC WAVE FIELD SOURCE
   DISTRIBUTION :
   $k_{\parallel}$ = 6-10 m^{-1} AND DOMINANT m = 0,±1,±2 EXCITATION

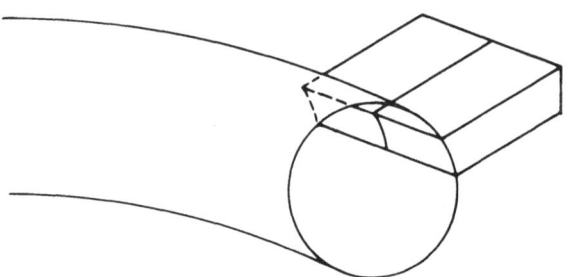

b) EXCITATION BY MEANS OF STANDARD WAVEGUIDES

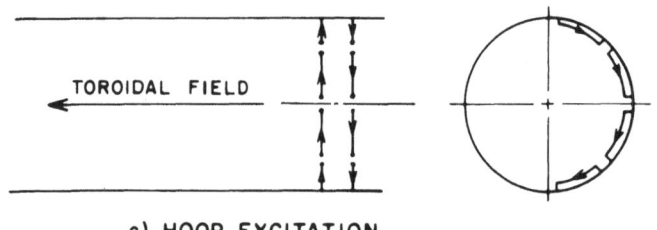

c) HOOP EXCITATION

Figure 28

Considerations for Magnetosonic Wave
Heating Launching Structure

An ideal system would have a spectrum that excites lower order poloidal m values. Higher order m values cause the finite ion gyroradius heating to peak further toward the plasma edge, leading to edge heating rather than core heating. In order to excite only low m numbers, the source distribution would have to cover most of the poloidal circumference. A further problem is caused by the fact that for large tokamaks such as UWMAK-III, the minor circumference is large, compared to a free-space wavelength. This introduces phase shifts of a wave traveling in the poloidal direction, which can contribute to a higher-order poloidal spectrum and reduce the efficiency. This may be overcome by introducing external circuit phase shifts that correct for the natural phase shift of the wave as it travels in the poloidal sense.

A standard rectangular waveguide launcher, flush mounted in the wall as shown in Figure 28b introduces new, special excitation problems, since its toroidal dimension must be larger than half a free space wavelength for propagation of the dominant TE_{10} mode. This can be overcome by using smaller waveguides, possibly filled with the high, dielectric ceramic that reduces the dimension of the guide and excites larger k values. If a small excursion of conducting material into the torus can be tolerated, a strip line located over ceramic dielectric inside a rectangular guide could be envisioned that does not have a cutoff limitation due to small aperture size. The conducting strip would have to be coupled to the torus by making a small loop that is connected electrically to the bottom of the guide.

The constraints on the design of the RF power system are as follows:[60] It is assumed that up to 200 MW are provided by as many as 12 separate feed systems placed in pairs located toroidally around the devices with a minimum cross-sectional area of the feed, so that the field strengths remain below the breakdown level. While the magnitude of electric field strengths for 60 MHz breakdown in a tokamak environment is not precisely known, several estimates are indicated in the literature. In UWMAK-III, we adopted the criterion that the RF field should be less than 3 kV/cm in the RF feeds during startup. It is preferable to have the startup neutral pressure outside the plasma confinement region below 10^{-5} torr. However, 10^{-4} torr is all that can be accommodated, given the large device size, pumping capacity and limited pumping area.

The design of the waveguide feed is obtained subject to
the following conditions: (1) each element of the launching
structure must be capable of supplying 17 MW of power with a
maximum electric field strength of 3 kV/cm; (2) the ratio of
wave frequency to cutoff frequency of the feed must be greater
than 1.2 in order that the attenuation in the guide due to
skin effect is not too large. If we require that the feed
be a vacuum region, then one can consider two types. In all
cases, due to a lower cost of fabrication, it is advisable
to consider a copper coaxial line running most of the distance from the RF source to the torus, with a 1 m diameter.
A coaxial-to-waveguide transition could be made near the
torus wall.

If the neutron and plasma flux to the launching aperture
during startup dictate a vacuum-filled waveguide, the following designs can be considered. A rectangular guide with a
ratio of wave frequency to cutoff frequency of $f/f_c = 1.2$
for the TE_{10} mode. This would require a toroidal inner
dimension of 3 m and a poloidal dimension of 1.5 m as shown
in Figure 29a. A waveguide of this size with the imposed
maximum field intensity of 3 kV/cm would have a theoretical
power injection energy capability of 500 MW and a wave impedance of 190 ohms.

For the design used in UWMAK-III, a ridged waveguide
has been considered in order to reduce the toroidal extent
of the guide for improved wave penetration and heating. This
reduces the mode conversion potential and provides a means
of determining the fraction of wave power deposited initially
in the ions versus the electrons. The H-shaped guide would
have the inner dimensions of 1 m high by 1 m wide, with a
gap width of 8.7 cm, which is 50 cm in extent as shown in
Figure 29b. The wave impedance of the dominant TE_{10} mode is
70 ohms and the theoretical power-carrying capability is 30 MW.

If neutron and plasma heating of the guide aperture during startup are not too severe, a guide that is partially or
fully filled with a dielectric medium can be considered.
Alumina ceramic (Al_2O_3) has a relative dielectric constant
of $\varepsilon_r = 8$, a loss tangent of 10^{-4}, and a maximum temperature
of operation of 1500-1900°C. A ridged waveguide that is filled
with ceramic, 50 cm wide by 50 cm high, with a 10 cm gap can

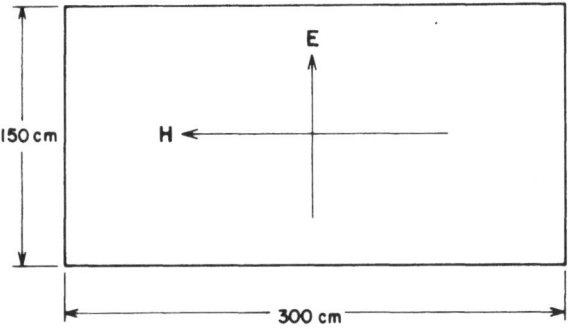

a) STANDARD RECTANGULAR WAVEGUIDE

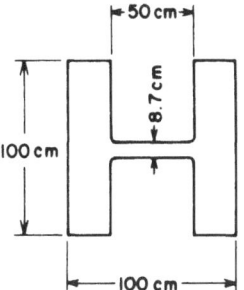

b) RIDGED WAVEGUIDE

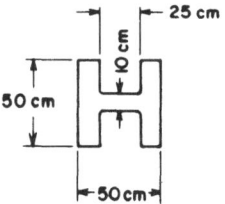

c) RIDGED WAVEGUIDE WITH CERAMIC ($\epsilon_r = 8$) FILLED DIELECTRIC

Figure 29

Size Comparison of Te_{10} Waveguide Feeds for 60 MH_3RF

be designed as shown in Figure 29c. The wave impedance is
40 ohms, and a theoretical power transfer of 60 MW can be
obtained.

It is evident from this discussion that several alternatives for the design of magnetosonic wave-launching structures exist. Much further analytical work in relating a given
wave source distribution on the torus wall to the heating
deposition profiles in the plasma is required; however, with
the analysis that has been performed to date, the recommen-
dations were as follows: [60] From a purely plasma point of
view, a wave source distribution is proposed that goes half-
way around the minor cross-section on the outside edge. A
second source, located very close to the first, but shifted
in phase by 180° is proposed in order to set up short wave
lengths $\lambda = 1$ m in the toroidal sense. While the above is
optimum for plasma physics consideration, it presents obvious
problems from the viewpoint of vertical field coil placements
and access. The added complexity led to the design approach
shown in Figure 30. This design was used as the basis for
locating the equilibrium vertical field coils used in the
MHD analysis described earlier.

The equilibrium plasma conditions during the plasma
burn for UWMAK-III have been studied by using both a global
plasma energy balance model and a more sophisticated space-
time tokamak code. [62,63] In both cases, thermal equilibria
are found that produce 5,000 MW_{th} of power. In the latter

case, however, the average ion and electron temperatures, the
plasma β_θ, and the particle confinement time are found to be
significantly lower than those predicted by the global model
because the temperature and density profiles are sharply
peaked at the axis of the torus.

The plasma and magnetic field parameters are listed in
Tables 6 and 7. The energy confinement time is 1.6 sec.,
while the particle confinement time is 0.55 sec. The fractional burnup is a very low 0.83% and the fueling rate for
D and T atoms is 1.43×10^{21}/sec. This low fractional burnup
and correspondingly high fueling rate result because the
diffusion of particles from the plasma is governed by the
diffusion coefficient at the plasma edge. The edge condition
associated with the use of a diverter implies a very low
density and therefore a quite collisionless plasma. In turn,
this leads to a large diffusion coefficient assuming micro-

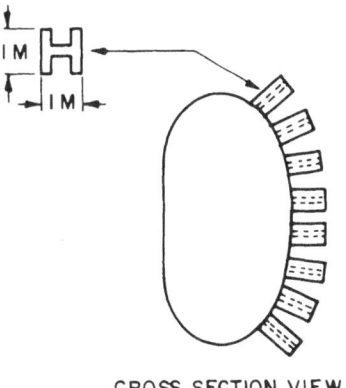

CROSS SECTION VIEW

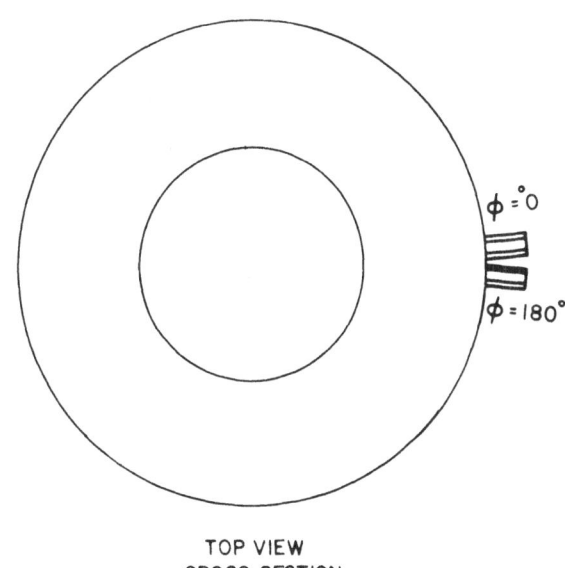

TOP VIEW
CROSS SECTION

Figure 30

Schematic of a Conceptual Magnetosonic Wave Launching System

Table 6

Plasma Parameters for UWMAK-III

	Space-Time	Point Model
Mean Ion Temperature (keV)	11.4	18.4
Mean Electron Temp. (keV)	11.9	22.9
Mean Ion Density (cm^{-3})	7.9×10^{13}	6.46×10^{13}
Mean Alpha Density (cm^{-3})	8.8×10^{11}	1.9×10^{12}
Plasma Current (MA)	15.6	15.7
Stability Factor at Plasma Edge, $q(a)$	2.7	2.69
Plasma Width (m)	2.7	2.7
Plasma Height (m)	5.4	5.4
Radius of Equivalent Volume Circular Plasma (m)	3.83	---
Major Radius (m)	8.1	8.1
Axial Toroidal Magnetic Field (kG)	40.0	40.0
Plasma Volume (m^3)	2370	2360
Electron Poloidal Beta, β_θ^e	0.69	1.05
Total Poloidal Beta, β_θ	1.65	2.3
Total Toroidal Beta, β_ϕ	.058	.083
Energy Content of Plasma (GJ)	1.32	1.63
Wall Surface Area (m^2)	1600	1600
Particle Confinement Time (s)	0.547	3.33
Energy Confinement Time (s)	1.64	1.66
$\bar{n}_e \tau_E$	1.33×10^{14}	1.14×10^{14}
Fractional burnup, f_b	0.83%	5.9%
Voltage Around Torus (volts)	0.059	

Table 7

Power Parameters

Energy per Fusion (MeV)	21.7
Thermal Power during Burn (MW)	5000
Average Power Density of Plasma (MW/m^3)	1.72
14 MeV Neutron Production Rate (s^{-1})	1.44×10^{21}
Power to Wall (MW)	71.3
Radiation (MW)	63.2
Particles (MW)	8.1
Power Wall loading (MW/m^2)	0.44
Power to Diverter (MW)	725

Fueling Parameters

Tritium (deuterium) Consumption Rate	$1.94 \times 10^{21}/s$
Tritium Burnup Rate	.62 kg/day
Deuterium Burnup Rate	.413 kg/day
Particle Leakage Rate (D+T+α)	$3.44 \times 10^{23}/s$
Fueling Rate:	
Deuterium	49.8 kg/day
Tritium	74.7 kg/day

instability scaling. The low value of τ_p in the space-dependent model means the amount of tritium in the refueling cycle will be much larger than heretofore expected on the basis of global model calculations alone. In this case, the fueling rate for tritium is 74.7 kg/day. A plot of the density and temperature profiles during the plasma burn is shown in Figure 31.

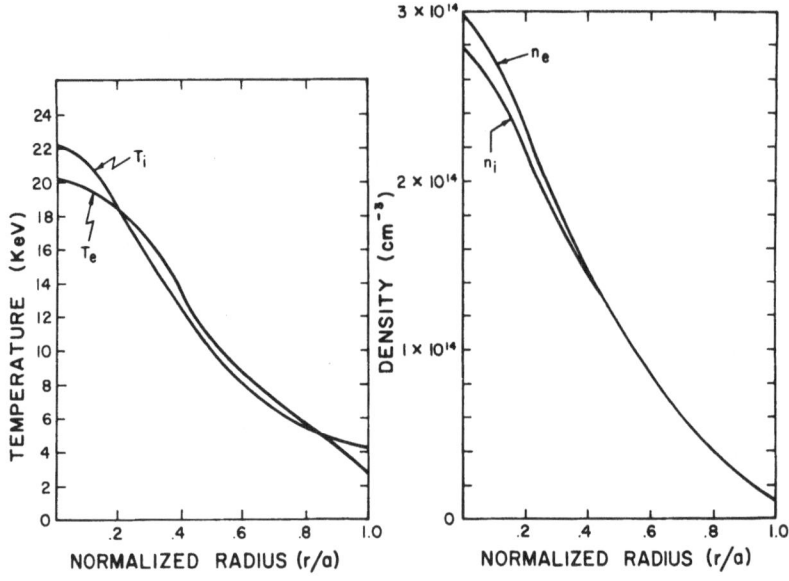

Figure 31

Spatial Profiles of Density and Temperature Calculated for
The UWMAK-III Plasma During the Burn Period

C.   Impurity Control and Vacuum Pumping

        Two forms of impurity control are included in the
UWMAK-III design: a magnetic diverter and a low Z liner.
The magnetic diverter is of the axisymmetric, poloidal double
null type such as we have used previously (1,5) and Figure 32
shows the location of the various coils, together with the
location of the plate onto which the plasma particles deposit
their energy.   The vacuum pumps and the baffle system are
also shown.

        The low Z liner consists of a thick graphite zone (25 cm)
on the inner portion of the chamber nearest the centerline,
and a thin carbon curtain,(64) as described in UWMAK-II(5)
mounted on the outer blanket first wall.   The operation of
the carbon curtain is similar to that previously described,

and the temperature during the plasma burn is 1300°C. As such, physical sputtering is expected to dominate and this is on the order of 0.01 (55). The thick graphite zone on the inner side of the chamber will operate with a much higher surface temperature in the neighborhood of 1800°C. Again, physical sputtering should be dominant, as estimates indicate the formation of acetylene only at somewhat higher temperatures ($\gtrsim$ 2000°C).

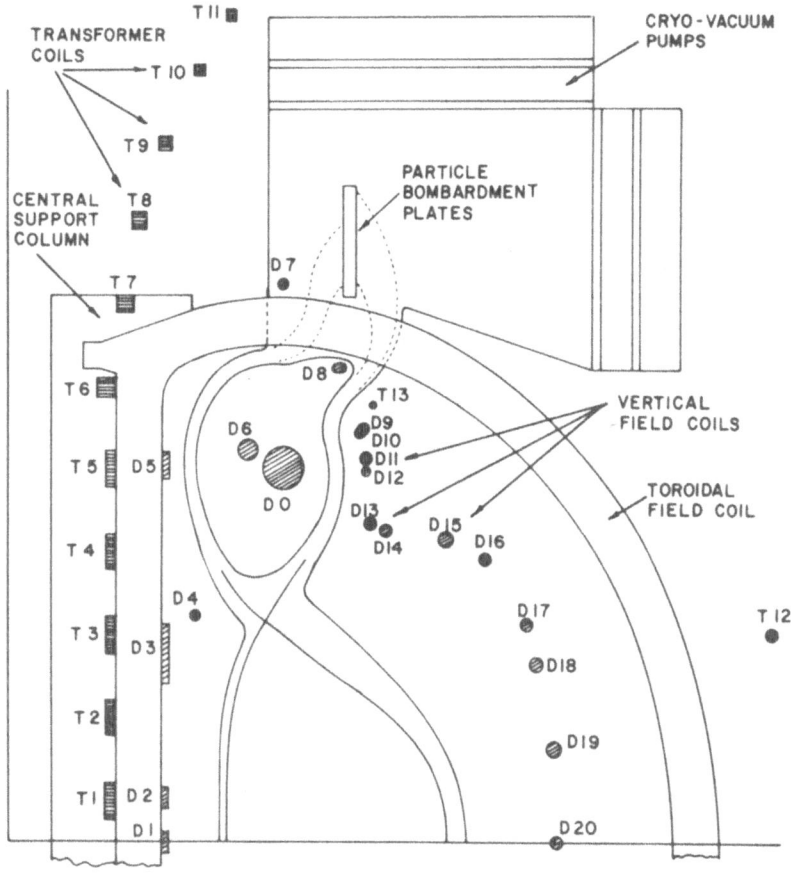

Figure 32

Transformer and Diverter (Vertical Field) Coil Locations
and Diverter Design

The effectiveness of the diverter in reducing the
charged particle flux to the wall has been estimated,
using the model previously described.[65] The plasma
density profile in the scrape-off region is determined by
cross-field diffusion and by plasma flow along the diverted
field lines to the particle bombardment plates. In the outer
diverter the plasma must pass through a region of higher B on
its way to the collector so magnetic mirroring can be important; however, the plasma would have a loss-cone distribution
and be subject to all the usual microinstabilities associated
with mirror machines. This has been reviewed in Reference 66.
As such, we have taken the diverter efficiency to be $\gtrsim$~99%
for collecting particles diffusing from the plasma.

Further, in both space-dependent and space-independent
calculations, the plasma is assumed to be essentially free
of impurities. This follows simply from the assumption that
in the presence of turbulence, high Z impurities will be
transported out of the plasma with the same diffusion coefficient as the plasma itself. With this assumption, one
can set the flux of high Z impurities leaving the plasma
equal to the impurity reflux:

$$-D\frac{\partial n_z}{\partial r} = -S(1-\eta)D\frac{\partial n_{DT}}{\partial r} + \frac{a\bar{S}}{2}n_{DT}^2 <\sigma v>$$
$$+ Sn_{DT}\dot{}<\sigma v>_{cx}\lambda_{cx}$$

(43)

where D is the diffusion coefficient, n_z is the density of
high Z impurities, n_{DT} is the density of plasma ions, n_n
is the density of neutrals at the plasma edge, S is the D-T
sputtering yield (of order .01), η is the diverter unload
efficiency (of order .99), $<\sigma v>_{cx}$ is the charge exchange
cross-section at the plasma edge, λ_{cx} is the charge exchange
mean free path at the plasma edge, and $\bar{S}$ is the sputtering
coefficient for neutrons ($\sim 10^{-5}$). The first term on the
right represents sputtering by plasma ions diffusing out
of the plasma core and it is the dominant term. The second
and third terms represent neutron and charge exchange neutral
sputtering, respectively. By setting

$$\frac{\partial n}{\partial r} \propto \frac{n}{a},$$ one finds

$$\frac{\bar{n}_z}{\bar{n}_{DT}} \simeq S(1-\eta) \simeq 10^{-3} \text{ to } 10^{-4} \tag{44}$$

Therefore, we have assumed that the plasma is essentially free of impurities.

The design of the diverter in UWMAK-III is unique in several ways; first, the particles bombard a thin, solid plate of TZM located *outside* the toroidal field magnets and the pumping and heat transfer problems are not coupled. Placing the bombardment plates outside the TF coils allows the field lines to be fanned out, thus reducing the power density of particles on the plates. Combining this with the relatively small flow conductance back through the slots to the plasma, one expects that there will be little backstreaming of neutrals and impurities.

The surfaces of the TF magnets will experience direct charged particle bombardment if the separatrix around the coils intercepts the surface or if the particles can cross-field drift (due to curvature and ∇B drifts) to field lines that intercept the surface. The latter effect is not considered to be important, since the particles can drift only a distance of the order of the gyroradius while traversing a distance along the field lines equal to the radius of curvature of field. The fields are strong (toroidal field = 4T, poloidal field ~1T) and the gyroradius is small (≤ 1 cm), so this effect can be ignored. Calculation of the location of separatrix requires extensive 3D field line computer calculations. This has not yet been done for UWMAK-III. Similar calculations for UWMAK-I[1] and UWMAK-II[5] showed the coils were protected except near the inner leg of the "D". This leads us to believe the method described will work for UWMAK-III. If the coils suffer some charged particle bombardment, one can place particle collectors on the surface of the magnets or at other strategic locations, to protect the magnets from erosion and heat input.

The location of the particle collector plates is also shown in Figure 32. This is based on the assumption that the coils are protected by the magnetic field against bombardment. This arrangement requires two collectors back-to-back to receive the flux from the inner and outer scrape-off zones. The particles from the inner scrape-off zone pass by the sup-

ports for the portion of the blanket and shield directly above
and below the plasma. These supports are guarded against
charged particle bombardment by two parallel currents. The
line currents create a separatrix and thereby divert the field
around the supports. The required current to guard the supports is 1.9 MA per support. Because of the location behind
the blanket and shield, this can be a superconducting current.

The particle collection plates in UWMAK-I and II consis-
ted of a flowing lithium film to absorb both the D,T particles
and the associated thermal energy. This design is not suitable
for UWMAK-III because of the very large thermal load carried
on the plates. The average thermal load in UWMAK-I was
1.0 MW/m^2. This load increased to 3 MW/m^2 in UWMAK-II, with
the peak load estimated to be around 30 MW/m^2. The lithium
temperature can vary only from 186°C (its M.P.) to 350°C
(the B.P. at the plasma pressure 10^{-5} torr). With this very
small coolant temperature rise available, and with the in-
tense thermal load, the heat transfer problem becomes almost
insurmountable.

The design philosophy for the particle collector plates
in UWMAK-III has been completely changed. The pumping and
heat transfer problems have been decoupled. The vacuum pumping capacity has been increased by more than one order of
magnitude to compensate for the loss of pumping by the lithium
film. The sole purpose of the collector plates now is to
slow down the particles and convert their kinetic energy to
thermal energy. A coolant is now required to transfer the
thermal energy to a power conversion system and sodium is
used.

The main structure of the collector plate will consist
of a bank of tubes that form the coolant passages. To avoid
excessive sputtering due to incident high-energy particles,
a thin sheet of TZM is used to cover the main plate. Good
thermal contact between the sheet and the plate is maintained
by means of a liquid lithium film that fills the void between
them. The main sheet of TZM acts as a sacrificial wall and
will be designed to allow frequent replacement. One of the
most important questions then, is how often does the sacrificial wall have to be replaced? Two schematic drawings of
the particle collector plate are shown in Figures 33 and 34.

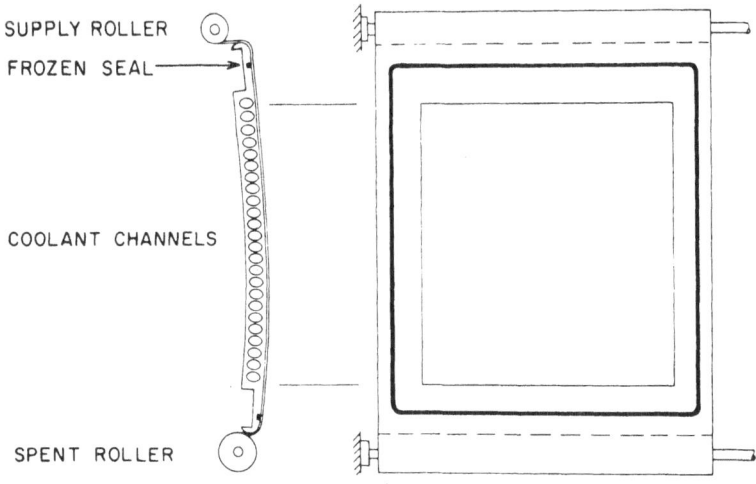

Figure 33

Schematic Drawing of Collector Plate Design for UWMAK-III

The location of the collector plate is outside the toroidal magnet as shown in Figure 32, at a major radius of 7 m. There are 36 collector plates, each 1.14 m wide and 3 m high, giving a total surface area of 123 m^2. The sacrificial wall is 0.1 mm thick, weighing a total of 125 Kg per replacement.

For the interaction of D,T particles with TZM, the sputtering yield is about .005. The particle flux is 5.2 x 10^{23}, so the erosion rate from the sacrificial wall is .041 cm^3/sec, or 1.490 kg/hr.

Since the sacrificial wall must be replaced frequently, a periodic automatic replacement process is envisaged that is shown schematically in Figures 33 and 34. The thin TZM sheet is fed from a supply roller and taken up on a spent roller. A frozen lithium seal is provided on the periphery of the plate to prevent lithium from leaking out. After a TZM sheet has been in service for 17 hours, the peripheral seal is melted and the sheet advanced until new material covers the plate. The seal is then allowed to solidify again. This process can be carried out during a burn cycle with no down time for the reactor.

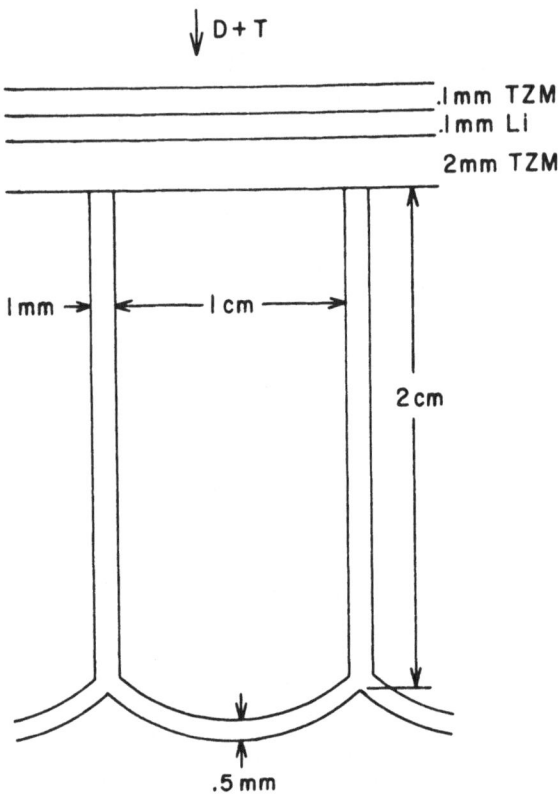

Figure 34

Cross-Section of Particle Collector Plate
In the Diverter of UWMAK-III

The TZM plate is cooled with sodium and a calculation of
the MHD effects on pressure drop from both Hartmann and end-
of-loop effects has been made. The total pressure drop is
100 psi and the sodium exit temperature is 600°C. This sodium
is then used to feed a steam generator system so the 687 MW$_t$

(average) associated with the diverter cycle can contribute to the electrical output. This is discussed in the section on the power cycle.

The particles coming off the TZM plate must be collected by the primary vacuum system; thus, the main function of the primary vacuum system is to remove the unburned fuel during the burn cycle while maintaining the plasma chamber surrounding the plasma confinement zone at a pressure of 1×10^{-4} torr or better. A secondary function, but equally as important, is to pump the chamber down to a relatively high vacuum between burn cycles in order to clean and condition the chamber surfaces.

A pumping scheme for a fusion reactor is complicated by the fact that twice as many pumps as needed in normal operation must be provided to allow one-half of them to be recycled while the other half is on line. This implies low conductance gate valves, heaters, backup pumps and a maze of ducts and vacuum lines. Nevertheless, a system such as UWMAK-III lends itself rather nicely to a simple design. We envisage that the cryo pumps will be rectangular in shape and will cover about 65% of the available space. The total pumping area will be $1600\,m^2$ with only one-half of it in use at one time.

The pumps will be arranged in even numbers, covering the horizontal and vertical space available in the vacuum plenum. Sliding plates will act as gate valves and will be sealed with conventional elastomer seals. The seals will be shielded from neutrons but will suffer damage from scattered gammas, and will have to be replaced periodically. Each sliding plate will serve a pair of pumps, one on-line and the other being regenerated. During changeover, the plate is moved to seal the saturated pump while exposing a freshly regenerated one to the chamber plenum. The back end of the pumps will have a similar gate plate arrangement operating in reverse of the front gate plate. The rear plate isolates the pump on line from the common pumpout plenum while opening the pump being regenerated to it. This scheme will provide essentially no pumping impedance, will be space saving, and requires no extrapolation of present-day vacuum technology. The refrigeration system requires 65,000 l/day of liquid helium and 555,000 l/day of liquid nitrogen. The power required is, however, relatively small, 7.3 kW at 4.2 K and 1 MW at 77 K. The combined power needed at 300 K is 6.2 MW.

D.   Magnet Systems

  1.   Toroidal Field Coils

        The toroidal field coils have a modified constant
tension "D" shape, corrected for the field variation due to
the finite number (18) of coils. The embedded conductor in
a solid "D" shaped alloy aluminum former is used to minimize
the material requirements while providing maximum strength
and stiffness. This construction was described in detail in
UWMAK-I report (1) and the cold central structure and secondary vacuum wall used in the UWMAK-II (5) design are retained.
The superconductor is TiNb superconducting filaments processed
in an OFHC copper matrix for convenience, but stabilized with
high-purity aluminum in the case of the toroidal field magnets. The coils are fully cryogenically stabilized and produce a maximum field of 8.75 T and an axial field at R =
8.1 m of 4.05 T. A detailed description is given by Boom,
Moses and Young. [67]

The vertically-elongated plasma and a new blanket removal scheme, coupled with only 18 toroidal field magnets
instead of the 24 used in UWMAK-II permits the use of much
smaller TF magnets. The inside surface of the outer leg is
at 16.5 m, as shown in Figure 35. The shape of the constant
tension "D" naturally accommodates a vertically-elongated
plasma. In this case the plasma, blanket, shield and diverters fit within a "D" shape based on a 1/R field variation
without modifications; however, the local fields near the
18 individual windings are now higher because of field perturbations caused by the discreteness of the coils. Thus,
the specific "D" shape produced by assuming a simple 1/R
variation of the field becomes vertical at a position on the
inside leg that is not high enough to clear the interior
structure. Figure 35 shows the shape of the modified "D"
chosen to give additional height. From point A to point C,
the magnet has a constant tension shape.

        For a uniformly wound toroid, the shape A to C would be
described by a radius of curvature ρ proportional to the radial distance from the axis of the toroid, as

$$\rho = KR. \qquad (45)$$

For discrete toroidal coils such as those used in UWMAK-III,
one should correct for the local perturbations of the magnetic field due to the flux leakage between the finite number

of finite coils. An analytic expression for the radius of curvature ρ from work by Moses and Young[68] is:

$$\rho = KR\left[\cfrac{1}{1 + \cfrac{1}{N}\left(\cos\phi + \cfrac{R}{\rho}\ell n\cfrac{1.284R}{CN}\right)}\right] \tag{46}$$

where:

K = a constant directly proportional to local toroidal field intensity

R = distance from major toroid

N = number of discrete magnets in the toroidal field system

ϕ = angle from midplane to a normal vector

C = radius of TF coil cross-section if the "D" magnet has a round cross-section or it is approximately $W/\sqrt{\pi}$ if the cross-section is a square of width W.

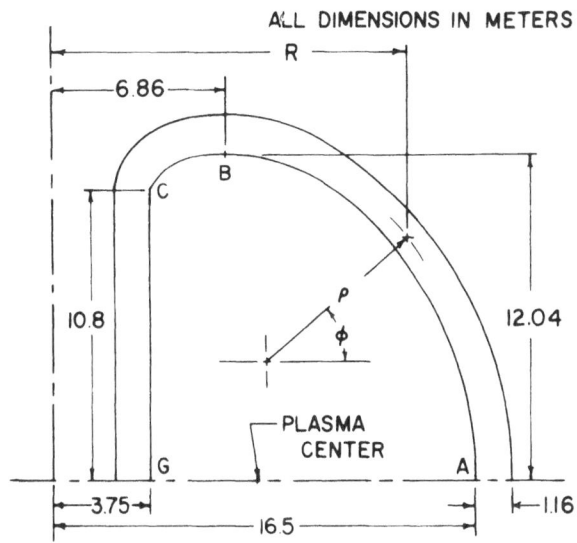

Figure 35

UWMAK-III Toroidal Field Magnet Design

Table 8

Specifications of Toroidal Field Magnets for UWMAK-III

a. Field Specifications:

    On-axis field, B_o 4.05 @ 8.1m

    Field at inside turn of far edge (point A)   1.99T @ 16.5m

    Maximum field at the conductor, B_m (point C)   8.75T @ 3.75m

    Field Ripple along the far edge of the plasma

    Total electromagnetic energy stored in TF coils   30 MWh

b. Design Stress:

    Maximum design stress in 316 stainless steel at
    4.2K for UWMAK-I and II or in 2219-T8T aluminum $\dfrac{311 \text{ MPa}}{(45,000 \text{ psi})}$
    at 4.2K for UWMAK-III

    Design strain   ≤ 0.0038

c. Magnet Components:

    Number of magnets   18

    Number of discs per magnet   16

    Number of conductor turns per disc. In UWMAK-III,
    the two end discs have no turns on outer surface   64 (32 each side)

    Conductor current   9,545A @ 16 discs
                                    10,909A @ 14 discs

    Disc cross-section (see III-A-2)   5 x 116 cm

Superconductor composite: TiNb in copper matrix, filament diameter = 0.038 cm, twisted every 30 cm	182 filaments max.
Stabilizer cross-section, copper in UWMAK-I and II, aluminum in UWMAK-III	1.5 x 1.5 cm
Insulation between conductor and stainless steel disc	0.05 cm reinforced epoxy
Spacer between discs	
Aluminum alloy bolts to fasten discs together against slipping, 2.54 cm diameter	1250
Lateral support between magnet and dewar: reinforced epoxy struts on each side of cross-section	3500 cm^2
Each magnet weight	174 Mt

d. Total Materials Required in Magnet System:

TiNb alloy superconductor	76 Mt
Stabilizer material	717 Mt Al
Structural disc material	2290 Mt
Dewars	730 Mt (aluminum) 119 Mt (SS)
Reinforced epoxy in discs	43 Mt
Micarta spacers between discs	108 Mt
Reinforced epoxy struts in dewars	10 Mt
Aluminum alloy bolts to assemble discs	34 Mt
Cryogenic insulation	503,000 m^2
Liquid helium inside dewars	144,000 liters

Solving for a square cross-section gives:

$$\rho = \frac{R[K - \frac{1}{N}\ln\frac{2.276R}{WN}]}{1 + \frac{\cos\phi}{R}}.$$ (47)

This expression, although approximate, gives results that
agree within a few percent to values found by much more dif-
ficult-to-use iterative procedures, and this result was used
in the toroidal field coil design.

A unique feature of the UWMAK-III TF coils is the use
of an Al conductor and an Al alloy structure. The use of
316 stainless steel for the structural material in the toroidal field magnets in the UWMAK-I and UWMAK-II designs was
predicated on a limiting maximum strain of 0.002 in the copper stabilizer and a disired high-stress level to minimize
structural volume. Recent work on alloy aluminum-reinforced
high-purity aluminum stabilizers at the University of
Wisconsin-Madison by H. Segal[57] leads to the possibility
of cycling the high-purity aluminum stabilizer to a strain
level of 0.0038, with little degradation to its resistivity
ratio. Work by Cameron[1,5] has raised serious questions
regarding the future availability of nickel and chromium, and
had cast some doubt on the advisability of committing so much
stainless steel to the construction of a fusion power plant.
With these two factors as principal motivation, although
recognizing the problems created by the extra volume (lower
stress level) required for the alloy aluminum toroidal field
magnet structure, the decision was made to use aluminum
alloy 2219 in the UWMAK-III design. This alloy was selected
for its strength, ability to be welded and heat treated, and
its ability to be formed at a reasonable cost. A complete
list of specifications for the toroidal field magnets is given
in Table 8.

The radiation level on the magnet components is about
four orders of magnitude less than for UWMAK-I. In that case,
at 3×10^{-4} dpa/year (dpa = displacements per atom), we allowed
for extra copper and for extra cooling surface, and planned
for room-temperature anneals every 10 years. For UWMAK-III,
in contrast, the dpa rate on the front surface of the magnet

is 1.6×10^{-8} dpa; therefore, we expect up to 40 years of operation without any anneals. The damage to insulators, which
was not considered a problem for UWMAK-I, is negligibly small
for UWMAK-III.

2.   VF and OH Windings

        The vertical field coils are constructed inside
the toroidal field magnets as they were in UWMAK-II. The
maximum stored energy in these coils is less than in UWMAK-II,
due primarily to the increased magnetic field utilization for
noncircular plasmas. A four-element conductor introduced by
Cornish[69] has been modified to accommodate the requirements
of both the vertical field magnets and the ohmic heating magnets. The improved cooling afforded by this design has
allowed an increase in the current density, and this fact,
coupled with smaller diameter coils, has resulted in a less-
expensive set of magnets. The vertical field and ohmic-
heating magnets are cold-worked OFHC copper for both stabilizer and structure, with some stainless steel also used in the
more highly-stressed vertical field magnets. As with the
toroidal field coils, all magnets are cryogenically stabilized.

        To reduce the AC losses, the superconducting filaments
are extruded with cupronickel in an OFHC copper matrix, drawn
to size, twisted, and finally formed into a square cross-
section. All conductor sections containing the superconduc-
tors will have the same size in all the coils, but the number
of TiNb filaments will depend upon the current being carried
and the maximum ambient field at the given coil location.
The general specifications of the vertical field and ohmic
heating coils are given in Table 9.

E.   Neutron and Photon Transport Analysis of
Blanket and Shield

        The philosophy behind the blanket design was described
in Part A of this section; a schematic illustration of the
overall design is given in Figure 24. The neutronic behavior
of this design has been analyzed thoroughly, using one- and
two-dimensional multigroup S_n neutron and photon transport
codes (70,72). The data used have been discussed previously
(73). For these calculations, there were 46 neutron groups
and 21 gamma groups.

Table 9

General Specifications
For Vertical Field and Ohmic Heating Coils

1. Conductor TiNb + CuNi sleeve + Copper Stabilizer,
    fully cryogenically stabilized

2. Conductor current in Type 1 = 14,160 to 14,520 A
    Conductor current in Type 2 = 10,390 to 11,110 A
    Conductor current in Type 3 = 4,340 to 6,750 A

3. Conductor current density in Type 1 = 4880 to 5320 A/cm^2
    Conductor current density in Type 2 = 3120 to 3850 A/cm^2
    Conductor current density in Type 3 = 1410 to 3580 A/cm^2
    (based on gross coil area)

4. Conductor size range in Type 1 = 1.6 cm x 1.6 cm
    Conductor size range in Type 2 = 1.6 cm x 1.6 cm
    Conductor size range in Type 3 = 1.6 cm x 0.951 cm to
    1.6 cm x 2.09 cm (this includes the bonded stainless
    steel reinforcement in Types 2 and 3)

5. TiNb - CuNi - Cu composite section: 0.80 cm x 0.80 cm
    with 34 to 182, 0.038 cm diameter filaments, twisted
    every 30 cm

6. Copper to superconductor ratio = 6.03 to 33.2

7. Design stress at 4.2 K: 124 MPa (18,000 psi) in 10%
    cold reduced OFHC copper, 211 MPa (30,600 psi) in
    316 stainless steel in conductor types 1 and 2

    207 MPa (30,000 psi) in 10% cold reduced OFHC, 351 MPa
    (50,900 psi) in 316 cold stainless steel in conductor
    type 3

8. Spiral wrapped with 0.1 cm thick fibreglas reinforced
    epoxy over 50% of the surface; alternate conductors
    wrapped to produce a spacing of 0.1 cm in both vertical
    and radial directions as the coils are layer wound

9. Design limit temperatures to 5.2 K in a 4.2 K bath
    for all normal situations

10. Total TiNb superconductor required = 64.5 Mt

11. Total copper required = 1331 Mt

12. Total 316 SS reinforcement required = 436 Mt

13. Total insulation for wrapping conductors = 26 Mt

The first set of calculations is two-dimensional and based on using x-y geometry to approximately represent a vertical cut through the system as shown in Figure 36. Toroidal effects are not included here. The inside of a torus in the poloidal direction is the most difficult part of a tokamak to replace because there is little access space. Thus, a thick graphite wall (region 12 in Figure 36) is introduced in front of the vacuum wall (region 13 in Figure 36) which acts to soften, reflect and attenuate the neutrons incident from the plasma in the manner described in Reference 56. This ISSEC (Internal Spectral Shifter and Energy Converter) zone reduced both the displacement and helium production rates in the first structural wall (region 13 in Figure 36) and thus increases the wall life for this zone. The ISSEC zone is *not* actively cooled but radiates heat to the first structural wall behind it and also to the first wall on the outer portion of the plasma chamber. The inner blanket is further simplified by use of helium cooling. The absence of flowing liquid lithium eliminates the problem of the high pressure drop that results from using a liquid metal coolant in the high magnetic field region. The absence of lithium also alleviates tritium removal problems by eliminating the need for this function in the inner blanket. Finally, the inner blanket and shield is consciously designed to be as thin as feasible to allow for a lower aspect ratio and to allow higher magnetic fields in the plasma for the same magnetic field at the coil.

The ISSEC cannot be extended around the chamber since it would result in problems in achieving adequate tritium production. For this reason, and because one has relatively easy access to the outer blanket and shield (zones 1, 2, 3, 4, 7, 8, 9, 10 of Figure 36), the outer blanket system is designed for convenient replacement. Experience with the UWMAK-I and UWMAK-II designs indicates that resource limitations make use of Be as a neutron multiplier undesirable. This leads to the choice of a liquid lithium-cooled system for the outer blanket to achieve adequate breeding without a contribution from the inner blanket. The choice of zone thickness is made on the basis of results from a series of one-dimensional calculations. The final structural details are chosen on the basis of heat transfer and mechanical design.

Both the plasma and the chamber are noncircular and the magnetic axis that serves at the plasma axis is not at the geometrical center of the plasma.

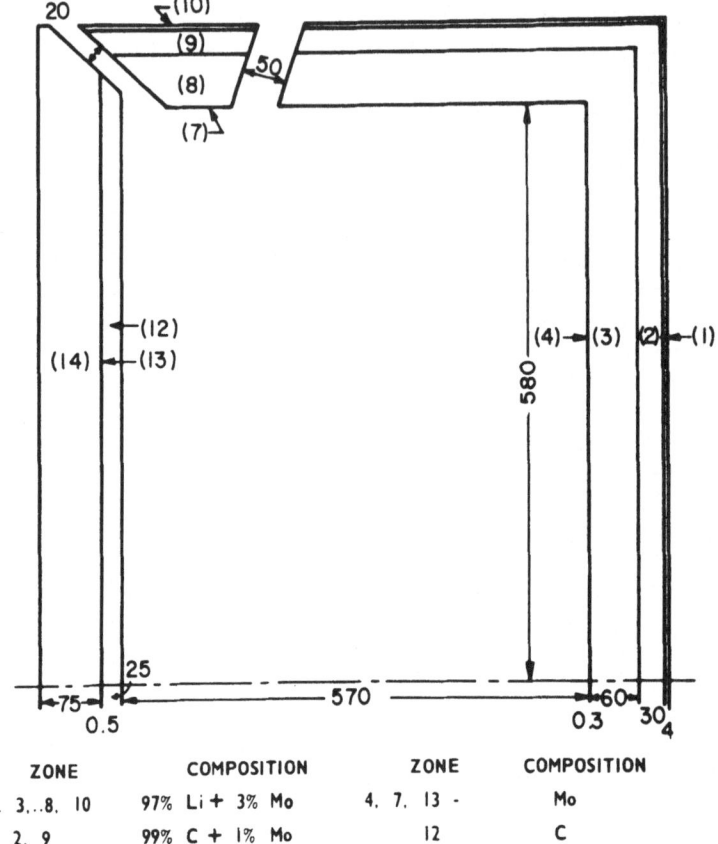

Figure 36

Blanket Model Used for Two-Dimensional Neutronics
Analysis of UWMAK-III (74)

The neutron source distribution is proportional to the
square of the ion density and has been approximately represented as a function of x and y in a vertical plane through
the torus as

$$S(x,y) = S(0,0) \ [1 - (\frac{x - 50}{200})^2]^3 \ [1 - (\frac{y}{580})^2]^3 \ ,$$

$$- 150 \leq x \leq 250;$$

and

$$S(x,y) = S(150,y) \; [0.01 \; x + 2.50], \; - 250 \leq x \leq -150.$$

The source has a center at R = 9.0 m, shifted out from the geometric center of the plasma by finite β effects as predicted by the MHD calculations.

The major results of the calculation are given in Table 10. The breeding ratio exceeds 1 and the total energy per fusion is 21.7 MeV. Also, the ISSEC performs its function well, since the neutron radiation damage indicators, such as dpa (displacements per atom) and helium production rate are substantially reduced in the first wall behind the ISSEC (zone 13 on Figure 36) compared with the unprotected outer blanket first wall (zone 4 on Figure 36).

Table 10

Nuclear Performance of UWMAK-III Blanket Design
(x-y geometry representation)

Breeding Ratio	1.074
Energy per Fusion	21.7 MeV
Energy Attenuation:	
of Inner Blanket and Shield	3×10^{-7} MeV/MeV
of Outer Blanket and Shield	3×10^{-7} MeV/MeV
DPA Rate per Year:	
Inner Blanket First Wall	2.9
Outer Blanket First Wall	18.8
Helium Production Rate (Approx./yr):	
Inner Blanket First Wall	10.8
Outer Blanket First Wall	157.0

At low aspect ratio, toroidal geometry effects can be
very important. This is especially true in systems that
are not poloidally symmetric with respect to functions like
tritium breeding; thus, the geometry is based on a horizontal cut through the torus midplane as illustrated in Figure 37.
This is quite reasonable for vertically elongated systems
like UWMAK-III and the model is illustrated in detail in
Figure 38. We now find a breeding ratio of 1.27 and a ratio

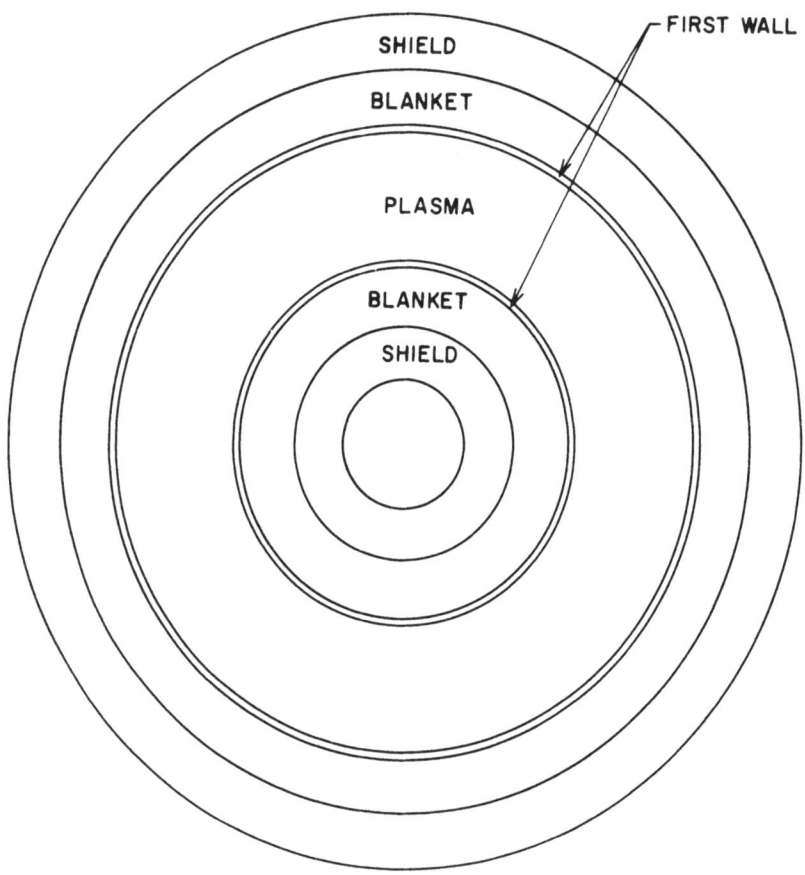

Figure 37

Horizontal Cut Through the Midplane of a Tokamak.
A one-dimensional cylindrical representation of this type
automatically includes the effects of toroidal geometry.

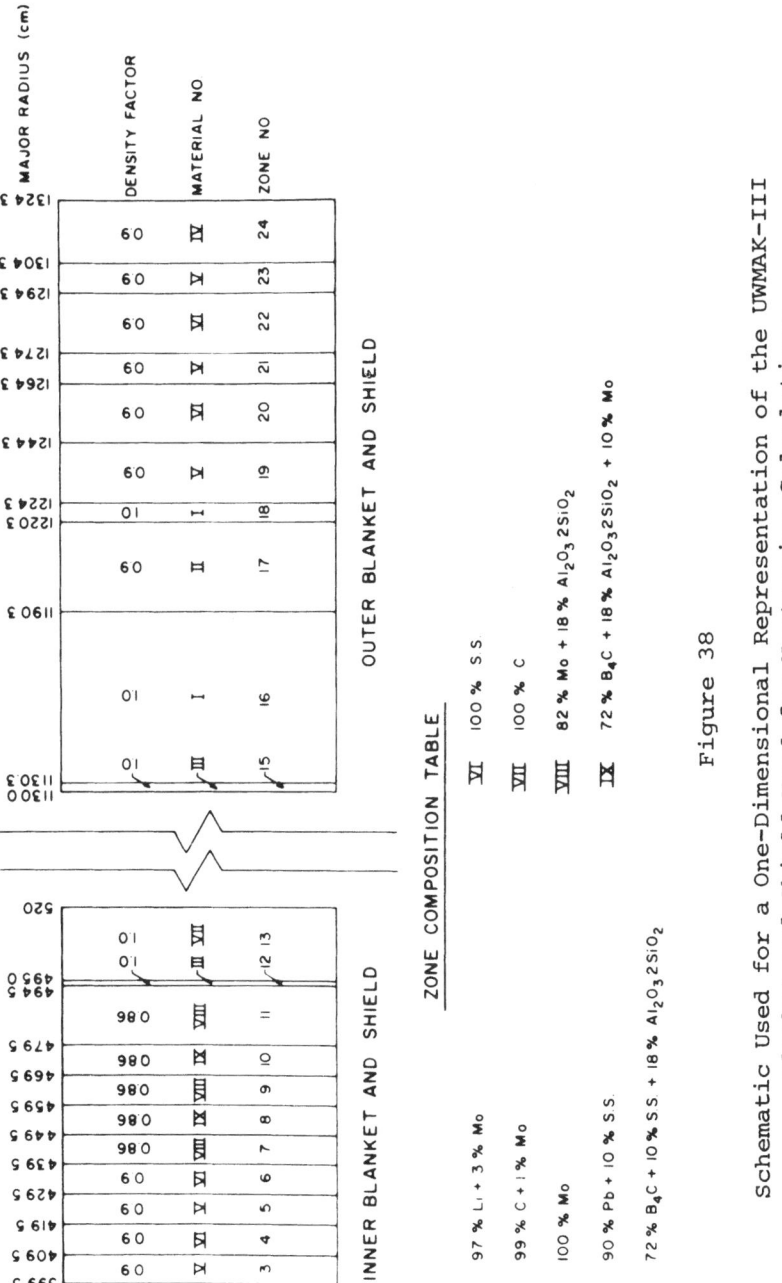

Figure 38

Schematic Used for a One-Dimensional Representation of the UWMAK-III
Blanket and Shield Used for Neutronics Calculations

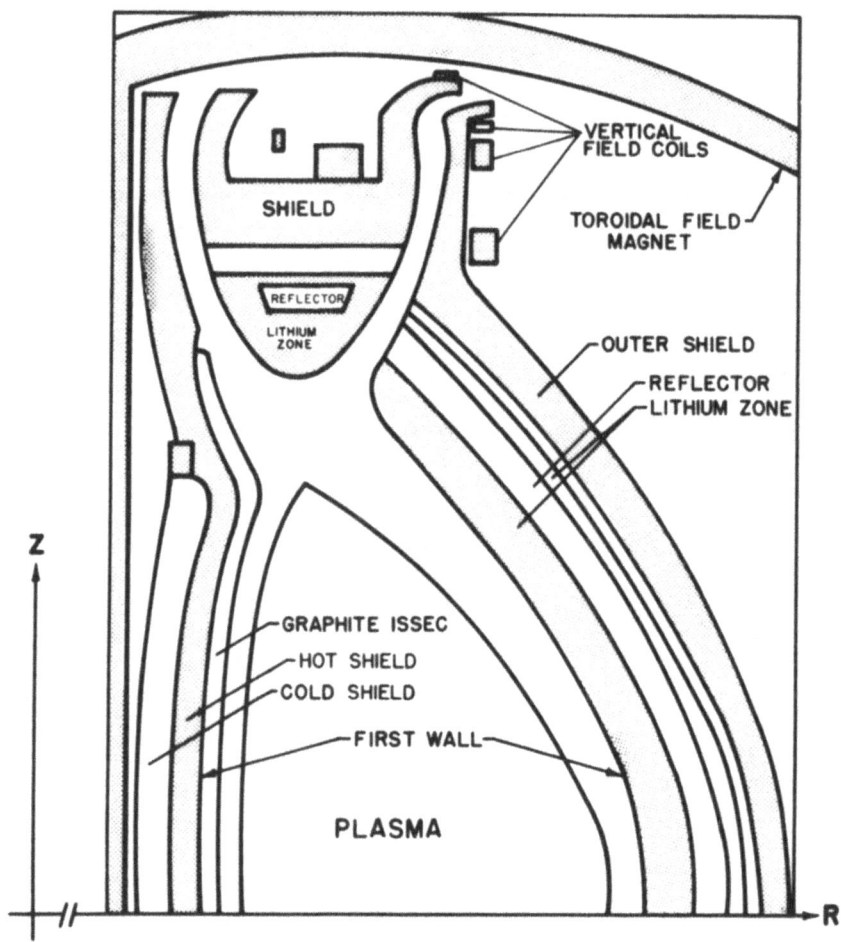

Figure 39

Geometrical Representation of UWMAK-III Used for
Two-Dimensional, R-Z, Toroidal Neutronics Calculations

of dpa rates for the outer blanket first wall to the inner
blanket first structural wall of 9 to 1.

Finally, a calculation using R-Z geometry to represent
the toroidal configuration was undertaken (75). There are
also some other minor geometry and composition changes that

are included for the inner blanket and outer shield. The
calculation has been performed using the DOT program (71),
using S_6 (30 directions) with a P_3 approximation of the
scattering cross-section.

The neutron source distribution has been represented as
a function of R and Z as

$$S(R,Z) = S(R,0) \ [1 - (Z/Zi)^2]^3$$

$$S(R,0) = S(0,0) \ [1 - ((R-900)/180)^2]^3 \ 1080 > R > 900$$

$$S(R,0) = S(0,0) \ [1 - ((900-R)/360)^2]^3 \ 540 < R < 900$$

This source represents a source peaked at R = 9 m (i.e.,
shifted 90 cm toward the outside blanket from the plasma
center at R = 810 cm) but symmetrical about the midplane.
Zi is the vertical boundary of the source distribution.
The shift of 90 cm rather than 50 cm, as in the x-y calculations, is based on more recent MHD calculations.   The
geometry model used in this calculation is shown in Figure 39, where the source distribution, diverter slots, and
the different coil positions are determined from the plasma
physics analysis.

Table 11 presents the primary results of the analysis.
The total breeding ratio is 1.25, which is about 16% higher
than predicted by the x-y calculation. In fact, this increase in the breeding ratio is not completely due to the
correction from the toroidal effect. Careful comparison
with the tritium production rates on a per zone basis shows
the toroidal effect increases the breeding by about 9%.
The other 7% increase is due to the change in the composition
of the shielding and to a change in the thickness of the last
zone of the outer blanket. Also, the first zone of the outer
shield has been changed from B_4C and stainless steel to carbon. ·This change has the effect of reflecting more low energy neutrons to increase the ^6Li (n,t)α reaction rate.

A summary of the results for the heat deposition per
zone in the blanket region is given in Tables 12 and 13.
The R-Z calculation indicates that 17.6 MeV per D-T neutron
is deposited in the blanket and shield. It can be noted
from Table 13 that the heat deposition in the outer blanket
is four times that in the inner blanket (including the ISSEC).
This was only a factor of 2 in the x-y calculations.

Table 11

Nuclear Performance of UWMAK-III Blanket Design
(Based on R-Z Neutronics)

Breeding Ratio   1.252

Energy per Fusion   21.09 MeV

Energy Attenuation:

    of Inner Blanket and Shield[*] 6.0×10^{-7} MeV/D-Tn

    of Outer Blanket and Shield[*] 1.0×10^{-6} MeV/D-Tn

Power in Blankets   3595

Power in Shields   90

DPA Rate per Year:

    Inner Blanket First Wall   3.9[*]

    Outer Blanket First Wall   16.5[*]

Helium Production Rate(approx./yr):

    Inner Blanket First Wall   12.5[*]

    Outer Blanket First Wall   103[*]

---

[*]   At midplane.

Figure 40 shows the variation of the neutron wall loading as a function of poloidal angle. As a result of the shifted neutron source, the wall loading has a maximum at the outer wall at $\theta = 0°$. The angle θ varies from $0°$ where the midplane intersects the outer blanket to $180°$, where the midplane intersects the inner blanket. This shows the reason the tritium breeding ratio of 1.25 is so much larger than the value from the x-y calculations. Neutrons tend to preferentially go toward the outer breeding blanket such that

Table 12

Neutron and Gamma Heating Rates [*]
By Zone from R-Z Toroidal Calculation in UWMAK-III Blanket

Zone	Neutron Heating	Gamma Heating	Total Heating (per zone)
2	1.3481	0.6775	2.0256
3	0.0697	1.5886	1.6583
4	0.0841	1.0409	1.1250
5	0.0117	0.1781	0.1898
6	8.0960	1.7348	9.8308
10	0.3711	0.0809	0.4520
7	0.7874	0.1815	0.9689
8	0.3136	0.4979	0.8115
9	0.0300	0.0384	0.0684
36	0.0166	0.0424	0.0590
Total:	11.1283	6.0610	17.1893

[*]   All the heating rates are in MeV per D-T neutron.

excessive parasitic absorptions in the inner blanket are
avoided. The wall loading can be obtained at essentially
no expense by representing the source in (R-Z) coordinates,
using one neutron group, and plotting the flux at the
points where the first wall would be located. This simple
approach gives results that match more elaborate and
expensive calculations (76) for the same cases. There is
now, however, no restriction on the shape of the plasma
cross-section or on the neutron source distribution. From

Table 13

Power Distribution for UWMAK-III

Zone	Percent	Power[*] (MW)
ISSEC	9.6	480
Inner Blanket (hot shield)	7.9	395
Outer Blanket	57.9	2895
Diverter	6.1	305
Diverter Blades (α particles)	16.7	835
Inner Shield (cold shield)	0.9	45
Outer Shield	0.6	15
Diverter Shield	0.3	30

----

[*]   Total thermal output, 5000 MW_{th} during plasma burn.

the figure, one can see that the wall loading varies substantially at different locations around the first wall.

Perhaps more relevant than wall loading is the variation
of response functions like the atom displacement rate (in
dpa/yr) and the helium production rate (in approx. He/yr) in
the first structural wall of TZM. These are shown in Figure 41, and are the result of the full 46 neutron group, 21
gamma group, R-Z calculation. Again, the variations in θ.

are substantial. The relatively low values that occur when $\theta \leq 84^{o}$ are caused by the graphite zone in front of the hot inner shield. It is interesting to note that the dpa and He production rates peak at $\theta = 50^{o}$, whereas the maximum wall loading occurs at $\theta = 0^{o}$. The reason is that the wall loading is related to the neutron current, whereas reaction rates are proportional to the neutron flux. Therefore, the variation of wall loading with poloidal angle does not necessarily indicate the point of maximum reaction rate.

In summary, the representation of low aspect ratio toroidal CTR systems using (R) or (R-Z) coordinates allows one to adequately represent these systems using existing one and two-dimensional neutronics codes. Neutron wall

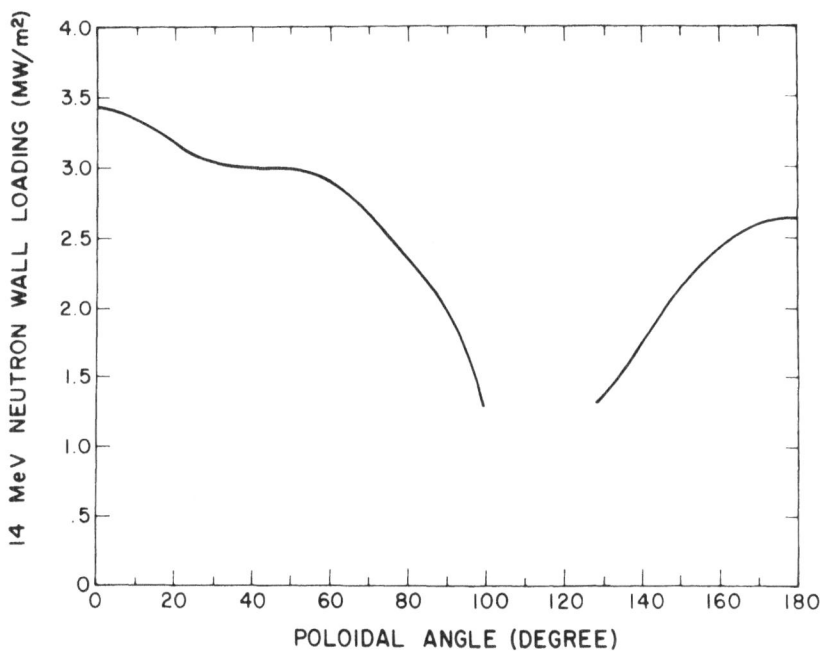

Figure 40

Variation of Neutron Wall Loading
as a Function of Poloidal Angle

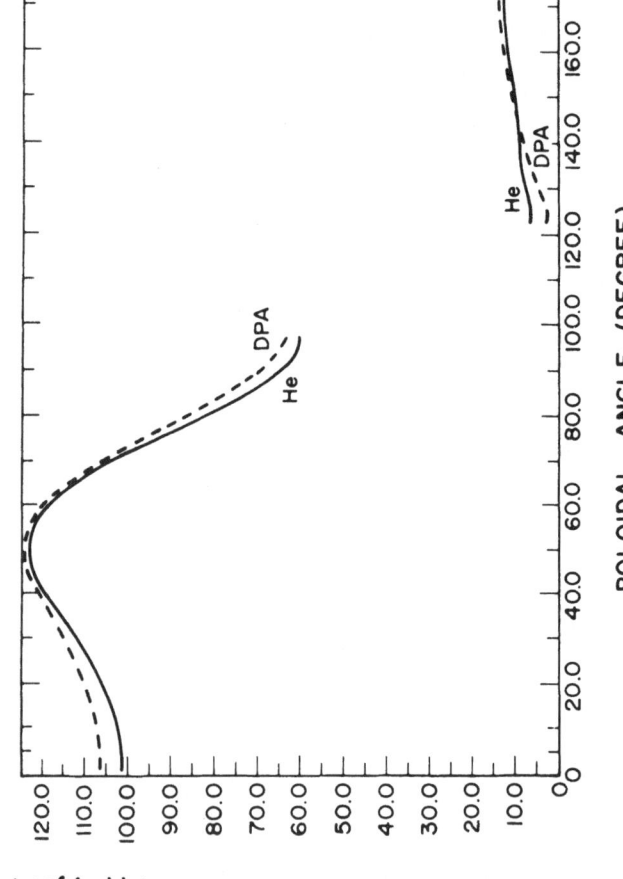

Figure 41

Displacement (dpa) and Helium Generation Rates in the First
TZM Wall as a Function of Poloidal Angle

loading distributions can be trivially calculated using one
neutron group. In general, the inclusion of toroidal effects
can have an important impact on the final results.

F. Thermal Hydraulics and Mechanical Design Aspects
   of the Blanket System

The blanket design described in the previous section
has three distinct components from a heat-transfer viewpoint;
the hot inner shield, the partial ISSEC of 25 cm graphite,
and the outer breeding blanket. The neutronics calculations
give the local neutron and gamma heating rates required as
input to the heat-transfer analysis.

The maximum allowable temperature in the graphite is not
yet accurately defined. This temperature limitation may be
set by any one of the following processes:

1. evaporation, causing the carbon vapor to be
   condensed on the panels in the cryopumps,
   causing impurity buildup in the plasma

2. high temperatures inside the ISSEC, causing
   carbon transport

3. Formation of hydrocarbons, particularly
   acetylene, at temperatures above 1500°C.

In this study and for reasons described in Section IVB, it is
assumed that the maximum allowable temperature is 2000°C.

The heat deposited in the partial ISSEC zone will be
transported to the front and back surfaces and radiated away.
Thus, the maximum temperature will occur inside the graphite
rather than at its surface. The heat transfer calculations
based on the heating rates generated in the x-y geometry neutronics calculations show the temperatures in the ISSEC are
within the specified limits if the graphite has a thermal
conductivity comparable to 809S graphite. The surface temperature is < 2000°C, and the maximum temperature in the
interior is ~2300°C. The toroidal geometry calculations,
using R as the primary coordinate, show that the heat deposition is reduced by about 25%, compared with the x-y analysis.
As such, the temperature within the graphite drops accordingly,
and nuclear grade graphite could be used without exceeding the
specified temperature limits. The graphite curtain has no

temperature problems as calculations give its operating
temperature as 1300°C.

The thermal energy in the ISSEC will be radiated to the
first structural walls. The surface heating loads are calculated to be 350 kW/m^2 on the TZM first wall of the inner
hot shield and 250 kW/m^2 on the outer first wall.

Two of the most severe problems in the design of a
lithium-cooled tokamak blanket are MHD problems and thermal
recirculation problems (77,78). The MHD problems are caused
by the circulation of electrically-conducting lithium across
magnetic field lines. The thermal recirculation problems
are caused by the proximity of hot and cold lithium streams.
The major feature of the outer blanket design in UWMAK-III
is to minimize these two effects.

The basic shape of a heat-transfer unit is U-shaped.
This heat-transfer cell design gives the lowest possible
$\underline{V} \times \underline{B}$ value within the blanket and thus minimizes MHD
effects; however, this shape must be modified to prevent
heat transfer between adjacent legs of the cell. Figure 42
shows a heat transfer cell for the UWMAK-III blanket. The
width of the cell is 37 cm, and there is a 10 cm wide static
lithium zone in the middle of the cell, which serves as
thermal insulation. The high temperature in the middle of
the static zone prevents heat transfer from the hot toward
the cold region. The calculated MHD pressure drop within
the blanket is only 0.4 MPa. The important dimensions of
the outer blanket are given in Table 14.

The lithium coolant pressure drop has been calculated,
including both effects described above and using the local
magnetic field, $\underline{B}$, consisting of poloidal and toroidal components. The maximum coolant pressure on the first wall is
only 0.5 MPa and the pumping power required is just 2 MW.
This shows the advantage of using the liquid metal coolant
for the outer blanket only.

The lithium velocity in the buffer zone shown in Figure 42 is enough for tritium recovery but can be considered
static from a heat transfer viewpoint. A two-dimensional
finite difference method has been used to calculate the
temperature profile for a typical cell as shown in Figure 43
(79). The maximum temperature of the TZM structure is
1000°C, and the exit temperature of the lithium to the power
cycle is 980°C.

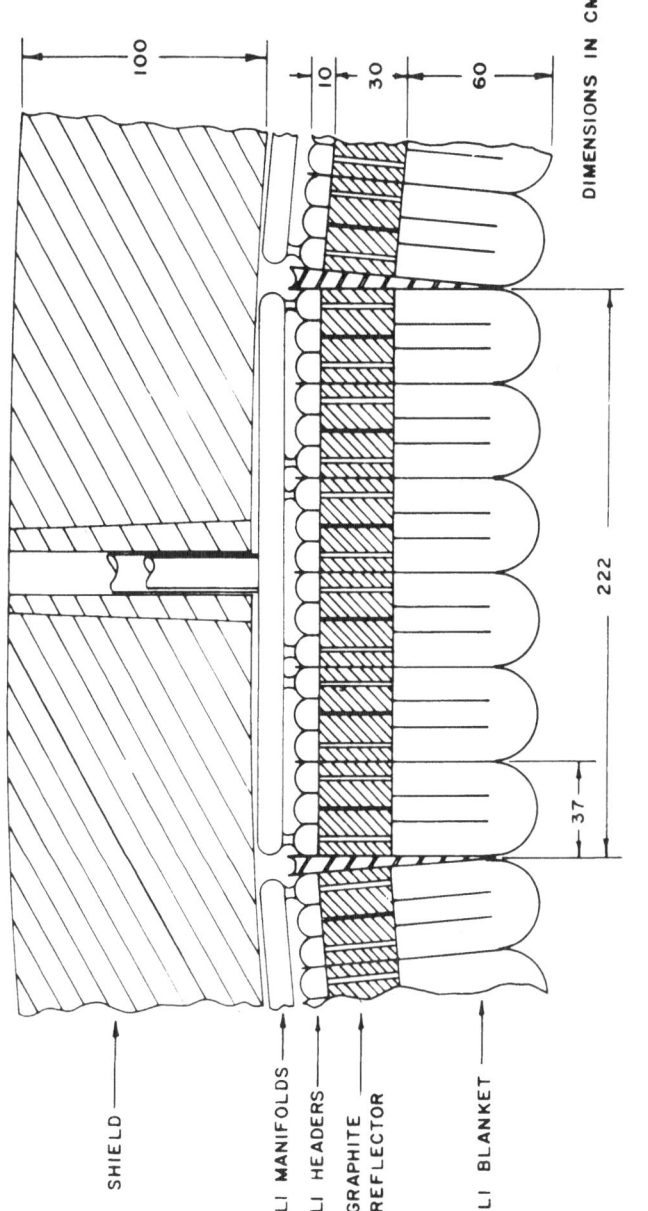

Figure 42

Cross-Sectional View of an Outer Blanket Looking in the Toroidal Direction

Table 14

Important Dimensions of the Outer Blanket

Thickness of the First Wall	1.5 mm
Width of the U-Cell	37 cm
Width of the Lithium Gap	10 cm
U-Cell Wall Thickness	6 mm
Baffle Wall Thickness	2 mm
First Breeding Zone Thickness	60 cm
Refractory Zone Thickness	30 cm
Second Breeding Zone Thickness	10 cm
Percentage Structure in the Blanket	3%
Diameter of the Header	20 cm
Wall Thickness of the Header	2 mm
Diameter of the Feed Tube	50 cm
Wall Thickness of the Header	5 mm

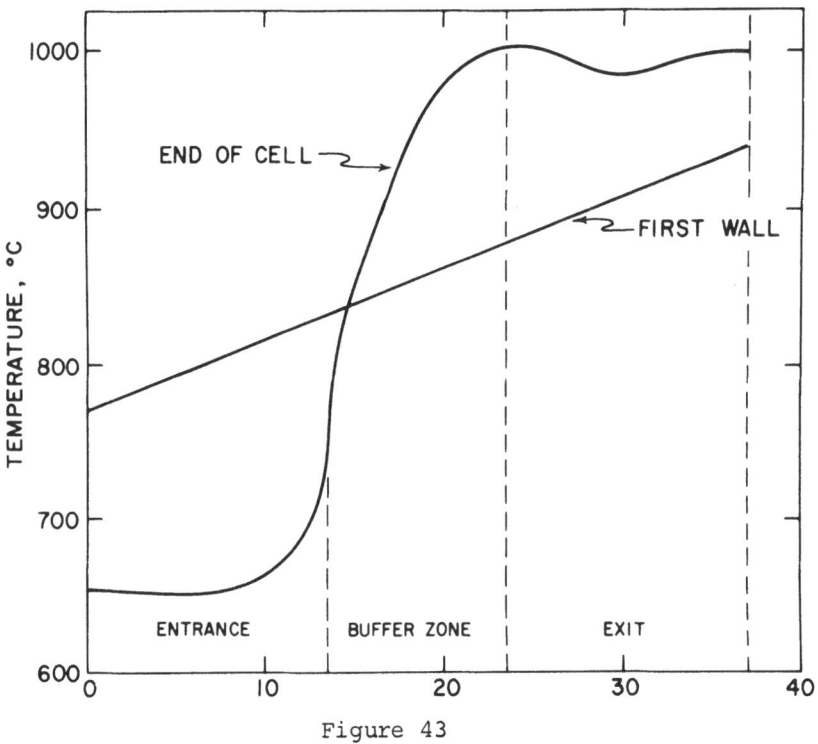

Figure 43

Temperature in a U-bend Cell of the Outer Blanket

The inner hot shield, or inner blanket, is cooled with helium. A 0.5 m diameter feed tube is introduced between the TF magnets and connected to a manifold running in the toroidal direction. The surface heating load on the inner blanket is 350 kW/m^2, and together with the space-dependent volumetric heating from the neutronics analysis, leads to a coolant inlet temperature of 488°C and an outlet temperature of 870°C at 6.7 MPa pressure. These coolant conditions match the temperature of the helium coming from the secondary heat exchanger of the outer blanket cycle such that identical turbines and thermal energy storage can be used. This is discussed in Part J of this section. The distribution of coolant tubes within the inner blanket is arranged to be inversely proportional to the heating rate so that the heat load on each coolant is approximately the same (see Figure 44).

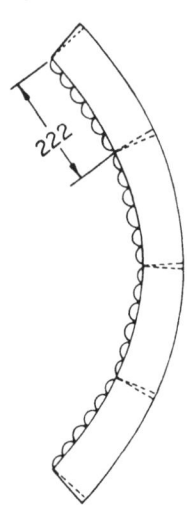

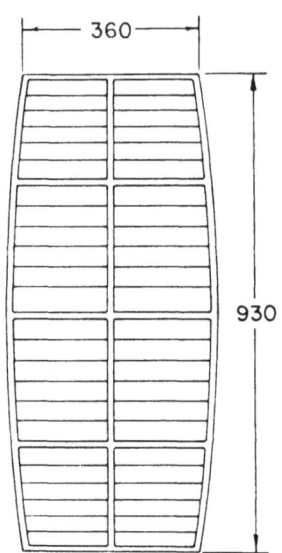

DIMENSIONS IN CM

(a)   (b)

Figure 44

Skeletal Frame for Region A of Outer Blanket
(a) side view in the toroidal direction
(b) radial view from outside the blanket

A large amount of Mo is used in the inner blanket for
shielding reasons and most of this can be in the much less
expensive form of powder; however, because both powdered
Mo and B_4C are poor thermal conductors, we find that approximately 35% of the total Mo must be in the form of fine
sheet, oriented to provide a conduction path to the TZM
coolant tubes. The maximum allowable temperature in the
B_4C and nonstructural Mo is assumed to be 1250°C.

Turning now to the mechanical design aspects, the blanket is divided into 18 equal modules. This allows a blanket
module to be taken out between each of the 18 toroidal
field magnets. Each module is further divided vertically
into two regions. The region subtended by the angles ± 60°
as measured from the plasma center will be called region A.
The rest of the outer blanket, consisting of the part extending to the first diverter slot and the central region between
the diverter slots will be called region B. Regions A and B
are taken to have lifetimes limited by radiation damage of
1.5 and 3.0 years, respectively.

Each A region of an outer blanket module will have a
skeletal frame shown in Figure 45. This frame is divided
into eight trapezoidal cavities. An assembled section of
the blanket, consisting of six cells, fits into each cavity
and is sealed to fit. This design has the flexibility of
allowing the removal of a single section or the removal
of the whole frame with eight sections. Region B will
have a similar frame, but the sections will be only 180 cm
wide and will have six 30 cm wide cells.

Looking radially out from the plasma, the first 60 cm
of the blanket contains lithium with a minimum amount of
structure, the next 30 cm contains TZM clad graphite reflector and the last 10 cm contains the lithium distribution headers. Holes in the graphite block allow the
passage of lithium from the headers to the front wall and
back. A toroidal view of an outer blanket section is
shown in Figure 42, and a poloidal view in Figure 46. The
heavy wedge-shaped material is part of the skeletal frame.
It is wedge shaped to allow for poloidal and toroidal curvature. Blanket sections themselves are flat for ease of
fabrication.

The entire outer blanket structural material is TZM,
with the possible exception of the skeletal frame. TZM
displays superior properties in thin sheet form and is
difficult to fabricate in heavy sections. The front wall
material is 1.5 mm thick, the cell frame material is 2.5 mm
thick, the partition and header material is 1.0 mm thick,
and the graphite cladding material is 0.5 mm thick. The
skeletal frame thickness tapers from 5 cm thick at the
heaviest section down to 0.5 cm, and is about 80 cm high.

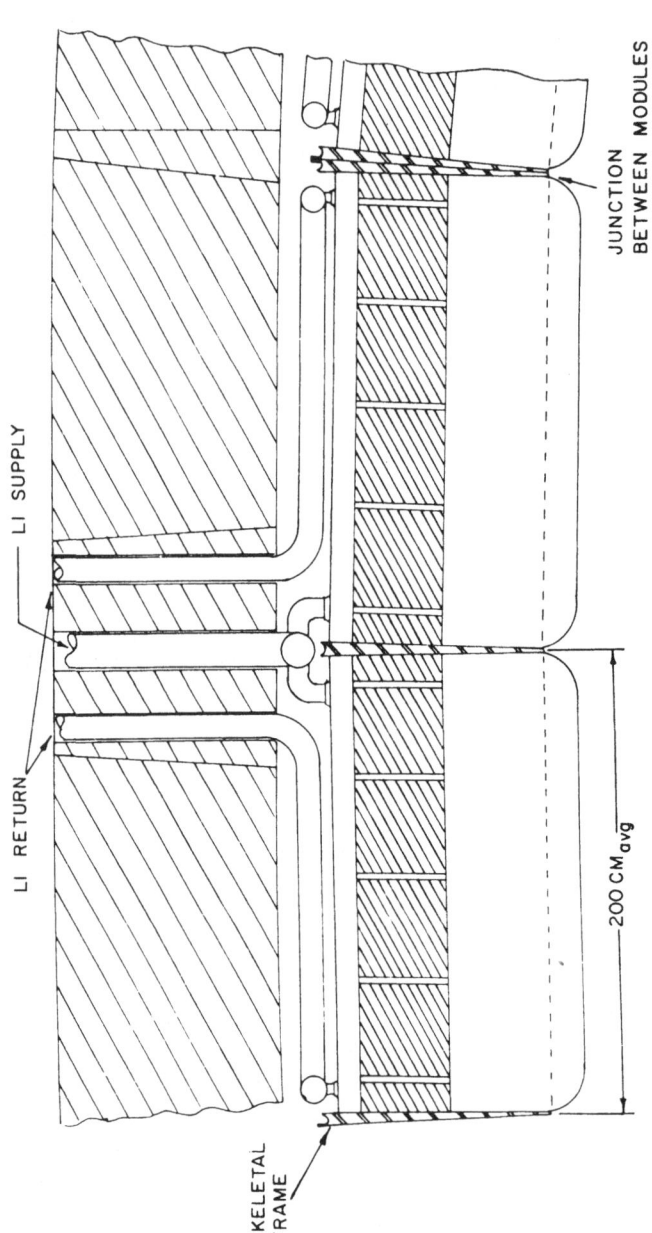

Figure 45

Cross-section of Outer Blanket and Shield
Viewed in the Poloidal Direction

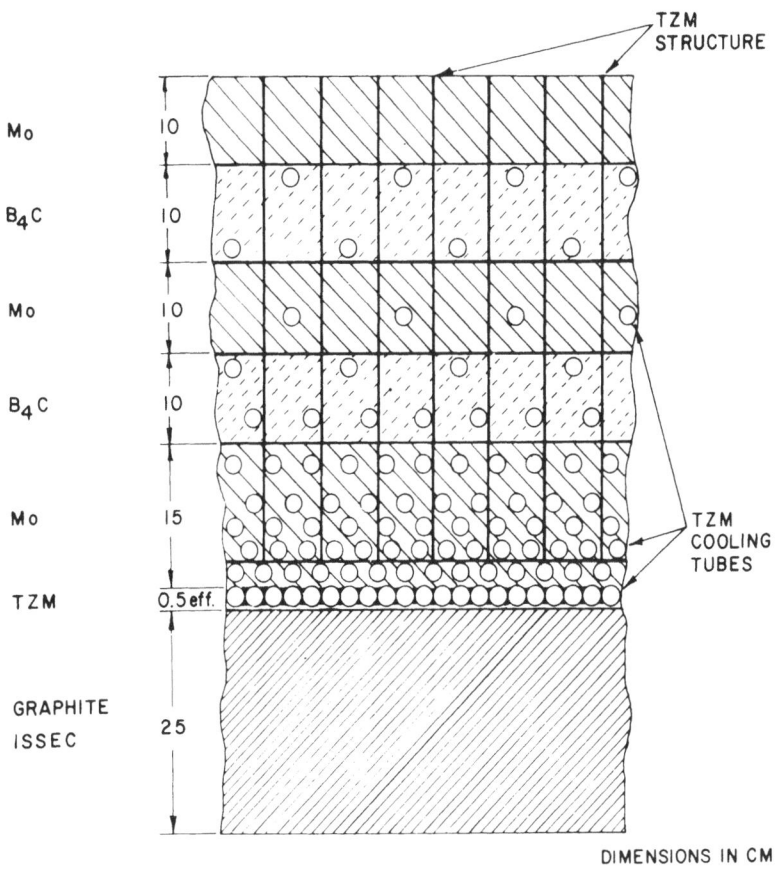

Figure 46

Cross-sectional View
Looking in the Toroidal Direction of the Inner Blanket

TZM can be welded, but the weld zone is recrystallized metal and its ductility and strength are affected. A preferable joining method is by brazing. Several high-temperature brazing materials have been developed for TZM which have proven to be extremely good. For example, a 25% Ti, 3% Cr, Be brazing material has a liquidous temperature of 1110°C and a remelt temperature of 1700°C. Such a brazing material would be ideal for building up a blanket section, since certain parts can be brazed separately. The whole assembly can

then be brazed together without affecting the preassembled
joints. The final seal between the section and the frame
can be either a braze or a weld joint. This region is not
highly stressed and suffers little radiation damage.

The skeletal frame material can be either TZM or niobium.
The latter is easier to fabricate and weld in heavy sections
but is not as strong. If the frame is made of niobium, the
joint between the TZM section and the frame may present problems because of the difference in their coefficients of expansion.

The design of the inner hot blanket is quite different
from that of the outer blanket. A bank of TZM tubes, cooled
with helium at 5 MPa is joined together and essentially provides the seal for the plasma chamber on the inner wall. These
tubes are 2 cm in diameter and extend for 12 m in the vertical
direction. Additional cooling tubes are provided within the
inner blanket. These tubes are embedded in layers of Mo and
B_4C. Figure 46 shows a section through the inner blanket.
Helium gas is supplied to the tubes from a manifold located
between the inner diverter slot and the TF coil at the top
and exhausts to a similar manifold at the bottom. The stresses
in the front wall have been determined by simply adding the
thermal and pressure stresses together. No attempt has been
made to determine the effect of swelling or creep. A more
thorough analysis using finite element codes can be made
when more information is available on the swelling of TZM
and other effects. Stress-relieved TZM rolled sheet has a
low value for tensile strength at 1000°C of 500 MPa, while
for recrystallized TZM sheet material at 1000°C, it is
200 MPa. Note also that a stress of 188 MPa will produce
0.2% creep in the material at 1000°C in three years. Based
on this information, it was decided that a design stress of
70 MPa would be satisfactory. This corresponds to a thickness of 1.5 mm. Even if 0.5 mm is lost to sputtering and
blistering after three years, the stress will rise only to
97 MPa.

The inner shield is essentially a continuous cylinder
subtending the inner legs of the toroidal field coils and is
composed of Mo and SS honeycomb structure filled with various
materials. The weight of the inner shield is 2780 tonnes.
It will be supported entirely on columns protruding between

the TF coils and built into the floor of the containment
building.

The central shield and blanket sections will be supported on structures that also penetrate between the TF coils.
Upper sections will be supported on the reactor containment
building ceiling beams and the lower sections on the floor.
Each of these blanket and shield sections weighs 900 tonnes
or 50 tonnes per support.

The outer blanket module frames will be attached to reinforced plates that also comprise the wall of the outer diverter slots. These plates are attached to a shield support ring.
At the same time, the blanket skeletal frame also will be
supported on the shield as previously mentioned.

Although expansion joints for the blanket would be very
beneficial, it is difficult to visualize an expansion joint
that can subtend the entire plasma chamber including the vacuum plenum. At a temperature of $1000°C$, the outer radius
of the chamber will increase approximately 7 cm. This would
present difficulties, especially if the blanket is attached
to the shield that does not get as hot. An alternative might
be to design expansion joints into the shield without sacri-
ficing efficiency. Some imaginative engineering will be
needed to solve this problem.

In order to make the blanket accessible for servicing,
the following procedures must be followed: (1) the plasma
chamber and toroidal enclosure are brought to atmospheric
pressure; (2) the lateral support structure between TF coils
must be removed, opening up a passage ± 8 m from the central
plane and simultaneously draining the blanket of lithium;
(3) if there are RF waveguides in the module in question,
they must be disconnected and extracted; (4) the VF coils
must be retracted (they are supported on mechanical jacks);
(5) the outer lithium manifolds must be disconnected and
removed; (6) the outer shield must be opened and firmly
secured.

It appears that for a system like UWMAK-III, one would
discard an entire spent module upon removal. The spent module
would be crushed, compacted and stored in special rad-waste
areas. The amount of material thus disposed of per year, averaged over three years, is 164 tonnes of TZM and 472 tonnes of

graphite. Replacement of the blanket modules would consist of the insertion and weldment of new modules into place following the functions enumerated in the foregoing paragraph in reverse order.

G.   The Tritium Systems in UWMAK-III

Details of the tritium system for both UWMAK-II and UWMAK-III have been described by Clemmer, Larsen and Wittenberg.[80] There are at least four aspects to the tritium problem in UWMAK-III: the recovery and reprocessing of tritium in the diverter and fueling cycle; the extraction of tritium from the breeding material and other heat-transfer fluids; the recovery of tritium from the reactor building; and loss of tritium to the environment in gaseous or aqueous phases. The routes along which tritium flows are shown schematically in Figure 47.

The reactor building will have an inert atmosphere contained within an impermeable metal shell. The tritium cleanup system for this atmosphere is specified to have capacity for removing 100 curies (3.7 TBq) per day* at a tritium pressure of 0.1 µPa. This amount represents the maximum permissible leak rate from all the systems within the shell.

The largest amount of tritium is handled in the fueling and reprocessing cycle and is directly related to the small particle confinement time and low fractional burnup. The temperature of the piping and containment vessels, except within the torus, is very low and the tritium loss problem will be minimal so long as the many joints and valves perform satisfactorily; however, the outer blanket power cycle operates at a very high temperature. Here, the tritium bred in the lithium diffuses through the heat exchanger walls to the sodium and then to the helium. The pressure of tritium in the helium loop ultimately becomes that of the reactor building as long as the cleanup system in the sodium loop maintains the tritium concentration at or below the minimally required level. Losses to the building environment at this point should be low (<1 curie/day). In turn, the tritium in the lithium must be maintained at a concentration such that diffusion into the sodium does not result in a tritium loss rate that exceeds the sodium cleanup system capacity. The tritium concentration in the lithium also must be low enough to prevent substantial leakage through the piping and valves.

---

*1 curie (Ci) = 37 x 10^9 becquerel (Bq).

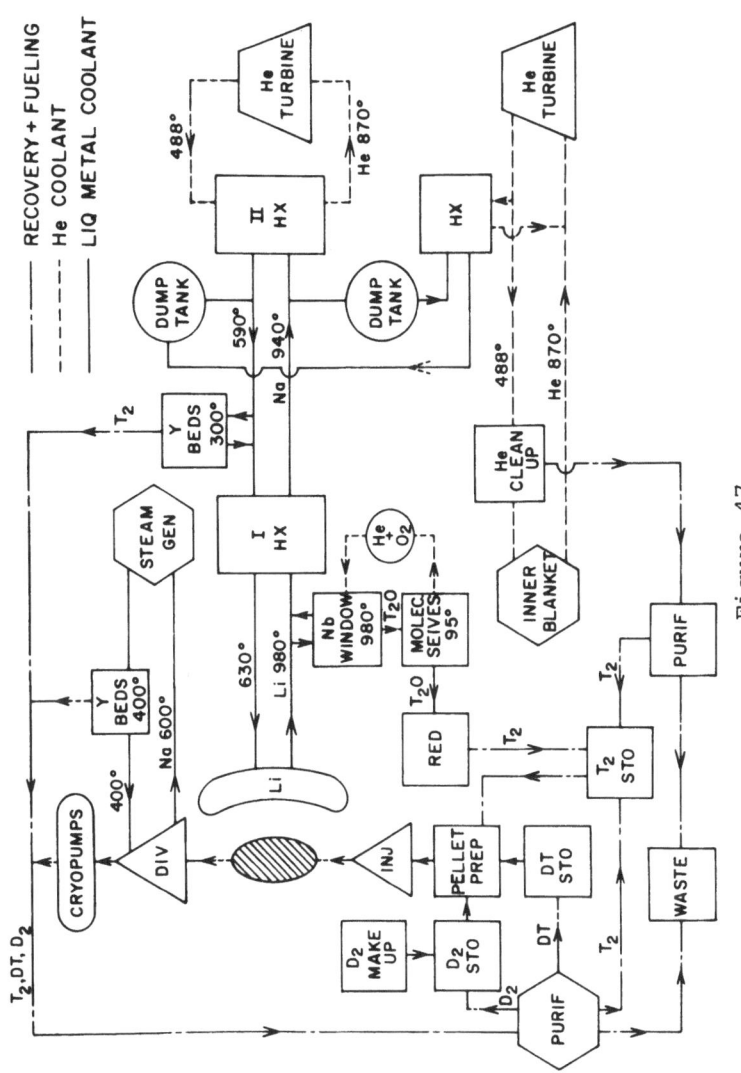

Figure 47

Schematic Flow Diagram for Tritium in UWMAK-III

A system has been designed to meet these requirements. This system would operate with a steady-state tritium inventory in the lithium of 1 kg (2.08 ppm) and a leak rate into the sodium of 1256 curies (45 TBq) per day. The sodium cleanup system will process this amount of tritium pressure over the sodium of 0.1 μPa.

Fueling the plasma is by the injection of solid D-T pellets at a rate of 1.63 kg of tritium and 1.085 kg of deuterium on a twenty-four hour basis. Of this quantity, only 0.620 kg of tritium and 0.413 kg of deuterium are consumed. The unburned fuel and product helium ions are directed to the diverter plates. The particles are collected over a five-hour period in cryopumps containing molecular sieves at cryogenic temperatures. The total inventory of tritium in the cryopumps after 5 hours is 15.5 kg. Two sets of cryopumps operate such that the one set collects hydrogen isotopes and helium while the second set is regenerated. The regenerated gaseous hydrogen isotopes and helium are freed from the molecular sieves by heating to 30 K. This hydrogen isotope stream is joined by other tritium streams and is fed to a cryogenic fractional distillation apparatus, based on a design by W. R. Wilkes.[81] Four sets of cryogenic distillation columns are required, and one should note that these will require a considerable scale-up and redesign to accommodate the 31 kg inventory of the cryopumps.

Tritium is extracted by diverting about 5% of the liquid lithium in the outer blanket to a niobium window.[82] Niobium is particularly well suited for this purpose because it has a high permeability to hydrogen at these temperatures[83], and it allows us to reduce the tritium concentration in the lithium to a level compatible with subsequent tritium losses to the environment. Other methods for the extraction of tritium from lithium, such as yttrium beds, molten salt extraction, liquid alloy getters and fractional distillation, were considered but were not found to be as attractive.

The niobium window is not without its own problems, however. The oxygen level in the niobium must be less than 100 ppm or the lithium will erode the surface, even going up grain boundaries to react with oxygen. Also, carbon transport from the TZM will occur, causing embrittlement of the niobium. In addition, at 980°C, niobium will be subject to attack by oxygen, and therefore, some protective barrier is

required. Palladium is the barrier of choice, since it has
a high hydrogen permeability. A palladium thickness of
0.001 cm, deposited by electrodeposition, will give a co-
herent layer of suitable durability. This will require a
total of 170 kg of palladium for the calculated surface area
of the window. Also, the mutual solubility of Nb and Pd is
quite high, and it is expected that the palladium will diffuse
into the niobium, though no firm data exist for the rates of
interdiffusion; therefore, an intermediate yttrium layer is
proposed that should minimize the rates of diffusion so that
a practical "window" lifetime can be achieved. Yttrium was
chosen because it, along with the other IIIa elements and
some of the rare earth metals, is among the few metals that
do not form solid solutions with niobium. The required
yttrium layer thickness is estimated to be 0.25 μm. This
will require a total of 2.0 kg of this element for the calculated surface area of the window.

The high temperature in UWMAK-III creates a situation
in which large amounts of tritium could leak through pipes.
For example, the high temperature side of the lithium system
is at 980°C and has a tritium pressure of 130 μPa. The approximate pipe dimensions are 200 m^2 and 5 mm thick. Assuming
that the downstream pressure is zero, the leak through such
piping is 92 curies (3.5 TBq) per day; however, to protect
the TZM from oxidation, the pipe is double walled, using the
outer pipe of stainless steel lined with ceramic insulating
material. Since the temperature of the jacket is much lower,
and the permeability of the ceramic is very low, the release
of tritium to the reactor is vanishingly small.

The plant inventory of tritium is estimated to be 35.8
kg (Table 15). It should be noted that of this, only 1.7 kg
is in the breeding-recovery cycle. The 34.1 kg in the refueling cycle includes 18.6 kg in reserve storage. The reason the
refueling cycle inventory is so large is that more realistic,
space-dependent plasma calculations were used in this study.
The result is a fractional tritium burnup of only 0.8%, compared to 7.2% and 2.8% burnup calculated on the basis of the
simpler plasma modes used in UWMAK-I (1) and UWMAK-II (5).
Assuming 5% burnup, the refueling cycle tritium inventory
would be only 5.5 kg, and the total plant inventory would be
7.2 kg.

Table 15

Tritium Inventories (kg)

Cryopumps	7.75
Reprocessing	7.75
Storage	18.6
Outer Blanket-Li	1.0
Recovery-Li	0.666
Outer Blanket-Na	2.0×10^{-4}
Outer Blanket-Na (Y beds)	1.6×10^{-4}
Diverter-Na	6.0×10^{-4}
Diverter-Y beds	7.5×10^{-3}
Total:	35.8 kg

H.   Radioactivity and Afterheat

Induced radioactivity and afterheat are important aspects of fusion systems based on a neutron-producing fuel cycle and the analysis for UWMAK-III is particularly interesting because of the use of TZM rather than stainless steel as the primary structural material. The procedure for calculating radioactivity in the UWMAK-III system follows very closely that developed in previous work.[1,5,84,85] Modifications have been made to account for the two-dimensional characteristic of the UWMAK-III blanket, as well as to incorporate more second-order reactions.

The buildup of the total activity in the outer blanket is shown in Figure 48. The rapid initial buildup is due to the production and saturation of the short half-life isotopes, ^{99}Mo ($t_{1/2}$ = 66 hr), ^{99m}Tc ($t_{1/2}$ = 6 hr) and ^{91}Mo ($t_{1/2}$ = 16 m). The activity approaches a constant value after one to two months. If the blanket were to be left in for a long period of time (> ~5 year), a slight decrease in activity would result because of the burnup of some of the parent nuclei overriding the slow buildup of long-lived nuclei. Figure 48 also shows the buildup in the first breeding zone due to the large amount of material contained therein; however, the highest specific activity is, of course, in the first wall. The first wall and the first breeding zone are the major contributors with the reflector contributing only ~5% of the total.

The activity in the blanket after shutdown is of considerable interest not only from the point of view of radiation levels during the blanket changeout or other maintenance, but also because of its implication in accident potential, recycling and long-term radioactive waste. Figure 49 shows the activity of the outer blanket and the inner blanket first wall following two years of operation. The activity levels at shutdown are about the same as in UWMAK-I and II; however, the behavior of the activity following shutdown is quite different for TZM than for 316 stainless stell. In particular, after a slow decay for the first day or so, there is a rapid decay where the activity drops by a factor of 100 in only one week. This is in contrast to stainless steel systems where the time is measured in years for this to take place. After three years, the short-lived activities in the TZM have died away and the activity has assumed its long-term value, which is almost four orders of magnitude below the shutdown value. The long-term activity is dominated by ^{93}Mo ($t_{1/2}$ = 3 x 10^3 yr) and ^{93m}Nb ($t_{1/2}$ = 12 yrs), with a small contribution from ^{99}Tc ($t_{1/2}$ = 2 x 10^5 yr). This residual activity is higher than would be associated with a SS system, but its effect is put in better perspective in the discussion of afterheat. It is seen that the first wall contributed about 25% of the activity at shutdown.

The buildup of the afterheat is shown in Figure 50. As was the case with the radioactivity, saturation is seen to

occur after about three weeks of operation. The major portion
of the afterheat comes from the first breeding zones as expected.

The decay of afterheat following shutdown is shown in
Figure 51 for the case of two years of operation. Tritium
is neglected. The sharp initial drop from 0.31% to 0.24%
immediately following shutdown is due to the decay of very
short-lived ^{6}He ($t_{1/2}$ = 0.8 sec). The decay heat decreases
more rapidly than the activity and has gone down a factor of
100 after one month. By the end of one year, however, the
afterheat is approaching its long-term value of approximately
10^{-6}%, or a total heating of only ~50 watts. Although the
activity is dominated by ^{93m}Nb and ^{93}Mo, because of the
larger amount of energy in its decay, the afterheat at long
times is primarily due to ^{99}Tc. This level of afterheat is
greater than that expected for a SS system, but less than
that of an Nb based structure.

We have also calculated the "Biological Hazard Potential"
or BHP for UWMAK-III. BHP is defined as the activity in
curies/kW$_{th}$ divided by the maximum permissible concentration
(MPC) for general public exposure in curies/km^3 of air,
as taken from 10CFR20, Appendix B, Table II, Column 2. The
BHP is the volume of air required to dilute the activity per
thermal kilowatt to MPC. It has proven a useful if not
definitive way of making comparisons between systems and of
making initial assessments of potential hazards of catas-
trophic events.

The BHP of the total blanket at shutdown after two years
of operation is shown in Table 16. As is expected, the values
are higher and are due primarily to the outer blanket. The
BHP after 200 years of decay is also shown, and these values
are reduced by a factor of 500 from the values at shutdown.
It is clear that the major contribution comes from five
relatively short-lived isotopes and that the total BHP is of
the same order as for the stainless steel first wall of
UWMAK-I. It should be noted that contributions for isotopes
not listed explicitly in 10CFR20 have not been included because of the unrealistically high values that would result if
the general MPC value for unspecified isotopes was applied.

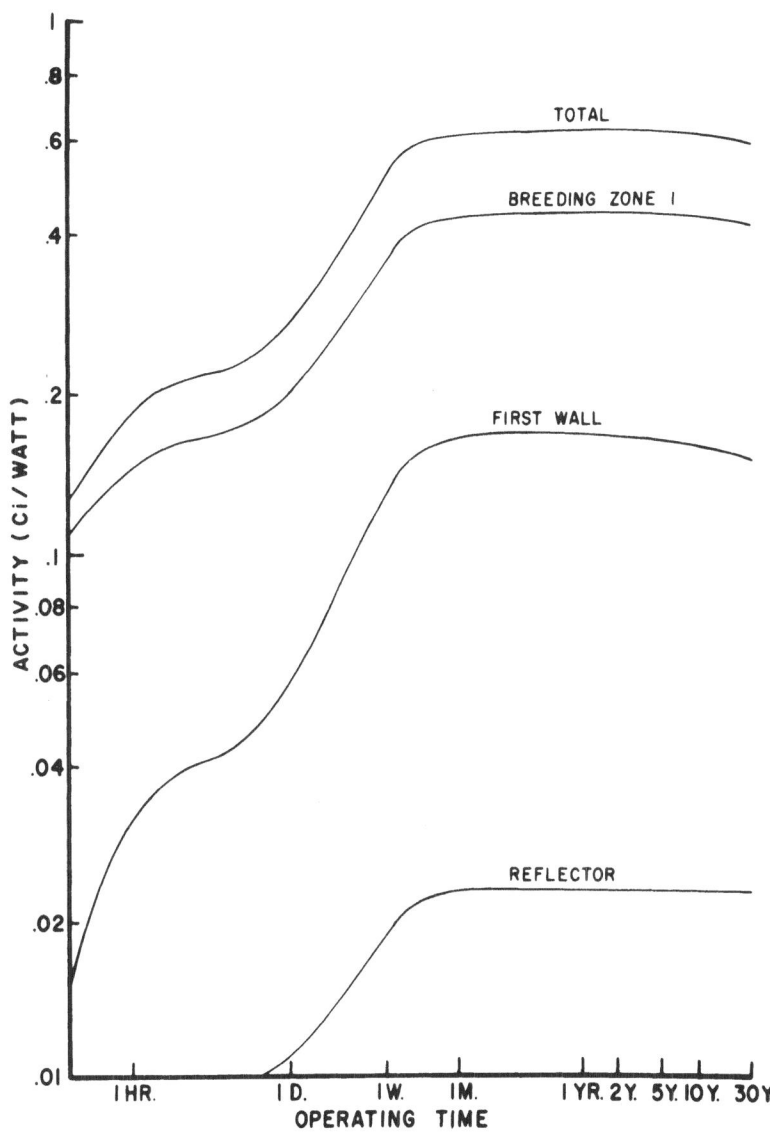

Figure 48
UWMAK-III Outer Blanket Activity
At Shutdown vs Operating Time

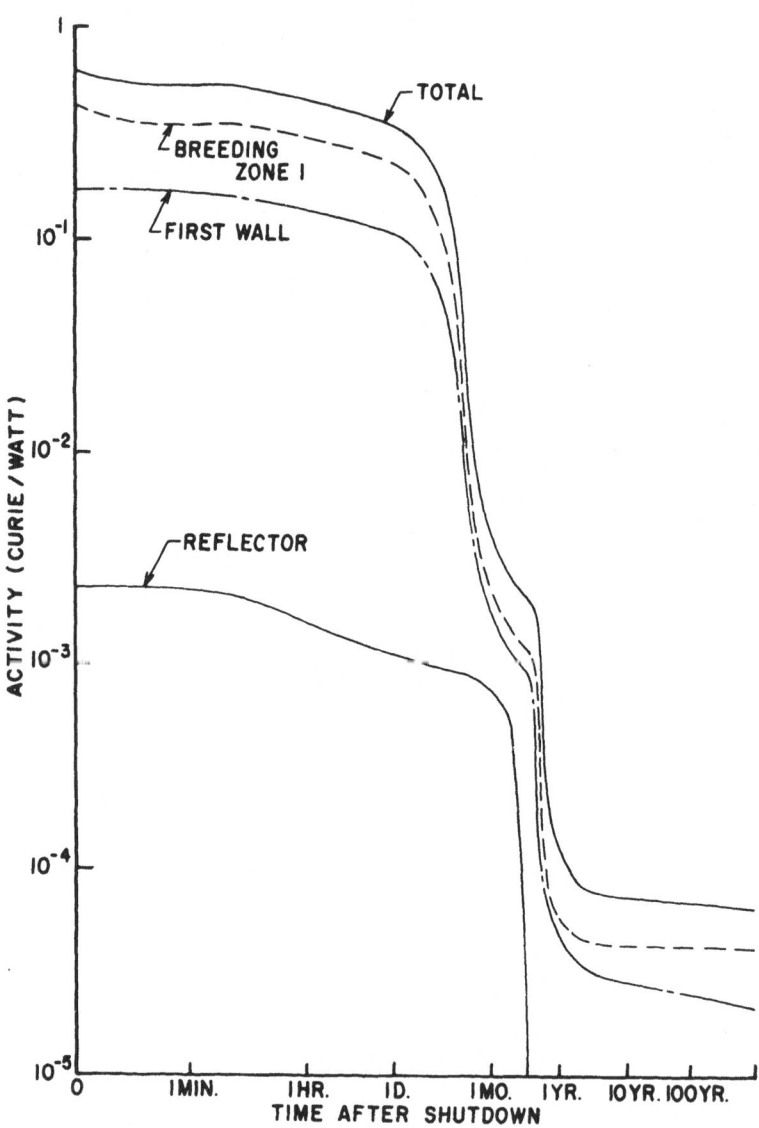

Figure 49

Decay of Radioactivity in Outer Blanket
After Two Years of Operation

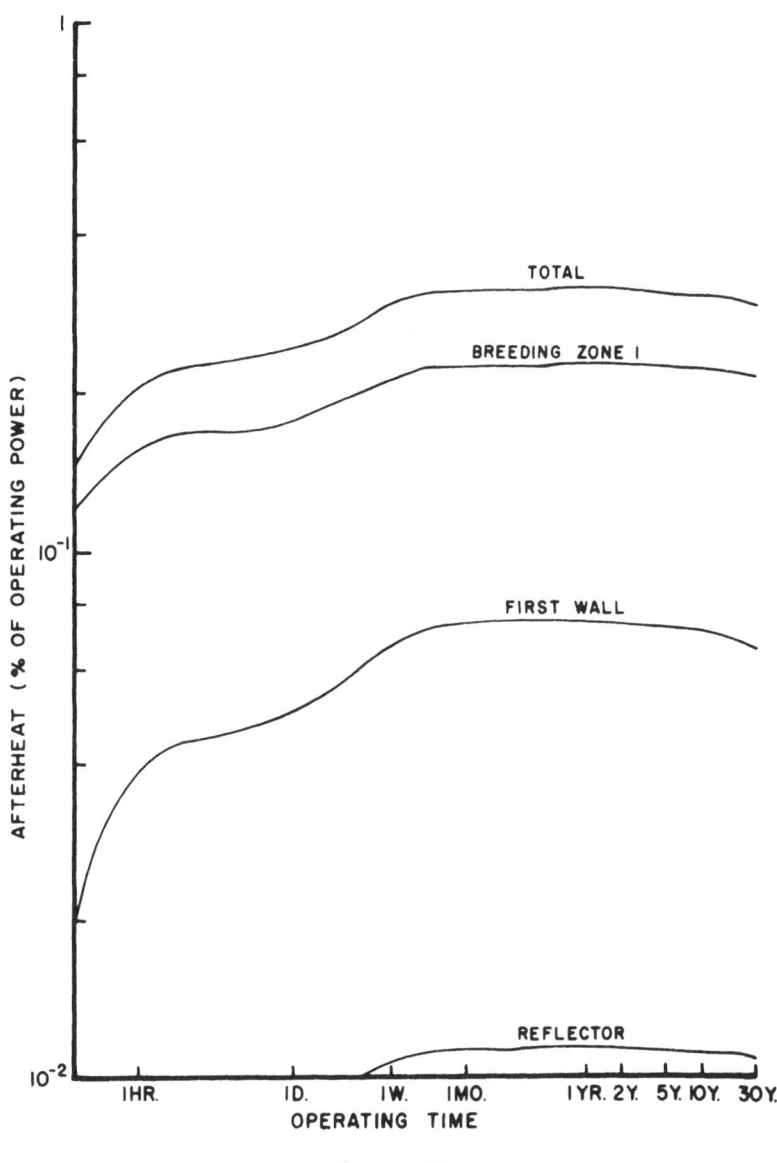

Figure 50

UWMAK-III Outer Blanket Afterheat at
Shutdown vs Operating Time

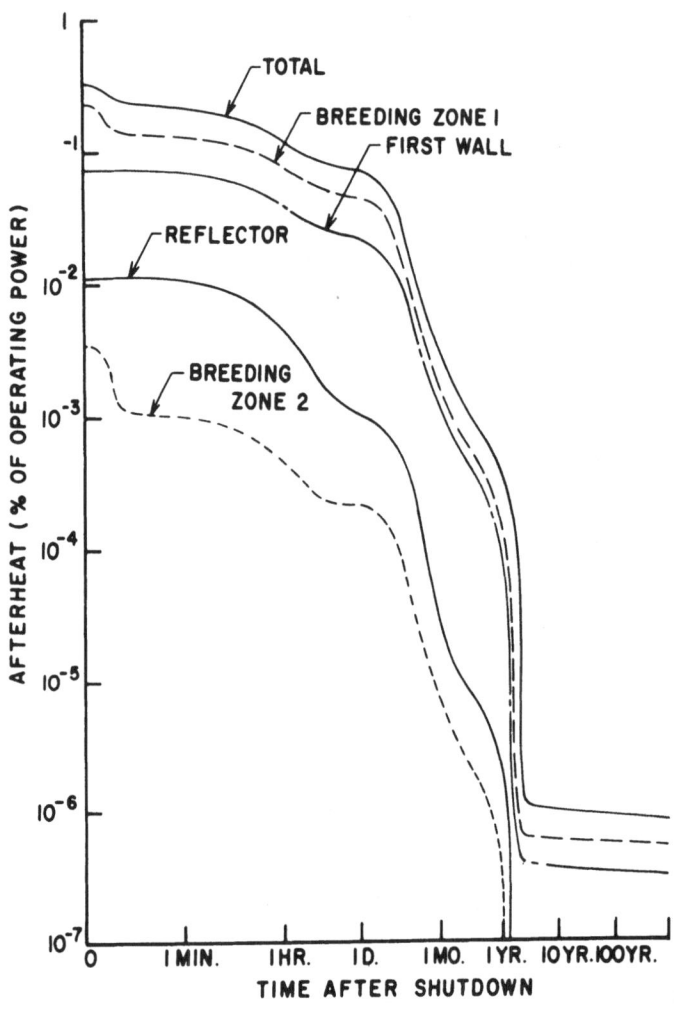

Figure 51

Decay of Afterheat Level as a Function of Time After
Shutdown for Outer Blanket Following Two Years' Operation

Table 16

BHP of Dominant Isotopes at Shutdown in UWMAK-III
Two-Year Operating Time

BHP (km^3 air/kW_{th} Operating Power)

Isotope	Outer Blanket	Hot Shield	Total
Nb 92^m	44.43	7.11	51.54
Zr 89	38.27	5.66	43.93
Nb 95^m	30.24	4.60	34.84
Mo 99	28.63	17.27	45.91
Nb 96	18.64	2.60	21.24
Nb 91^m	9.81	1.25	11.06
Tc 101	1.46	1.49	2.95
Mo 101	1.46	1.57	3.03
Mo 95	1.60	.24	1.84
Zr 95	1.02	.16	1.18
Mo 91	1.01	.12	1.13
Total (including other isotopes)	181.38	43.66	225.04

BHP of Dominant Isotopes 200 Years after Shutdown
Two-Year Operating Time

Mo 91	.32	.15	.47
Nb 93^m	.01	~ 0	.01
Total:	.33	.15	.49

I.   Materials

The choice of the molybdenum-based alloy, TZM (99.4% Mo, 0.5% Ti, 0.1% Zr, 0.01% C), is governed in part by the basic philosophy of the UWMAK-III study; i.e., an advanced high-temperature reactor, and in part on the assessment that TZM is the most attractive of the materials with the potential for high-temperature (~1000°C) operation. On the other hand, it is fully anticipated that near-term fusion systems and most likely the first demonstration reactors will use an austenitic stainless steel as the basic structural material. The reasons for this are presented in previous work (1,5), where 316 stainless steel was the structural material chosen. Such systems have upper limits to the temperature that can be achieved such that the overall efficiency of the plant will be in the 35 to 39% range. To achieve still higher plant efficiencies, it is necessary to go to higher temperatures, and this implies the use of either refractory metals (V, Nb, Mo or Ta), or nonmetals such as carbon or silicon carbide. The use of nonmetals as structural materials seems very difficult, and the choice of TZM was based on a comparison of the metals and alloys that potentially can meet the necessary criteria. A detailed exposition on the case for Mo-based alloys over other high-temperature materials has been given by Kulcinski, Davis and Schmunk.[86]

At least 10 different properties must be considered in deciding on a structural material for a fusion reactor. These are physical properties, mechanical properties, thermal properties, fabricability, neutronic properties, chemical compatibility with coolants, radiation damage , materials availability, interaction with tritium, and cost. Obviously, no metal or alloy will satisfy all these criteria, and the ultimate choice of a CTR structural material will depend upon the compromise between all of the 10 properties mentioned above. At the present time, there are only five metal systems seriously considered for CTR application, and each has its own temperature range of importance. These are summarized below:

Low temperature (~200°C)	– Aluminum alloys Austenitic steels
Medium temperature (200°-600°C)	– Austenitic steels Vanadium alloys

High temperature (600°-1000°C)	– Vanadium alloys
	  Nb alloys
	  Mo or Mo alloy

Three physical properties of metals are of major importance in fusion reactor technology. These are density, electrical resistivity, and vapor pressure. Table 17 summarizes these properties for the five systems considered here. The mass density is important with respect to dead weight loads of massive structures. This load is a complex function of yield or creep strength for a given internal and external pressure load, and the mass of the structure depends roughly on the ratio of $\rho(T)/\sigma(T)$, where $\rho(T)$ is the density at a given temperature and low ρ/σ are desirable. In the low temperature (<200°C) range, the Al alloys are best with Nb alloys clearly much worse. In the intermediate range (400°-600°C), aluminum alloys are no longer applicable, and V-20Ti is clearly the best, with 316 SS and TZM next. Finaly, in the high temperature range (600°-1000°C), V and Mo alloys are the best, while Nb-1Zr is clearly much worse.

There are five main mechanical properties that are of interest to fusion reactor designers: yield strength; high-strain rate ductility; creep strength; fatigue life; and fracture toughness. These properties have been compared for Mo, TZM, Nb, 316 SS and aluminum; it is found that over the temperature range of anticipated operation, molybdenum (or its alloys) has the highest yield strength, creep strength, and fatigue life of the metals compared; however, its ductility is low and its fracture toughness (ability to resist propagation of existing cracks or those produced by fatigue or thermal cycling of the system) is about the same, or slightly worse than for the other bcc metals.

As for thermal properties, a high value of thermal conductivity is required to minimize induced stresses in rigidly bound blanket components. Molybdenum is clearly superior over the entire temperature range whereas the extremely high coefficients of Al and steel could present some problems.

A large heat capacity is desirable to minimize wide thermal fluctuation for given energy inputs and also to act as an energy reservoir in the event of an accident situation. Aluminum has the highest heat capacity for the low temperature region, with Mo taking that honor for the intermediate and high-temperature regimes. A summary of selected thermal properties for the materials under discussion is given in Table 18.

Table 17

Selected Physical Properties
of Potential CTR Structural Material

Property	Al	Steel	V	Nb	Mo
Atomic Weight	27	56	50.9	92.9	95.9
Melting Point (°C)	660	1430	1900	2468	2620
Density (Mg/m^3)	2.7	7.98	6.1	8.6	10.2
Atomic Density (10^{28} atoms/m^3)	6.0	8.5	7.2	5.6	6.4
Electrical Resistivity: (μΩ-cm) - 100°C	3.9	78	24	17	8
400°C	9	100	44	31	11
800°C		122	66	43	18
Vapor Pressure at Temperature (°C): 10^{-7} torr	800	900	1200	1920	1830
10^{-4} torr	1150	1100	1810	2250	2350
10^{-1} torr	1660	1500	2050	3200	

The important thermal figure of merit to consider (86) is $\alpha E/k(1-\nu)$, where α is the thermal expansion coefficient, E is Young's modulus, k is the thermal conductivity, and ν is Poisson's ratio. These are listed in Table 19 for the materials considered here. The materials fall into three distinct classes: those that have very low values of the figure of merit (Al and its alloys), those with intermediate values (Mo, Nb, and V), and those with very high values (stainless steel). This comparison shows that the stresses induced in Al are much lower than in the other systems (a factor of 3 lower than for steel). Within the refractory

Table 18

Selected Thermal Properties
of Importance for CTR Materials (86)

Property	Al	Steel	V	Nb	Mo
Thermal Conductivity (W/mK):					
100°C	2.34	.17	0.3	0.5	1.3
400°C	2.30	.21	0.4	0.6	1.3
800°C		.27	0.4	0.8	1.2
Coefficient of Thermal Expansion (10^{-6}/K):					
100°C	25	16.5	9	7	5
400°C	30	17.5	10	7	5
800°C		19	11	9	6
Heat Capacity (J/kg K):					
100°C	920	460	500	290	670
400°C	1090	540	540	335	710
800°C		630	630	335	795

system, the order of increasing stresses is V, Mo, and Nb, but the difference is not more than 15%.

Kulcinski, Davis and Schmunk (86) have also reviewed the area of fabricability for the various structural materials. This property is one of the most important and yet one of the most difficult to measure quantitatively. The final proof of whether a material can be fabricated is the ability to construct both large and small components of varying shapes and thicknesses. Such structures should be crack free and should be able to maintain vacuum of high-pressure loads.

Table 19

Figure of Merit for Thermal Stresses
in Potential CTR First Wall Materials

(Temperature $^{\circ}$C)

Material	25	200	400	600	800	1000
Mo	3620	3450	3250	3270	3250	2980
Nb	3810	3460	3230	3260	3580	3620
V	3490	3440	3430	3270	3100	
316 SS	27,700	24,300	21,400	18,100	16,300	
Al	1100	1200	1300	1400		

$(\alpha E/k(1-\nu))$, newton per meter-watt. A number as low as
possible is desired.

small components of varying shapes and thicknesses. Such
structures should be crack free and should be able to maintain vacuum of high-pressure loads.

There is no need to discuss the fabrication and joining
of Al alloys and stainless steels. It is even relatively
easy to fabricate *pure* V and Nb into complex shapes and tubes.
However, joining of these last two metals is not easy, and
must be performed under closely controlled conditions to avoid
interstitial pickup. The fabrication of V-Ti alloys is not
easily accomplished and must be carefully controlled to avoid
cracking in large angle bends.

Much of the development of the refractory metals, and in
particular, the Mo alloys, was initiated in the mid-to-late
1950's. At that time, the refractory metal industry was in
the process of developing new alloys. At the same time the

aerospace industry was beginning to consider their use as
primary structures. The original accent was on strength at
temperature, and not on fabricability. Alloying elements
such as carbon, zirconium and tungsten were used in ever
increasing amounts, until some of the alloys were so brittle
that they could not be easily fabricated. It was at this
point that most of the reports concerning the difficulty of
fabricating refractory alloys began to emerge, and many programs designed to use these alloys were modified or stopped.

Many improvements have been made in the last 15 years,
and many of the brittle and difficult-to-fabricate alloys
have been eliminated. New, improved processing and casting
techniques are now used, and the metals in general have lower
ductile-to-brittle transition temperatures (DBTT) as a result of lower interstitial content. This trend can be seen
in the molybdenum industry where alloys such as Mo-0.5 Zr,
Mo-0.5 Ti and TZM were created. For one reason or another,
most of these alloys have been eliminated from major application with the exception of TZM. This alloy initially was
difficult to fabricate, and most of the forming was performed
at elevated temperatures; however, as the metal suppliers
began to understand more about the metallurgy of the alloy,
the forming temperature began to drop. Then, with the advent
of the Navy Sheet Rolling Program[87], significant improvements were made in the formability of the alloy. As improvements were made in the alloy, the formation temperature
began to drop until it is now at room temperature.

The TZM produced now is the product of technology developed as part of the Navy Sheet Rolling Program (87). This
material was evaluated in the structural sheet evaluation
program conducted at McDonnell-Douglas (88) and structures
have been fabricated that have sharp angles of over 120°,
and where the panels are joined by welding; thus, many
molybdenum and TZM structures have been fabricated and
operated successfully over the past 15 years. While welding
large components of Mo alloys in the field will be difficult
(as will the welding of Nb and V alloys), it will not be
serious enough to remove Mo or TZM from consideration as
viable CTR structural material.

The neutronics aspects of a reactor material are critical
if breeding ratios of one or greater are to be achieved. Five
cross-sections (or sets of cross-sections) are important, and
these are:

absorption (n,γ), neutron multiplication (e.g. $(n,2n)$)
hydrogen isotope gas production, and inelastic scattering.

These reactions can lead to induced radioactivity, afterheat,
and chemical changes in the host material.

Particular problems for Mo are its large number of isotopes and the fact that Mo has not been used as a major fission
reactor material, and therefore, received little attention
until now. The problem is that not all of the above cross-
sections have been measured for seven Mo isotopes and one must
rely to a large degree on theoretical calculations.

The general conclusions that can be drawn from an examination of the available nuclear data are: (1) aluminum is
most desirable in the low-energy neutron range (E < 0.1 MeV).
This element is followed by 316 SS, V, Mo and Nb in order of
increasing cross-section. In fact, the Nb parasitic absorption cross-section is two orders of magnitude higher than Al
in this range; however, alloying elements may modify this
conclusion; (2) the order of desirability is the same in the
intermediate (0.1 to 1-3MeV) range with the exception that V
now has a lower parasitic cross-section than 316 SS; and (3)
in the high-energy regime (3-14 MeV), Mo actually has the
lowest parasitic cross-section, followed by Nb, V, Al and
316 SS in that order. The high-energy cross-sections are
dominated by (n,p) and (n,α) reactions that are particularly
bad for the Al and 316 SS system where they exceed the Li
(n,t) values by factors of ~5. This competition not only
removes a neutron that could have produced another tritium
atom, but it does so at an energy where the Li-7 (n,t) reaction could give another neutron back to the system, along
with a tritium atom.

The $(n,2n)$ reaction partially compensates for parasitic
absorption and analysis indicates that the final breeding
ratio is generally highest with V, about the same for Mo and
stainless steel, and lowest for Nb when the same volume percent structure is used in a liquid lithium-cooled blanket.[89]

The gas production cross-sections of all the refractory
metals (V, Nb, Mo) are low, compared with Al or stainless
steel. This is advantageous, particularly for the case of
helium production, because this can lead to severe embrittlement.

The ability of structural materials to withstand the
corrosive environment of high-temperature coolants is essential. For fusion, only three coolants appear to be feasible
at the present time: helium gas, liquid lithium, and lithium
salts. It is found that aluminum can be used only with helium gas, and it is severely attacked by liquid metals.
Steel can be used up to the creep limit with helium and
Li-Be-F, but studies have recently shown that excessive cor-
rosion takes place in dynamic liquid lithium above 500°C.

Vanadium, niobium and molybdenum are all immune to
liquid lithium for temperatures approaching half the melting
point of the structure; however, V and Nb are susceptible to
attack by the fluorine in Flibe salts. Vanadium and Nb are
both limited to 700 °C and 800 °C, respectively. Since the
solubilities of 0_2, N_2 or C are so low in Mo, this metal can
be safely used to 1000°C. In summary, the Mo or TZM system
represents the best system from a standpoint of chemical
compatibility. This is also important with regard to the
transport of radioactive elements around the coolant loops.

Tritium containment within the blanket is clearly important, and this means that the metal should have a low diffusivity for hydrogen isotopes at the operating temperature
of the metal. Figure 52 shows the variation of hydrogen
diffusivity with temperature in five metal systems. It is
clear that Mo and Al make effective diffusion barriers to
tritium at their expected operating temperatures. The 316 SS
has a diffusivity at 600°C about equal to that in Mo at
1000°C; however, the hydrogen diffusivity in Nb and V are
much too high to allow tolerable leakage rates.

Taking all of the above factors into account, it appears
that Mo (or TZM) is superior to other metals for very
high temperature fusion reactor applications, mechanical
strength, compatibility, availability, and cost. Mo or TZM
is approximately the same as V and Nb with respect to physical, thermal, neutronic and radiation damage considerations.
The biggest drawback to Mo is its difficult (but not impossible) joining problems, particularly in the field. If
this problem were to be solved in the next 10-20 years, an
Mo system certainly would look to be the most favorable candidate for high-temperature applications.

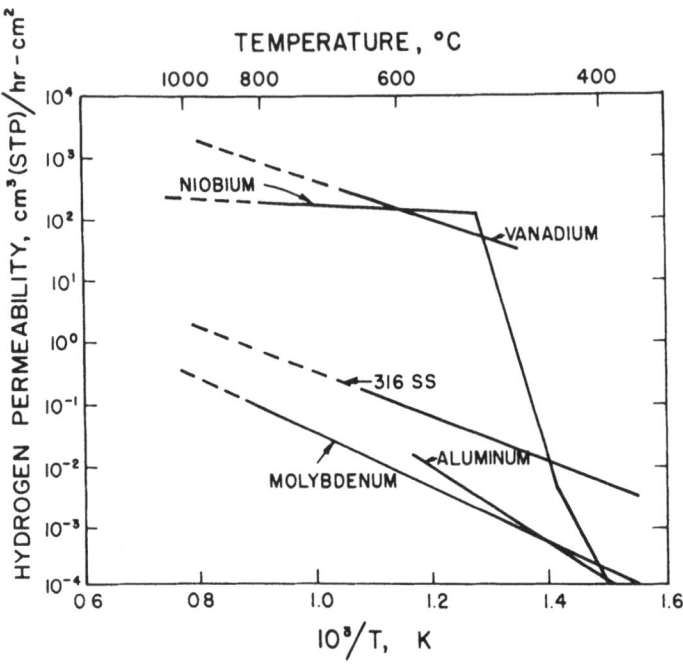

Figure 52

Tritium Permeability as a Function of Temperature
In Several Candidate CTR Materials

Up to this point we have discussed, primarily, material
properties without considering the effects of neutron irradiation damage. The neutron and gamma environment is especially
severe in a D-T fusion system because the 14 MeV neutron is
not only quite energetic, but is also above the threshold
energies where gas production reactions like (n,p) and (n,α)
become important (this threshold is related to the nuclear
Coulomb barrier and typically is about 4.5 MeV). Thus, atoms
will be displaced from their lattice sites while large amounts
of gas are being generated in the material. In Part E of
this section, we reported the values for the helium production rate (in appm/yr) and atomic displacement rate (in dpa/yr)
in the first wall of TZM. The problem now is to determine the
effect of this radiation damage on the properties of the materials. For UWMAK-III, the relevant materials are TZM and

carbon, shielding materials such as boron carbide, lead and 316 stainless steel, and magnet materials such as mylar (superinsulation) and NbTi.

A summary of all known irradiation studies on Mo up to 1975 has been given by Schmunk and Kulcinski.(91) Kulcinski, in Reference 8, has attempted to interpret these data and deduce the effects of radiation damage on the TZM in UWMAK-III. Much of the data are for pure Mo rather than for TZM, and extensive data are not available in the high-temperature range from 700°C to 1000°C. For reference, one year of operation in UWMAK-III corresponds to about 20 dpa and about 150 appm of He in the first wall exposed directly to the 14 MeV neutrons. Unfortunately, there are no data with such high helium content at these dpa levels. From the limited data on swelling, it is deduced that the swelling might vary from as little as 0.4% to as much as 12% in the 10-15 dpa range at 700-1000°C.

Irradiation also affects mechanical properties such as tensile strength, creep, hardness and fatigue. It appears that irradiation of Mo to greater than 5 dpa at less than 400°C causes an increase in the DBTT from the range of -80° to 20°C up to 550-650°C. Irradiation at temperatures above 800°C appears to have much less effect on the DBTT, and actually increases the post-irradiation creep life. It is also found that the fatigue life increases by a factor of two, and this is important since the number of burn pulses per year in UWMAK-III is about 12,500, at 80% plant factor. Injection of helium up to 10-20 appm and subsequent testing increases fatigue life at about 400°C and has little effect on the ductility at 800°C.

For these reasons, it is tentatively concluded that the mechanical properties of Mo and TZM will not be severely degraded at 800°-1000°C for dpa levels up to 15-25. Clearly, however, the data base upon which to make a final judgment is far from sufficient.

Graphite is used in three forms in UWMAK-III. A thick zone is employed to moderate neutrons in front of the first TZM wall on the inner portion of the torus.(56) A thin carbon curtain is used in front of the first wall on the outer portion of the torus to protect the plasma from high Z impurities sputtered from the first wall.(64) Finally, graphite

is used behind the tritium breeding zone of the outer
blanket as a neutron reflector. The situation regarding
radiation damage to graphite at temperatures less than
1400°C has been reviewed by Gray and Morgan (92), and also
in Reference 5 by Kulcinski. Initially, the graphite shrinks
on the order of 1-3% in volume before there is the onset of
swelling. For temperatures less than 1400°C, the shrinkage
is reversed, and growth begins at a neutron fluence of 10^{24}
neutrons/m^2; however, at temperatures in excess of 1300°C
to 1400°C, the shrinkage phase appears to last much longer
although this can be dependent on the type of graphite. Van
den Berg et al. (93) have recently discussed irradiations
up to 1400°C, and in general, find that the reversal of the
shrinkage phase is much slower than at lower temperatures
and remains in the -1% to -2% range for fission fluences up
to 1.5 x 10^{26} n/m^2. Morgan, Woodruff and Gray (94) attribute
differences in behavior up to 1300°C to differences in graphite
morphology so that it appears, with the appropriate choice of
graphite, that one would be able to use it at temperatures
greater than 1400°C. Such temperatures would be reached and
exceeded in the ISSEC zone discussed in parts A, E and F of
this section. Another interesting aspect of the use of
graphite in a 14 MeV neutron environment is the large production rate for helium. Helium is produced by reactions such
as ^{12}C(n,n'3α) and ^{12}C(n,α)^{9}Be, and the generation rate can
be as high as 3000 appm He in graphite for a 14 MeV neutron
wall loading of 1 MW/m^2; however, several recent experiments (95)
have shown that the helium either implanted or generated in
situ is essentially totally released at temperatures above
1300°-1400°C. Therefore, one expects little impact from the
high helium generation rates in both the ISSEC and the graphite curtain. In the graphite reflector, the neutron flux is
greatly reduced, and the spectrum is much softer, so that
the lifetime of the graphite there will be much longer than
that of the TZM.

Surface effects such as physical and chemical sputtering,
photodesorption and blistering have been reviewed by Behrisch
(55) and by McCracken (95). For carbon, the maximum physical
sputtering yield for hydrogen on carbon is about 10^{-2} carbon
atoms per incident hydrogen atom, with the peak occurring at
about 300 eV. In certain temperature ranges, one can expect
enhanced carbon removal due to the formation of hydrocarbons,
particularly methane (CH_4) and acetylene (C_2H_2). Erents, Braganza and McCracken (97) have investigated methane production

during hydrogen ion bombardment as a function of the temperature of pyrocarbon samples. The ion energy (H^+ or D^+) was 20 keV. They find that the methane production is less than 0.2% of the beam level at temperatures below 175 °C or above 925 °C. In between, the methane production rate peaks at a temperature of about 550 °C and at an equivalent carbon removal rate of 6-8 x 10^{-2} carbon atoms per incident hydrogen atom. Effectively, no acetylene was observed for temperatures up to 1000 °C. Balooch and Olander[98] have investigated the interaction of thermal hydrogen (H and He) with graphite at temperatures approaching 2250 °C. They also find significant production at 400 °-600 °C, and that this drops to very low levels above 800 °C. Acetylene formation begins to be observed above 1000 °C, and increases with increasing temperature, and extending their data, it appears that another maximum may exist at about 2000 °C, but the equivalent yield would still be about 10^{-2}. Vacuum properties have been shown to be good[64,99], so the use of graphite as a thin curtain or as a thick moderating zone (ISSEC) appears promising.

The last component to be discussed from the viewpoint of radiation effects is superconducting magnets. The main elements here are the superconductor itself, the insulation (typically mylar), epoxy electrical insulation, the stabilizer (usually Cu, but Al in UWMAK-III), and the magnet structural material. The most important, or limiting item, appears to be radiation damage to the superinsulation. The threshold for damage is at about 30 Mrads (0.30 Mgy)* and the mylar is expected to lose about 25% of its tensile strength at 120 Mrads (1.20 Mgy). This latter value typically is used to design radiation shields. Other effects are less severe and include resistivity changes in the conductor stabilizer and decreases in the critical current density for the superconductor.

J. Power Conversion System

The power conversion system for UWMAK-III has two distinct components; the power cycle associated with the heat deposited in the inner hot shield and the outer blanket, and the heat deposited in the diverter by particles diffusing from the plasma. The advanced power conversion system designed for the high-temperature coolants of the inner hot shield and the outer blanket is based on a closed-cycle helium gas turbine system and yields a total power blanket output of 1755 MW_e. The diverter power cycle is based on a conventional Rankine

---

*1 rad = 0.01 gray (gy), J/kg.

steam power cycle to utilize the 687 MW_{th} (average) in the
diverter. The sodium coolant of the diverter plates leaves
the reactor at 600°C and the total power output is 295 MW_e.
The gross electrical power output is 2050 MW_e and the net
electrical output is 1985 MW. This gives a net plant efficiency of 4.19%. A summary of system parameters is given in
Table 20, and an extensive description has been given by
Conn and Kuo.[100]

The closed-cycle helium gas turbine offers the potential
of high performance, low cost and less environmental effect,
and would appear most suitable for integration (directly or
indirectly) with advanced-design fusion reactors. In addition to the attractive thermal efficiency, the helium gas
turbine will be smaller in size (about one-quarter) compared to the steam turbine of the same output capacity,
primarily because of a low overall expansion ratio (about
3:1 for the helium turbine, compared to approximately
3000:1 for the steam turbine). Also, the closed-cycle helium
turbine power system will be more adaptable economically to
dry cooling because heat injection takes place over a temperature range much higher than the condenser temperature in
a Rankine steam power system.

From the viewpoint of power utility, a fusion power plant
must be capable of delivering uninterrupted constant-output,
50 or 60 hertz AC power under normal operating conditions.
This requires that the turbomachines operate at 3000 or 3600
rpm, and that the operational characteristics of the power
conversion systems must be adaptable to the existing power
grids. Based on the power plant output level mentioned above,
the two-zone blanket design, and the need to provide a reasonable system redundancy, turbomachinery unit capacity of approximately 500 to 600 MW_e would be required.

The decision to have a helium-cooled inner blanket without tritium breeding, and a lithium-cooled outer blanket with
tritium breeding would mean that use of a primary heat exchanger to enable the containment and primary extraction of
tritium within the lithium loop will be imperative (see
Figure 53 left and top diagram). If no intermediate loop is
used, then the secondary loop of the primary heat exchanger
and the inner blanket can be connected in series (Figure 53
right and top), or in parallel (not shown), or separately
(Figure 53 right and bottom) with helium gas turbine power

Table 20

Power Cycle Parameters for UWMAK-III

Outer Blanket

Thermal Power	2668 MW
Power Cycle	Li-Na-He
Li-Na IHX Area	5750 m^2
Na Temperatures	540-940 °C
Na-He IHX Area	12,300 m^2
He Temperature	488-870 °C
Power Output	1170 MW

Inner Blanket

Thermal Power	1334 MW
Power Cycle	He-Direct
Helium Temperatures	488-870 °C
Helium Pressure	7 MPa
Na-He HX Thermal Storage	6150 m^2
Power Output	585 MW

Diverter Cycle

Thermal Power	686 MW
Power Cycle	Na-Steam
Na Temperatures	400-600 °C
Steam Generator Area	6176 m^2
Steam Temperature	538 °C
Steam Pressure	16.5 MPa
Power Output	295 MW

Plant Power System

Thermal Power (Continuous)	4735 MW
Gross Electrical Power	2050 MW
Net Electrical Power	1985 MW
Net Plant Efficiency	41.9%

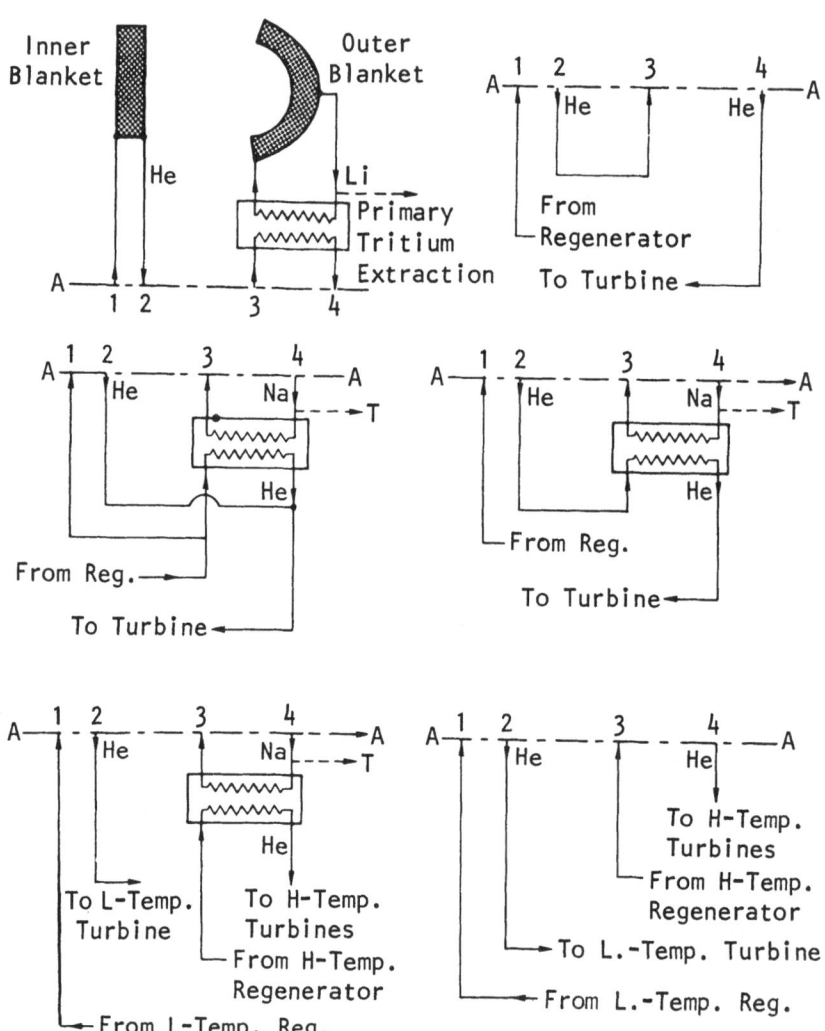

Figure 53

Various Integration Schemes
For UWMAK-III Power Conversion System

systems. However, the need for secondary extraction and/or
stringent containment of tritium and the need to store thermal energy using a liquid metal would require an intermediate
sodium loop at a penalty of increased temperature drop between
the lithium coolant and the helium turbine inlet. In this
case again, the inner and outer blankets can be connected in
series, in parallel, or separately with the helium gas turbine
power systems. Since the inner blanket does not involve tritium breeding, a separate connection scheme would be generally
preferable. In this case, the power split between the inner
and outer blanket loading must be controlled such that the
number of gas turbine loops and their unit capacity would be
acceptable from the economic and operational points of view.

The final reference design conditions for the baseline
cycle configuration were selected, based on interface problems and preliminary trade-off considerations. The temperature and pressure for the major cycle points are identified
in the cycle flow diagram shown in Figure 54. The turbine
inlet temperature of 870°C was selected, based essentially
on projected blanket capabilities with consideration given to
the heat exchanger temperature drops and projected turbomachinery technology. As pointed out earlier, the limiting
factor is the blanket outlet temperature, and thus, the
selection of turbine inlet temperature involved more system
constraints than trade-offs. The compressor inlet temperature of 45°C was selected, based on the dry cooling tower
requirement and reasonable precooler and intercooler designs;
despite the dry cooling tower, use of intercooling for more
than three points gain in cycle efficiency is justifiable,
considering the potential saving in power generation cost
over the lifetime of the fusion plant versus capital cost
of the intercooler.

Selection of the turbine inlet pressure is based largely
on component sizes and heat-transfer considerations. As will
be mentioned later, use of 7 MPa inlet pressure was mandatory
for a turbomachinery unit capacity in the 500 to 600 MW_e
range. The overall pressure ratio of 2.8 (or 2.6 for the
turbine) was selected for a maximum cycle efficiency, and the
intercooling pressure was selected so required compressor work
will be at a minimum. The regenerator effectiveness of 88%
represents a compromise to accommodate the integration constraints, yet generally provide a balance between the system
performance and costs. Based on the operating conditions

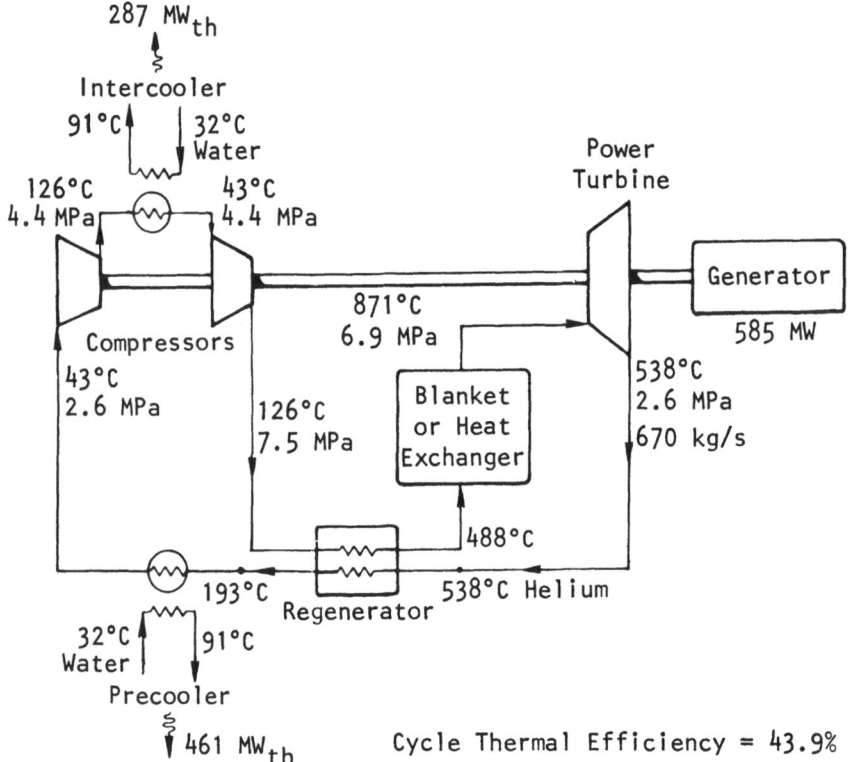

Figure 54

Reference-Design, Closed-Cycle, Helium Gas Turbine System
For UWMAK-III Fusion Plant. Three Loops Needed.

specified in Figure 54, the cycle efficiency was estimated
to be 43.9%, thus allowing an overall plant efficiency of
greater than 40%.

As for the power system arrangement, a first glance at
the thermal power rating of the UWMAK-III fusion reactor at
5000 MW$_{th}$ would suggest that between three and six units of
helium turbomachinery would seem appropriate. Further examination of the inner (helium) and outer (lithium) blanket
temperature capability and possible thermal power split would
narrow down the number of inner and outer blanket turbo-
machineries to 1 + 2, 1 + 3, 2 + 4, and 2 + 6, using a direct-
cycle for the inner blankets. After more detailed examination
on blanket and turbomachinery design characteristics, the
1 + 2 arrangement using three identical 585 MW$_e$ power conversion systems was selected, primarily based on the projected
turbomachinery technology and economic considerations.

The finalized system arrangement for the UWMAK-III fusion
power plant is shown in Figure 55. During each thirty-minute
"burn time", the Numbers 1 and 2 power conversion systems will
draw thermal energy from the outer blanket through the primary
(lithium to sodium) and the secondary (sodium to helium) heat
exchangers at a turbine inlet temperature of 870°C, while
the No. 3 power conversion system will be directly integrated
with the inner blanket, thus also operating at the same 870°C
turnine inlet temperature. During this firing time, the sodium
thermal storage system will be "charged" to store sufficient
thermal energy to operate all three power conversion systems
during the subsequent 1.5 minute "down time". During the
down time, thermal storage systems will be "discharged" through
the same sodium-helium heat exchanger to operate the No. 2
power loops, and a separate sodium-helium heat exchanger to
operate the No. 3 unit. Although extraction of tritium will
take place primarily in the lithium loop, any stringent restriction on the tritium level in the power converstion systems
could require additional tritium extraction in the sodium
intermediate loop. Operating temperatures and flow rates for
key points in the inner and outer blanket transfer loops are
also indentified in Figure 55.

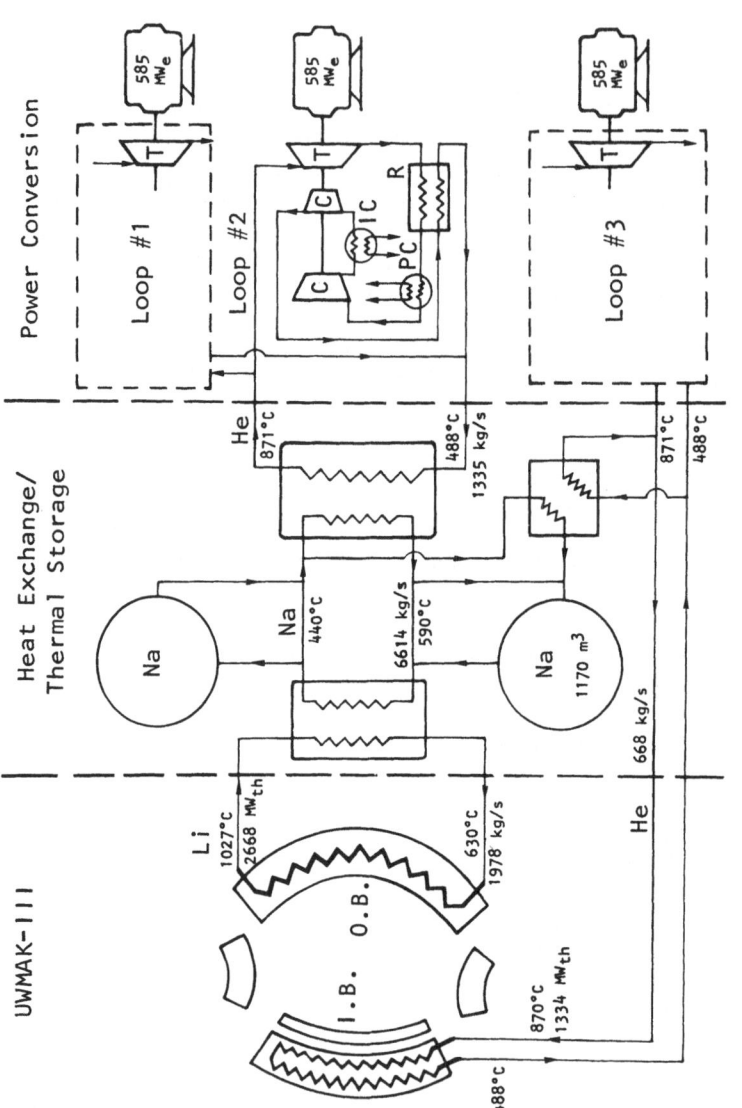

Figure 55

Schematic of Final System Arrangement for
Closed-Cycle, Helium-Gas, Power Conversion System on UWMAK-III

Based on the foregoing discussions, the reference design, 585-MW$_e$ turbomachinery (three needed) for the UWMAK-III
fusion power plant was based on a single-shaft, single-flow
configuration at 3600 rpm shaft speed. This single-shaft
turbomachinery can be coupled directly to the alternater to
avoid overspeed, and generally represents a fairly compact
design with serious penalties in the performance of the
compressors. Generally speaking, the turbomachinery can be
designed for either maximum efficiency of maximum output
(per unit weight or volume) or somewhere in between. The
current design is for maximum efficiency based on 50% reaction at the mean diameter and a flow coefficient of approximately 0.5. Based on the unit capacity and specified rpm,
the current design, using one intercooler, is regarded as the
best compromise for satisfying the power plant requirements
and constraints.

Design data for the turbomachinery are summarized in
Table 21; Figure 56 depicts the turbomachinery structure
including the arrangement of compressors and turbine, bearings
and seals, and the overall dimensions. While conventional
seals are regarded as sufficient for internal sealing, her-
metic seals may be needed for the end seals. Because of the
large flow area required at the turbine exit, an axial-flow
flange design was adopted. The turbomachinery has the unique
cost-saving feature of employment of the same constant hub
diameter for turbine and compressors, all on the single shaft.
The maximum diameter and blade lengths are 2.36 and 0.35 m,
respectively, and the stresses resulting from these configurations are regarded as well within the projected material capabilities.

In addition to the turbomachinery, similar detailed conceptual designs have been developed for the regenerator, precooler and intercooler. A summary of the design details is
given in Table 22.

The diverter power cycle, to convert the heat carried to
the collector plates by particle diffusion from the plasma, is
based on a conventional Rankine steam power cycle, and uses the
sodium coolant leaving the plate at 600°C. The steam conditions are 16.5 MPa at 538°C with 538°C reheat. The sodium re-
turns at 400°C, and the net output from this cycle is 295 MW$_e$.
This is a substantial improvement over previous designs[1,5]
in which the use of liquid lithium to collect the particles

limited the lithium outlet temperature to only 325°C. This
limited the efficiency of the diverter cycle to only 25%.
As such, the design here is much more compatible with the
overall high performance of UWMAK-III

Table 21

Closed-Cycle Gas Turbine Turbomachinery Summary

Reference Design
585 MW$_e$ Unit Output (Three Units Needed)

	Low Compressor	High Compressor	Turbine
Inlet Temperature, °C	43	43	871
Outlet Temperature, °C	126	126	538
Inlet Pressure, MPa	2.6	4.4	70
Outlet Pressure, Mpa	4.4	7.5	2.6
Pressure Ratio	1.68	1.69	2.63
Adiabatic Efficiency, %	89	89	91
Number of Stages	10	10	9
Work per Stage, kJ/kg	8.9	8.9	35.9
Helium Flow, kg/s	667	667	667
Hub Diameter, m	1.65	1.65	1.65
Inlet Tip Diameter, m	2.02	1.81	2.03
Outlet Tip Diameter, m	1.93	1.82	2.36
Wheel Speed, Hub, m/s	311	311	311
Maximum Tip Speed, m/s	381	353	444
Shaft Speed, rpm	3600	3600	3600

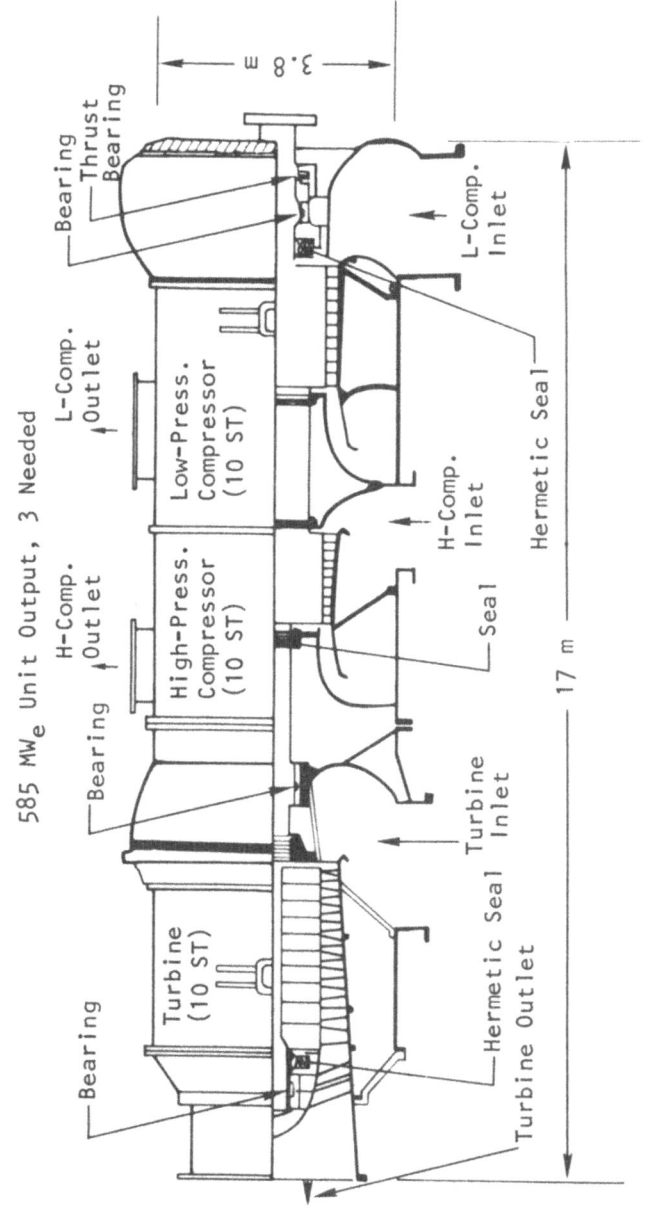

Figure 56

Closed-Cycle Helium Gas Turbines
For UWMAK-III Fusion Plant

Table 22

Heat Exchanger Design Data Summary
Reference Design

	Regenerator	Precooler	Intercooler
Heat Rate, MW	1263	400	287
Flow Rate, kg/s			
Hot Side	670 (helium)	670 (helium)	670 (helium)
Cold Side	670 (helium)	189 (water)	1177 (water)
Pressure Loss, $\Delta P/P$ = %:			
Hot Side	2.8	1.3	1.1
Cold Side	1.0	---	---
Effectiveness, %	88	92	88
Overall Heat Transfer Coefficient, kW/m^2K	1.95	2.5	1.97
Heat Transfer Area, m^2	1.3×10^3	6.5×10^3	5.5×10^3
No. Heat Exchange Elements	150	150	150
Effective Tube Length, m	17.5	11	9.5
Overall Shell Length, m	23	14.5	13
Approximate Weight, tonnes	185	115	100

K.   Plant Design for UWMAK-III

    With the ever-increasing successes in fusion develop-
ment programs worldwide, and with the hope of building commercial fusion power plants, the requirements of integrated
plant design have begun to receive serious attention. The
importance of the balance-of-plant should not be underesti-
mated. For fission plants, the balance-of-plant can amount
to over 85% of the plant capital cost. For fusion plants,
this share may decrease somewhat. The fact remains, however,
that the design of the balance-of-plant and construction planning have significant impact upon the economic viability of
fusion power plants. The conceptual design and corresponding
cost estimate play a major role in identifying the high-cost
areas of the balance-of-plant, and form the basis for development of design approaches for greater cost-effectiveness.
Therefore, a balance-of-plant study was carried out for
UWMAK-III by the Bechtel Corporation.[101]

    The arrangement of equipment and plant is based on the
concept of identifying system functions and functional
interrelationships and assigning these into modules. This
forms the framework for definition of major buildings and
functional requirements. These are: a reactor containment
building containing the reactor and other components of the
nuclear island; a hot cell for handling irradiated blanket
segments; a reactor auxiliary building containing safety-
related equipment and other equipment closely associated with
the nuclear island; a heat exchanger building containing the
heat exchange and transport equipment (it also contains the
thermal storage equipment); a turbine building containing the
energy-conversion equipment; a control building containing
the diesel generators; a radwaste building containing the
equipment for storing high-pressure helium; and an inductive
energy storage building containing equipment associated with
magnetic energy supply systems. The general plant layout is
shown in Figure 57.

    The plant heat rejection system provides the heat sink
for the power cycles, various components of the nuclear island
and other plant equipment, and rejection of this waste heat to
the atmosphere. Guidelines for the heat rejection system design are based on the concept of "zero water use". Consistent
with this concept, dry cooling towers were selected for rejec-
ting the plant heat to the atmosphere.

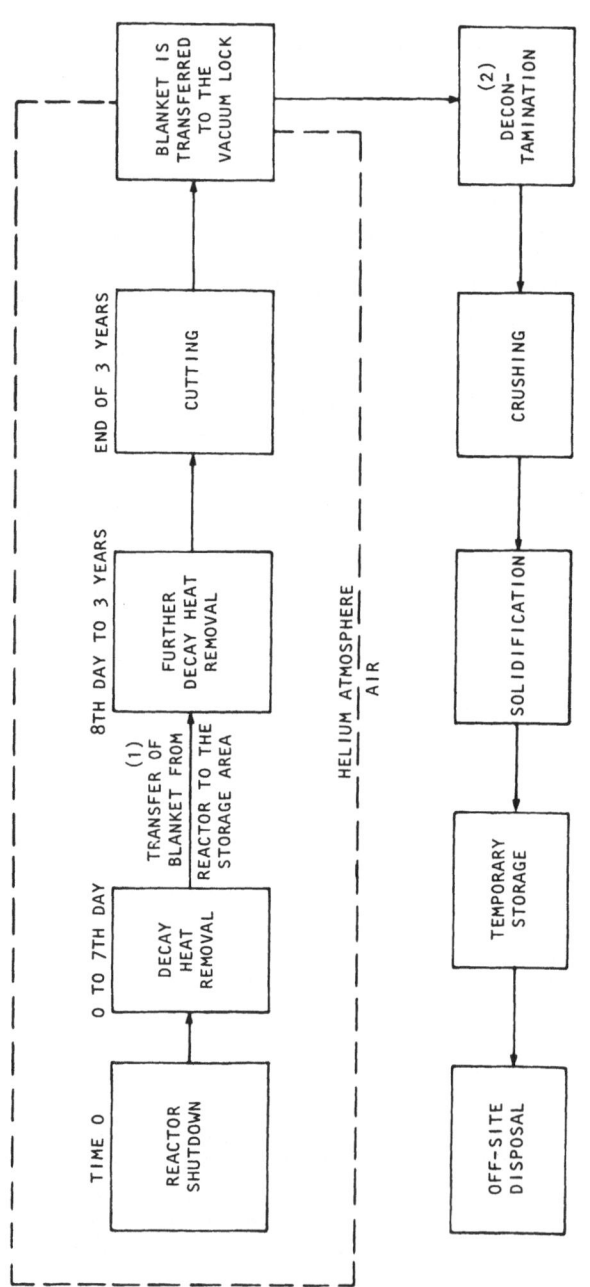

Note: (1) Transfer of the largest blanket piece takes about 8 hours
      (2) Any lithium left in the blanket is converted into LiOH and then removed

Figure 57

Block Diagram Outlining Steps Involved
In Changing Out Blanket Modules No Longer Serviceable Because of Radiation Damage

The use of dry cooling towers also imposes some constraints on the performance of the energy conversion cycles.
The minimum temperature obtained in the cycle with a dry
tower is considerably higher than that obtained by a wet
tower; the reason being that the dry tower approaches the wet
bulb temperature. Also, for economic reasons, the temperature
drive for the dry tower is larger than for the wet tower.
This has significant deleterious effects on the cycle efficiency; however, this sacrifice probably can be justified in
locations where availability of water is scarce.

The replacement of the blanket segments from the outer
blanket generates a large volume of solid radioactive wastes.
A preliminary concept was developed for the removal, processing, storing, and off-site disposal of the wastes. The
guideline used to establish the concept is to reduce the specific activity of the blanket to a sufficiently low level
(7.5 EBq/m^3) so that the shielding requirement during the
off-site transportation is reduced to a minimum.

The process block diagram is shown in Figure 57, and the
main considerations are as follows: The decay heat from the
blanket at the time of the reactor shutdown is about 9 MW_{th};
it is therefore necessary to circulate coolant to remove the
decay heat. The decay heat is reduced by a factor of about
40 in seven days, at which time gas cooling can take over.
The lithium is drained from the cooling system. The modules
are then transferred to the storage area for further decay.

The blanket will be held in the storage area for three
years before processing for disposal; however, the frame
attached to the blanket can be reused, and it will be separated from the blanket during the first six months in storage
so the frame can be welded to a new blanket.

At the end of three years, the used blanket is transferred to the hot work shop where the large pieces will be
cut into small pieces (1m x 2m) using a laser. The small
pieces are then passed through a vacuum lock to the decon-
tamination area which has an air atmosphere.

The annual blanket waste volume is 1500 m^3 and weighs
1000 tonnes. Solidification is performed by mixing the
blanket fragments with cement. The waste will be shipped in
3 m^3 containers with 5 cm steel shielding. There will be

530 containers required annually. It may be more economical
to use larger containers, but shielding requirements and
activity contents of the waste may limit the container size;
however, no optimization was performed in this study.

A first-order consideration has been given to the plant
electrical system to identify the major equipment associated
with the main generation system and the system leading to
the ultra-high voltage (700 KV) switchyard. The study is
also intended to define the major plant load centers and to
identify the essential (Class 1E) loads related to plant
safety.

Conceptual design studies were performed for the plant
buildings and structures, and for the plant, equipment and
piping arrangement. The structural design effort was confined to the major structures; viz., the containment building, hot cell, radwaste building, auxiliary building, heat
exchanger building, and the turbine building. Only super-
ficial consideration could be given to the plant equipment
and piping arrangement.

The general plant arrangement and internal arrangement
of the buildings are patterned, wherever possible, after
the generic concept developed by the Bechtel Corporation
for fission reactor plants. The greatest effort was devoted to those areas unique to UWMAK-III, such as the containment building, heat exchanger building, turbine and
hot cell. Other buildings such as the control building,
radwaste building, etc., are similar to those in other
nuclear plants, and their arrangements can be adapted to
this plant readily.

Major features of the arrangement include the following:
The containment building is centrally located relative to the
outer building; it is a reinforced concrete stucture lined
with stainless steel, and has an inert environment to prevent
lithium or sodium fire; the heat exchanger building, radwaste
building, auxiliary building, hot cell and maintenance building are arranged around the containment building. This
arrangement seems to have an advantage in handling blanket
sections and radioactive waste, and in arranging the reactor
primary coolant piping between the containment building and
the heat exchanger building. This arrangement makes the plant
layout very compact, with maximum utilization of space around

the containment, by locating the reactor support systems and facilities close to the nuclear island. This is also expected to offer some economic advantage for piping, cabling and structural considerations. The peninsular arrangement of the Turbine Building orients the axes of the turbine generators such that any missiles generated by the machines will not strike the containment building. Access corridors run the length of the turbine building and continue in the control building and the auxiliary building to provide accessibility to all equipment.

L.   Survey of Material Requirements and Resource Implications

It is of interest to investigate the material require- ments of UWMAK-III type reactors to determine whether there might be major problems with regard to either reserves or procurement of specific elements. This has been done by Cameron[102] as part of the UWMAK-III study.[8] The reduced overall size of UWMAK-III, compared with earlier studies[1,5] makes the requirements as summarized in Tables 23 and 24 correspondingly less. United States reserves and resources, together with U.S. 1974 production, are summarized in Table 25. Also listed are the requirements for a 10^6MW$_e$ economy (500 reactors) of UWMAK-III type reactor systems. Of course, the requirements for a smaller economy of between five and fifty reactors will be proportionately smaller.

Of the many elements listed, some of the key elements include aluminum, lithium, molybdenum, niobium, chromium, copper and manganese. Each has a facet of their reserve and resource picture worth noting.

The United States has very limited bauxite reserves, and imports over 90% of its bauxite supply. An adequate domestic supply of aluminum ore would require development of resources of high-alumina clays, anorthosite, alunite and dawsonite, singly or in combination. The resources of aluminum in such materials are very large. The necessary technology is avail- able for extraction from clays, anorthosite and alunite, but energy cost and dollar cost would be above those of extraction of aluminum from bauxite. Mining of these three materials would be open cast, but no unusual environmental problems would be involved. Dawsonite production would be involved with oil shale production and would share the environmental and other problems facing the oil shale effort.

Table 23

Material Requirements - UWMAK-III Nuclear Island (Metric Tonnes)

Elements	Blanket Initial Structure(a)	Blanket Partial 1st Wall Replacement 19 Times,30 Yrs(b)	Blanket Bal.1st Wall Replacement 9 Times,30 Yrs(c)	Bal. of Plasma Chamber & Vac.Plenum	Shield Inner & Outer Shield	Shield ISSEC Replacement 9 Times,30 Yrs(d)
Al					167.73	
B					942.26	
C	1039	9,080.5	5049.35	50	494.33	2098.8
Cr					1128.63	
Nb						
Cu						
He						
Fe				165	3724.49	
Pb					3611.06	
Li	395					
Ti	1.97	17.22	5.46	0.55	0.48	
Mn				5	112.86	
Mo	335.6	3,107.95	987.56	99.33	1049.34	
V						
Ni				30	677.18	
Zr	0.43	3.76	0.93	0.12	0.10	
TOTAL:	1792	12,209.4	6043.3	350	11,908.5	2098.8

COMPONENT WEIGHTS (MT)

	Initial Structure	Partial 1st Wall	Bal.1st Wall	Bal. of Plasma	Inner & Outer Shield	
$Al_6Si_2O_{13}$					441.4	
304 SS	358			250	5643.17	
TZM		3,198.92	944.22	100	86.67	
Mo					963.25	
B_4C					1203.4	

(a) 18 modules plus 3 spares; (b) 18 partial modules replaced every 1 1/2 years plus 3 spares; (c) 18 bal. modules replaced every 3 years plus 3 spares; (d) ISSEC graphite replaced every 3 years plus 3 spares on hand.

Elements	Magnets		Shield, Blanket & Coil Support Structure	Vac.Pumps	Total 30 Yrs	Tonnes/MWe (Based on 1955 MWe)
	T.F.Coils Dewars & Central Struct.	VF & OH Coils & Dewars				
Al	4623.56		300	20	4,811.3	2.42
B					942.3	0.47
C					17,762.0	8.95
Cr	23.8	71.4		35.2	1,609.0	0.61
Nb	50	33			83.0	0.04
Cu	382.34	889		14.3	1,285.6	0.65
He	18	13			31.0	0.02
Fe	78.54	236	900	116.6	5,310.2	2.65
Pb					3,611.1	1.82
Li					395	0.20
Ti	30.23	17			72.9	0.04
Mn	16.93	7.14	30	3.52	175.5	0.09
Mo					5,599.8	2.62
V	6.35				6.4	0.003
Ni	14.28	43	180	21.12	965.6	0.49
Zr	10.58				15.9	0.01
TOTAL	5254.6	1309.5	1500	210.3	42,676.6	21.50

COMPONENT WEIGHTS (MT)

Elements	T.F.Coils Dewars & Central Struct.	VF & OH Coils & Dewars	Shield, Blanket & Coil Support Structure	Vac.Pumps	Total 30 Yrs	Tonnes/MWe (Based on 1955 MWe)
$Al_6Si_2O_{13}$			1500	176	441.4	0.22
304 SS	119	357			8,045.2	4.05
TZM					4,667.8	2.35
Mo					963.3	0.49
Al alloy 2219	4233			20	4,253	2.14
Hi Purity Al	716.5				716.5	0.36
Al					1,203.4	0.61
B_4C				14.3	998.3	0.50
Carb.St. 1020	95	889			126	0.06
Cu	76	50				
NbTi						

Table 24

Material Requirements – UWMAK-III Balance of Plant (Metric Tonnes)

Element	Heat Exchangers Li-Na (a)	Heat Exchangers Na-He (b)	Heat Exchangers Na-Steam (c)	Piping Li,Na He & Steam	Tritium Extraction System (d)	Thermal Flywheel Na Tanks	Li & Na Dump Tanks (e)	Li & Na Pumps	Containment Building Liner
Al									
Cr		79	60		0.4			120	222
Nb					9.26				
Cu									
He		19		1.0					
Fe		261	197	33,796	1.3			396	733
Mn		8	6	204				12	22
Mo	99.4	423		993	1.0	49.7	178.8	745	
Ni		47	36		0.24			72	133
Na		400	500	180		900			
Ti	0.50		2.3	5.5	0.02	0.27	1.0	4.1	
Y									
Zr	0.10		0.5	1.2		0.06	0.22	1.0	
Pd					0.50				
Li	20			50					
TOTAL	120	1237	801.8	35,230.7	12.72	950	180	1350	1110
304 SS		395	298		2.0			600	
Carb.St. 1020				34,000					
TZM	100	426		1,000	1.0	50	180	750	

(a) 2 heat exchangers
(b) 6 heat exchangers
(c) 2 heat exchangers
(d) Material needed for 30 years operation
(e) Li dump tanks and diverter Na dump tanks; thermal flywheel tanks are used as dump tanks for the intermediate Na loop.

Element	Buildings	Energy Storage Unit	Turbine Machinery (f)	Generator & Exciters (g)	Regenerator Precooler Intercooler He-steam	Misc.	Total	Tonnes MWe Based 1985
Al			2.96				2.96	.001
Cr		81	30.72	50	32		675.1	0.34
Nb		7.5					16.5	.008
Cu		362		295			657	0.33
He		15				37.5 (h)	72.5	0.037
Fe	78,486	269	230.9	1980	319.3		116,670	58.78
Mn	474	8	1.32		4.5		739.8	0.37
Mo			7.42	20			2,517.3	1.27
Ni		49	68.56		19.2		425	0.21
Na							1,980	1.00
Ti		4	3.56				21.2	0.01
Y							0.02	
Zr							3.08	.001
Pd							0.50	
Li			13.9				70	.035
Co							13.9	.007
Concrete 304,000m³								
TOTAL	78,960	795.5	359.3	2345	370.5	37.5	123,865	62.4
304 SS					160		2,565	1.29
316 SS		400					400	0.20
Carb.St.	78,960				215		113,175	57
1020							2,507	1.26
TZM							362	0.18
Cu		382						0.006
NbTi		11.5					11.5	
Concrete 394m999m³								

(f) Turbine made of blades-UDIMET-700; Casing-INCONEL 718; disc-A-286; compressors(2) made of blades-PWA-IN-4017.

(g) Silicon steel 495 MT; carbon steel 500; copper 50; Cr.Mo.V. stainless steel 163; insulation 25.

(h) Liquid He in storage.

Table 25

United States Production, Reserves and Resources
of Metals Required for UWMAK-III
(metric tonnes x 10^6)

Metal	Requirement for $10^6 MW_e$	United States		
		Production, 1974 [1]	Reserves [2]	Resources [3]
Al	2.42	.40	~ 2.0	Very large
B	.47	.18	10.1	Very large
Cr	1.08	0	--	1.53
Nb	.06	0	.006	.07
Cu	.98	~ 1.59	85.0	100.0
Fe	60.28	50.0 [4]	2,000.0 [5]	10,000.0
Pb	1.82	.68	59.0	Large
Li	0.24	~ .002	1.5	1.7
Mn	0.46	0	--	10.0
Mo	4.08	.06		
Ni	.67	.015	.20	16.0
Na	1.0	.200 [6]	Unlimited	
Ti	.05	.29	34.0	29.0
Y	$<10^{-5}$	.00004	Probably large	
Zr	0.01	.07	10.0	No data
Pd	0.001	0	0	Large [7]
Co	< 0.003	1,000	Included in 800,000 resources	

(1) Primary metal only.
(2) Metal in ores minable under current economic conditions.
(3) Metal in material judged available at prices up to three times present prices.
(4) Fe content of domestically mined ores.
(5) Fe content of ore minable under current economic conditions.
(6) Metal only; approximately 1% of total production of sodium minerals.
(7) Assumes development of newly discovered deposits in the Stillwater Complex, Montana.

Data from U.S.G.S. (Prof. Paper 820, 1973), U. S. Bureau of Mines (Commodity Data Summaries, 1975), and J. D. Vine (Lithium, Trans. American Nuclear Society, 22, 1975).

United States reserves·of molybdenum are estimated by the U. S. Geological Survey at 3×10^6 tonnes, and by the U. S. Bureau of Mines at 4×10^6 tonnes. The survey estimates identified marginal resources at 17.5×10^6 tonnes. These figures suggest that at prices somewhat higher than present prices, the UWMAK-III, 10^6 MW$_e$ requirements could be met, but information concerning possible rates of production of molybdenum beyond the end of the century is not available. The current rate of production is only about 55,000 metric tonnes per year; thus, obtaining molybdenum could become a serious problem because of procurement difficulties rather than of resources.

If the chromite resources in the Stillwater Complex, Montana, are not used for other purposes, the UWMAK-III requirement could be met from that source, at prices probably not very much higher than present prices; otherwise, the requirement might be met from domestic sources only by paying high prices for domestic chromite concentrations, probably at three times the present prices, or more.

Niobium is a key element of the superconductor for the UWMAK-III coils. The requirement for niobium is approximately equal to the amount of niobium in ore blocked out by exploration of the Powderhorn Complex, Colorado, the only sizeable domestic resource. This is not a comfortable situation. Stockpiling of niobium in advance of use appears to be the only answer to the problem of procurement unless new deposits can be discovered in the United States. The situation would be considerably more difficult if niobium were used as the structural material.

The copper requirement for UWMAK-III is small, relative to the estimated reserves of 85×10^6 tonnes; however, these reserves will be about half-depleted by the end of the century at present rates of production. Even now, domestic production is not quite equal to domestic consumption. The use of copper in fusion reactors will have to compete with conventional uses from available domestic supply, and it is not certain that present rates of domestic copper production can be maintained to the end of the century. Much will depend upon success in discovering new deposits of copper during the next 25 years.

If it becomes necessary to draw on foreign sources of copper, the outlook for such supplies is fairly favorable.

At least seven countries should be producing large amounts
of copper during the first quarter of the 21st century; Peru,
Mexico, Chile, Panama, Zambia, Zaire, and the Papua-New Guinea.
Copper from the deep sea nodules should also have entered the
market by that time.

World reserves of manganese are enormous, but the United
States has no manganese reserves. At three times the present
prices, the UWMAK-III requirement probably could be met from
off-grade resources in Maine, Arizona and Minnesota. The
nodules of the deep sea basins are another potential source.
Presumably, the present dispute over jurisdiction of deep-sea
mining will have been resolved.

Current reserves of nickel in the United States are too
small to meet the UWMAK-III requirement, and will be virtually
exhausted by the end of the century; however, nickel deposits
in the Duluth Complex, Minnesota, should be in production by
the end of the century and should be able to supply the necessary amounts. Nickel won from the manganese nodules of the
deep sea floors could be an alternate source.

Finally, the United States Geological Survey[103] states
that the identified resources of lithium in the United States
likely to reach the marketplace by the Year 2000 are scarcely
adequate to meet the anticipated growth of the industry for
conventional use and are grossly inadequate for the anticipated
requirement for new energy-related uses. A search for new deposits is needed. The brines of western saline basins and certain types of clay deposits appear to offer the most promise.

M.   Economic Analysis

    1.   Role of the Study

The role of an economic study for an early conceptual
design such as UWMAK-III is to provide us with an indication
of where major cost areas might be in tokamak reactors. This
helps guide future tokamak reactor studies, and indicates
areas wherein increased research and design effort can pay
the greatest dividends. On the other hand, many parts of the
UWMAK-III system, by the nature and scope of the study, remain
undefined; this means that an accurate determination of absolute cost is extremely difficult. This is particularly true
if one tries to compare fusion with power systems like fission
reactors where the designs are known in much greater detail;

therefore, we place greater emphasis on the relative cost of
key items in UWMAK-III than on any absolute figures.

It is also important to recognize that the economic analysis for UWMAK-III is *specific* to this system. That is, many
of the items that have a strong impact on the results are
governed by certain design choices peculiar to UWMAK-III.
Different design choices can have very different effects on
cost, so that a generalization regarding the cost of tokamak
reactors is not possible at this time (as a specific example,
one will note in the following sections that the estimated
cost of high-temperature refractory metal piping was a substantial cost driver; these costs would be sharply reduced
if stainless steel has been chosen as the structural material).
The results reported here were obtained as part of a joint
study by the Bechtel Corporation and the University of
Wisconsin.[104]

Turning to the estimating procedure itself, it is normal
estimating practice, when engineering is well defined, to derive the cost estimate by building up the cost, item by item.
In a less well-defined study, where perhaps only 60% of the
equipment is identified, this method can still be used, but
a large portion of the cost must be based on allowances and
factors; however, in a preconceptual estimate, such as the
UWMAK-III study where less definition is available, a different approach is necessary. The plant must be viewed as a
series of systems that should be similar in content (but not
necessarily in function) to systems in existing analogous
plants, and the cost obtained by inductive reasoning, using
the costs of the analogous known systems. The estimate is
thus derived by a macro, rather than micro approach. There
is still considerable margin for variation in the unknown
systems, but the potential variation is dimished by using
this composite systems approach.

Existing nuclear power plants are the best comparable
source of pertinent data, and the cost estimate and schedule
for UWMAK-III rely heavily on data from recent Bechtel Corporation power plant experience. The basis for the estimate
is discussed and the resultant capital cost for the entire
plant is summarized in Part 2 of this section.

Parts 3 and 4 of this section are important in that
we attempt here to place the estimate of plant economics for

UWMAK-III in perspective by providing some comparison with
estimates of electric power cost for other methods of generating electricity. We have asked how the cost developed for
UWMAK-III compares with cost estimates developed by the
group for other, non-fuel limited approaches to electric power
generation (it is important to maintain consistency by using
the estimates of a particular group, since we find estimates
by various groups, including costs developed with Federal
government guidelines to vary widely). Other approaches to
electric generation include fission breeder reactors, solar
power, fuel cells and coal-fired MHD plants. Also included
for reference are consistent estimates for light water reactors, coal-fired steam plants and gas turbines. These systems, particularly LWR's and gas turbines, truly are not
comparable because the fuel supply no longer will be available at the time that advanced technologies such as breeder
reactors and fusion are ready to play a substantial role.

Of course our main goal was to determine the key design
decisions and cost items that most affect the economics of
UWMAK-III. A discussion is given of design choices made at
the beginning of the study that turned out, after the fact,
to have more strongly affected the estimated cost than had
been anticipated. From these considerations, one can draw
conclusions regarding directions for future work.

2. Cost Estimate

The basis for the cost estimate involves many items.
The facility under discussion is assumed to be a first-
generation model designed in a time when the major equipment
functions have been successfully demonstrated. At that time,
industry will have developed the capability to produce the
equipment, a capability that does not necessarily exist today.
Development and first-of-a-kind costs are, therefore, not
considered.

In general, the cost of the reactor plant equipment was
developed by the University of Wisconsin, and the Balance
of Plant cost was determined by the Bechtel Corporation. The
cost estimates are for the fourth quarter 1975 dollars, with
no allowance for escalation.

The basis of the estimate is the plot plan, plan and
elevation of the power block buildings (the main structures,
housing the reactor and power generation facilities), and a

single-line electrical diagram. All of these were used as guides, and with the exception of building volumes and some quantities of concrete and lengths of main heat-transfer piping, no quantity takeoffs were made. Fuller understanding of the engineering scope, extent and type of systems was provided in informal discussions with the engineering team.

The capital cost estimate is composed of field cost, engineering cost, a contingency allowance, owner's cost and interest during construction. The largest category, that of field cost, comprises the direct cost of permanent plant equipment and indirect cost of temporary construction materials, supervision, etc., that are to be distributed across the entire facility.

Direct field costs are: costs for permanent plant equipment, materials, labor and subcontract items; these form the bulk of field cost. These are derived from two basic sources: bulk materials and equipment. In general, both are available from current and historic sources in a system and category format.

The major items of reactor equipment have been estimated by the University of Wisconsin, and these have been included without comment. Items that lie in the Balance of Plant category; e.g., cooling towers, have been estimated, using current price guidelines from major manufacturers and current pricing data from within Bechtel Corporation. The cost of other equipment has been established as a function of the appropriate building.

The bulk of the mechanical equipment peculiar to UWMAK-III has been priced by the University of Wisconsin. Peripheral equipment such as pumps, condensers, and heating, ventilating and air-conditioning equipment has been estimated by eliminating from an LWR mechanical account all those items related to a fission reactor and scaling up the remainder of the cost to the increased capacity.

The cooling tower was treated separately. Its duty corresponds to that of an equivalent 800 MW_e LWR plant except that the circulating flow is one-half and the temperature is much higher. Its price is estimated to be 40% higher than that of an equivalent wet cooling tower.

Remote handling equipment was also treated separately and an allowance included, although little is known beyond the general function; certainly, no equipment of the sort exists today.

There is a close correlation between the station load of a power plant and the cost of the electrical equipment. On a net basis, UWMAK-III is more efficient than an LWR facility, but its equipment is more sophisticated. In sum, on a per-megawatt basis, UWMAK-III's electrical equipment is assessed to be 90% of that in an LWR plant, excluding the unique equipment for a fusion plant, and was estimated on this basis. To this base figure was added the estimated cost of the unique systems, the energy storage and inverter facilities, and a low-voltage, cryogenic transmission system from the energy storage facility to the containment building.

The wage rate included is based on a craft mix and wage agreement as experienced in the Pacific Northwest which is a representative area for the United States. Labor availability and consequent productivity is also affected by plant site choice. For this study, labor productivity is again based on the Pacific Northwest experience and is generally regarded as good. Also, it is normal for a part of the work to be subcontracted, examples being earth work, cooling towers, insulation and building finishings. In this respect, a fusion plant is regarded as being normal, and thus, standard power plant relationships have been used.

Indirect cost can be a controversial part of any cost estimate. Previous UWMAK studies[1,5] have used the WASH 1230[105] as a guide for determining indirect cost; however, they are probably understated in WASH 1230 (20% of direct labor costs compared with 90% to 110% in Bechtel's experience). The different bases of the cost scope make comparisons difficult. For UWMAK-III, an approach consistent with the Bechtel scope has been used.

Principal categories of indirect costs and their approximate percentages for a typical fission facility are:

. Temporary construction facilities   15%

. Miscellaneous construction services   18%
  (cleanup, maintenance of tools and
  equipment, material handling, guards,
  welder tests, etc.)

. Construction equipment, tools,
  supplies and utilities   19%

. Field office cost   42%
  (supervision, engineering, administration
  warehousing, field purchasing, medical
  and overhead)

. Other (insurance, taxes and miscellaneous)   6%

                                                           100%

In the UWMAK-III estimate, indirect costs are calculated
at 100% of direct labor cost. The labor cost is not identified separately, but generally is in the range of 20% to 25%
of the total direct cost, and an average of 23% is assessed
on the basis of a high-bulk material component cost, these
being labor intensive. The resulting indirect cost, therefore, is 23% of the total direct cost.

Engineering services provided include engineering cost,
home office cost and fee. Engineering includes both preliminary and final design, specifications, etc. Home office
cost, which comprises estimating and scheduling services,
procurement, startup, quality assurance and project management, lies in the 50% to 60% range of engineering cost. Fee
normally is a function of the total project cost, and there
are commonly-accepted guidelines on appropriate schedules.

The sum of these three categories falls into the 12% to
22% range, which is consistent historically. The actual value
is dependent on the complexity of the project. For Light
Water Reactor plants, 12% to 15% is normal. For this study,
a figure of 16% of the total field cost has been used for
engineering services because it is assumed that many of the
problems lie in equipment and material specifications and in
construction procedures, increasing the relative cost of
engineering services.

Contingency is the sum of money added to an estimate to
provide for items not specified in the engineering detail but
that are known from experience to be a necessary part of a
facility. Implicitly, it will be used, and it is not a draw-
down fund to compensate for overruns or changes and additions
to the project scope. Commonly-used ranges for large complex
facilities are from 15% of total construction cost for well-
defined projects to 30% or more for highly conceptual studies.

As UWMAK-III is a first-generation plant there will be many
problems that are unique to the facility; thus, the contingency allowance has been set at 30%.

It is important to recognize that contingency provides
for the unknown in the design only to the extent that the
engineering is understood, and experience shows that it is
quite difficult to assess the degree to which future processes
are understood relative to hardware. If the arrangement of
the plant components contains major uncertainties, or the
materials selected prove to be deficient, or the duty more
severe than anticipated, or if additional major subsystems
are required, then the scope of the engineering is inadequate;
this is not covered by contingency.

Two final categories of cost include "Other Owner Costs"
and "Interest during Construction". Other owner costs include the owner's general office and accounting cost, startup
and operator training, spare parts and inventory cost, and
environmental impact studies. These costs lie in a range of
4% to 12%, and a figure of 8% has been selected as typical.
This is based on the assumption that major equipment and
startup problems have been resolved, and that environmental
data acquisition and reports present no unusual concern.

As for interest during construction, we have taken the
center of gravity of cash flow at the point wherein two-
thirds of the construction is complete, estimated from a
proposed schedule to be three years from the plant fuel load
date. Interest during construction is calculated at 8% of
the total capital cost for three years. Escalation can be
a difficult and tricky cost to include, and it can distort
the meaning of final results. For purposes here, no pro-
vision is made for future escalation, and estimates are based
on fourth quarter 1975 dollars.

A summary of the estimated capital cost for UWMAK-III is
given in Table 26. The cost of electricity includes operating
and maintenance cost, fuel cost, and annual return on capital.
Operating and maintenance cost includes salaries, miscellaneous
supplies and equipment, outside support services, miscellaneous
and administrative cost, blanket rebuilding cost, and coolant
makeup. These are estimated to be 1.5 mils per kWh.

Table 26

Estimated Capital Cost for UWMAK-III
(1985 MW_e, Molybdenum Alloy TZM as Structural Material)

Direct Field Cost	Cost ($/kW_e$)
1. Land and Land Rights	0.5
2. Structures and Site Facilities	285.1
3. Reactor Plant Equipment	408.1
4. Turbine Plant Equipment	284.6
5. Electric Plant Equipment	141.0
6. Miscellaneous Plant Equipment	30.2
7. Special Materials	4.0
TOTAL DIRECT COST:	1,154

Indirect Cost	
8. Construction Facilities and Services	267.0
9. Engineering Services	226.7
10. Contingency	493.7
TOTAL ESTIMATED CONSTRUCTION COST:[*]	2,140

---

[*] A primary cost driver was the design choice of the Molybdenum based alloy, TZM, as the structural material. Costs are therefore specific to UWMAK-III, and can be expected to be less in a lower technology system; e.g., using 316 Stainless Steel and a Steam Cycle. See discussion in text.

The fuel cost is very small, estimated at 0.01 mills
per kWh. UWMAK-III operates on a mixture of 50% deuterium
and 50% tritium. The initial tritium inventory will be a
substantial capital cost that has not been taken into account;
however, after the start of routine operation, a surplus of
tritium will exist for which no credit has been allowed, so
it has been assumed that tritium is available at zero cost.
Deuterium is burned and lost in the operation of the plant,
and an allowance for recycling is made.

The largest component of the cost of electricity is the
annual return on capital. The annual return on capital depends on many factors, particularly the financial structure
of the utility and long-term interest rates. Consistent with
recent experience, a 15% fixed charge rate is used.

3.   A Perspective on Cost

To provide a perspective for evaluating the cost
estimate for UWMAK-III, we show in Figure 58 the estimated
cost of power for UWMAK-III in approximate relationship to
other base-load plants. The systems have been arranged into
three groups according to the time period during which they
will probably reach commercial operation. The figure has
been developed using Bechtel Corporation experience, and
published reports, adjusted where necessary to be comparable
to UWMAK-III; thus, all costs are in present-day dollars and
include interest during construction, but exclude future
escalation cost. The cost of power for each facility is
divided into capital cost, fuel cost, and operating and
maintenance cost. The ranges of cost include the various
types of facilities and differing site conditions. Coal was
priced at $23 per ton and oil at $12 per barrel.

Escalation was not considered in the calculation because
it would distort the basis of comparison, since its inclusion
would entail some arbitrary assessment of the time each type
plant is likely to be in commercial operation; however, excluding escalation introduces *another bias* resulting from
neglecting the effect of escalating fuel prices. For fission
plants scheduled for operation in 1982, the cost of fuel might
be from 6 to 12 mils per kilowatt hours on a fifteen-year level-
ized cost basis, depending on whether fuel costs escalate at
5% or nearer 10%. For coal-fired plants the cost could be
from 22 to 42 mils per kilowatt hours on the same basis. The
higher figures represent approximately 29% of total generating

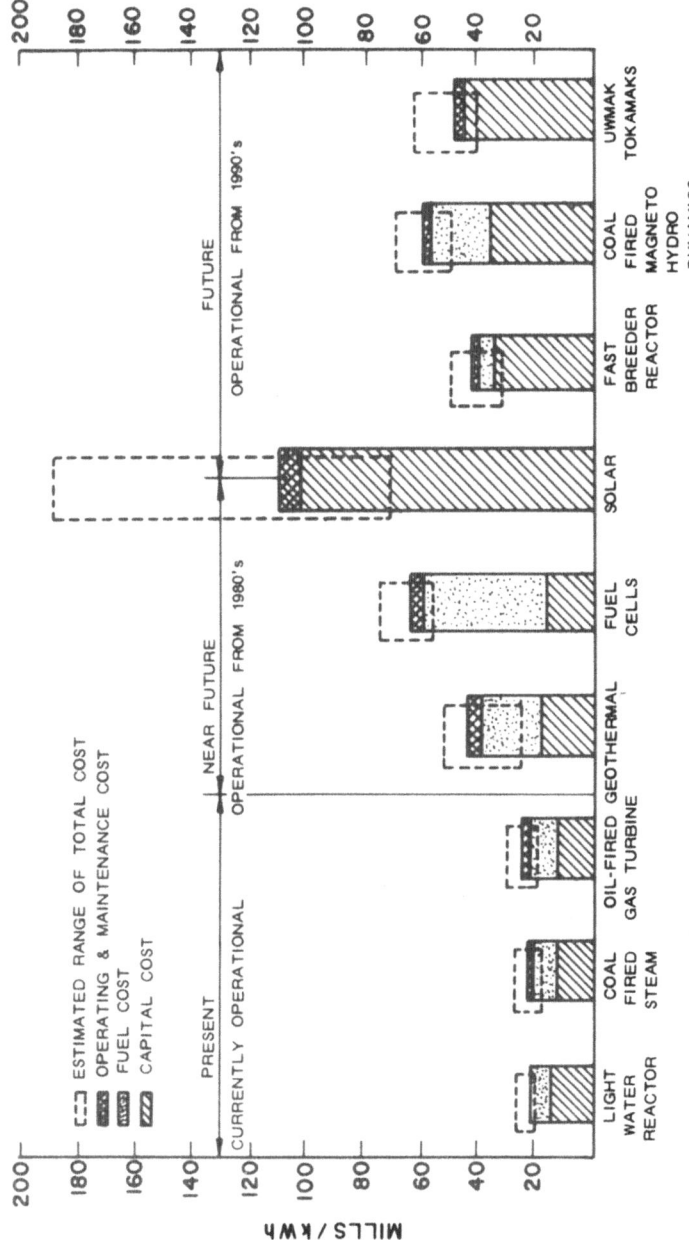

Figure 58

Approximations to the Cost of Electric Power (1975 Dollars)
from Various Advanced Energy Systems

costs for fission plants and about 56% for coal-fired plants
compared with roughly 21% and 36%, respectively, for units
starting up in 1978-79, but based on 1976 dollars. Further
extrapolation would be difficult, but it is possible that fuel
cost may continue to escalate faster than capital cost, gradually increasing the incentive for high capital/low fuel cost
plants such as UWMAK-III.

Peak shaving and standby power systems are excluded from
the comparison since criteria other than pure economic consideration govern their selection: reliability, simplicity,
or low capital costs are frequently the dominant factors.
Hydro power also is excluded, since the potential for a significant capacity increase in the continental United States
is limited.

The source for all costs included in Figure 58, except
those for fast breeder reactors, is the Bechtel Corporation.
These cost estimates were developed by Bechtel, based either
on direct experience or on a joint study. This provides the
results with a consistency that generally is not possible,
since the basis for a cost estimate can differ widely among
groups. The light water reactor and coal-fired steam costs
are based on Bechtel experience. The gas turbine estimate
is based on a joint GE-NASA study. In the lower range are
combined cycle gas turbine/steam plants, and in the higher
range are closed-cycle regenerative plants.

The cost estimate for geothermal power was developed by
Bechtel in conjunction with input from the Pacific Gas and
Electric Company. The lowest figures reflect Pacific Gas and
Electric experience with their Geysers plants; however, the
conditions are unique (relative cleanliness and high-temperature with superheat). The high capital cost is a reflection
of the relatively large steam turbine and cooling towers
required for a low-grade heat cycle. High-temperature brine
plants require more extensive equipment.

Cost figures for fuel cell and coal-fired MHD plants were
developed by Bechtel, based on joint studies by General Electric Company and NASA. The fuel source for fuel cells is
likely to be coal derived. Both high and low temperature

fuel cells fall within the same approximate range with the
former having the higher capital cost and lower fuel cost.
The cost for coal-fired MHD plants are for those that include
both open cycle, closed-cycle inert gas, and closed-cycle
liquid metal MHD plants. The open cycle is the least expensive, the liquid metal the most expensive, and the inert gas
is in the middle range.

Cost for central station solar power is based on studies
by the Bechtel Corporation. The types of plants considered
range from small solar ponds to large ocean thermal-gradient
powered plants. Also included at about the mid-range are
solar plants using heliostats focusing on a central tower
furnace that powers a steam plant. To be comparable with
other base-load plants, the central tower system includes a
heat storage system to accommodate night generation of power.

A major physical problem with land-based solar plants is
the space required. A system designed to run on solar power
for eight hours during the day and to have sufficient thermal
storage to carry a reduced night load would require 3 1/2 to
4 square miles for 100 MW$_e$. Further, unless summer loads are
considerably higher than winter loads, the peak seasonal demand and generating capacity will be out of phase.

Ocean thermal gradient plants have neither of these problems; however, their overall efficiency is very low (in the
order of 2% to 3%), and the environmental effects of trans-
posing 4,000 MW$_{th}$ of energy between the top and bottom of the
ocean to generate 100 MW$_e$ have yet to be analyzed.

The perspective provided by Figure 58 is that the electricity cost as developed for UWMAK-III appears to be comparable with estimates developed by Bechtel for several other advanced energy systems such as fuel cells and coal-fired MHD
plants. All costs are less than estimates for central station
solar power. The dashed boxes on Figure 58 indicate a range
of possible total cost, but such a range has not been developed for UWMAK-III. This should be done by changing several
of the key design choices such as the choice of the primary
structural material, and assessing the impact.

Fuel cells with a low capital cost but a high fuel cost appear to be the type system ideally suited for peaking plants Fusion and coal-fired MHD are considered as base-loaded plants because the capital cost makes up a greater part of the electricity cost. The capital cost associated with the coal-fired MHD system (approximately 40 mils per kilowatt-hour) is higher than for coal-fired steam because of the high-temperature materials and cryogenic magnets associated with the MHD power cycle. The fuel cost adds about a third to the electricity cost for both coal-fired steam and coal-fired MHD. In contrast, the fusion power cost is primarily related to capital cost, while fuel cost is only one to two mils per kilowatt-hour. This is important if fuel costs escalate faster than the capital cost. Past patterns of price behavior are a poor guide in predicting the future, but from an economics standpoint, if the scarcity of fuel is perceived to be greater than the scarcity of capital (not an unlikely event since fossil sources are finite whereas capital sources are not), the cost of fuel will indeed rise faster than capital cost.

The solar estimates, which range from 70 to 200 mills per kilowatt-hour, exceed those of all other systems. Further, solar power plants may not be feasible as base-loaded systems except in the southwestern part of the United States because of this high load requirement during winter. Providing the energy storage capacity necessary for a base-loaded plant in other parts of the country would roughly double the costs indicated in Figure 58 (this does not detract from the positive role to be played by solar energy for space heating on homes and other buildings).

The other potential base-loaded system not yet discussed is the Liquid Metal Fast Breeder Reactor (LMFBR). The LMFBR is the most highly developed of the nonfuel-limited power plants of the future, but consistent costs for this system were not available.

We shall therefore attempt an estimate for LMFBR's to develop a like range of cost; we begin with a discussion of the cost estimate for the Clinch River Breeder Reactor (CRBR) project.

The total estimated construction cost for CRBR was $891 million in 1974 (see ERDA Congressional Testimony, 1976). We have used fourth quarter 1975 dollar amounts in this report, and as discussed earlier, mid-1974 prices are to be escalated by 22%, based on Bechtel escalation figures and the Handy Whitman Index. This would bring the CRBR estimated construction cost to $1.087 million in the fourth quarter 1975 dollar amount, or $3,105 per kW_e. Adding owner cost and interest during construction on the basis applied earlier gives a total estimated capital cost for CRBR of $1.455 billion, or $4,157 per kilowatt electric. The contribution to the cost of electricity from return on capital based on a 15% fixed charge rate is 89 mils/kWh. These costs do not include development cost, fuel cost, operating cost, or escalation beyond 1975. The plant is scheduled to operate in 1982.

The above costs are *not* strictly comparable because the CRBR is a demonstration plant, and it is difficult to estimate consistently the cost of commercial plants from CRBR costs. This point has been made in a recent article by Levenson, Murphy and Zaleski (*106*); however, Levenson et al. do point out that demonstration plants such as Shippingport for light water reactors proved to be about twice as expensive as commercially built light water reactors like Dresden and Yankee. They also point out that "the cost of Shippingport, after normalization to 1974 conditions, is the same as the estimated cost of CRBR with a 15% contingency". If one infers that fast breeder reactors will experience a similar cost reduction, then a commercial LMFBR will cost about $200/kWe and will have electricity costs in the range of 30-55 mils/kWh (1975 dollar amounts). This result is in basic agreement with current reports by the French for the cost of Superphenix, a 1200 MWe LMFBR that will follow the French demonstration LMFBR, Phenix. The electricity cost for Superphenix is projected to be approximately double the cost of light water reactors. Lower costs than these will mean the commercial LMFBR's will have to be less than half the cost of demonstration fast breeder reactors. The cost of electricity from an LFMBR indicated on Figure 28 is 45 mils/kWh, with a range from 30-55 mils per kilowatt hours. The costs for CRBR are summarized in Table 27.

Table 27

Estimated Capital Cost for 350 MW$_e$
Clinch River Breeder Reactor
(4th Quarter, 1975 Dollar Amounts)

Total Estimated Construction Cost	(In Millions of Dollars)
($891 million[*] (1974 dollars) + 22% escalation to 1975	$1,087
Other Owner Costs	87
Interest During Construction	261
Total Estimated Capital Cost	$1,435
Total Construction Cost ($/kW$_e$)	3,105
Total Capital Cost ($/kW$_e$)	4,100

Estimated Electricity Cost for CRBR

	(mills/kWh)
A. Operation and Maintenance	1
B. Fuel	5
C. Return on Capital	89
Total Estimated Electricity Cost:	95

---

[*] 1976 ERDA Congressional Testimony

4.   Analysis of UWMAK-III Cost

The cost detail summarized in Part 2 has been analyzed to understand which items were keys to determine cost.
The unanticipated result is that the cost of high-temperature
refractory metal piping is critical, while cost of the toroidal field magnet system is relatively less important. This
last result is crucial because it is opposite from the accepted folklore that the magnet costs are the primary factor
in determining fusion plant cost.

Piping cost was high because it is necessary to protect
refractory metals from the atmosphere to avoid oxide formation and hydrogen embrittlement. Importantly, this is a
general characteristic of refractory metals and is not specific to the use of molybdenum-based alloys. The cost estimate
developed for such pipe in the primary system is approximately
$35,000 per foot. As such, about a quarter of the nuclear
island costs and about a third of the closed-cycle helium gas
turbine plant equipment cost were for pipe. It will be of
great interest to redesign UWMAK-III (a relatively small unit
physically, with R = 8m) using a lower technology structural
material such as 316 stainless steel. Clearly, the absolute
cost will be reduced significantly, and it is important to
determine just how much.

An important auxiliary result is the breakdown by major
category of the total cost. Direct cost is 40% of total cost.
Since all indirect cost is computed as a percentage of direct
cost, every saving of a dollar in direct cost means a saving
of $2.5 in total cost; therefore, a substantial reduction in
piping cost (part of the direct cost) will have a multiple
effect on total estimated cost.

The direct cost in UWMAK-III was divided as follows:
25% for buildings, 35% for the reactor plant equipment (nuclear island), 25% for the turbine plant, 12% for the electric
plant equipment, and 3% for remaining miscellaneous items;
therefore, the nuclear island is 35% of the direct cost, while
the balance of plant is 65%. The toroidal field magnets were
estimated to be 10% of the reactor plant equipment cost, or
just 3.5% of the total direct cost. This is much lower than
anticipated, and leads us to conclude that we can afford to
pay for higher magnetic field.

This result has some very important implications. In particular, it has an implication for the size of tokamak reactors. There are two reasons one would make a tokamak large. The first is to improve the confinement time of the plasma, and the second is to keep the neutron wall loading to a prescribed but low value. A wall-loading constraint at some level (high or low) is applicable to *all* magnetic approaches to fusion, and is *not* specific to tokamaks. On the other hand, increasing the size to improve confinement time is more peculiar to closed-line devices like tokamaks than it is to an open-ended system like a mirror machine; however, almost all theories, as well as experience, indicate the confinement time improves with magnetic field. Therefore, within a given wall-loading constraint, tokamaks can be made as small as possible by increasing the magnetic field. The cost penalty on the coils will not be overwhelming, whereas keeping the size small will lower the cost of buildings, piping, blanket systems, and so on.

Turning to the role of the structures in this design, the building cost was about a quarter of the direct cost and the reactor building was approximately a third of the total building cost. Future designs of the reactor building should therefore aim for a highly optimized design. A constraint is the apparent need for access to a tokamak from both the side and the top. This leads to a tall, large-diameter building for UWMAK-III; however, further examination may allow for improvement.

The choice of power conversion cycle also merits further careful study. The cost of a closed-cycle helium plant appears to be in the same range as a steam cycle (about \$140/kW$_e$ for the turbogenerator island). In this case, the determining factor in the choice of the cycle would be the power conversion efficiency and since capital cost is relatively high, the choice of an efficient power cycle is very important. There are indications from other studies that advanced steam cycles can produce comparable efficiencies at lower cost, compared to closed-cycle helium gas systems (see "Study of Advanced Energy Conversion Techniques for Utility Application Using Coal or Coal Derived Fuels," Oral Briefing on Results of Task II, General Electric Company, February 10, 1976).

Lastly, a comment concerning the intermediate loop. We have felt that the use of liquid lithium as the primary coolant would necessitate the use of an intermediate loop to isolate

the primary coolant from the turbine cycle. This entails
an added heat exchanger, an extra building, and added pipe
length. The intermediate sodium loop in UWMAK-III is about
15% of the nuclear island cost. The development of a blanket
design and cooling systems that can eliminate the need for
the intermediate loop can thus lead to a potential saving
in total direct cost. An example of such a blanket system
is the use of Li_2O particles in a moving bed similar to those
used in the chemical industry.(107) Combined with the lower
cost associated with a material like stainless steel, such a
system will be less expensive than the UWMAK-III design
employing TZM, and thus, should be investigated.

In summary, the economic analysis of UWMAK-III has shown
that cost can be determined by certain key design decisions
made early in a reactor study. For this system, the choice
of a refractory metal alloy caused piping cost to be a main
cost driver. It has been found that magnet cost is not dominant, and that it will make good sense to increase magnetic
fields to achieve the required plasma confinement time. One
can then design the reactor in a minimum size consistent with
other design constraints. It is also found that the need for
an intermediate loop results in increased cost. Yet this
loop is required only for certain choices of coolant and
breeder materials, and may be eliminated with alternate blanket designs; thus, there are many avenues open for reducing
the cost of tokamak systems below those of UWMAK-III. Finally,
a cost comparison shows UWMAK-III costs to be in the same
range as those projected for other advanced energy systems.
As noted earlier, directly comparable LMFBR costs were not
available for inclusion here, but estimates based on the
analyses of others indicate that these costs will be in
approximately the same range.

V.   FUTURE DIRECTIONS

The headings for the 12 specific parts of Section IV can
be taken as a list of the major technological areas important
for fusion reactors generally and tokamak reactors specifically. In carrying out conceptual designs, we have developed
a detailed and quantitative understanding of the problems in
these areas that subsequently has indicated the direction we
should follow in future fusion reactor research.

The early studies such as the studies of UWMAK-I (*1*) and
UWMAK-II (*5*) were based on relatively conservative assumptions
regarding plasma performance, radiation damage and materials
performance, toroidal field magnet design, and so on.   In
general, these assumptions led to relatively large reactors
(major radius, 13 m; plasma radius, 5 m).   The direction in
the UWMAK-III study was to extend the design parameters, such
as plasma beta and neutron wall loading, in an effort to reduce the size of the reactor.   This effort was basically
successful.   Nevertheless, as the neutron wall loading is
increased, the predictions of first wall lifetime as limited
by radiation damage are reduced.   This would require more
frequent shutdown to replace damaged blanket segments.   Predictions (*108*) of lifetime have been about five MW years/m^2
for existing materials like 316 stainless steel or TZM.   If
this remains, the wall loading in any fusion reactor will
probably be limited to 2.5 MW/m^2 and this limits the minimum
size of any system.   Some very recent experiments by Bloom
et al. (*109*) indicate that the wall lifetime in 316 stainless
steel may be as high as 16.5 MW yr/m^2; however, these data
are not definitive, particularly with respect to the ductility measurements.

A very interesting approach for future consideration is
to ask, "What are the minimum materials performance criteria
that are required to produce a reliable and economically competitive fusion reactor?"   The answer will provide a goal for
the fusion material development program, and it may require
the development of new materials just as the fission program
developed zircalloy.   In the same vein, we should ask a similar question about magnetic field strength, blanket performance characteristics, power cycle design, and plasma physics
performance.   An example here is the finding in the UWMAK-III
study that the cost of the toroidal field magnet set was a
relatively small part of the total direct cost.   One would
conclude from this that if we can develop a material that can
operate to higher fluence, we can design a smaller fusion
reactor and maintain the power output by raising the power
density in the plasma.   Since the power density varies as
$\beta^2 B^4$, our goal can be achieved by either increasing β, a
plasma physics parameter, or by increasing B, a technological
parameter (see equation 12).

An approach like this will bring fusion reactor design
and analysis into a second phase, which not only can produce

a priority list of technological problems, but can also generate a set of technological goals. Meeting these goals will
ensure the technical and commercial feasibility of power from
controlled fusion. The road to this endpoint will surely be
difficult, but the ultimate promise of this new energy source
deserves our best efforts.

REFERENCES

1. Badger, B., Abdou, M.A., Boom, R. W., Brown, R.G., Chang,
   T. E., Conn, R.W., Donhowe, J.M., El-Guebaly, L.A.,
   Emmert, G.A., Hopkins, G.R., Houlberg, W.A., Johnson, A.B.,
   Kamperschroer, J.H., Klein, D., Kulcinski, G.L., Loft,R.G.,
   McAlees, D.G., Maynard, C.W., Mense, A.T., Neil, G.R.,
   Norman, E., Sanger, P.A., Stewart, W.E., Sung, T.,
   Sviatoslavsky, I., Sze, D.K., Vogelsang, W.F., Wittenberg,
   L.J., Yang, T.F., Young, W.D., "UWMAK-I, A Wisconsin Tor-
   oidal Fusion Reactor Design", Nuclear Engineering Depart-
   ment Report FDM-68, The University of Wisconsin, Madison,
   1973 (Vol. 1); May 1975 (Vol. 2). See also, Kulcinski,
   G.L., Conn, R.W., "Conceptual Design of a 5000 MW(th) D-T
   Tokamak Reactor", in Fusion Reactor Design Problems, IAEA,
   Vienna, 1974, P. 51; and Conn, R.W., Kulcinski, G.L.,
   "Technological Implications for Tokamak Fusion Reactors
   of the UWMAK-I Conceptual Design", Proceedings First
   National Topical Conference on the Technology of Controlled
   Nuclear Fusion, CONF-740402, Vol. I, U.S.A.E.C., Page 56,
   1974.

2. Mills, R.G., Editor, "A Fusion Power Plant", Princeton
   Plasma Physics Laboratory Report MATT-1050, Princeton
   University, August 1974.

3. "An Engineering Design Study of a Reference Theta-Pinch
   Reactor (RTPR)", LA-5336 or ANL-8019, Joint Report by
   Los Alamos Scientific Laboratory and Argonne National
   Laboratory, 1974.

4. Werner, R.W., Carlson, G.A., Hovingh, J., Lee, J.D.,
   Peterson, M., "Progress Report No. 2 in the Design
   Considerations for a Low Power Experimental Mirror
   Fusion Reactor", Lawrence Livermore Laboratory Report
   UCRL-7405 4-2, 1974.

5. Badger, B., Conn, R.W., Kulcinski, G.L., Abdou, M.A.,
   Aronstein, R., Avci, H.I., Boom, R.W., Cheng, E.T., Davis,
   J., Donhowe, J.M., Emmert, G.A., Eyssa, Y., Ghoniam, N.M.,
   Ghose, S., Houlberg, W., Kesner, J., Lue, W., Maynard, C.W.,
   Mense, A., Mohan, N., Peterson, H.A., Sung, T.Y.,
   Sviatoslavsky, I., Sze, D.K., Vogelsang, W.F., Westerman,R.,
   Wittenberg, L.J., Yang, T.F., Young, J., Young, W.D.,
   "UWMAK-II, A Conceptual Tokamak Power Reactor Design",
   Nuclear Engineering Department Report FDM-112, University
   of Wisconsin, December 1975.

6. Ribe, F.L., Rev. Mod. Phys., 47,7, 1975. For a general
   view of worldwide research in fusion reactor design see:
   Fusion Reactor Design Problems, IAEA, Vienna, 1974; Plasma
   Physics and Controlled Nuclear Fusion Research 1974, IAEA,
   Vienna, 1975, Vol. III; Proc. First National Topical Conf.
   on the Technology of Controlled Nuclear Fusion, CONF-740402,
   Volumes I and II, U. S. Atomic Energy Commission, 1974.

7. Steiner, D., Nucl. Sci. Eng., 58, P. 102, 1975. See also,
   Nozawa, M., Steiner, D., "An Assessment of the Power Bal-
   ance in Fusion Reactors," Oak Ridge National Laboratory
   Report, ORNL-TM-4421, 1974.

8. Conn, R.W., Kulcinski, G.L., Maynard, C.W., Aronstein, R.,
   Avci, H.I., Blackfield, D., Boom, R., Bowles, A., Cameron,
   E., Cheng, E.T., Clemmer, R., Dalhed, S., Davis, J., Emmert,
   G.A., Ghoniem, N.M., Ghose, S., Gohar, Y., Kesner, J.,
   Kuo, S., Larsen, E., Ramer, E., Scharer, G., Schmunk, R.E.,
   Sung, T.Y., Sviatoslavsky, I., Sze, D.K., Vogelsang, W.F.,
   Yang, T.F., Young, W.D., "UWMAK-III, A High Performance,
   Noncircular Tokamak Power Reactor Design," Nuclear Engi-
   neering Department Report FDM-150, The University of Wis-
   consin, 1976.

9. Mills, R.G., "Catalyzed Deuterium Fusion Reactors," Princeton Plasma Physics Laboratory Report TM-259, 1971.

10. Duane, B.H., "Fusion Cross Section Theory", Battelle North-
    west Laboratory Report, BNWL-1685, 1972.

11. Miley, G.H., Towner, H., Ivich, N., "Fusion Cross-Sections
    and Reactivities", Nuclear Engineering Program, Report
    COO-2218-17, University of Illinois, 1974.

12. Lawson, J.D., Proceedings Physical Society, B-70, P.6 1957.

13. Meade, D., Nuclear Fusion 14, P. 289, 1974.

14. Conn, R.W., Kesner, J., Nuclear Fusion 15, P. 775, 1975.

15. Post, R.F., Annual Review of Nuclear Science, 20, P. 509, 1970.

16. Futch, A.H., Jr., Holdren, J.P., Killeen, J., Mirin, A. A., Plasma Physics, 14, P. 211, 1972.

17. Plasma Physics and Controlled Nuclear Fusion Research 1974, Fifth Conference Proceedings, Tokyo, IAEA, Vienna, 1975. See papers by D. E. Baldwin et al., Vol. I, P. 301, and M. E. Rensink et al., Vol. I, P. 311.

18. Gott, Yu.B., Ioffe, M.S., Telkovsky, V.C., Nuclear Fusion Supplement 3, P. 1045, 1962.

19. Ibid Ref. 17, paper by Coensgin, F.H., Cummins, F.W., Molvik, A.W., Nexsen, W.E., Simonen, T.C., Stallard, B.W., Vol. II, P. 323.

20. Post, R.F., Rosenbluth, M.N., Phys. Fluids, 9, P. 730, 1966.

21. Plasma Physics and Controlled Nuclear Fusion Research, Proceedings Fourth Conference, Madison, Wisconsin (IAEA, Vienna, 1971). See D.E. Baldwin et al., Vol. II, P. 735.

22. Molvik, A.W., Coensgen, F.H., Cummins, W.F., Nexsen, W.E., Simonen, T.C., Phys. Rev. Letts., 32, P. 1109, 1974.

23. Logan, B.G., "Two Component Experiments in 2XIIB", Lawrence Livermore Laboratory Report, UCID-16851, 1975.

24. Coensgen, F.H., et al., "Startup of a Neutral-Beam-Sustained Plasma in a Quasi-DC Magnetic Field", Lawrence Livermore Laboratory Report, UCRL-78057, 1976.

25. Ibid, Ref. 21, paper by Burnett, S.C., et al., Vol. III, P. 201.

26. Ibid, Ref. 17, Paper by Cantrell, E.L. et al., Vol. III, P. 13.

27. Artsimovich, L.A., Nuclear Fusion 12, P. 215, 1972.

28. Furth, H.P., Nuclear Fusion 15, P. 487, 1975.

29. Spitzer, L., Physics of Fully Ionized Gasses, J. Wiley, New York, Second Edition, 1962.

30. Gorbunov, E.P., et al.,"Controlled Fusion and Plasma Physics,"Proceedings 6th European Conference, Moscow, Vol. I., P. 1, 1973.

31. Ibid Ref. 19, D.Dimock et al., Vol. 1, P. 451.

32. Ibid Ref. 17, Equipe TFR, Vol. I, P.P. 127 and 135.

33. Bol, K., et al., Phys. Rev. Letts., 29, P. 495, 1972.

34. Ibid Ref. 17, Paper by Bol, K., et al., Vol. I, P. 83; also, Bol, K., et al., Phys., Rev. Letts., 32, P. 661, 1974.

35. Berry, L.A., Bulletin American Physical Society 20, P. 1332, 1975.

36. Dei Cas, R., TFR Group, Bulletin American Physical Society, 20, P. 1332, 1975.

37. Galeev, A.A., Sagdeer, R.Z., Soviet Physics JETP 32, P. 572, 1971.

38. Bickerton, R.J., Conner, J.W., Taylor, J.B., Nature, Phys. Science 229, P. 110, 1972.

39. Ibid Ref. 17, Meade, D.M., Furth, H.P., Rutherford, P.H., Seidl, F., Duchs, D.F., Vol. I, P. 605.

40. Ohkawa, T., Voorhies, H.G., Phys. Rev. Letts., 22, P. 1275, 1969; see also Ohkawa, T., Jensen, T.H., Plasma Physics 12, P. 789, 1970.

41. Ibid Ref. 17, Ohkawa, T., et al., Vol. I, P. 281.

42. Rosenbluth, M.N., Hazeltine, R.D., Hinton, F.L., Phys.
     Fluids, 15, P. 116, 1972.

43. Pfrisch, D., Schlüter, A., Max Planck Institute
     Report, MPI/PA/7/62, 1962.

44. Hinton, F.L., Rosenbluth, M.N., Phys. Fluids 16,
     P. 836, 1973.

45. Galeev, A.A., Sagdeev, R.Z., Zh. Eksp. Theo. Fiz. 53,
     P. 348, 1967; also, Soviet Phys. Dokl. 14, P.1198, 1970,
     and Sov. Phys. JETP 32, P. 572, 1971.

46. Berry, L.A., Clarke, J.F., Hogan, J.T., Phys. Rev.
     Letts., 32, P. 362, 1974.

47. Yoshikawa, S., Phys. Rev. Letts., 25, P. 353, 1970; also
     ibid Ref. 17, Yoshikawa, S., Christofilos, N., Vol. II,
     P. 357.

48. Artsimovich, L.A., J.E.T.P. Letters 13, P. 70, 1971.

49. Hazeltine, R.D., Hinton, F.L., Rosenbluth, M.N., Phys.
     Fluids 16, P. 1645, 1973.

50. Kadomtsev, B.B., Pogutse, O.P., Zh. Ex. sp. Teor. Fi.
     51, P. 1734, 1966; also, Sov. Phys. J.E.T.P. 24, P.
     1172, 1967.

51. Kadomtsev, B.B., Pogutse, O.P., Soviet Phys. Dokl. 14,
     P. 470, 1969.

52. Kadomtsev, B.B., Pogutse, O.P., Nuclear Fusion 11,
     P. 67, 1971.

53. Ibid Ref. 17, Horton, W., et al., Vol. I, P. 541, and
     Coppi, B., Pozzolo, R., Rewoldt, G., Schep, T.,
     Vol. I, P. 549.

54. Spano, A.H. (Compiler), Nuclear Fusion 15, P. 909, 1975.

55. Ibid Ref. 17, Behrisch, R., Kadomtsev, B.B., Vol. II,
     P. 229.

56. Conn, R.W., Kulcinski, G.L., Avci, H., Magraby, M. El,
     Nuclear Technology 26, P. 125, 1975.

57. Segal, H., Richards, T.G., "Low Temperature Resistance Studies on Cyclically Strained Aluminum", in *Advances in Cryogenic Engineering*, P. 21, in press.

58. Yang, T., Conn, R.W., Bulletin American Phys. Society 20, P. 1280, 1975; also, "MHD Equilibrium and Stability Calculations for a Noncircular Highβ_θ Tokamak Plasma", Nuclear Engineering Department Report, UWFDM-152, University of Wisconsin, 1975.

59. Ibid Ref. 17, Chance, M.S., 3t al., Vol. I, P. 463.

60. Scharer, J., Conn, R.W., Blackfield, D., "Study of RF and Neutral Beam Heating in Tokamaks", Electric Power Research Institute Report, EPRI ER-268, 1976.

61. Ibid Ref. 17, Adam, J. et al., Vol. I, P. 65.

62. Hogan, J.T., Meth. in Comp. Physics 16, 1976; also, Oak Ridge National Laboratory Report, ORNL-TM-5153, 1975.

63. Kesner, J., Conn, R.W., *Nuclear Fusion*, 16, P. 397, 1976.

64. Kulcinski, G.L., Conn, R.W., Lang, G., Nuclear Fusion 15, P. 327, 1975.

65. Mense, A.T., "Poloidal Diverters for Tokamak Reactors", *Ph.D. Thesis*, University of Wisconsin, 1977.

66. Mense, A.T., Emmert, G.A., Callen, J.D., Nuclear Fusion 15, P. 703, 1975.

67. Boom, R.W., Moses, R.W., Jr., Young, W.C., "Magnet Design of Toroidal Field Coils for the UWMAK-II and III Tokamak Systems," Nuclear Eng. and Design, 39, P. 99, 1976.

68. Moses, R., Young, W., "Analytic Expressions for Magnetic Forces on Sectored Toroidal Coils", 6th Symposium on Engineering Problems of Fusion Research, Paper D-3-4, November 1975; also, Nuclear Engineering Department Report UWFDM-143, University of Wisconsin, 1975.

69. Cornish, D., Lawrence Livermore Laboratory, private communication.

70. Engle, W., Jr., "A User's Manual for ANISN", Oak Ridge Gaseous Diffusion Plant Report, K-1693, 1967.

71. Mynatt, F.R., et al., "The DOT-III Two-Dimensional Discrete Ordinates Transport Code", Oak Ridge National Laboratory Report, ORNL-TM-4280, 1973.

72. Bell, G.E., Glasstone, S., Nuclear Reactor Theory, Van Nostrand-Reinhold, New York, 1970.

73. Abdou, M.A., Conn, R.W., Nuclear Science and Engineering, 55, P. 226, 1974.

74. Gohar, Y., University of Wisconsin, private communication.

75. Conn, R.W., Gohar, Y., Maynard, C.W., Trans. American Nuclear Society, 22, P. 16, 1976.

76. Chapin, D.L., Price, W., Jr., Trans. American Nuclear Society 21, P. 66, 1975.

77. Hoffman, M.A., and Carlson, G.A., "Calculation Techniques for Estimating the Pressure Losses for Conducting Fluid Flows in Magnetic Fields", Lawrence Livermore Laboratory Report UCRL-51010, 1971.

78. Sze, D.K., Stewart, W.E., "Lithium Cooling for a Low-β Tokamak Reactor", Proceedings, 1972 Symposium on the Technology of Controlled Thermonuclear Fusion Experiments and Engineering Aspects of Fusion Reactor, CONF-721111, AEC Symposium Series No. 31, USAEC, 1974.

79. Sze, D.K., Ibid Ref. 8, Chapter VI.

80. Clemmer, R.G., Larson, E.M., Wittenberg, L.W., "Tritium Handling, Breeding and Containment in Two Conceptual Fusion Reactor Designs: UWMAK-I and UWMAK-II," Nuclear Engineering and Design, 39, P. 85, 1976; also Ref. 8, Chapter XII.

81. Wilkes, W.R., Trans. American Nuclear Society, 19, P. 20, 1974.

82. Watson, J.S., "An Evaluation of Methods for Recovering Tritium from Blanket or Cooling Systems of Fusion Reactors", Oak Ridge National Laboratory Report ORNL-TM-3794, 1972.

83. Barrer, R.M., Diffusion In and Through Solids, Cambridge University Press, P. 168, 1941.

84. Vogelsang, W.F., Kulcinski, G.L., Lott, R.G., Sung, T.Y., Nuclear Technology 22, P. 379, 1974.

85. Conn, R.W., Sung, T.Y., Abdou, M.A., Nuclear Technology 26, P. 391, 1975.

86. Kulcinski, G.L., Davis, J., Schmunk, R.E., "The Case for Molybdenum Alloys in D-T Fusion Reactors", Nuclear Engineering Department Report UWFDM-142, The University of Wisconsin, 1975.

87. Bianchi, L.M., et al., Universal Cyclops Corporation Report (available from National Technical Information Service (NTIS)), Report No. AD-458529, 1964.

88. Davis, J.W., McDonnell-Douglas Astronautics-East, St. Louis, Missouri, private communication.

89. Cheng, E.T., Conn R.W., Trans. American Nuclear Society, 22, P. 44, 1975.

90. Gill, W.W., et al., AIChE 6, P. 139, 1960.

91. Schmunk, D., Kulcinski, G.L., "Survey of Irradiation Data on Molybdenum", University of Wisconsin Report UWFDM-161, September 1976.

92. Gray, W.J., Morgan, W.C., "High Temperature Graphite Irradiations: 550 to 1450°C", Battelle Pacific Northwest Laboratories Report, BNWL-1672, 1972.

93. Van den Berg, M., Everett, M.R., Kingsbury, A., "The Relationship Between Irradiation Temperature and Dimensional Changes of Nuclear Graphites", 12th Biennial Conference on Carbon, University of Pittsburg, PP 307-310, 1975.

94. Morgan, W.C., Woodruff, E.M., Gray, W.J., "Irradiation
     Behavior of Graphite at Very High Temperature", in
     2nd National Topical Conference on Controlled Fusion
     Technology, Richland, Washington, September, 1976, to be
     published.

95. Holt, J.B., Hosmer, D.W., Guinan, M.W., Condit, R.H.,
     Borg, R.J., "Helium Generation and Diffusion in Graphite",
     ibid.

96. McCracken, G.M., Rep. Prog. Phys. 38, P. 241, 1975.

97. Erents, S.K., Braganza, C.M., McCracken, G.M., "Methane
     Formation During the Interaction of Energetic Protons
     and Deuterons with Carbon", Journal Nuclear Material,
     in press.

98. Balooch, M., Olander, D.R., Journal Chem. Phys. 63,
     P. 4772, 1975.

99. Lang, G., Holmes, V.L., Nuclear Fusion 16, P. 162, 1976.

100. Conn, R.W., Kuo, S., "An Advanced Conceptual Tokamak Fu-
     sion Reactor Utilizing Closed Cycle Helium Gas Turbines,"
     Nuclear Engineering and Design, 39, P. 45, 1976.

101. Bechtel Corporation Scientific Development, "Balance of
     Plant and Cost Study for the Conceptual Fusion Reactor
     Design, UWMAK-III", Bechtel Corporation Report to the
     University of Wisconsin, February 1976.

102. Cameron, E., University of Wisconsin, Department of
     Geology, private communication; also, Ref. 8, Chapter XII.

103. Vine, J.D., Trans. American Nuclear Society 23, P. 55,
     1975.

104. Bowles, A., Von Fischer, E., Bechtel Corporation,
     Scientific Development, and Conn, R.W., Sviatoslavsky, I.,
     The University of Wisconsin; extensive details reported
     in Chapter XIII, Ref. 8, and in Ref. 101.

105. United Engineers and Contractors, "Pressurized Water
     Reactor Plant, 1000 MW$_e$ Central Station Power Plants-
     Investment Cost Study", AEC Report WASH-1230, Vol. I, 1971.

106. Levenson, M., Murphy, P.M., Zaleski, C.P.L., Nuclear News 19, P. 54, 1976.

107. Sze, D.K., Larsen, E.M., Cheng, E.T., Clemmer, R.G., Trans. American Nuclear Society 22, P. 21, 1975.

108. Kulcinski, G.L., Brown, R.G., Lott, R.G., Sanger, P.A., Nuclear Technology 22, P. 20, 1974.

109. Bloom, E.E., Wiffen, F.W., Moziasz, P.J., Stiegler, J.O., "Temperature and Fluence Limits for a Type 316 Stainless Steel CTR First Wall," Nuclear Technology, 31, P. 115, 1976.

ACKNOWLEDGEMENTS

The research on UWMAK-III was supported by the Electric Power Research Institute. That research involved many members of the Fusion Technology Program at the University of Wisconsin, and their work is represented in the final report on UWMAK-III. Their contributions are also cited in specific references throughout this article. I am indebted to them for the intellectually stimulating atmosphere which they have created.